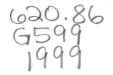

OCCUPATIONAL SAFETY AND HEALTH
for Technologists, Engineers, and Managers

Third Edition

DAVID L. GOETSCH

Prentice Hall

Upper Saddle River, New Jersey
Columbus, Ohio

Library of Congress Cataloging-in-Publication Data

Occupational safety and health for technologists, engineers, and managers / David L. Goetsch.—3rd ed.

p. cm.

Originally published: Industrial safety and health in the age of high technology. New York: Macmillan Pub. Co., 1993.

Includes bibliographical references and index.

ISBN 0-13-924085-3 (hc)

1. Industrial safety—United States. 2. Industrial hygiene—United States. I. Title

T55.G586 1999

658.4'08—dc21

97-44002
CIP

Editor: Stephen Helba
Production Editor: Patricia S. Kelly
Design Coordinator: Karrie M. Converse
Text Designer: Kip Shaw
Cover Designer: Karrie M. Converse
Production Manager: Deidra M. Schwartz
Electronic Text Management: Marilyn Wilson Phelps, Karen L. Bretz, Tracey B. Ward
Marketing Manager: Frank Mortimer, Jr.

This book was set in Clearface and Swiss by Prentice Hall and was printed and bound by R.R. Donnelley & Sons Company. The cover was printed by Phoenix Color Corp.

Earlier edition entitled *Industrial Safety and Health: In the Age of High Technology for Technologists, Engineers, and Managers,* © 1993 by Macmillan Publishing Company.

Printed in the United States of America

10 9 8 7 6 5 4 3

ISBN: 0-13-924085-3

Prentice-Hall International (UK) Limited, *London*
Prentice-Hall of Australia Pty. Limited, *Sydney*
Prentice-Hall of Canada, Inc., *Toronto*
Prentice-Hall Hispanoamericana, S. A., *Mexico*
Prentice-Hall of India Private Limited, *New Delhi*
Prentice-Hall of Japan, Inc., *Tokyo*
Pearson Education Asia Pte. Ltd., *Singapore*
Editora Prentice-Hall do Brasil, Ltda., *Rio de Janeiro*

PREFACE

BACKGROUND

The field of occupational safety and health and has undergone significant change over the past two decades. There are many reasons for this. Some of the more prominent include the following: technological changes that have introduced new hazards in the workplace; proliferation of health and safety legislation and corresponding regulations; increased pressure from regulatory agencies; realization by executives that a safe and healthy workplace is typically a more productive workplace; health care and workers' compensation costs; increased pressure from environmental groups and the public; a growing interest in ethics and corporate responsibility; professionalization of health and safety occupations; increased pressure from labor organizations and employees in general; rapidly mounting costs associated with product safety and other types of litigation; and increasing incidents of workplace violence.

All of these factors, when taken together, have made the job of the modern safety and health professional more challenging and more important than it has ever been. These factors have also created a need for an up-to-date book in workplace safety and health that contains the latest information needed by people who will practice this profession in the age of global competition.

WHY WAS THIS BOOK WRITTEN AND FOR WHOM?

This book was written in response to the need for an up-to-date practical teaching resource that focuses on the needs of modern safety and health professionals practicing in the workplace. It is intended for use in universities, colleges, community colleges, and corporate training settings that offer programs, courses, workshops, and/or seminars in occupational safety and health. Educators in such disciplines as industrial technology, manufacturing technology, industrial engineering, engineering technology, occupational safety, management, and supervision may find this book both valuable and easy to use.

The direct, straightforward presentation of material focuses on making the theories and principles of occupational safety and health practical and useful in a real-world setting. Up-to-date research has been integrated throughout in a down-to-earth manner.

ORGANIZATION OF THE BOOK

The text contains thirty chapters, each focusing on a major area of concern for modern safety and health professionals. The chapters are presented in an order that is compatible with the typical organization of a college-level safety and health course. A standard chapter format is used throughout the book. Each chapter begins with a list of major topics and ends with a comprehensive summary. Following the summary, each chapter contains end material including review questions, key terms and concepts, and endnotes. Within each chapter are case studies to prompt classroom discussion, as well as at least one safety fact or myth. These materials are provided to encourage review, stimulate additional thought, and provide opportunities for applying what has been learned.

HOW THIS BOOK DIFFERS FROM OTHERS

This book was written because in the age of global competition, safety and health in the workplace has changed drastically. Many issues, concerns, and factors relating *specifically* to modern workplace environments have been given more attention, greater depth of coverage, and more illumination than other textbooks. Some of the areas receiving more attention and specific occupational examples are the following:

- The OSHAct and OSHA
- Standards and Codes
- Laws and liability
- Stress-related problems
- Life safety
- Evolving roles of health and safety professionals
- Health and safety training
- Environmental issues and ISO 14000 standards
- Computers, robots, and automation
- Ethics and safety
- Bloodborne pathogens in the workplace
- Product safety and liability
- Ergonomics and safety
- Relationship between safety and quality
- Workplace violence
- Repetitive strain injuries (RSI)

NEW IN THIS EDITION

This third edition of *Occupational Safety and Health* contains much new and updated material, including the following:

- Chapters were combined and the order of chapters rearranged as requested by many of the book's users.
- Photographs were added to better illustrate selected topics explained in the text.
- At least one *safety fact* or *safety myth* (debunked) was added to every chapter to enhance student interest.
- A discussion case related directly to the content was added to each chapter to generate classroom discussion and additional research.
- Chapter 2 was updated to include new and additional material on repetitive strain/soft tissue injuries and on estimating the cost of accidents.
- Chapter 4 was updated to reflect the new OSHA guidelines for record keeping.
- Chapter 5 was updated to include a new section on cost cutting strategies for workers' compensation.
- Chapter 6 (formerly Chapter 30) was updated to include additional material on the ergonomics of aging and VDT hazards as well as a new section on repetitive strain injuries (RSI).
- Chapter 9 was updated to include new sections on the OSHA Fall Protection standard and forklift safety as well as new material on ladder safety, footwear/foot protection, and proper lifting techniques.
- Chapter 13 was updated to include new sections on Life Safety and flame resistant clothing.
- Chapter 14 (old Chapters 14 and 24 combined) was updated to contain the following new sections: NIOSH guidelines for respirators, ventilation and the sick building syndrome, OSHA's Process Safety Standard, and OSHA's Confined Space Standard.
- Chapter 15 was updated to include the most recent information concerning electromagnetic fields in the workplace.
- Chapter 16 was updated to include new information on hearing protection devices and a new section on other effects of noise hazards.
- Chapter 18 (formerly Chapter 20) was updated to include OSHA's Process Safety Standard.
- Chapter 20 (formerly Chapter 22) was updated to include a new section on teamwork in promoting safety.
- Chapter 24 (formerly Chapter 27) was updated to include a new section on Hepatitis B (HBV) in the workplace.
- Chapter 25 (formerly Chapter 28) was completely rewritten and now covers all of the previous environmental material plus the ISO 14000 environmental standard.
- Chapter 27 (formerly Chapter 6) was updated to include a section on the position of risk manager.

- Chapter 28 (formerly Chapter 31) was completely rewritten to introduce the concept of Total Quality Management or TSM (safety management according to the principles of TQM).
- Chapter 30 is a new chapter on the critical subject of workplace violence and the role of safety and health professionals in preventing it.

ABOUT THE AUTHOR

David L. Goetsch is Provost of the joint campus of the University of West Florida and Okaloosa-Walton Community College in Fort Walton Beach, Florida, and professor of safety and quality management. He also administers the state of Florida's Center for Manufacturing Competitiveness that is located on this campus. In addition, Dr. Goetsch is President of The Management Institute, a private consulting firm dedicated to the continual improvement of organizational competitiveness, safety, and quality. Dr. Goetsch is co-founder of The Quality Institute, a partnership of the University of West Florida, Okaloosa-Walton Community College, and the Okaloosa Economic Development Council. He currently serves on the executive board of the Institute.

ACKNOWLEDGMENTS

The author acknowledges the invaluable assistance of the following people in developing this book: Harvey Martin, health and safety manager of Metric Systems Corporation in Fort Walton Beach, Florida, for providing up-to-date research material; Faye Crawford, for word processing of the manuscript; and the following reviewers for their invaluable input: Albert S. Kirk, University of Southern Maine; Susan B. Meyer, University of Minnesota, Duluth; and Stephen G. Mraz, Pensacola Junior College.

CONTENTS

CHAPTER SIX
Ergonomic Hazards and Repetitive Strain Injuries   144

CHAPTER SEVEN
Stress and Safety   178

Safety versus Health

The title of this book purposely includes the words *safety* and *health*. Throughout the text, the titles "safety and health professional" and "safety and health manager" are used. This, too, is done by design. This approach underscores the point that the field of occupational safety has been broadened to encompass both safety and health. Consequently, managers, technical personnel, and engineers in this field must be knowledgeable about safety and health and be prepared to oversee a corporate program that encompasses both areas of responsibility.

Safety and health, although closely related, are not the same. One view is that safety is concerned with injury-causing situations, whereas health is concerned with disease-causing conditions. Another view is that safety is concerned with hazards to humans that result from sudden severe conditions; health deals with adverse reactions to prolonged exposure to dangerous, but less intense, hazards. Both of these views are generally accurate in portraying the difference between safety and health. However, the line between these two concepts is not always clearly marked.

For example, stress is a hazard that can cause both psychological and physiological problems over a prolonged period of time. In this case, it is a health concern. On the other hand, an overly stressed worker might be more prone to unintentionally forget safety precautions and thus might cause an accident. In this case, stress is a safety concern.

Since managers in this evolving profession are likely to be responsible for both health and safety, it is important that they have a broad academic background covering both. This book attempts to provide that background.

This broadening of the scope of the profession does not mean that specialists in safety and health are not still needed. They are. Chapter 27 shows how today's safety and health manager is a generalist who often heads a team of specialists such as safety engineers, health physicists, industrial hygienists, occupational nurses, occupational physicians, and risk managers. In order to manage a team of specialists in these various areas, safety and health managers must have the broad and comprehensive background that this book provides.

Safety and Health Movement, Then and Now

From its meager beginnings in a tiny Chicago office in 1913, the National Safety Council (NSC) developed into a comprehensive, broad-based safety organization with thousands of members from business, industry, agriculture, education, labor, and government. The development of the **safety movement** in the United States has paralleled that of the NSC. In the early 1900s, industrial accidents were commonplace in this country; for example, in 1907 over 3,200 people were killed in mining accidents. During this period, legislation, precedent, and public opinion all favored management. There were few protections for workers' safety.

Working conditions for industrial employees today have improved significantly. The chance of a worker being killed in an industrial accident is less than half of what it was 60 years ago.[1] According to the NSC, the current death rate from work-related injuries is approximately four per 100,000, or less than a third of the rate 50 years ago.[2]

Improvements in safety until now have been the result of pressure for legislation to promote safety and health, the steadily increasing costs associated with accidents and injuries, and the professionalization of safety as an occupation. Improvements in the future are likely to come as a result of greater awareness of the cost effectiveness and resultant competitiveness gained from a safe and healthy workforce.

This chapter examines the history of the safety movement in the United States and how it has developed over the years. Such a perspective will help practicing and prospective safety professionals form a better understanding of both their roots and their future.

DEVELOPMENTS BEFORE THE INDUSTRIAL REVOLUTION

It is important for students of industrial health and safety to first study the past. Understanding the past can help safety and health professionals examine the present and future with a sense of perspective and continuity. Modern developments in health and safety are neither isolated nor independent. Rather, they are part of the long continuum of developments in the safety and health movement.

The continuum begins with the days of the ancient Babylonians. During that time, circa 2000 B.C., their ruler Hammurabi developed his **Code of Hammurabi.** The code encompassed all of the laws of the land at that time, showed Hammurabi to be a just ruler, and set a precedent followed by other Mesopotamian kings. What is significant about the code from the perspective of safety and health is that it contained clauses dealing with injuries, allowable fees for physicians, and monetary damages assessed against those who injured others.[3] This clause from the code illustrates Hammurabi's concern for the proper handling of injuries: "If a man has caused the loss of a gentleman's eye, his own eye shall be caused to be lost."[4]

This movement continued and emerged in later Egyptian civilization. As evidenced from the temples and pyramids that still remain, the Egyptians were an industrious people. Much of the labor was provided by slaves, and there is ample evidence that slaves were not treated well, that is, unless it suited the needs of the Egyptian taskmasters.

One such case occurred during the reign of Rameses II (circa 1500 B.C.), who undertook a major construction project, the Rameuseum. To ensure the maintenance of a workforce sufficient to build this huge temple bearing his name, Rameses created an industrial medical service to care for the workers. They were required to bathe daily in the Nile and were given regular medical examinations. Sick workers were isolated.[5]

The Romans were vitally concerned with safety and health, as can be seen from the remains of their construction projects. The Romans built aqueducts, sewerage systems, public baths, latrines, and well-ventilated houses.[6]

As civilization progressed, so did safety and health developments. In 1567, Philippus Aureolus produced a treatise on the pulmonary diseases of miners. Entitled *On the Miners' Sickness and Other Miners' Diseases,* the treatise covered diseases of smelter workers and metallurgists and diseases associated with the handling of and exposure to mercury. Around the same time, Georgius Agricola published his treatise *De Re Metallica,* emphasizing the need for ventilation in mines and illustrating various devices that could be used to introduce fresh air into mines.[7]

The eighteenth century saw the contributions of Bernardino Ramazzini, who wrote *Discourse on the Diseases of Workers.* Ramazzini drew conclusive parallels between diseases suffered by workers and their occupations. He related occupational diseases to the handling of harmful materials and to irregular or unnatural movements of the body. Much of what Ramazzini wrote still has relevance today.[8]

The Industrial Revolution changed forever the methods of producing goods. According to LaDou, the changes in production brought by the Industrial Revolution can be summarized as follows:

- Introduction of **inanimate power** (i.e., steam power) to replace people and animal power
- Substitution of machines for people
- Introduction of new methods for converting raw materials
- Organization and specialization of work, resulting in a division of labor[9]

These changes necessitated a greater focusing of attention on the safety and health of workers. Steam power increased markedly the potential for life-threatening injuries, as did machines. The new methods used for converting raw materials also introduced new risks of injuries and diseases. Specialization, by increasing the likelihood of boredom and inattentiveness, also made the workplace a more dangerous environment.

MILESTONES IN THE SAFETY MOVEMENT

Just as the United States traces its roots to Great Britain, the safety movement in this country traces its roots to England. During the Industrial Revolution, child labor in factories was common. The hours were long, the work hard, and the conditions often unhealthy and unsafe. Following an outbreak of fever among the children working in their cotton mills, the people of Manchester, England, began demanding better working conditions in the factories. Public pressure eventually forced a government response, and in 1802 the Health and Morals of Apprentices Act was passed. This was a milestone piece of legislation: It marked the beginning of governmental involvement in workplace safety.

When the industrial sector began to grow in the United States, hazardous working conditions were commonplace. Following the Civil War, the seeds of the safety movement were sown in this country. Factory inspection was introduced in Massachusetts in 1867. In 1868, the first barrier safeguard was patented. In 1869, the Pennsylvania legislature passed a mine safety law requiring two exits from all mines. The Bureau of Labor Statistics (BLS) was established in 1869 to study industrial accidents and report pertinent information about those accidents.

The following decade saw little new progress in the safety movement until 1877, when the Massachusetts legislature passed a law requiring safeguards for hazardous machinery. This year also saw passage of the Employer's Liability Law, establishing the potential for **employer liability** in workplace accidents. In 1892, the first recorded safety program was established in a Joliet, Illinois, steel plant in response to a scare caused when a flywheel exploded. Following the explosion, a committee of managers was formed to investigate and make recommendations. The committee's recommendations were used as the basis for the development of a safety program that is considered to be the first safety program in American industry.

Around 1900, Frederick Taylor began studying efficiency in manufacturing. His purpose was to identify the impact of various factors on efficiency, productivity, and profitability. Although safety was not a major focus of his work, Taylor did draw a connection between lost personnel time and management policies and procedures. This connection between safety and management represented a major step toward broad-based safety consciousness.

In 1907, the U.S. Department of the Interior created the Bureau of Mines to investigate accidents, examine health hazards, and make recommendations for improvements. Mining workers definitely welcomed this development since over 3,200 of their fellow workers were killed in mining accidents in 1907 alone.[10]

One of the most important developments in the history of the safety movement occurred in 1908 when an early form of **workers' compensation** was introduced in the United States. Workers' compensation actually had its beginnings in Germany. The practice soon spread throughout the rest of Europe. Workers' compensation as a concept made great strides in the United States when Wisconsin passed the first effective workers' compensation law in 1911. In the same year, New Jersey passed a workers' compensation law that withstood a court challenge.

The common thread among the various early approaches to workers' compensation was that they all provided some amount of compensation for on-the-job injuries regardless of who was at fault. When the workers' compensation concept was first introduced in the United States, it covered a very limited portion of the workforce and provided only minimal benefits. Today, all 50 states have some form of workers' compensation that requires the payment of a wide range of benefits to a broad base of workers. Workers' compensation is examined in more depth in Chapter 5.

The Association of Iron and Steel Electrical Engineers (AISEE), formed in the early 1900s, pressed for a national conference on safety. As a result of the AISEE's efforts, the first meeting of the **Cooperative Safety Congress (CSC)** took place in Milwaukee in 1912. What is particularly significant about this meeting is that it planted the seeds for the eventual establishment of the National Safety Council. A year after the initial meeting of the Cooperative Safety Congress, the **National Council of Industrial Safety (NCIS)** was established in Chicago. In 1915, this organization changed its name to the National Safety Council. It is now the premier safety organization in the United States.

From the end of World War I (1918) through the 1950s, safety awareness grew steadily. During this period, the federal government encouraged contractors to implement and maintain a safe work environment. Also during this period, industry in the United States arrived at two critical conclusions: (1) there is a definite connection between quality and safety, and (2) off-the-job accidents have a negative impact on productivity. The second conclusion became painfully clear to manufacturers during World War II when the call-up and deployment of troops had employers struggling to meet their labor needs. For these employers, the loss of a skilled worker due to an injury or for any other reason created an excessive hardship.[11]

The 1960s saw the passage of a flurry of legislation promoting workplace safety. The Service Contract Act of 1965, the Federal Metal and Non-metallic Mine Safety Act, the Federal Coal Mine and Safety Act, and the Contract Workers and Safety Standards Act all were passed during the 1960s. As their names indicate, these laws applied to a limited audience of workers. According to the Society of Manufacturing Engineers (SME), more significant legislation than that enacted in the 1960s was needed because

> Generally, the state legislated safety requirements only in specific industries, had inadequate safety and health standards, and had inadequate budgets for enforcement. . . . The injury and death toll due to industrial mishaps was still . . . too high. In the late 1960s, more than 14,000 employees were killed annually in connection with their jobs. . . . Work injury rates were taking an upward swing.[12]

These were the primary reasons behind passage of the Occupational Safety and Health Act of 1970 **(OSHA)** and the Federal Mine Safety Act of 1977. These federal laws, particularly OSHA, represent the most significant legislation to date in the history of the safety movement. Figure 1–1 summarizes some significant milestones in the development of the safety movement in the United States.

ROLE OF ORGANIZED LABOR

Organized labor has played a crucial role in the development of the safety movement in the United States. From the outset of the Industrial Revolution in this country, organized labor has fought for safer working conditions and appropriate compensation for workers injured on the job. Many of the earliest developments in the safety movement were the result of long and hard-fought battles by organized labor.

Although the role of unions in promoting safety is generally acknowledged, there is a school of thought that takes the opposite view. Proponents of this dissenting view hold that union involvement actually slowed the development of the safety movement. Their theory is that unions allowed their demands for safer working conditions to become entangled with their demands for better wages and, as a result, they met with resistance from management. Regardless of the point of view, there is no question that working conditions in the earliest years of the safety movement were often reflective of an insensitivity to safety concerns on the part of management.

Among the most important contributions of organized labor to the safety movement was their work to overturn antilabor laws relating to safety in the workplace. These laws were the fellow servant rule, the statutes defining contributory negligence, and the concept of assumption of risk.[13] The **fellow servant rule** held that employers were not liable for workplace injuries that resulted from the negligence of other employees. For example, if Worker X slipped and fell, breaking his back in the process, because Worker Y spilled oil on the floor and left it there, the employer's liability was removed. In addition, if the actions of employees contributed to their own injuries, the employer was absolved of any liability. This was the doctrine of **contributory negligence.** The concept of **assumption of risk** was based on the theory that people who accept a job assume the risks that go with it. It says employees who work voluntarily should accept the consequences of their actions on the job rather than blaming the employer.

Since the overwhelming majority of industrial accidents involve negligence on the part of one or more workers, employers had little to worry about. Therefore, they had little incentive to promote a safe work environment. Organized labor played a crucial role in bringing sometimes deplorable working conditions to the attention of the general

1867	Massachusetts introduces factory inspection.
1868	Patent awarded for first barrier safeguard.
1869	Pennsylvania passes law requiring two exits from all mines and the Bureau of Labor Statistics is formed.
1877	Massachusetts passes law requiring safeguards on hazardous machines and the Employer's Liability Law is passed.
1892	First recorded safety program is established.
1900	Frederick Taylor conducts first systematic studies of efficiency in manufacturing.
1907	Bureau of Mines created by U.S. Department of the Interior.
1908	Concept of worker's compensation is introduced in the United States.
1911	Wisconsin passes the first effective worker's compensation law in the United States and New Jersey becomes the first state to uphold a worker's compensation law.
1912	First Cooperative Safety Congress meets in Milwaukee.
1913	National Council of Industrial Safety is formed.
1915	National Council of Industrial Safety changes its name to National Safety Council.
1916	Concept of negligent manufacture is established (product liability).
1936	National Silicosis Conference convened by the U.S. Secretary of Labor.
1970	Occupational Safety and Health Act passes.
1977	Federal Mine Safety Act passes.
1986	Superfund Amendments and Reauthorization Act
1990	Amended clean Air Act of 1970
1996	Introduction of Total Safety Management (TSM) concept.

Figure 1–1
Milestones in the safety movement.

public. Public awareness and, in some cases, outrage eventually led to these **employer-biased laws** being overturned in all states except one. In New Hampshire, the fellow servant rule still applies.

ROLE OF SPECIFIC HEALTH PROBLEMS

Specific health problems that have been tied to workplace hazards have played significant roles in the development of the modern safety and health movement. These health problems contributed to public awareness of dangerous and unhealthy working conditions which, in turn, led to legislation, regulations, better work procedures, and better working conditions.

DISCUSSION CASE

What Is Your Opinion?

John Andrews and Andrea Washington are both safety and health professionals. Both hold safety-related college degrees from universities with strong safety and health programs. Both have 10 years of experience, and both are highly respected by their colleagues. They know each other well and often meet over lunch to discuss common problems, issues, and concerns. They are good friends. But on one issue, they always disagree. That issue is the role of organized labor in the safety movement. John Andrews thinks that labor's role has been insignificant. Andrea Washington thinks that it has been significant. What is your opinion?

Lung disease in coal miners was a major problem in the 1800s, particularly in Great Britain, where much of the Western world's coal was mined at the time. Frequent contact with coal dust led to a widespread outbreak of anthrocosis among Great Britain's coal miners. Also known as the black spit, this disease persisted from the early 1800s, when it was first identified, until around 1875 when it was finally eliminated by such safety and health measures as ventilation and decreased work hours.

In the 1930s, Great Britain saw a resurgence of lung problems among coal miners. By the early 1940s, British scientists were using the term *coal-miner's pneumoconiosis,* or CWP, to describe a disease from which many miners suffered. Great Britain designated CWP a separate and compensable disease in 1943. However, the United States did not immediately follow suit, even though numerous outbreaks of the disease had occurred among miners in this country.

The issue was debated in the United States until Congress finally passed the Coal Mine Health and Safety Act in 1969. The events that led up to passage of this act were tragic. An explosion in a coal mine in West Virginia in 1968 killed 78 miners. This tragedy focused attention on mining health and safety, and Congress responded by passing the Coal Mine Health and Safety Act. The act was amended in 1977 and again in 1978 to broaden the scope of its coverage.

Over the years, the diseases suffered by miners were typically lung diseases caused by the inhalation of coal dust particulates. However, health problems were not limited to coal miners. Other types of miners developed a variety of diseases, the most common of which was silicosis. Once again, it took a tragic event—the Gauley Bridge disaster—to focus attention on a serious workplace problem.

As part of a project to bring hydroelectric power to a remote part of West Virginia in the 1930s, a large tunnel had to be dug through the mountains. As it turned out, the tunnel cut through pure silica. Exposure to this substance led to numerous deaths which, in turn, provoked a public outcry. Following in the footsteps of *Uncle Tom's Cabin,* written by Harriet Beecher Stowe before the Civil War, a fictitious account of the Gauley Bridge disaster entitled *Hawk's Nest* by Hubert Skidmore whipped the public outcry into a frenzy, forcing Congress to respond.

Congress held a series of hearings on the matter in 1936. Also in 1936, representatives from business, industry, and government attended the National Silicosis Conference, convened by the U.S. Secretary of Labor. Among other outcomes of this conference was a finding that silica dust particulates did, in fact, cause silicosis.

Mercury poisoning is another health problem that has contributed to the evolution of the health and safety movement by focusing public attention on unsafe conditions in the workplace. The disease was first noticed among the citizens of a Japanese fishing village in the early 1930s. A disease with severe symptoms was common in Minamata, but extremely rare throughout the rest of Japan. After much investigation into the situation, it was determined that a nearby chemical plant periodically dumped methyl mercury into the bay that was the village's primary source of food. Consequently, the citizens of this small village ingested hazardous dosages of mercury every time they ate fish from the bay.

Mercury poisoning became an issue in the United States after a study was conducted in the early 1940s that focused on New York City's hat-making industry. During that time, many workers in this industry displayed the same types of symptoms that had been seen among the citizens of Minamata, Japan. Since mercury nitrate was used in the production of hats, enough suspicion was aroused to warrant a study. The study linked the symptoms of workers with the use of mercury nitrate. As a result, the use of this hazardous chemical in the hat-making industry was stopped, and a suitable substitute—hydrogen peroxide—was found.

Another important substance in the evolution of the modern health and safety movement has been asbestos. At one time, asbestos was considered a "wonder" material. Consequently, it was widely used, particularly in the building construction industry. By the time it was determined that asbestos is a hazardous material, the fibers of which can cause lung cancer (mesothelioma), thousands of buildings contained the substance. As these buildings began to age, the asbestos, particularly that used to insulate pipes, began to break down. As asbestos breaks down, it releases dangerous microscopic fibers into the air. These fibers are so hazardous that removing asbestos from old buildings has become a highly specialized task requiring special equipment and training.

DEVELOPMENT OF ACCIDENT PREVENTION PROGRAMS

In the modern workplace, there are many different types of **accident prevention** programs ranging from the simple to the complex. Widely used accident prevention techniques include failure minimization, fail-safe designs, isolation, lockouts, screening, personal protective equipment, redundancy, timed replacements, and many others. These techniques are individual components of broader safety programs. Such programs have evolved since the late 1800s.

In the early 1800s, employers had little concern for the safety of workers and little incentive to be concerned. Consequently, organized safety programs were nonexistent, a situation that continued for many years. However, between World War I and World War II, industry discovered the connection between quality and safety. Then, during World War II, troop call-ups and deployments created severe labor shortages. Faced with these shortages, employers could not afford to lose workers to accidents or for any other reason. This realization created a greater openness toward giving safety the serious consideration that it deserved. For example, according to the Society of Manufacturing Engineers, around this time industry began to realize the following:

■ Improved engineering could prevent accidents.

■ Employees were willing to learn and accept safety rules.

■ Safety rules could be established and enforced.

■ Financial savings from safety improvement could be reaped by savings in compensation and medical bills.[14]

With these realizations came the long-needed incentive for employers to begin playing an active role in creating and maintaining a safe workplace. This, in turn, led to the development of organized safety programs sponsored by management. Early safety programs were based on the **"Three E's of safety"**: Engineering, Education, and Enforcement (see Figure 1–2). The engineering aspects of a safety program involve making design improvements to both product and process. By altering the design of a product, the processes required to manufacture it can be simplified and, as a result, made less dangerous. In addition, the processes used to manufacture products can be engineered in ways that decrease potential hazards associated with the processes.

The education aspect of a safety program ensures that employees know how to work safely, why it is important to do so, and that safety is expected by management. Safety education typically covers the what, when, where, why, and how of safety.

The enforcement aspect of a safety program involves making sure that employees abide by safety policies, rules, regulations, practices, and procedures. Supervisors and fellow employees play a key role in the enforcement aspects of modern safety programs.

DEVELOPMENT OF SAFETY ORGANIZATIONS

Today, numerous organizations are devoted in full or at least in part to the promotion of safety and health in the workplace. Figure 1–3 lists professional organizations with

---

SAFETY FACT

Safety Movement and War

World War II actually had a positive effect on the modern safety and health movement. During the war, there was a shortage of able-bodied, skilled workers in factories supporting the war effort because most of these workers were in the armed services. Consequently, preserving the safety and health of the relatively few skilled workers still available was paramount. The law of supply and demand suddenly made workplace safety a significant issue, which it still is today. The military war is over, but the economic war still rages. To be competitive in this international conflict, employers today must follow the lead of their predecessors during World War II and protect their employees.

Figure 1–2
Three E's of safety.

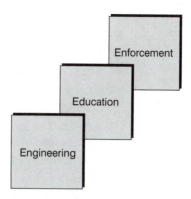

workplace safety as part of their missions. Figure 1–4 lists governmental agencies concerned with safety and health. These lists are extensive now, but this has not always been the case. Safety organizations in this country had humble beginnings.

The grandfather of them all is the National Safety Council. The Society of Manufacturing Engineers traces the genesis of this organization as follows:

> The Association of Iron and Steel Electrical Engineers was organized in the first decade of the 20th century and devoted much attention to safety problems in its industry. In 1911, a request came from this association to call a national industrial safety conference. The first

Figure 1–3
Professional organizations concerned with workplace safety.

Alliance for American Insurers

American Board of Industrial Hygiene

American Council of Government Industrial Hygienists

American Industrial Hygiene Association

American Insurance Association

American National Standards Institute

American Occupational Medical Association

American Society of Mechanical Engineers

American Society of Safety Engineers

American Society for Testing and Materials

Chemical Transportation Emergency Center

Human Factors Society

National Fire Protection Association

National Safety Council

National Safety Management Society

Society of Automotive Engineers

System Safety Society

Underwriters Laboratories

Figure 1–4
Government agencies concerned with workplace safety.

American Public Health Association

Bureau of Labor Statistics

Bureau of National Affairs

Commerce Clearing House

Environmental Protection Agency

National Institute for Standards and Technology (formerly National Bureau of Standards)

National Institute of Occupational Safety and Health

Occupational Safety and Health Administration

Superintendent of Documents, U.S. Government Printing Office

U.S. Consumer Product Safety Commission

Cooperative Safety Congress met in Milwaukee in 1912. A year later, at a meeting in New York City, the National Council of Industrial Safety was formed. It began operation in a small office in Chicago. At its meeting in 1915, the organization's name was changed to the National Safety Council (NSC).[15]

Today, the NSC is the largest organization in the United States devoted solely to safety and health practices and procedures. Its purpose is to prevent the losses, both direct and indirect, arising out of accidents or from exposure to unhealthy environments. Although it is chartered by an act of Congress, the NSC is a nongovernmental, not-for-profit, public service organization.

The Occupational Safety and Health Administration (OSHA) is the government's administrative arm for the Occupational Safety and Health Act. Formed in 1970, OSHA sets and revokes safety and health standards, conducts inspections, investigates problems, issues citations, assesses penalties, petitions the courts to take appropriate action against unsafe employers, provides safety training, provides injury prevention consultation, and maintains a database of health and safety statistics.

Another governmental organization is the National Institute of Occupational Safety and Health **(NIOSH).** This organization is part of the Centers for Disease Control of the Department of Health and Human Services. NIOSH is required to publish annually a comprehensive list of all known toxic substances. NIOSH will also provide on-site tests of potentially toxic substances so that companies know what they are handling and what precautions to take.

SAFETY AND HEALTH MOVEMENT TODAY

The safety and health movement has come a long way since the Industrial Revolution. Today, there is widespread understanding of the importance of providing a safe and healthy workplace. The tone was set during and after World War II when all of the various practitioners of occupational health and safety began to see the need for cooperative efforts. These practitioners included safety engineers, safety managers, industrial hygienists, occupational health nurses, and physicians.

One of the earliest and most vocal proponents of the cooperative or integrated approach was H. G. Dyktor. He proposed the following objectives of integration:

- Learn more through sharing knowledge about health problems in the workplace, particularly those caused by toxic substances.
- Provide a greater level of expertise in evaluating health and safety problems.
- Provide a broad database that can be used to compare health and safety problems experienced by different companies in the same industry.
- Encourage accident prevention.
- Make employee health and safety a high priority.[16]

Integrated Approach to Safety and Health

This integrated approach has become the norm that typifies the health and safety movement of today. By working together and drawing on their own respective areas of expertise, safety and health professionals are better able to identify, predict, control, and correct safety and health problems.

OSHA reinforces the integrated approach by requiring companies to have a plan for doing at least the following: (1) providing appropriate medical treatment for injured or ill workers, (2) regularly examining workers who are exposed to toxic substances, and (3) having a qualified first-aid person available during all working hours.

Smaller companies may contract out the fulfillment of these requirements. Larger companies often maintain a staff of safety and health professionals. According to Hamilton and Hardy, the health and safety staff in a modern industrial company might include the following positions:

- *Industrial hygiene chemist and/or engineer.* Companies that use toxic substances may employ industrial hygiene chemists periodically to test the work environment and the people who work in it. In this way, unsafe conditions or hazardous levels of exposure can be identified early and corrective/preventive measures can be taken. Dust levels, ventilation, and noise levels are also monitored by individuals serving in this capacity.
- *Radiation control specialist.* Companies that use or produce radioactive materials employ radiation control specialists who are typically electrical engineers or physicists. These specialists monitor the radiation levels to which workers may be exposed, test workers for levels of exposure, respond to radiation accidents, develop company-wide plans for handling radiation accidents, and implement decontamination procedures when necessary.
- *Industrial safety engineer or manager.* Individuals serving in this capacity are safety and health generalists with specialized education and training. In larger companies,

they may be devoted to safety and health matters. In smaller companies, they may have other duties in addition to safety and health. In either case, they are responsible for developing and carrying out the company's overall safety and health program including accident prevention, accident investigation, and education/training.[17]

Other professionals who might be part of a company's safety and health team include occupational nurses, physicians, psychologists, counselors, educators, and dietitians.

New Materials, New Processes, and New Problems

The job of the safety and health professional is more complex than it has ever been. The materials out of which products are made have become increasingly complex and exotic. Engineering metals now include carbon steels, alloy steels, high-strength low-alloy steels, stainless steels, managing steels, cast steels, cast irons, tungsten, molybdenum, titanium, aluminum, copper, magnesium, lead, tin, zinc, and powdered metals. Each of these metals requires its own specialized processes.

Nonmetals are more numerous and have also become more complex. Plastics, plastic alloys and blends, advanced composites, fibrous materials, elastomers, and ceramics also bring their own potential hazards to the workplace.

In addition to the more complex materials being used in modern industry and the new safety and health concerns associated with them, modern industrial processes are also becoming more complex. As these processes become automated, the potential hazards associated with them often increase. Computers; lasers; industrial robots; nontraditional processes such as explosive welding, photochemical machining, laser beam machining, ultrasonic machining, and chemical milling; automated material handling; expert systems; flexible manufacturing cells; and computer-integrated manufacturing have all introduced new safety and health problems in the workplace and new challenges for the safety and health professional.

Chapter 22 is devoted to coverage of the special safety and health problems associated with computers, robots, and automation. In addition, coverage of specific aspects of these problems is provided in different chapters throughout this book.

Rapid Growth in the Profession

The complexities of the modern workplace have made safety and health a growing profession. Associate and baccalaureate degree programs in industrial technology typically include industrial safety courses. Some engineering degree programs have safety and health tracks. Some colleges and universities offer full degrees in occupational safety and health.

The inevitable result of the increased attention given to industrial safety and health is that more large companies are employing safety and health professionals and more small companies are assigning these duties to existing employees. This is a trend that is likely to continue as employers see their responsibilities for safety and health spread beyond the workplace to the environment, the community, the users of their products, and the recipients of the by-products and waste.

SUMMARY

1. Safety and health awareness has a long history. There is evidence of occupational safety and health efforts as far back as the time of the Egyptian pharaohs. The Code of Hammurabi, circa 2000 B.C., contained clauses that could be interpreted as early attempts at workers' compensation. There is also evidence of concern for safety and health during the time of the Romans.
2. Milestones in the development of the safety movement in the United States include the following: first recorded safety program in 1892, creation of the Bureau of Mines in 1907, passage of the first effective workers' compensation law in the United States in 1911, and passage of OSHA in 1970.
3. Organized labor has played a crucial role in the development of the safety movement in the United States. Particularly important was the work of unions to overturn antilabor laws inhibiting safety in the workplace.
4. Specific health problems associated with the workplace have contributed to the development of the modern safety and health movement. These problems include lung diseases in miners, mercury poisoning, and lung cancer tied to asbestos.
5. Widely used accident prevention techniques include failure minimization, fail-safe designs, isolation, lockouts, screening, personal protective equipment, redundancy, and timed replacements.
6. The development of the safety movement in the United States has been helped by the parallel development of safety organizations. Prominent among these are the National Safety Council, the National Safety Management Society, the American Society of Safety Engineers, and the American Industrial Hygiene Association.
7. The safety and health movement today is characterized by professionalization and integration. The safety and health team of a large company might include an industrial chemist or engineer, radiation control specialist, industrial safety engineer or manager, occupational nurse, counselor, psychologist, and dietitian. New materials and processes are introducing new safety and health problems, making the integrated approach a practical necessity and promoting growth in the profession.

KEY TERMS AND CONCEPTS

Accident prevention	Industrial safety engineer
Assumption of risk	Industrial safety manager
Code of Hammurabi	National Council of Industrial Safety
Contributory negligence	NIOSH
Cooperative Safety Congress	Organized labor
Employer-biased laws	OSHA
Employer liability	Radiation control specialist
Fellow servant rule	Safety movement
Inanimate power	Three E's of safety
Industrial hygiene chemist	Workers' compensation

REVIEW QUESTIONS

1. To what cause(s) can the improvements in workplace safety made to date be attributed?
2. Explain the significance of the Code of Hammurabi in terms of the safety movement.
3. Describe the circumstances that led to the development of the first organized safety program.
4. What is Frederick Taylor's connection to the safety movement?
5. Explain the development of the National Safety Council.
6. What impact did labor shortages in World War II have on the safety movement?
7. Explain the primary reasons behind the passage of OSHA.
8. Summarize briefly the role that organized labor has played in the advancement of the safety movement.
9. Define the following terms: fellow servant rule, contributory negligence, and assumption of risk.
10. Explain the Three E's of safety.
11. Explain the term *integration* as it relates to modern safety and health.

ENDNOTES

1. Society of Manufacturing Engineers. *Manufacturing Management* (Dearborn, MI: Society of Manufacturing Engineers, 1989), p. 12-1.
2. National Safety Council. *Accident Facts,* 1990 ed.
3. LaDou, J. (ed.). *Introduction to Occupational Health and Safety* (Chicago: National Safety Council, 1986), p. 28.
4. Ibid., p. 28
5. Soubiran, A. "Medical Services under the Pharaohs," *Abbottempo,* 1963, 1, pp. 19–23.
6. LaDou, J. (ed.). *Introduction to Occupational Health and Safety* (Chicago: National Safety Council, 1986), p. 31.
7. Ibid., p. 34.
8. Ibid., p. 35.
9. Ibid., p. 37.
10. Society of Manufacturing Engineers. *Manufacturing Management,* p. 12-1.
11. Ibid., p. 12–2.
12. Ibid., p. 12–2.
13. Ibid., p. 12–1.
14. Ibid., p. 12–2.
15. Ibid., p. 12–1.
16. Dyktor, H. G. "Integration of Industrial Hygiene with Industrial Medicine," *Industrial Medicine,* 1940, Vol. 9, No. 4, p. 193.
17. Hamilton, A. and Hardy, H. *Industrial Toxicology* (Boston: John Wright Publishing Sciences Group, 1983).

Accidents and Their Effects

There is a long history of debate in this country concerning the effect of **accidents** on industry (the workers and the companies) and the cost of preventing accidents. Historically, the prevailing view was that **accident prevention** programs were too costly. The more contemporary view is that accidents are too costly and that accident prevention makes sense economically. As a result, accident prevention, which had been advocated on a moral basis, is now justified in economic terms.

Accidents are the fourth leading cause of death in this country after heart disease, cancer, and strokes. This ranking is based on all types of accidents including motor vehicle accidents, drownings, fires, falls, natural disasters, and work-related accidents.

Although deaths from **natural disasters** tend to be more newsworthy than workplace deaths, their actual impact is substantially less. For example, the most publicized natural disasters in the United States in 1989 were Hurricane Hugo (26 deaths) and the

100 deaths/year on avg

San Francisco/Oakland earthquake (61 deaths). During that same year, there were *more than* 10,400 accidental workplace deaths in this country.[1] The following quote from the National Safety Council puts **workplace accidents** and deaths in the proper perspective, notwithstanding their apparent lack of newsworthiness.

> While you make a 10-minute speech—2 persons will be killed and about 170 will suffer a disabling injury. Costs will amount to $2,800,000. On the average, there are 11 accidental deaths and about 1,030 disabling injuries every hour during the year.[2]

This chapter provides prospective and practicing **safety and health professionals** with the information they need to have a full understanding of workplace accidents and their effect on industry in the United States. Such an understanding will help professionals play a more effective role in keeping both management and labor focused appropriately on safety and health in the workplace.

COSTS OF ACCIDENTS

The following news brief appeared in the *Occupational Health & Safety Letter:*

> An Illinois contracting firm was fined $750,000 last month on charges that it willfully ignored federal safety rules and caused the death of three workers in a 1988 sewer tunnel explosion in Milwaukee.[3]

To gain a proper perspective on the economics of workplace accidents, we must view them in the overall context of all accidents. The overall cost of accidents in the United States is approximately $150 billion. These costs include such factors as **lost wages, medical expenses, insurance administration, fire-related losses,** motor vehicle **property damage,** and **indirect costs.**

Figure 2–1 breaks down this overall amount by categories of accidents. Figure 2–2 breaks them down by cost categories. Notice in Figure 2–1 that workplace accidents rank second behind motor vehicle accidents in cost. Figure 2–2 shows that the highest cost category is wages lost by workers who are either injured or killed. The category of indirect losses from work accidents consists of costs associated with responding to accidents (i.e., giving first aid, filling out accident reports, handling production slowdowns).

Clearly accidents on and off the job cost U.S. industry dearly. Every dollar that is spent responding to accidents is a dollar that could have been reinvested in modernization, research and development, facility upgrades, and other competitiveness-enhancing activities.

Figure 2–1
Accident costs by accident type (in billions, in a typical year).

Motor vehicle accidents	$722.0
Workplace accidents	48.5
Home accidents	18.2
Public accidents	12.5

Figure 2–2
Accident costs by categories
(in billions, in a typical year).

Wages lost	$37.7
Medical expenses	23.7
Insurance administration	28.4
Property damage (motor vehicle)	26.8
Fire losses	9.4
Indirect losses for work accidents	22.5

ACCIDENTAL DEATHS IN THE UNITED STATES

On March 3, 1991, an explosion occurred at an oil refining company in Louisiana, killing four workers. The explosion occurred as three gasoline synthesizing units were being brought back on line after having been down for maintenance. Workplace deaths caused by explosions are not unheard of, but neither are they common.

Accidental deaths in the United States result from a variety of causes, including motor vehicle accidents, falls, poisoning, drowning, fire-related injuries, **suffocation** (ingested object), firearms, medical complications, air transport accidents, machinery, mechanical suffocation, and the impact of falling objects. The National Safety Council periodically computes death totals and death rates in each of these categories. The statistics for a typical year are as follows:

■ *Motor vehicle accidents.* Motor vehicle accidents are the leading cause of accidental deaths in the United States each year. They include deaths resulting from accidents involving mechanically or electrically powered vehicles (excluding rail vehicles) that occur on or off the road. In a typical year, there are approximately 47,000 such deaths in the United States.

■ *Falls.* This category includes all deaths from **falls** except those associated with transport vehicles. For example, a person who is killed as the result of falling while boarding a bus or train would not be included in this category. In a typical year, there are approximately 13,000 deaths in the United States from falls.

■ *Poisoning.* The **poisoning** category is divided into two subcategories: (1) poisoning by solids and liquids, and (2) poisoning by gases and vapors. The first category includes deaths that result from the ingestion of drugs, medicine, widely recognized solid and liquid poisons, mushrooms, and shellfish. It does not include poisoning from spoiled food or salmonella. The second category includes death caused by incomplete combustion (for example, gas vapors from an oven or unlit pilot light) or from carbon monoxide (for example, exhaust fumes from an automobile). In a typical year, there are approximately 6,000 deaths in the first category and 1,000 in the second.

■ *Drowning.* This category includes work-related and nonwork-related **drownings** but excludes those associated with floods or other natural disasters. In a typical year, there are approximately 5,000 deaths from drowning in the United States.

■ *Fire-related injuries.* This category includes deaths from burns, asphyxiation, falls, and those that result from falling objects in a fire. In a typical year, there are over 4,000 fire-related deaths in the United States.

■ *Suffocation (ingested object).* This category includes deaths from the ingestion of an object that blocks the air passages. In many such deaths, the ingested object is food. In a typical year, there are approximately 4,000 suffocation deaths in the United States.

■ *Firearms.* This category includes deaths that result when recreational activities involving firearms or household accidents involving firearms result in death. For example, a person killed in the home while cleaning a firearm would be included in this category. However, a person killed in combat would not be. In a typical year, there are approximately 2,000 deaths in this category.

■ *Others.* This category includes deaths resulting from **medical complications** arising out of mistakes made by health care professionals, air transport injuries, interaction with machinery, **mechanical suffocation,** and the impact of falling objects. In a typical year, there are over 14,000 deaths in these subcategories.[4]

ACCIDENTS VS. OTHER CAUSES OF DEATH

Although there are more deaths each year from **heart disease, cancer,** and **strokes** than from accidents, these causes tend to be concentrated among people at or near retirement age. Among people 37 years of age or younger—prime working years—accidents are the number one cause of death. Figure 2–3 summarizes the causes of death for persons from 25 to 44 years of age. Notice that the leading cause is accidents.

Figure 2–3 shows that accidents represent a serious detriment to productivity, quality, and competitiveness in today's workplace. Yet accidents are the one cause of death and injury that companies can most easily control. Although it is true that companies might have some success in decreasing the incidence of heart disease and stroke among

Figure 2–3
Causes of accidents (ages 25 to 44 years in a typical year).

Accidents	27,484
Motor vehicle	16,405
Poison (solid, liquid)	2,649
Drowning	1,516
Falls	1,138
Fire-related	899
Cancer	20,305
Heart disease	15,874

their employees through such activities as corporate wellness programs, their impact in this regard will be limited. However, employers can have a significant impact on preventing accidents.

WORK ACCIDENT COSTS AND RATES

Workplace accidents cost employers millions every year. Consider the following examples from 1990. Arco Chemical Company was ordered to pay $3.48 million in fines as a result of failing to protect workers from an explosion at its petrochemical plant in Channelview, Texas. The steel-making division of USX paid a $3.25 million fine to settle numerous health and safety violation citations. BASF Corporation agreed to pay a fine of $1.06 million to settle OSHA citations associated with an explosion at a Cincinnati chemical plant that caused two deaths and seventeen injuries.

These examples show the costs of fines only. In addition to fines, these employers incurred costs for safety corrections, medical treatment, survivor benefits, death/burial costs, and a variety of indirect costs. Clearly, work accidents are expensive. However, the news is not all bad. The trend in the rate of accidents is downward.

Work **accident rates** in this century are evidence of the success of the safety movement in the United States. As the amount of attention given to workplace safety and health has increased, the accident rate has decreased. According to the National Safety Council,

> Between 1912 and 1993, accidental work deaths per 100,000 population were reduced 81 percent, from 21 to 4. In 1912, an estimated 18,000 to 21,000 workers' lives were lost. In 1993, in a workforce more than triple in size and producing 11 times the goods and services, there were approximately 10,000 work deaths.[5]

As was shown in Figure 2–1, the cost of these 10,000 work deaths and work injuries was $48.5 billion. This translates into a cost of $420 per worker in the United States, computed as the value-added required per worker to offset the cost of work injuries. It translates further into $610,000 per death and $18,000 per disabling injury.[6]

Although statistics are not available to document the supposition, many safety and health professionals believe that the major cost of accidents and injuries on the job results from damage to morale. Employee morale is a less tangible factor than documentable factors such as lost time and medical costs. However, it is widely accepted among management professionals that few factors affect productivity more than employee morale. Employees with low morale do not produce up to their maximum potential. This is why so much time and money are spent every year to help supervisors and managers learn different ways to help improve employee morale.

Since few things are so detrimental to employee morale as seeing a fellow worker injured, accidents can have a devastating effect on morale. Whenever an employee is injured, his or her colleagues silently think, "That could have been me," in addition to worrying about the employee. Morale is damaged even more if the injured employee is well liked and his or her family is known to other employees.

SAFETY FACT

Safety Disasters Have Long-Term Effects

A safety disaster can continue to haunt an organization for years, no matter how well it responds. A case in point is the chemical release that occurred on December 3, 1984, at a Union Carbide plant in Bhopal, India. Thousands of employees and nearby residents were killed as a cloud of toxic chemical mist escaped from the plant and was carried on the wind through a surrounding community. Many years later, human rights and environmental groups are still attempting to close Union Carbide down. These groups continue to ask New York's Attorney General to take the legal action necessary to revoke Union Carbide's corporate charter.

TIME LOST BECAUSE OF WORK INJURIES[7]

An important consideration when assessing the effect of accidents on industry is the amount of **lost time** due to **work injuries.** According to the National Safety Council, approximately 35,000,000 hours are lost in a typical year as a result of accidents. This is actual time lost from disabling injuries and does not include additional time lost for medical checkups after the injured employee returns to work. Accidents that occurred in previous years often continue to cause lost time in the current year.

DEATHS IN WORK ACCIDENTS

Deaths on the job have decreased markedly over the years. However, they still occur. For example, in a typical year, there are 10,400 work deaths in the United States. The causes of death in the workplace vary. They include those related to motor vehicles, falls, electric current, drowning, fires, air transport, poison, water transport, machinery, falling objects, rail transports, and mechanical suffocation.[8] Figure 2–4 gives a complete breakdown of the percentages for the various categories of causes.

WORK INJURIES BY TYPE OF ACCIDENT

Work injuries can be classified by the type of accident from which they resulted. The most common causes of work injuries are

- Overexertion
- Impact accidents
- Falls
- Bodily reaction (to chemicals)

Figure 2–4
Work deaths by cause for a typical year.

Motor vehicle related	37.2%
Falls	12.5
Electric current	3.7
Drowning	3.2
Fire related	3.1
Air transport related	3.0
Poison (solid, liquid)	2.7
Water transport related	1.6
Poison (gas, vapor)	1.4
Other	31.6

- Compression
- Motor vehicle accidents
- Exposure to radiation/caustics
- Rubbing or abrasions
- Exposure to extreme temperatures

Overexertion, the result of employees working beyond their physical limits, is the leading cause of work injuries. According to the National Safety Council, almost 31 percent of all work injuries are caused by overexertion. **Impact accidents** involve a worker being struck by or against an object. The next most prominent cause of work injuries is falls.[9] The remaining accidents are distributed fairly equally among the other causes just listed.

DISCUSSION CASE

What Is Your Opinion?

Mack Jones has been a safety engineer at Zumwalt Processing Company for almost 25 years. His son David is a recent college graduate who has been the assistant safety director at another company for just six months. Over supper last night, they had a discussion about work injuries. During the discussion, Mack said he thought back injuries were still the problem that safety professionals should worry about the most. David disagreed. He said repetitive-motion/soft tissue injuries such as carpal tunnel syndrome were a bigger problem in the modern computerized workplace. What is your opinion?

DEATH RATES BY INDUSTRY[10]

A variety of agencies and organizations including the Bureau of Labor Statistics, the National Center for Health Statistics, and the National Safety Council collect data on **death rates** within industrial categories. Such information can be used in a variety of ways, not the least of which is in assigning workers' compensation rates. The most widely used industrial categories are agriculture, including farming, forestry, and fishing; mining/quarrying, including oil and gas drilling and extraction; construction; manufacturing; transportation/public utilities; trade, both wholesale and retail; services, including finance, insurance, and real estate; and federal, state, and local government.

When death rates are computed on the basis of the number of deaths per 100,000 workers in a given year, the industry categories rank as follows (from highest death rate to lowest):

1. Mining/quarrying
2. Agriculture
3. Construction
4. Transportation/public utilities
5. Government
6. Manufacturing
7. Services
8. Trade

The rankings sometimes change slightly from year to year. For example, agriculture and mining/quarrying may exchange the first and second ranking in any given year. This is also true at the low end of the rankings with services and trade. However, generally, the typical ranking is as listed.

PARTS OF THE BODY INJURED ON THE JOB

In order to develop and maintain an effective safety and health program, it is necessary to know not only the most common causes of death and injury but also the parts of the body most frequently injured. According to the National Safety Council,

Disabling work injuries in the entire nation totaled approximately ~~1.8~~ 1.75 million in ~~1993~~ 1998. Of these, about 10,400 were fatal and 60,000 resulted in some permanent impairment. Injuries to the back occurred most frequently, followed by thumb and finger injuries and leg injuries.[11]

Typically, the most frequent injuries to specific parts of the body are as follows (from most frequent to least):

1. Back
2. Legs and fingers
3. Arms and multiple parts of the body
4. Trunk
5. Hands
6. Eyes, head, and feet
7. Neck, toes, and body systems

The back is the most frequently injured part of the body. Legs and fingers are injured with approximately the same frequency, as are arms and multiple parts of the body; the hands are next in frequency, followed by the eyes, the head, and feet; and neck, toes, and body systems. This ranking shows that one of the most fundamental components of a safety and health program should be instruction on how to lift without hurting the back (see Chapter 9).

CHEMICAL BURN INJURIES

Chemical burn injuries are a special category with which prospective and practicing safety professionals should be familiar. According to the Bureau of Labor Statistics, of the various industry categories, the greatest incidence of chemical burns (approximately one-third) occurs in manufacturing.[12] Other high incidence industries are services, trade, and construction.

The chemicals that most frequently cause chemical burn injuries include acids and alkalies; soaps, detergents, and cleaning compounds; solvents and degreasers; calcium hydroxide (a chemical used in cement and plaster); potassium hydroxide (an ingredient in drain cleaners and other cleaning solutions); and sulfuric acid (battery acid). Almost 46 percent of all chemical burn injuries occur while workers are cleaning equipment, tools, and vehicles.[13]

What is particularly disturbing about chemical burn injuries is that a high percentage of them occur in spite of the use of personal protective equipment, the provision of safety instruction, and the availability of treatment facilities. In some cases, the personal protective equipment is faulty or inadequate. In others, it is not properly used in spite of instructions.

Preventing chemical burn injuries presents a special challenge to safety and health professionals. The following strategies are recommended:

- Familiarize yourself, the workers, and their supervisors with the chemicals that will be used and their inherent dangers.
- Secure the proper personal protection equipment for each type of chemical that will be used.
- Provide instruction on the proper use of personal protection equipment and then make sure that supervisors confirm that the equipment is used properly every time.
- Monitor that workers are wearing personal protection equipment and replace it when it begins to show wear.

HEAT BURN INJURIES

Heat burn injuries present a special challenge to safety and health professionals in the modern workplace. Almost 40 percent of all such injuries occur in manufacturing every year. The most frequent causes are flame (this includes smoke inhalation injuries), molten metal, petroleum asphalts, steam, and water. The most common activities associated with heat burn injuries are welding, cutting with a torch, and handling tar or asphalt.[14]

Following are several factors that contribute to heat burn injuries in the workplace. Safety and health professionals who understand these factors will be in a better position to prevent heat burn injuries.

- Employer has no health and safety policy regarding heat hazards.
- Employer fails to enforce safety procedures and practices.
- Employees are not familiar with the employer's safety policy and procedures concerning heat hazards.
- Employees fail to use or improperly use personal protection equipment.
- Employees have inadequate or worn personal protection equipment.
- Employees work in a limited space.
- Employees attempt to work too fast.
- Employees are careless.
- Employees have poorly maintained tools and equipment.[15]

These factors should be considered carefully by safety and health professionals when developing accident prevention programs. Employees should be familiar with the hazards, should know the appropriate safety precautions, and should have and use the proper personal protection equipment. Safety professionals should monitor to ensure that safety rules are being followed, that personal protection equipment is being used correctly, and that it is in good condition.

REPETITIVE STRAIN/SOFT TISSUE INJURIES

Repetitive strain injury (RSI) is a broad and generic term that encompasses a variety of injuries resulting from cumulative trauma to the soft tissues of the body, including tendons, tendon sheaths, muscles, ligaments, joints, and nerves. Such injuries are typically associated with the soft tissues of the hands, arms, neck, and shoulders.

Carpal tunnel syndrome (CTS) is the most widely known repetitive strain injury. There are also several other repetitive strain injuries to the body's soft tissues. The carpal tunnel is the area inside the wrist through which the median nerve passes. It is formed by the wrist bones and a ligament. CTS is typically caused by repeated and cumulative stress on the median nerve. Symptoms of CTS include numbness, a tingling sensation, and pain in the fingers, hand, and/or wrist.

Stress placed on the median nerve typically results from repeated motion while the hands and fingers are bent in an unnatural position. However, sometimes the stress results from a single traumatic event such as a sharp blow to the wrist.

Evidence suggests a higher incidence of CTS among women than men. Scientists at the University of Washington found that the incidence rate for CTS among women was 1.96 per 1,000 full-time equivalent (FTE) workers and 1.58 per 1,000 FTE for men. According to their study, the overall incidence rate for CTS is increasing at a rate of more than 15 percent per year.[16]

A common misconception about RSI is that it is synonymous with carpal tunnel syndrome. It isn't. In fact, carpal tunnel syndrome is relatively rare among RSI patients. Following is a list of the broad classifications of RSI and the types commonly associated with each classification:

Muscle and Tendon Disorders
- Tendinitis
- Muscle damage
- Tenosynovitis
- Stenosing tenosynovitis
    DeQuervain's disease
    Flexor tenosynovitis (trigger finger)
    Shoulder tendinitis
    Forearm tendinitis

Cervical Radioculapathy
- Epicondylitis
- Ganglion cysts

Tunnel Syndromes
- Carpal tunnel syndrome
- Radial tunnel syndrome

Ulnar Nerve Disorders
- Sulcus ulnaris syndrome
- Cubital tunnel syndrome
- Guyon's canal syndrome

Nerve and Circulation Disorders
- Thoracic outlet syndrome
- Raynaud's disease

Other Associated Disorders
- Reflex sympathetic dysfunction
- Focal dystonia (writer's cramp)
- Degenerative joint disorder (osteoarthritis)
- Fibromyalgis
- Dupuytreu's contracture[17]

RSI and its manifestations are explained in greater detail in Chapter 6.

ESTIMATING THE COST OF ACCIDENTS

Even decision makers who support accident prevention must consider the relative costs of such efforts. Clearly, accidents are expensive. However, to be successful, safety and health professionals must be able to show that accidents are more expensive than prevention. To do this, they must be able to estimate the cost of accidents. The procedure for estimating costs set forth in this section was developed by Professor Rollin H. Simonds of Michigan State College working in conjunction with the Statistics Division of the National Safety Council.

Cost Estimation Method

Professor Simonds states that in order to have value, a cost estimate must relate directly to the specific company in question. Applying broad industry cost factors will not suffice. To arrive at company-specific figures, Simonds recommends that costs associated with an accident be divided into *insured* and *uninsured* costs.[18]

Determining the insured costs of accidents is a simple matter of examining accounting records. The next step involves calculating the uninsured costs. Simonds recommends that accidents be divided into the following four classes:

- *Class 1 accidents.* Lost workdays, permanent partial disabilities, and temporary total disabilities.
- *Class 2 accidents.* Treatment by a physician outside of the company's facility.
- *Class 3 accidents.* Locally provided first aid, property damage of less than $100, or the loss of fewer than eight hours of work time.
- *Class 4 accidents.* Injuries that are so minor they do not require the attention of a physician, result in property damage of $100 or more, or cause eight or more work hours to be lost.[19]

Average uninsured costs for each class of accident can be determined by pulling the records of all accidents that occurred during a specified period and sorting the records according to class. For each accident in each class, record every cost that was not covered by insurance. Compute the total of these costs by class of accident and divide by the total number of accidents in that class to determine an average uninsured cost for each class, specific to the particular company.

Figure 2–5 is an example of how the average cost of a selected sample of Class 1 accidents might be determined. In this example, there were four Class 1 accidents in the pilot test. These four accidents cost the company a total of $554.23 in uninsured costs, or an average of $138.56 per accident. Using this information, accurate estimates of the cost of an accident can be made as can accurate predictions.

Other Cost Estimation Methods

The costs associated with workplace accidents, injuries, and incidents fall into broad categories such as the following:

Class of Accident	Accident Number							
Class 1	1	2	3	4	5	6	7	8
Cost A	16.00	6.95	15.17	3.26				
Cost B	72.00	103.15	97.06	51.52				
Cost C	26.73	12.62	—	36.94				
Cost D	—	51.36	—	38.76				
Cost E	—	11.17	—	24.95				
Cost F	—	—	—	−13.41				
Cost G	—	—	—	—				
Total	114.73	185.25	112.23	142.02				

Grand Total: $554.23

Average Cost per Accident: $138.56 (grand total ÷ number of accidents)

Signature: _____ Date: _____

Figure 2–5
Uninsured costs worksheet.

- Lost work hours
- Medical costs
- Insurance premiums and administration
- Property damage
- Fire losses
- Indirect costs

Calculating the direct costs associated with lost work hours involves compiling the total number of lost hours for the period in question and multiplying the hours times the applicable loaded labor rate. The loaded labor rate is the employee's hourly rate plus benefits. Benefits vary from company to company but typically inflate the hourly wage by 20 to 35 percent. A sample cost-of-lost-hours computation follows:

$$\text{Employee Hours Lost (4th quarter)} \times \text{Average Loaded Labor Rate} = \text{Cost}$$
$$386 \times \$13.48 = \$5,203.28$$

In this example, the company lost 386 hours due to accidents on the job in the fourth quarter of its fiscal year. The employees who actually missed time at work formed a pool of people with an average loaded labor rate of $13.48 per hour ($10.78 average hourly wage plus 20 percent for benefits). The average loaded labor rate multiplied times the 386 lost hours reveals an unproductive cost of $5,203.28 to this company.

By studying records that are readily available in the company, a safety professional can also determine medical costs, insurance premiums, property damage, and fire losses for the time period in question. All of these costs taken together result in a subtotal cost. This figure is then increased by a standard percentage to cover indirect costs to determine the total cost of accidents for a specific time period. The percentage used to calculate indirect costs can vary from company to company, but 20 percent is a widely used figure.

SUMMARY

1. The approximate cost of accidents in the United States is $150 billion annually. This includes the direct and indirect costs of accidents that occur on and off the job.
2. The leading causes of accidental deaths in the United States are motor vehicle accidents, falls, poisoning, drowning, fire-related injuries, suffocation, firearms, medical complications, air transport accidents, machinery-related injuries, mechanical suffocation, and the impact of falling objects.
3. The leading causes of death in the United States are heart disease, cancer, and stroke. However, these causes are concentrated among people at or near retirement age. Among people 37 years of age and younger, accidents are the number one cause of death.
4. Between 1912 and 1993, the number of accidental work deaths per 100,000 population declined by 81 percent, from 21 to 4.
5. In 1989, 35,000,000 work hours were lost as a result of accidents. This is actual time lost from disabling injuries and does not include additional time lost to medical checkups after the injured employee returns to work.
6. The leading causes of death in work accidents are motor vehicle-related, falls, electric current, drowning, fire-related, air transport-related, poison, and water transport-related.
7. The leading causes of work injuries are overexertion, impact accidents, falls, bodily reaction, compression, motor vehicle accidents, exposure to radiation/caustics, rubbing or abrasions, and exposure to extreme temperatures.
8. When death rates are computed on the basis of number of deaths per 100,000 workers, the industry categories are ranked as follows (from highest death rate to lowest): mining/quarrying, agriculture, construction, transportation/public utilities, government, manufacturing, services, and trade.
9. Typically, the ranking of injuries to specific parts of the body are as follows (from most frequently injured to least): back; legs and fingers; arms and multiple parts of body; trunk; hands; eyes, head, and feet; and neck, toes, and body systems.
10. The chemicals most frequently involved in chemical burn injuries include acids and alkalies; soaps, detergents, and cleaning compounds; solvents and degreasers; calcium hydroxide; potassium hydroxide; and sulfuric acid.

11. The most frequent causes of heat burn injuries are flame, molten metal, petroleum, asphalt, steam, and water.
12. Carpal tunnel syndrome (CTS) is an injury to the median nerve in the wrist that typically results from repeated stress placed on the nerve. Symptoms of CTS include numbness, a tingling sensation, and pain in the hand and/or wrist.
13. Repetitive strain injury (RSI) is a broad and generic term that encompasses a variety of injuries resulting from cumulative trauma to the soft tissues of the body, including tendons, tendon sheaths, muscles, ligaments, joints, and nerves. Such injuries are typically associated with the soft tissues of the hands, arms, neck, and shoulders.

KEY TERMS AND CONCEPTS

Accident prevention	Lost wages
Accident rate	Mechanical suffocation
Accidents	Medical complications
Carpal tunnel syndrome	Medical expenses
Chemical burn injuries	Natural disasters
Death rates	Overexertion
Drowning	Poisoning
Falls	Property damage
Fire-related losses	Repetitive strain injury
Heat burn injuries	Safety and health professional
Impact accidents	Suffocation
Indirect costs	Work injuries
Insurance administration	Workplace accidents
Lost time	

REVIEW QUESTIONS

1. What are the leading causes of death in the United States?
2. When the overall cost of an accident is calculated, what elements make up the cost?
3. What are the five leading causes of accidental deaths in the United States?
4. What are the leading causes of death in the United States of people between the ages of 25 and 44?
5. Explain how the rate of accidental work deaths now compares with the rate in the early 1900s.
6. What are the five leading causes of work deaths?
7. What are the five leading causes of work injuries by type of accident?

8. When death rates are classified by industry type, what are the three leading industry types?
9. Rank the following body parts according to frequency of injury from highest to lowest: neck, fingers, trunk, back, and eyes.
10. Name three chemicals that frequently cause chemical burns in the workplace.
11. Identify three factors that contribute to heat burn injuries in the workplace.
12. Explain the difference between RSI and carpal tunnel syndrome.

ENDNOTES

1. *Accident Facts,* 1991 ed. (Chicago: National Safety Council), p. 37.
2. Ibid., p. 25.
3. Business Publishers, Inc. *Occupational Health & Safety Letter,* April 17, 1991, Vol. 21, No. 8, p. 66.
4. *Accident Facts,* 1994–1997 eds. (Chicago: National Safety Council), pp. 4–5.
5. Ibid., p. 34.
6. Ibid., p. 35.
7. Ibid., p. 35.
8. Ibid., p. 36.
9. Ibid., p. 36.
10. Ibid., p. 37.
11. Ibid., p. 38.
12. U.S. Bureau of Labor Statistics. *Chemical Burn Injuries,* Bulletin 2353 (Washington, D.C.: U.S. Government Printing Office, October 1989).
13. *Accident Facts,* 1997 ed. (Chicago: National Safety Council), p. 40.
14. Ibid., p. 41.
15. Ibid., p. 41.
16. Business Publishers, Inc. *Occupational Health & Safety Letter,* May 29, 1991, Vol. 21, No. 11, p. 89.
17. Pascarelli, Emil, and Quilter, Deborah. *Repetitive Strain Injury* (New York: John Wiley & Sons, Inc., 1994), pp. 51–52.
18. *Accident Prevention Manual for Industrial Operations: Administration and Programs,* 9th ed. (Chicago: National Safety Council, 1988), p. 158.
19. Ibid., p. 158.

CHAPTER THREE

Theories of Accident Causation

- The Domino Theory of Accident Causation
- The Human Factors Theory of Accident Causation
- The Accident/Incident Theory of Accident Causation
- The Epidemiological Theory of Accident Causation
- The Systems Theory of Accident Causation
- The Combination Theory of Accident Causation

almost 50

Each year, work-related accidents cost the United States ~~over~~ $48 billion.[1] This figure includes costs associated with lost wages, medical expenses, insurance costs, and indirect costs. The number of persons injured in **industrial place accidents** in a typical year is 7,128,000, or 3 per 100 persons per year.[2] In the workplace, there is one accidental death approximately every 51 minutes and one injury every 19 seconds.[3]

Why do accidents happen? This question has concerned safety and health decision makers for decades, because in order to prevent accidents we must know why they happen. Over the years, several theories of accident causation have evolved that attempt to explain why accidents occur. Models based on these theories are used to predict and prevent accidents.

The most widely known theories of accident causation are the domino theory, the human factors theory, the accident/incident theory, the epidemiological theory, the systems theory, and the combination theory. This chapter provides practicing and prospective safety professionals with the information they need to understand fully and apply these theories.

& behavioural theory

THE DOMINO THEORY OF ACCIDENT CAUSATION

An early pioneer of accident prevention and industrial safety was Herbert W. Heinrich, an official with the Travelers Insurance Company. In the late 1920s, after studying the reports of 75,000 industrial accidents, Heinrich concluded that

33

- 88 percent of industrial accidents are caused by unsafe acts committed by fellow workers.
- 10 percent of industrial accidents are caused by unsafe conditions.
- 2 percent of industrial accidents are unavoidable.[4]

Heinrich's study laid the foundation for his **Axioms of Industrial Safety** and his theory of accident causation, which came to be known as the **domino theory.** So much of Heinrich's theory has been discounted by more contemporary research that it is now considered outdated. However, since some of today's more widely accepted theories can be traced back to Heinrich's theory, students of industrial safety should be familiar with his work.

Heinrich's Axioms of Industrial Safety

Heinrich summarized what he thought health and safety decision makers should know about industrial accidents in ten statements he called Axioms of Industrial Safety. These axioms can be paraphrased as follows:

1. Injuries result from a completed series of factors, one of which is the accident itself.
2. An accident can occur only as the result of an unsafe act by a person and/or a physical or mechanical hazard.
3. Most accidents are the result of unsafe behavior by people.
4. An unsafe act by a person or an unsafe condition does not always immediately result in an accident/injury.
5. The reasons why people commit unsafe acts can serve as helpful guides in selecting corrective actions.
6. The severity of an accident is largely fortuitous and the accident that caused it is largely preventable.
7. The best accident prevention techniques are analogous with the best quality and productivity techniques.
8. Management should assume responsibility for safety since it is in the best position to get results.
9. The supervisor is the key person in the prevention of industrial accidents.
10. In addition to the direct costs of an accident (i.e., compensation, liability claims, medical costs, and hospital expenses) there are also hidden or indirect costs.[5]

According to Heinrich, these axioms encompass the fundamental body of knowledge that must be understood by decision makers interested in preventing accidents. Any accident prevention program that takes all ten axioms into account is more likely to be effective than a program that leaves one or more axioms out.

Heinrich's Domino Theory

Perhaps you have stood up a row of dominoes, tipped the first one over, and watched as each successive domino topples the one next to it. This is how Heinrich's theory of accident causation works. According to Heinrich, there are five factors in the sequence of events leading up to an accident. These factors can be summarized as follows:

1. *Ancestry and social environment.* Negative character traits that might lead people to behave in an unsafe manner can be inherited **(ancestry)** or acquired as a result of the **social environment.**
2. *Fault of person.* Negative character traits, whether inherited or acquired, are why people behave in an unsafe manner and why hazardous conditions exist.
3. *Unsafe act/mechanical or physical hazard.* **Unsafe acts** committed by people and **mechanical or physical hazards** are the direct causes of accidents.
4. *Accident.* Typically, accidents that result in injury are caused by falling or being hit by moving objects.
5. *Injury.* Typical injuries resulting from accidents include lacerations and fractures.[6]

Heinrich's theory has two central points: (1) Injuries are caused by the action of **preceding factors;** and (2) removal of the **central factor** (unsafe act/**hazardous condition**) negates the action of the preceding factors and, in so doing, prevents accidents and injuries.

Domino Theory in Practice

Construction Products Company (CPC) is a distributor of lumber, pipe, and concrete products. Its customers are typically small building contractors. CPC's facility consists of an office in which orders are placed and several large warehouses. Contractors place their orders in the office. They then drive their trucks through the appropriate warehouses to be loaded by CPC personnel.

Because the contractors are small operations, most of their orders are also relatively small and can be loaded by hand. Warehouse personnel go to the appropriate bins, pull out the material needed to fill their orders, and load the materials on customers' trucks. Even though most orders are small enough to be loaded by hand, many of the materials purchased are bulky and cumbersome to handle. Because of this, CPC's loaders are required to wear such personal protection gear as hard hats, padded gloves, steel-toed boots, and lower-back-support belts.

For years CPC's management team had noticed an increase in minor injuries to warehouse personnel during the summer months. Typically, these injuries consisted of nothing worse than minor cuts, scrapes, and bruises. However, this past summer had been different. Two warehouse workers had sustained serious back injuries. These injuries have been costly to CPC both financially and in terms of employee morale.

An investigation of these accidents quickly identified a series of events and a central causal behavior that set up a *domino effect* that, in turn, resulted in the injuries. What the investigation revealed is that during the summer months, CPC's warehouses become so hot that personal protection gear is uncomfortable. As a result, warehouse personnel simply discard it. Failure to use appropriate personal protection gear in the summer months had always led to an increase in injuries. However, since the injuries were minor in nature, management had never paid much attention to the situation. It was probably inevitable that more serious injuries would occur eventually.

To prevent a recurrence of the summer-injury epidemic, CPC's management team decided to remove the causal factor—failure of warehouse personnel to use their personal protection gear during the summer months. To facilitate the removal of this factor, CPC's management team formed a committee consisting of one executive manager, one warehouse supervisor, and three warehouse employees.

The committee made the following recommendations: (1) provide all warehouse personnel with training on the importance and proper use of personal protection gear; (2) require warehouse supervisors to monitor the use of personal protection gear more closely; (3) establish a company policy that contains specific and progressive disciplinary measures for failure to use required personal protection gear; and (4) implement several heat reduction measures to make warehouses cooler during the summer months.

CPC's management team adopted all of the committee's recommendations. In doing so, it removed the central causal factor that had historically led to an increase in injuries during the summer months.

THE HUMAN FACTORS THEORY OF ACCIDENT CAUSATION

The **human factors theory** of accident causation attributes accidents to a chain of events ultimately caused by **human error.** It consists of the following three broad factors that lead to human error: overload, inappropriate response, and inappropriate activities (see Figure 3–1). These factors are explained in the following paragraphs.

SAFETY FACTS

Pregnancy and Work

Strenuous physical work and pregnancy can be a dangerous combination. Too much strenuous labor can result in a miscarriage. The types of work to be avoided by pregnant employees include the following:

- Standing for more than three hours per day
- Operating industrial machinery
- Lifting more than 22 pounds
- Working in extremes of hot or cold

Shift work and work stations that require awkward postures can also put pregnant employees at risk. The third trimester is the most risk-intensive time during pregnancy.

Figure 3–1
Factors that cause human errors.

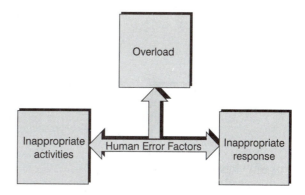

Overload

Overload amounts to an imbalance between a person's capacity at any given time and the load that person is carrying in a given state. A person's capacity is the product of such factors as his or her natural ability, training, state of mind, fatigue, stress, and physical condition. The load that a person is carrying consists of tasks for which he or she is responsible and added burdens resulting from **environmental factors** (noise, distractions, and so on), **internal factors** (personal problems, emotional stress, worry), and **situational factors** (level of risk, unclear instructions, and so on). The state in which a person is acting is the product of his or her motivational and arousal levels.

Inappropriate Response/Incompatibility

How a person responds in a given situation can cause or prevent an accident. If a person detects a hazardous condition but does nothing to correct it, he or she has responded inappropriately. If a person removes a safeguard from a machine in an effort to increase output, he or she has responded inappropriately. If a person disregards an established safety procedure, he or she has responded inappropriately. Such responses can lead to accidents. In addition to **inappropriate responses**, this component includes workstation incompatibility. The incompatibility of a person's workstation with regard to size, force, reach, feel, and similar factors can lead to accidents and injuries.

Inappropriate Activities

Human error can be the result of **inappropriate activities.** An example of an inappropriate activity would be a person undertaking a task he or she doesn't know how to do. Another example would be a person misjudging the degree of risk involved in a given task and proceeding based on that misjudgment. Such inappropriate activities can lead to accidents and injuries. Figure 3–2 summarizes the various components of the human factors theory.[7]

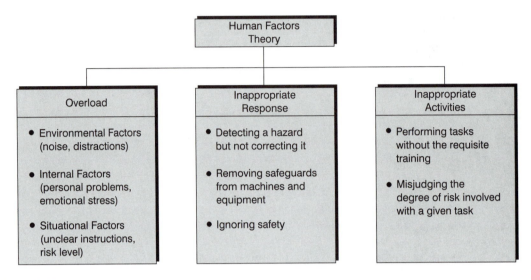

Figure 3–2
Human factors theory.

Human Factors Theory in Practice

Kitchenware Manufacturing Incorporated (KMI) produces aluminum kitchenware for commercial settings. After ten years of steady, respectable growth in the U.S. market, KMI suddenly saw its sales triple in less than six months. This rapid growth was the result of KMI's successful entry into European and Asian markets.

The growth in sales, although welcomed by both management and employees, quickly overloaded and, before long, overwhelmed the company's production facility. KMI responded by adding a second shift of production personnel and approving unlimited overtime for highly skilled personnel. Shortly after the upturn in production, KMI began to experience a disturbing increase in accidents and injuries. During his accident investigations, KMI's safety manager noticed that human error figured prominently in the accidents. He grouped all of the human errors identified into three categories: (1) overload, (2) inappropriate response, and (3) inappropriate activities.

In the category of *overload,* he found that the rush to fill orders was pushing production personnel beyond their personal limits in some cases, and beyond their capabilities in others. Stress, insufficient training of new employees, and fatigue all contributed to the overload. In the category of *inappropriate response,* the safety manager determined that many of KMI's production personnel had removed safeguards from their machines in an attempt to speed up production. All of the machines involved in accidents had had safeguards removed.

In the category of *inappropriate activities,* the safety manager found that new employees were being assigned duties for which they weren't yet fully trained. As a result, they often misjudged the amount of risk associated with their work tasks.

With enough accident investigations completed to identify a pattern of human error, the safety manager prepared a presentation containing a set of recommendations for corrective measures for KMI's executive management team. His recommendations were designed to prevent human-error-oriented accidents without slowing production.

THE ACCIDENT/INCIDENT THEORY OF ACCIDENT CAUSATION

The **accident/incident theory** is an extension of the human factors theory. It was developed by Dan Petersen and is sometimes referred to as the Petersen accident/incident theory.[8] Petersen introduced such new elements as **ergonomic traps,** the decision to err, and systems failures, while retaining much of the human factors theory. A model based on his theory is shown in Figure 3–3.

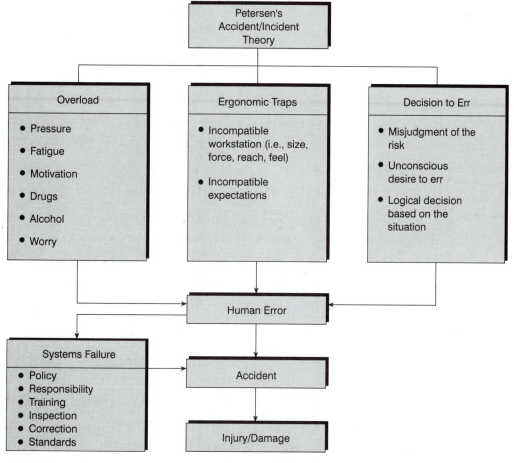

Figure 3–3
Accident/incident theory.

In this model, overload, ergonomic traps, and/or a decision to err lead to human error. The decision to err may be conscious and based on logic, or it may be unconscious. A variety of pressures such as deadlines, peer pressure, and budget factors can make a person decide to **behave in an unsafe manner.** Another factor that can influence such a decision is the "It won't happen to me" syndrome.

The systems failure component is an important contribution of Petersen's theory. First, it shows the potential for a **causal relationship** between management decisions/management behavior and safety. Second, it establishes management's role in accident prevention as well as the broader concepts of safety and health in the workplace.

Following are some of the different ways that systems can fail: Management does not establish a comprehensive safety policy. Responsibility and authority with regard to safety are not clearly defined. Safety procedures such as measurement, inspection, correction, and investigation are ignored or given insufficient attention. Employees do not receive a proper orientation. Employees are not given sufficient safety training. These are just some of the many types of systems failures that might occur, according to Petersen's theory.

Accident/Incident Theory in Practice

Poultry Processing Corporation (PPC) processes chickens and turkeys for grocery chains. Poultry processing is a labor-intensive enterprise involving a great deal of handwork. A variety of different knives, shears, and cleavers are used. Much of the work is monotonous and repetitive. Selected parts of the overall process must be done in cold conditions.

PPC has gone to great lengths to ensure that workstations are ergonomically sound, that personal protection gear is used as appropriate, and that adequate precautions are taken to prevent illness and injuries. As a result, PPC is an award-winning company in the area of workplace safety and health.

Consequently, the poultry processing industry was shocked when a class action lawsuit was filed against PPC on behalf of over 50 employees, all of whom claimed to be suffering from carpal tunnel syndrome. Because of PPC's excellent safety and health record, most observers felt sure that the company would be vindicated in the end.

The company's policies and procedures relating to safety and health were investigated thoroughly by consultants brought in by both PPC and the attorney for the plaintiffs. Over 100 witnesses gave depositions, and several preliminary hearings were held. By the time the trial finally rolled around, both sides had accumulated mountains of paper and filing cabinets full of evidence. Then, suddenly and without advance notice, PPC offered a substantial financial settlement, which the plaintiffs accepted.

It was one of PPC's outside consultants who discovered what had caused the outbreak of carpal tunnel syndrome. The company had always used a centralized approach to managing safety and health. Responsibility for such tasks as measurement, inspection, correction, and investigation was assigned to the safety manager, Joe Don Huttle. Huttle had an excellent record during his 20 years in the poultry-processing industry, with the last five spent at PPC. In fact, he was so well respected in the industry that his peers had elected him president of a state-wide safety organization. This, as it turned out, is where PPC's troubles began.

When Huttle took it over, the safety organization had experienced a three-year decline in membership and was struggling to stay afloat financially. He had been elected as "the man who could save the organization." Intending to do just that, Huttle went right to work. For months at a time he worked seven days a week, often spending as much as two weeks at a time on the road. When he was in his office at PPC, Huttle was either on the telephone or doing paperwork for the safety organization.

Within six months, he had reversed the organization's downhill slide, but not without paying a price at home. During the same six-month period, his duties at PPC were badly neglected. Measurement of individual and group safety performance had come to a standstill. The same was true of inspection, correction, investigation, and reporting.

It was during this time of neglect that the outbreak of carpal tunnel syndrome occurred. Safety precautions that Huttle had put in place to guard against this particular problem were no longer observed properly once the workers saw that he had stopped observing and correcting. Measurement and inspection might also have prevented the injuries had Huttle maintained his normal schedule of these activities.

PPC's consultant, in a confidential report to executive managers, cited the *accident/incident theory* in explaining his view of why the injuries occurred. In this report, the consultant said that Huttle was guilty of applying "It won't happen here" logic when he made a conscious decision to neglect his duties at PPC in favor of his duties with the professional organization. Of course, the employees themselves were guilty of not following clearly established procedures. However, since Huttle's neglect was also a major contributing factor, PPC decided to settle out of court.

THE EPIDEMIOLOGICAL THEORY OF ACCIDENT CAUSATION

Traditionally, safety theories and programs have focused on accidents and the resulting injuries. However, the current trend is toward a broader perspective that also encompasses the issue of industrial hygiene. **Industrial hygiene** concerns environmental factors that can lead to sickness, disease, or other forms of impaired health.

This trend has, in turn, led to the development of an epidemiological theory of accident causation. Epidemiology is the study of causal relationships between environmental factors and disease. The **epidemiological theory** holds that the models used for studying and determining these relationships can also be used to study causal relationships between environmental factors and accidents or diseases.[9]

Figure 3–4 illustrates the epidemiological theory of accident causation. The key components are **predisposition characteristics** and **situational characteristics.** These characteristics, taken together, can either result in or prevent conditions that might result in an accident. For example, if an employee who is particularly susceptible to peer pressure (predisposition characteristic) is pressured by his co-workers (situational characteristic) to speed up his operation, the result will be an increased probability of an accident.

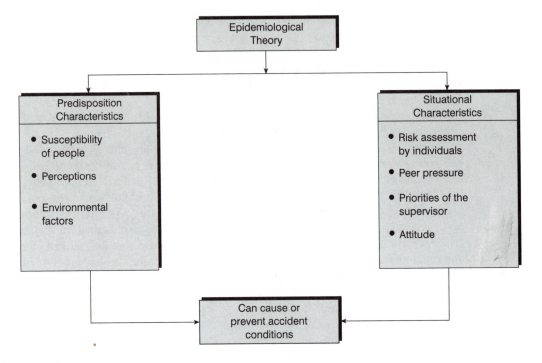

Figure 3–4
Epidemiological theory.

Epidemiological Theory in Practice

Jane Andrews was the newest member of the loading unit for Parcel Delivery Service (PDS). She and the other members of her unit were responsible for loading 50 trucks every morning. It was physically demanding work, and she was the first woman ever selected by PDS to work in the loading unit. She had gotten the job as part of the company's upward mobility program. She was excited about her new position because within PDS, the loading unit was considered a springboard to advancement. Consequently, she was anxious to do well. The responsibility she felt toward other female employees at PDS only served to intensify her anxiety. Andrews felt that if she failed, other women might not get a chance to try in the future.

Before beginning work in the loading unit, employees must complete two days of training on proper lifting techniques. The use of back-support belts is mandatory for all loading dock personnel. Consequently, Andrews became concerned when the supervisor called her aside on her first day in the unit and told her to forget what she had learned in training. He said, "Jane, nobody wants a back injury, so be careful. But the key to success in this unit is speed. The lifting techniques they teach in that workshop will just slow you down. You've got the job, and I'm glad you're here. But you won't last long if you can't keep up."

Andrews was torn between following safety procedures and making a good impression on her new supervisor. At first, she made an effort to use proper lifting techniques. However, when several of her co-workers complained that she wasn't keeping up, the supervisor told Andrews to "keep up or get out of the way." Feeling the pressure, she started taking the same shortcuts she had seen her co-workers use. Positive results were immediate, and Andrews received several nods of approval from fellow workers and a "Good job" from the supervisor. Before long, Andrews had won the approval and respect of her colleagues.

However, after two months of working in the loading unit, she began to experience persistent lower-back pain. Andrews felt sure that her hurried lifting techniques were to blame, but she valued the approval of her supervisor and fellow workers too much to do anything that might slow her down. Finally, one day while loading a truck, Andrews fell to the pavement in pain and could not get up. Her back throbbed with intense pain, and her legs were numb. She had to be rushed to the emergency room of the local hospital. By the time Andrews checked out of the hospital a week later, she had undergone major surgery to repair two ruptured discs.

Jane Andrews' situation can be explained by the *epidemiological theory* of accident causation. The predisposition factor was her susceptibility to peer pressure from her co-workers and supervisor. The applicable situational factors were peer pressure and the priorities of the supervisor. These factors, taken together, caused the accident.

THE SYSTEMS THEORY OF CAUSATION

A system is a group of regularly interacting and interrelated components that together form a unified whole. This definition is the basis for the **systems theory** of accident causation. This theory views a situation in which an accident might occur as a system comprised of the following components: person (host), machine (agency), and **environment**.[10] The likelihood of an accident occurring is determined by how these components interact. Changes in the patterns of interaction can increase or reduce the probability of an accident occurring.

For example, an experienced employee who operates a numerically controlled five-axis machining center in a shop environment might take a two-week vacation. Her tem-

---

DISCUSSION CASE

What Is Your Opinion?

"All accidents, one way or another, are the result of human error." "No, accidents are the result of a combination of things. I like the Combination Theory." "You're both wrong. Accidents are best explained by the Domino Theory." And so the debate went in Dr. Jameson's class at Burton State University. What is your opinion concerning the various theories of accident causation?

porary replacement might be less experienced. This change in one component of the system (person/host) increases the probability of an accident. Such a simple example is easily understood. However, not all changes in patterns of interaction are this simple. Some are so subtle that their analysis might require a team of people, each with a different type of expertise.

The most widely used systems model is that developed by R. J. Firenzie. The primary components of this model are the person/machine/environment, information, decisions, risks, and the task to be performed.[11] Each of the components has a bearing on the probability that an accident will occur. The model is illustrated in Figure 3–5.

As this model shows, even as a person interacts with a machine within an environment, three activities take place between the system and the task to be performed. Every time a task must be performed, there is the risk that an accident might occur. Sometimes the risks are great; at other times, they are small. This is where information collection and decision making come in.

Based on the information that has been collected by observing and mentally noting the current circumstances, the person weighs the risks and decides whether to perform the task under existing circumstances. For example, say, a machine operator is working on a rush order that is behind schedule. An important safety device has malfunctioned on his machine. Simply taking it off will interrupt work for only five minutes, but it will also increase the probability of an accident. However, replacing it could take up to an hour. Should the operator remove the safety guard and proceed with the task or take the time to replace it? The operator and his supervisor might assess the situation (collect information), weigh the risks, and make a decision to proceed. If their information was right and their assessment of the risks accurate, the task will probably be accomplished without an accident.

However, the environment in which the machine operator is working is unusually hectic and the pressure to complete an order that is already behind schedule is intense.

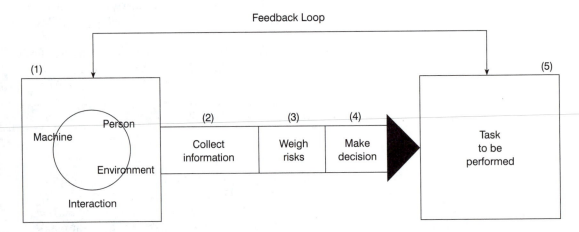

Figure 3–5
Systems theory model.

These factors are **stressors** that can cloud the judgment of those collecting information, weighing risks, and making the decision. When stressors are introduced between points 1 and 3 in Figure 3–5, the likelihood of an accident increases.

For this reason Firenzie recommends that five factors be considered before beginning the process of collecting information, weighing risks, and making a decision:

1.   Job requirements
2.   The workers' abilities and limitations
3.   The gain if the task is successfully accomplished
4.   The loss if the task is attempted but fails
5.   The loss if the task is not attempted[12]

These factors can help a person achieve the proper perspective before collecting information, weighing risks, and making a decision. It is particularly important to consider these factors when stressors such as noise, time constraints, or pressure from a supervisor might tend to cloud one's judgment.

Systems Theory in Practice

Precision Tooling Company (PTC) specializes in difficult orders that are produced in small lots, and in making corrections to parts that otherwise would wind up as expensive rejects in the scrap bin. In short, PTC specializes in doing the types of work that other companies cannot, or will not, do. Most of PTC's work comes in the form of subcontracts from larger manufacturing companies. Consequently, living up to its reputation as a *high-performance, on-time company* is important to PTC.

Because much of its work consists of small batches of parts to be reworked, PTC still uses several manually operated machines. The least experienced machinists operate these machines. This causes two problems. The first problem is that it is difficult for even a master machinist to hold to modern tolerance levels on these old machines. Consequently, apprentice machinists find holding to precise tolerances quite a challenge. The second problem is that the machines are so old that they frequently break down.

Complaints from apprentice machinists about the old machines are frequent. However, their supervisors consider time on the old "ulcer makers" to be one of the rites of passage that upstart machinists must endure. Their attitude is, "We had to do it, so why shouldn't you?" This was where things stood at PTC when the company won the Johnson contract.

PTC had been trying for years to become a preferred supplier for H.R. Johnson Company. PTC's big chance finally came when Johnson's manufacturing division incorrectly produced 10,000 copies of a critical part before noticing the problem. Simply scrapping the part and starting over was an expensive solution. Johnson's vice-president for manufacturing decided to give PTC a chance.

PTC's management team was ecstatic! Finally, they had an opportunity to partner with H.R. Johnson Company. If PTC could perform well on this one, even more lucrative contracts were sure to follow. The top managers called a company-wide meeting of all employees. Attendance was mandatory. The CEO explained the situation as follows:

Ladies and gentlemen, we are faced with a great opportunity. I've just signed a contract with H.R. Johnson Company to rework 10,000 parts that their manufacturing folks produced improperly. The rework tasks are not that complicated, but every part has got to go through several manual operations at the front end of the rework process. This means our manual machining unit is going to have to supply the heroes on this job. I've promised the manufacturing VP at Johnson that we would have his parts ready in 90 days. I know that's a lot to do in so short a period of time, but Johnson is in a real bind here. If we can produce on this one, they won't forget us in the future.

This put PTC's apprentice machinists on the spot. If PTC didn't perform on this contract, it would be their fault. They cursed their old machines and got to work. The CEO had said the rework tasks would not be "that complicated," but, as it turned out, the processes weren't that simple either. The problem was tolerances. Holding to the tolerances specified in the Johnson contract took extra time and a special effort on every single part. Before long, the manual machining unit was behind schedule, and management was getting nervous. The situation was made even worse by the continual breakdowns and equipment failures experienced. The harder the unit supervisor pushed, the more stressed the employees and machines became.

Predictably, it wasn't long before safety procedures were forgotten, and unreasonable risks were being taken. The pressure from management, the inexperience of the apprentice machinists, and the constant equipment failures finally took their toll. In a hurry to get back on schedule, and fearing that his machine would break down again, one machinist got careless and ran his hand into the cutter on his milling machine. By the time the machine had been shut down, his hand was badly mutilated. In the aftershock of this accident, PTC was unable to meet the agreed-upon completion schedule. Unfortunately, PTC did not make the kind of impression on H.R. Johnson's management team that it had hoped.

This accident can be explained by the systems theory. The *person-machine-environment* chain has a direct application in this case. The person involved was relatively inexperienced. The machine involved was old and prone to breakdowns. The environment was especially stressful and pressure packed. These three factors, taken together, resulted in this serious and tragic accident.

THE COMBINATION THEORY OF ACCIDENT CAUSATION

There is often a degree of difference between any theory of accident causation and reality. The various models presented with their corresponding theories in this chapter attempt to explain why accidents occur. For some accidents, a given model might be very accurate. For others, it might be less so. Often the cause of an accident cannot be adequately explained by just one model/theory. Thus, according to the **combination theory,** the actual cause may combine parts of several different models. Safety personnel should use these theories as appropriate both for accident prevention and accident investigation. However, they should avoid the tendency to try to apply one model to all accidents.

Combination Theory in Practice

Crestview Grain Corporation (CGC) maintains ten large silos for storing corn, rice, wheat, barley, and various other grains. Since stored grain generates fine dust and gases, ventilation of the silos is important. Consequently, all of CGC's silos have several large vents. Each of these vents uses a filter similar to the type used in home air conditioners that must be changed periodically.

There is an element of risk involved in changing the vent filters because of two potential hazards. The first hazard comes from unvented dust and gases that can make breathing difficult, or even dangerous. The second hazard is the grain itself. Each silo has a catwalk that runs around its inside circumference near the top. These catwalks give employees access to the vents that are also near the top of each silo. The catwalks are almost 100 feet above ground level, they are narrow, and the guardrails on them are only knee high. A fall from a catwalk into the grain below would probably be fatal.

Consequently, CGC has well-defined rules that employees are to follow when changing filters. Because these rules are strictly enforced, there had never been an accident in one of CGC's silos; not, that is, until the Juan Perez tragedy occurred. Perez was not new to the company. At the time of his accident, he had worked at CGC for over five years. However, he was new to the job of silo maintenance. His inexperience, as it turned out, would prove fatal.

It was time to change the vent filters in silo number 4. Perez had never changed vent filters himself. He hadn't been in the job long enough. However, he had served as the required "second man" when his supervisor, Bao Chu Lai, had changed the filters in silos 1, 2, and 3. Since Chu Lai was at home recuperating from heart surgery, and might be out for another four weeks, Perez decided to change the filters himself. Changing the filters was a simple enough task, and Perez had always thought the "second-man" concept was overdoing it a little. He believed in taking reasonable precautions as much as the next person, but in his opinion, CGC was paranoid about safety.

Perez collected his safety harness, respirator, and four new vent filters. Then he climbed the external ladder to the entrance/exit platform near the top of silo number 4. Before going in, Perez donned his respirator and strapped on his safety harness. Opening the hatch cover, he stepped inside the silo onto the catwalk. Following procedure, Perez attached a lifeline to his safety harness, picked up the new vent filters, and headed for the first vent. He changed the first two filters without incident. It was while he was changing the third filter that tragedy struck.

The filter in the third vent was wedged in tightly. After several attempts to pull it out, Perez became frustrated and gave the filter a good jerk. When the filter suddenly broke loose, the momentum propelled Perez backwards and he toppled off the catwalk. At first it appeared that his lifeline would hold, but without a second person to pull him up or call for help, Perez was suspended by only the lifeline for over 20 minutes. He finally panicked, and in his struggle to pull himself up, knocked the buckle of his safety harness open. The buckle gave way, and Perez fell over 50 feet into the grain below. The impact knocked his respirator off, the grain quickly enveloped him, and Perez was asphyxiated.

The accident investigation that followed revealed that several factors combined to cause the fatal accident—the combination theory. The most critical of these factors were as follows:

1. Absence of the supervisor
2. Inexperience of Perez
3. A conscious decision by Perez to disregard CGC's safety procedures
4. A faulty buckling mechanism on the safety harness
5. An unsafe design (only a knee-high guardrail on the catwalk).

SUMMARY

1. The domino theory of accident causation was one of the earliest developed. The theory posits that injuries result from a series of factors, one of which is an accident. The theory is operationalized in ten statements called the Axioms of Industrial Safety. According to this theory, there are five factors in the sequence of events leading to an accident: ancestry/social environment, fault of person, unsafe act/mechanical or physical hazard, accident, and injury.
2. The human factors theory of accident causation attributes accidents to a chain of events ultimately caused by human error. It consists of three broad factors that lead to human error: overload, inappropriate response, and inappropriate activities.
3. The accident/incident theory of accident causation is an extension of the human factors theory. It introduces such new elements as ergonomic traps, the decision to err, and systems failures.
4. The epidemiological theory of accident causation holds that the models used for studying and determining the relationships between environmental factors and disease can be used to study causal relationships between environmental factors and accidents.
5. The systems theory of accident causation views any situation in which an accident might occur as a system with three components: person (host), machine (agency), and environment.
6. The combination theory of accident causation posits that no one model/theory can explain all accidents. Factors from two or more models might be part of the cause.

KEY TERMS AND CONCEPTS

Accident/incident theory

Ancestry

Axioms of Industrial Safety

Causal relationships

Central factor

Combination theory

Domino theory

Environment

Environmental factors

Epidemiological theory

Ergonomic traps

Fault of person

Hazardous condition

Human error

Human factors theory

Inappropriate activities

Inappropriate response

Industrial hygiene

Industrial place accidents

Internal factors

Mechanical or physical hazard

Overload

Preceding factors

Predisposition characteristics

Situational characteristics

Situational factors

Social environment

Stressors

Systems theory

Unsafe act

Unsafe behavior

REVIEW QUESTIONS

1. Explain the domino theory of accident causation, including its origin and its impact on more modern theories.
2. What were the findings of Herbert W. Heinrich's 1920s study of the causes of industrial accidents?
3. List five of Heinrich's Axioms of Industrial Safety.
4. Explain the following concepts in the domino theory: preceding factor; central factor.
5. What are the three broad factors that lead to human error in the human factors theory? Briefly explain each.
6. Explain the systems failure component of the accident/incident theory.
7. What are the key components of the epidemiological theory? How does their interaction affect accident causation?
8. Explain the systems theory of accident causation.
9. What impact do stressors have in the systems theory?
10. List five factors to consider before making workplace decisions that involve risk.

ENDNOTES

1. *Accident Facts,* 1988–1991 eds. (Chicago: National Safety Council), p. 23.
2. Ibid., p. 24.
3. Ibid., p. 26.
4. Heinrich, H. W., Petersen, D., and Roos, N. *Industrial Accident Prevention,* 5th ed. (New York: McGraw-Hill Book Company, 1980).
5. Ibid.
6. Ibid.
7. Ibid.
8. Ibid.

9. Colling, D. A. *Industrial Safety: Management & Technology* (Upper Saddle River, NJ: Prentice Hall, 1990), p. 33.
10. Firenzie, R. J. *The Process of Hazard Control* (New York: Kendall/Hunt, 1978), p. 56.
11. Ibid.
12. Ibid.

The OSHAct, Standards, and Liability

Since the early 1970s, the amount of legislation passed—and the number of subsequent regulations—concerning workplace safety and health have increased markedly. Of all the legislation, by far the most significant has been the Occupational Safety and Health Act of 1970, called here the OSHAct. Prospective and practicing safety and health profession-als must be knowledgeable about the OSHAct and the agency established by it—the Occupational Safety and Health Administration, or OSHA. This chapter provides students with the information they need about the OSHAct, OSHA, and other pertinent federal legislation and agencies.

THE RATIONALE FOR THE OSHAct

Perhaps the most debilitating experience one can have on the job is to be involved in, or exposed to, a work-related accident or illness. Such an occurrence can be physically and psychologically incapacitating for the victim, psychologically stressful for the victim's fellow workers, and extraordinarily expensive for the victim's employer. In spite of this, until 1970, laws governing workplace safety were limited and sporadic. Finally, in 1970, Congress passed the Occupational Safety and Health Act **(OSHAct)** with the following stated purpose: " . . . to assure so far as possible every working man and woman in the nation safe and healthful working conditions and to preserve our human resources."[1]

According to the U.S. Department of Labor, in developing this comprehensive and far-reaching piece of legislation, Congress considered the following statistics:

- Each year, an average of 14,000 deaths were caused by **workplace accidents**.
- Each year, 2.5 million workers were disabled in workplace accidents.
- Each year, approximately 300,000 new cases of **occupational diseases** were reported.[2]

Clearly, a comprehensive, uniform law was needed to help reduce the incidence of work-related injuries, illnesses, and deaths. The OSHAct of 1970 addressed this need. It is contained in Title 29 of the Code of Federal Regulations, Parts 1900 through 1910. The act also establishes the Occupational Safety and Health Administration **(OSHA),** which is part of the U.S. Department of Labor and is responsible for administering the OSHAct.

OSHA'S MISSION AND PURPOSE

According to the Department of Labor, OSHA's mission and purpose can be summarized as follows:

- Encourage employers and employees to reduce workplace hazards.
- Implement new safety and health programs.
- Improve existing safety and health programs.
- Encourage research that will lead to innovative ways of dealing with workplace safety and health problems.
- Establish the rights of employers regarding the improvement of workplace safety and health.
- Establish the rights of employees regarding the improvement of workplace safety and health.

- Monitor job-related illnesses and injuries through a system of reporting and record keeping.
- Establish training programs to increase the number of safety and health professionals and to improve their competence continually.
- Establish mandatory workplace safety and health standards and enforce those standards.
- Provide for the development and approval of state-level workplace safety and health programs.
- Monitor, analyze, and evaluate state-level safety and health programs.[3]

OSHAct COVERAGE

The OSHAct applies to most employers. If an organization has even one employee, it is considered an employer and must comply with applicable sections of the act. This includes all types of employers from manufacturing and construction to retail and service organizations. There is no exemption for small businesses, although organizations with ten or fewer employees are exempted from OSHA inspections and the requirement to maintain injury/illness records.

Although the OSHAct is the most comprehensive and far-reaching piece of safety and health legislation ever passed in this country, it does not cover all employers. In general, the OSHAct covers employers in all 50 states, the District of Columbia, Puerto Rico, and all other territories that fall under the jurisdiction of the U.S. government. Exempted employers are

- Persons who are self-employed
- Family farms that employ only immediate members of the family
- Federal agencies covered by other federal statutes (in cases where these other federal statutes do not cover working conditions in a specific area or areas, OSHA standards apply)
- State and local governments (except to gain OSHA's approval of a state-level safety and health plan, states must provide a program for state and local government employees that is at least equal to its private sector plan)
- Coal mines (coal mines are regulated by mining-specific laws)

Federal government agencies are required to adhere to safety and health standards that are comparable to and consistent with OSHA standards for private sector employees. OSHA evaluates the safety and health programs of federal agencies. However, OSHA cannot assess fines or monetary damages against other federal agencies as it can against private sector employers.

There are many OSHA requirements to which employers must adhere. Some apply to all employers—except those exempted—whereas others apply only to specific types of employers. These requirements cover such areas of concern as the following:

- Fire protection
- Electricity

- Sanitation
- Air quality
- Machine use, maintenance, and repair
- Posting of notices and warnings
- Reporting of accidents and illnesses
- Maintaining written compliance programs
- Employee training

In addition to these, other more important and widely applicable requirements are explained later in this chapter.

OSHA STANDARDS

The following statement by the U.S. Department of Commerce summarizes OSHA's responsibilities relating to standards:

> In carrying out its duties, OSHA is responsible for promulgating legally enforceable standards. OSHA standards may require conditions, or the adoption or use of one or more practices, means, methods, or processes reasonably necessary and appropriate to protect workers on the job. It is the responsibility of employers to become familiar with standards applicable to their establishments and to ensure that employees have and use personal protective equipment when required for safety.[4]

The general duty clause of the OSHAct requires that employers provide a workplace that is free from hazards that are likely to harm employees. This is important because the general duty clause applies when there is no specific OSHA standard for a given situation. Where OSHA standards do exist, employers are required to comply with them as written.

How Standards Are Developed

OSHA develops standards based on its perception of need and at the request of other federal agencies, state and local governments, other standards-setting agencies, labor organizations, or even individual private citizens. OSHA uses the committee approach for developing standards, both standing committees within OSHA and special ad hoc committees. Ad hoc committees are appointed to deal with issues that are beyond the scope of the standing committees.

OSHA's standing committees are the National Advisory Committee on Occupational Safety and Health **(NACOSH)** and the Advisory Committee on Construction Safety and Health. NACOSH makes recommendations on standards to the secretary of health and human services and to the secretary of labor. The Advisory Committee on Construction Safety and Health advises the secretary of labor on standards and regulations relating specifically to the construction industry.

The National Institute for Occupational Safety and Health **(NIOSH),** like OSHA, was established by the OSHAct. Whereas OSHA is part of the Department of Labor, NIOSH is part of the Department of Health and Human Services. NIOSH has an education and research orientation. The results of this agency's research are often used to assist OSHA in developing standards.

How Standards Are Adopted, Amended, or Revoked

OSHA can adopt, amend, or revise standards. Before any of these actions can be undertaken, OSHA must publish its intentions in the *Federal Register* in either a notice of proposed rule making or an advance notice of proposed rule making. The **notice of proposed rule making** must explain the terms of the new rule, delineate proposed changes to existing rules, or list rules that are to be revoked. The advance notice of proposed rule making may be used instead of the regular notice when it is necessary to solicit input before drafting a rule.

After publishing notice, OSHA must conduct a public hearing if one is requested. Any interested party may ask for a public hearing on a proposed rule or rule change. When this happens, OSHA must schedule the hearing and announce the time and place in the *Federal Register.*

The final step, according to the Department of Labor, is as follows:

> After the close of the comment period and public hearing, if one is held, OSHA must publish in the Federal Register the full, final text of any standard amended or adopted and the date it becomes effective, along with an explanation of the standard and the reasons for implementing it. OSHA may also publish a determination that no standard or amendment needs to be issued.[5]

How to Read an OSHA Standard

OSHA standards are typically long and complex and are written in the language of lawyers and bureaucrats, making them difficult to read. However, reading OSHA standards can be simplified somewhat if one understands the system.

OSHA standards are part of the Code of Federal Regulations (C.F.R.) published by the Office of the Federal Register. The regulations of all federal-government agencies are published in the C.F.R. Title 29 contains all of the standards assigned to OSHA. Title 29 is divided into several parts, each carrying a four number designator (such as Part 1901, Part 1910). These parts are divided into sections, each carrying a numerical designation. For example, 29 C.F.R. 1910.1 means *Title 29, Part 1910, Section 1, Code 1 of Federal Regulations.*

The sections are divided into four different levels of subsections, each with a particular type of designator as follows:

First Level:	Alphabetically using lowercase letters in parentheses: (a) (b) (c) (d)
Second Level:	Numerically using numerals in parentheses: (1) (2) (3) (4)
Third Level:	Numerically using roman numerals: (i) (ii) (iii) (iv)
Fourth Level:	Alphabetically using uppercase letters in parentheses: (A), (B), (C), (D)

Occasionally, the standards go beyond the fourth level of subsection. In these cases, the sequence just described is repeated with the designator shown in parentheses underlined. For example: (a), (1), (i), (A).

Understanding the system used for designating sections and subsections of OSHA standards can guide readers more quickly to the specific information needed. This helps to reduce the amount of cumbersome reading needed to comply with the standards.

Temporary Emergency Standards

The procedures described in the previous section apply in all cases. However, OSHA is empowered to pass **temporary emergency standards** on an emergency basis without undergoing normal adoption procedures. Such standards remain in effect only until permanent standards can be developed.

To justify passing temporary standards on an emergency basis, OSHA must determine that workers are in imminent danger from exposure to a hazard not covered by existing standards. Once a temporary standard has been developed, it is published in the *Federal Register*. This step serves as the notification step in the permanent adoption process. At this point, the standard is subjected to all of the other adoption steps outlined in the preceding section.

How to Appeal a Standard

After a standard has been passed, it becomes effective on the date prescribed. This is not necessarily the final step in the **appeals process,** however. A standard, either permanent or temporary, may be appealed by any person who is opposed to it.

An appeal must be filed with the U.S. Court of Appeals serving the geographic region in which the complainant lives or does business. Appeal paperwork must be initiated within 60 days of a standard's approval. However, the filing of one or more appeals does not delay the enforcement of a standard unless the court of appeals handling the matter mandates a delay. Typically, the new standard is enforced as passed until a ruling on the appeal is handed down.

Requesting a Variance

Occasionally, an employer may be unable to comply with a new standard by the effective date of enforcement. In such cases, the employer may petition OSHA at the state or federal level for a variance. Following are the different types of variances that can be granted.

Temporary Variance

When an employer advises that it is unable to comply with a new standard but may be able to if given additional time, a **temporary variance** may be requested. OSHA may grant such a variance for up to a maximum of one year. To be granted a temporary variance, employers must demonstrate that they are making a concerted effort to comply and taking the steps necessary to protect employees while working toward compliance.

Application procedures are very specific. Prominent among the requirements are the following: (1) identification of the parts of the standard that cannot be complied with; (2) explanation of the reasons why compliance is not possible; (3) detailed explanations of the steps that have been taken so far to comply with the standard; and (4) explanation of the steps that will be taken to fully comply.

According to the U.S. Department of Labor, employers are required to keep their employees informed. They must " . . . certify that workers have been informed of the variance application, that a copy has been given to the employees' authorized representative, and that a summary of the application has been posted wherever notices are normally posted. Employees also must be informed that they have the right to request a hearing on the application."[6]

Variances are not granted simply because an employer cannot afford to comply. For example, if a new standard requires employers to hire a particular type of specialist but there is a shortage of people with the requisite qualifications, a temporary variance might be granted. However, if the employer simply cannot afford to hire such a specialist, the variance will probably be denied. Once a temporary variance is granted, it may be renewed twice. The maximum period of each extension is six months.

Permanent Variance

Employers who feel they already provide a workplace that exceeds the requirements of a new standard may request a **permanent variance.** They present their evidence, which is inspected by OSHA. Employees must be informed of the application for a variance and notified of their right to request a hearing. Having reviewed the evidence and heard testimony (if a hearing has been held), OSHA can award or deny the variance. If a permanent variance is awarded, it comes with a detailed explanation of the employer's on-going responsibilities regarding the variance. If, at any time, the company does not meet these responsibilities, the variance can be revoked.

Other Variances

In addition to temporary and permanent variances, an experimental variance may be awarded to companies that participate in OSHA-sponsored experiments to test the effectiveness of new health and safety procedures. Variances also may be awarded in cases where the secretary of labor determines that a variance is in the best interest of the country's national defense.

When applying for a variance, employers are required to comply with the standard until a decision has been made. If this is a problem, the employer may petition OSHA for an interim order. If granted, the employer is released from the obligation to comply until a decision is made. In such cases, employees must be informed of the order.

Typical of OSHA standards are the confined space and hazardous waste standards. Brief profiles of these standards provide an instructive look at how OSHA standards are structured and the extent of their coverage.

Confined Space Standard[7]

This standard was developed in response to the approximately 300 work-related deaths that occur in confined spaces each year. The standard applies to a broad cross section of industries that have employees working in spaces with the following characteristics: limited openings for entry or exit, poor natural ventilation, and a design not intended to accommodate continuous human occupancy. Such spaces as manholes, storage tanks, underground vaults, pipelines, silos, vats, exhaust ducts, boilers, and degreasers are typically considered confined spaces.

The key component in the standard is the *permit requirement*. Employers are required to develop an in-house program under which employees must have a permit to enter confined spaces. Through such programs, employers must do the following:

- Identify spaces that can only be entered by permit.
- Restrict access to identified spaces to ensure that only authorized personnel may enter.
- Control hazards in the identified spaces through engineering, revised work practices, and other methods.
- Continually monitor the identified spaces to ensure that any known hazards remain under control.

The standard applies to approximately 60 percent of the workers in the United States. Excluded from coverage are federal, state, and local government employees; agricultural workers; maritime and construction workers; and employees of companies with ten or fewer workers.

Hazardous Waste Standard

This standard specifically addresses the safety of the estimated 1.75 million workers who deal with hazardous waste: hazardous waste workers in all situations including treatment, storage, handling, and disposal; firefighters; police officers; ambulance personnel; and hazardous materials response team personnel.[8]

Occupational Hazards magazine summarizes the requirements of this standard:

- Each hazardous waste site employer must develop a safety and health program designed to identify, evaluate, and control safety and health hazards, and provide for emergency response.
- There must be preliminary evaluation of the site's characteristics prior to entry by a trained person to identify potential site hazards and to aid in the selection of appropriate employee protection methods.
- The employer must implement a site control program to prevent contamination of employees. At a minimum, the program must identify a site map, site work zones, site communications, safe work practices, and the location of the nearest medical assistance. Also required in particularly hazardous situations is the use of the buddy system so that employees can keep watch on one another and provide quick aid if needed.

- Employees must be trained before they are allowed to engage in hazardous waste operations or emergency response that could expose them to safety and health hazards.

- The employer must provide medical surveillance at least annually and at the end of employment for all employees exposed to any particular hazardous substance at or above established exposure levels and/or those who wear approved respirators for thirty days or more on site.

- Engineering controls, work practices, and personal protective equipment, or a combination of these methods, must be implemented to reduce exposure below established exposure levels for the hazardous substances involved.

- There must be periodic air monitoring to identify and quantify levels of hazardous substances and to ensure that proper protective equipment is being used.

- The employer must set up an informational program with the names of key personnel and their alternates responsible for site safety and health, and the requirements of the standard.

- The employer must implement a decontamination procedure before any employee or equipment leaves an area of potential hazardous exposure; establish operating procedures to minimize exposure through contact with exposed equipment, other employees, or used clothing; and provide showers and change rooms where needed.

- There must be an emergency response plan to handle possible on-site emergencies prior to beginning hazardous waste operations. Such plans must address personnel roles; lines of authority, training and communications; emergency recognition and prevention; safe places of refuge; site security; evacuation routes and procedures; emergency medical treatment; and emergency alerting.

- There must be an off-site emergency response plan to better coordinate emergency action by local services and to implement appropriate control actions.[9]

RECORD KEEPING AND REPORTING

One of the breakthroughs of the OSHAct was the centralization and systematization of **record keeping**. This has simplified the process of collecting health and safety statistics for the purpose of monitoring problems and taking the appropriate steps to solve them.

In January 1997, OSHA made substantial changes to its record keeping and reporting requirements. Employers had complained for years about the mandated injury and illness record-keeping system. Their complaints can be summarized as follows:

- Original system was cumbersome and complicated.
- OSHA record-keeping rule had not kept up with new and emerging issues.
- There were too many interpretations in many of the record-keeping documents.
- Record-keeping forms were too complex.
- Guidelines for record-keeping were too long and difficult to understand.

In response to these complaints, OSHA initiated a dialogue among stakeholders to improve the record-keeping and reporting process. Input was solicited and received from employers, unions, trade associations, record keepers, OSHA staff, state occupational safety and health personnel, and state consultation program personnel. OSHA's goals for the new record-keeping and reporting system were as follows:

- Simplify all aspects of the process
- Improve the quality of records
- Meet the needs of a broad base of stakeholders
- Improve access for employees
- Minimize the regulatory burden
- Reduce vagueness—give clear guidance
- Promote the use of data from the new system in local safety and health programs

In recording and reporting occupational illnesses and injuries, it is important to have common definitions. The Department of Labor uses the following definitions for record-keeping and reporting purposes:

> An occupational injury is any injury such as a cut, fracture, sprain, or amputation that results from a work-related accident or from exposure involving a single incident in the work environment. An occupational illness is any abnormal condition or disorder other than one resulting from an occupational injury caused by exposure to environmental factors associated with employment. Included are acute and chronic illnesses or diseases which may be caused by inhalation, absorption, ingestion, or direct contact with toxic substances or harmful agents.[10]

Reporting Requirements[11]

All occupational illnesses and injuries must be **reported,** if they result in one or more of the following:

- Death of one or more workers
- One or more days away from work
- Restricted motion or restrictions to the work that an employee can do
- Loss of consciousness of one or more workers
- Transfer of an employee to another job
- Medical treatment beyond in-house first aid (if it is not on the first-aid list, it is considered medical treatment)
- Any other condition listed in Appendix B of the rule.

Record-Keeping Requirements[12]

Employers are required to keep injury and illness records for each location where they do business. For example, an automobile manufacturer with plants in several states must keep records at each individual plant for that plant. Records must be maintained on an annual basis using special forms prescribed by OSHA. Computer or electronic copies can replace paper copies. Records are not sent to OSHA. Rather, they must be maintained locally for a minimum of three years. However, they must be available for inspection by OSHA at any time.

All records required by OSHA must be maintained on only the following forms:

- *Log and Summary of Occupational Injuries and Illnesses (OSHA Form 300).* All recordable injuries and illnesses must be recorded on this form (see Figure 4-1) within six working days of when the employer first becomes aware of the situation. To accommodate the use of computers and word processing systems, employers may create a computer version of Form 300 as long as it is clearly a detailed facsimile. Totals recorded must be posted wherever employee notices are usually posted.

- *Supplementary Record of Occupational Injuries and Illnesses (OSHA Form 301).* Form 300 is a plant-wide summary; Form 301 is a more detailed form for each individual injury (see Figure 4-2). Workers' compensation forms or other forms that contain sufficient detail may be substituted for Form 301.

- *Annual Survey (OSHA Form 200S).* This is a special form provided only to employees selected as participants in OSHA's annual statistical survey. Shortly after the end of the year, employers with more than 11 workers receive Form 200S for reporting on the immediate past year. Employers with 11 or fewer workers receive their form at the beginning of the year to be reported on.

Reporting/Record-Keeping Summary

Reporting and record-keeping requirements appear as part of several different OSHA standards. Not all of them apply in all cases. Following is a summary of the most widely applicable OSHA reporting and record-keeping requirements.

- *29 C.F.R. 1903.2(a) OSHA Poster.* OSHA Poster 2203, which advises employees of the various provisions of the OSHAct, must be conspicuously posted in all facilities subject to OSHA regulations.

- *29 C.F.R. 1903.16(a) Posting of OSHA Citations.* Citations issued by OSHA must be clearly posted for the information of employees in a location as close as possible to the site of the violation.

- *29 C.F.R. 1904.2 Injury/Illness Log.* Employers are required to maintain a log and summary of all *recordable* (Form 300) injuries and illnesses of their employees.

- *29 C.F.R. 1904.4 Supplementary Records.* Employers are required to maintain supplementary records (OSHA Form 301) that give more complete details relating to all recordable injuries and illnesses of their employees.

- *29 C.F.R. 1904.5 Annual Summary.* Employers must complete and post an annual summary of all recordable illnesses and injuries. The summary must be posted from February 1 to March 1 every year.

- *29 C.F.R. 1904.6 Lifetime of Records.* Employers are required to keep injury/illness records on file for three years.

- *29 C.F.R. 1904.7 Access to Records.* Employers are required to give employees, government representatives, and former employees and their designated representatives access to their own individual injury/illness records.

Establishment Name _____

Establishment Name _____

Mailing Address if Different _____

Industry Description and Standard Industrial Classification (SIC) if known _____

Calendar Year _____

Page ___ of ___

A. Employee's Name (e.g., Doe, Jane)	B. Case #	C. Date of Injury or Illness (m/d)	D. Department	E. Job Title	F. Description of Injury or Illness	G. Death (X)	H. Involving Days Away (X) (X Days)	I. Without Days Away Restricted Work Activity (X)	Other (X)	J. Employer Use
						CASE IDENTIFICATION / CASE DESCRIPTION / CASE CLASSIFICATION / OTHER				

Year End Totals _____
Annual Average Number of Employees _____
Total Hours Worked by All Employees _____

I have examined this Log and Summary and certify its accuracy and completeness. X _____

Title _____ Phone (___) _____ Date _____

Figure 4-1
OSHA Injury and Illness Log and Summary based on OSHA Form 300.

Employee

1. Last Name _____ First Name _____ MI___
2. Male ☐   Female ☐
3. Date of Birth ___/___/___
4. Home Address
5. Date hired ___/___/___

Health Care Provider

6. Name of health care provider _____
7. If treatment off-site, facility name and address _____
8. Hospitalized overnight as in-patient?
   (If emergency room only, mark *No*)   Yes ☐   No ☐

Illness or Injury

9. Specific injury or illness
   (e.g., Second degree burn or Toxic hepatitis) _____
10. Body part(s) affected (e.g., Lower right forearm)_____
11. Date of injury or illness: ___/___/___
12. If employee died, date of death: ___/___/___
13. If the case involved days away from work or restricted work activity, enter the date the employee returned to work at full capacity: ___/___/___
14. Time of Event: 15.   Time employee began work:
    (Specify a.m. or p.m.) _____   (Specify a.m. or p.m.) _____
16. All equipment, materials, or chemicals employee was using when the event occurred. (e.g., Acetylene cutting torch, metal plate)
    _____
    _____

17. Specify activity the employee was engaged in when the event occurred (e.g., Cutting metal plate for flooring). Indicate if activity was part of normal duties.
    _____
    _____
    _____

18. How injury or illness occurred. Describe the sequence of events and include any objects or substances that directly injured or made the employee ill. (e.g., Worker stepped back to inspect work and slipped on some scrap metal. As she fell, worker brushed against the hot metal.)
    _____
    _____
    _____

Employer Use
_____
_____
_____
_____
_____
_____

Figure 4-2
OSHA Injury and Illness Record based on OSHA Form 301.

- *29 C.F.R. 1904.8 Major Incident Report.* A major incident is the death of one employee or the hospitalization of five or more employees in one incident. All such incidents must be reported to OSHA.
- *29 C.F.R. 1904.11 Change of Ownership.* When a business changes ownership, the new owner is required to maintain the OSHA-related records of the previous owner.
- *29 C.F.R. 1910.20(d) Medical/Exposure Records.* Employers are required to maintain medical and/or exposure records for the duration of employment plus 30 years (unless the requirement is superseded by another OSHA standard).
- *29 C.F.R. 1910.20(e) Access to Medical/Exposure Records.* Employers that keep medical and/or exposure records are required to give employees access to their own individual records.
- *29 C.F.R. 1910(g)(1) Toxic Exposure.* Employees who will be exposed to toxic substances or other harmful agents in the course of their work must be notified when first hired and reminded continually on at least an annual basis thereafter of their right to access their own individual medical records.
- *29 C.F.R. 1910.20(g)(2) Distribution of Materials.* Employers are required to make a copy of OSHA's *Records Access Standard* (29 C.F.R. 1910.20) available to employees. They are also required to distribute to employees any informational materials provided by OSHA.

OSHA makes provisions for awarding record-keeping variances to companies that wish to establish their own record-keeping systems (see variance regulation 1905). Application procedures are similar to those described earlier in this chapter for standard variances. To be awarded a variance, employers must show that their record-keeping system meets or exceeds OSHA's requirements.

Incidence Rates

Two concepts can be important when completing OSHA 200/300 forms: *incidence rates* and *severity rates*. On occasion, it is necessary to calculate the total injury/illness incident rate of an organization in order to complete an **OSHA Form 200/301**. This calculation must include fatalities and all injuries requiring medical treatment beyond mere first aid.

The formula for determining the total injury/illness incident rate is as follows:

$$IR = N \times 200,000 \div T$$

IR = Total injury/illness incidence rate
N = Number of injuries, illnesses, and fatalities
T = Total hours worked by all employees during the period in question

The number 200,000 in the formula represents the number of hours that 100 employees work in a year (40 hours per week x weeks = 2,000 hours per year per employee). Using the same basic formula with only minor substitutions, safety managers can calculate the following types of incidence rates:

1. Injury rate
2. Illness rate

3. Fatality rate
4. Lost workday cases rate
5. Number of lost workdays rate
6. Specific hazard rate
7. Lost workday injuries rate

The *number of lost workdays rate,* which does not include holidays, weekends, or any other days that employees would not have worked anyway, takes the place of the old *severity rate* calculation.

Record Keeping and Reporting Exceptions

Among the exceptions to OSHA's record-keeping and reporting requirements, the two most prominent are as follows:

■ Employers with ten or fewer employees (full or part-time in any combination)
■ Employers in one of the following categories: real estate, finance, retail trade, or insurance

There are also partial exceptions to OSHA's record-keeping and reporting requirements. Most businesses that fall into Standard Industrial Classifications (SIC) codes 52–89 are exempt from all record-keeping and reporting requirements except in the case of a fatality or an incident in which five or more employees are hospitalized. Within this range of SIC codes, exemptions do *not* apply to the following classifications:

52:	Building materials/garden supplies
53/54:	General merchandise food stores
70:	Hotels/lodging establishments
75/76:	Repair services
79:	Amusement/recreation services
80:	Health services

KEEPING EMPLOYEES INFORMED

One of the most important requirements of the OSHAct is *communication.* Employers are required to keep employees informed about safety and health issues that concern them. Most of OSHA's requirements in this area concern the posting of material. Employers are required to post the following material at locations where employee information is normally displayed:

■ OSHA Poster 2203, which explains employee rights and responsibilities as prescribed in the OSHAct. The state version of this poster may be used as a substitute.
■ Summaries of variance requests of all types.
■ Copies of all OSHA citations received for failure to meet standards. Unlike other informational material, citations must be posted near the site of the violation. They

must remain until the violation is corrected or for a minimum of three days, whichever period is longer.

■ The summary page of OSHA Form 200/300 (Log and Summary of Occupational Injuries and Illnesses). Each year the new summary page must be posted by February 1 and must remain posted until March 1.

In addition to the posting requirements, employers must also provide employees who request them with copies of the OSHAct and any OSHA rules that may concern them. Employees must be given access to records of exposure to hazardous materials and medical surveillance that has been conducted.

WORKPLACE INSPECTIONS

One of the methods OSHA uses for enforcing its rules is the **workplace inspection.** OSHA personnel may conduct workplace inspections unannounced, and except under special circumstances, giving an employer prior notice is a crime punishable by fine, imprisonment, or both.

When OSHA compliance officers arrive to conduct an inspection, they are required to present their credentials to the person in charge. Having done so, they are authorized to enter, at reasonable times, any site, location, or facility where work is taking place. They may inspect, at reasonable times, any condition, facility, machine, equipment, materials, and so on. Finally, they may question, in private, any employee or other person formally associated with the company.

Under special circumstances, employers may be given up to a maximum of 24 hours' notice of an inspection. These circumstances are

■ When imminent danger conditions exist

■ When special preparation on the part of the employer is required

■ When inspection must take place at times other than during regular business hours

■ When it is necessary to ensure that the employer, employee representative, and other pertinent personnel will be present

■ When the local area director for OSHA advises that advance notice will result in a more effective inspection

Employers may require that OSHA have a judicially authorized warrant before conducting an inspection. The U.S. Supreme Court handed down this ruling in 1978 *(Marshall v. Barlow's Inc.).*[13] However, having obtained a legal warrant, OSHA personnel must be allowed to proceed without interference or impediment.

The OSHAct applies to approximately six million work sites in the United States. Sheer volume dictates that OSHA establish priorities for conducting inspections. These priorities are as follows: imminent danger situations, catastrophic fatal accidents, employee complaints, planned high-hazard inspections, and follow-up inspections.

After being scheduled, the inspection proceeds in the following steps:

■ The OSHA compliance officer presents his or her credentials to a company official.

- The compliance officer conducts an **opening conference** with pertinent company officials and employee representatives. The following information is explained during the conference: why the plant was selected for inspection, the purpose of the inspection, its scope, and applicable standards.

- After choosing the route and duration, the compliance officer makes the **inspection tour.** During the tour, the compliance officer may observe, interview pertinent personnel, examine records, take readings, and make photographs.

- The compliance officer holds a **closing conference,** which involves open discussion between the officer and company/employee representatives. OSHA personnel advise company representatives of problems noted, actions planned as a result, and assistance available from OSHA.

CITATIONS AND PENALTIES

Based on the findings of the compliance officer's workplace inspections, OSHA is empowered to issue citations and/or assess penalties. A citation informs the employer of OSHA violations. Penalties are typically fines assessed as the result of citations. The types of citations and their corresponding penalties, as quoted from OSHA 2056, 1991 (Revised), are as follows:

- *Other-than-serious violation.* A violation that has a direct relationship to job safety and health, but probably would not cause death or serious physical harm. A **proposed penalty** of up to $7,000 for each violation is discretionary. A penalty for an **other-than-serious violation** may be adjusted downward by as much as 95 percent, depending on the employer's good faith (demonstrated efforts to comply with the act), history of previous violations, and size of business.

- *Willful violation.* A violation that the employer intentionally and knowingly commits. The employer either knows that what he or she is doing constitutes a violation or is aware that a hazardous condition exists and has made no reasonable effort to eliminate it. Penalties of up to $70,000 may be proposed for each **willful violation,** with a minimum penalty of $5,000 for each violation. A proposed penalty for a willful violation may be adjusted downward, depending on the size of the business and its history of previous violations. Usually, no credit is given for good faith. If an employer is convicted of a willful violation of a standard that has resulted in the death of an employee, the offense is punishable by a court-imposed fine or by imprisonment for up to six months, or both. A fine of up to $250,000 for an individual or $500,000 for a corporation may be imposed for a criminal conviction.

- *Repeat violation.* A violation of any standard, regulation, rule, or order where, upon reinspection, a substantially similar violation is found. **Repeat violations** can bring a fine of up to $70,000 for each such violation. To be the basis of a repeat citation, the original citation must be final; a citation under contest may not serve as the basis for a subsequent repeat citation.

■ *Failure to correct prior violation.* Failure to correct a prior violation may bring a civil penalty of up to $7,000 for each day that the violation continues beyond the prescribed abatement date.[14]

In addition to the citations and penalties described in the preceding paragraphs, employers may also be penalized by additional fines and/or prison if convicted of any of the following offenses: (1) falsifying records or any other information given to OSHA personnel; (2) failing to comply with posting requirements; and (3) interfering in any way with OSHA compliance officers in the performance of their duties.

On March 1, 1991, OSHA's penalty maximum was increased from $10,000 for a single violation to $70,000, before the egregious multiplier is applied. According to Patrick R. Tyson,

> The egregious multiplier allows OSHA to assess a separate penalty for each instance of an OSHA violation. As a result, OSHA penalties, which in the past were often no more than a nuisance, have become serious business. And looking ahead, there is every indication that the number of these large penalties, and the penalties themselves, will continue to grow.[15]

Companies that had penalties pending on March 1, 1991, were subject to having their fines increased substantially. As a result, several companies moved to clear up their penalties before this date. This led to some of the largest settlements in OSHA's history. Tyson cites the following examples:

> USX Corp. settled its major citations (which originally totaled $7.3 million) and, as part of the settlement agreement, made a commitment to undertake significant safety and health correction actions, and agreed to pay a penalty of $3.25 million. In addition, ARCO also settled the citation that followed a disastrous explosion in Texas resulting in the deaths of 17 employees. ARCO agreed to pay the full penalty of $3.48 million. These two settlements are the largest and second largest penalties paid in OSHA's history, far surpassing the previous record high of $1.6 million. The ARCO settlement follows a similar settlement by BASF Corp. in which they also agreed to pay 100 percent of the $1,061,100 penalty imposed by OSHA, after the explosion in Cincinnati which resulted in the death of two employees.[16]

With OSHA penalties increasing in dollar amounts, there is more incentive for employers to appeal or even contest them through litigation. This trend was discussed by Edwin G. Foulke, Jr., director of the OSHA Review Commission, in October 1991.[17] In this interview, Foulke is quoted as saying, "Litigation is often based on whether a case can be settled more cheaply than if it were tried. As the fines increase, the incentive to settle decreases."[18] Clearly, OSHA's appeal process is becoming increasingly important. The process is explained in the following section.

THE APPEALS PROCESS

Employee Appeals

Employees may not contest the fact that citations were or were not awarded or the amounts of the penalties assessed. However, they may appeal the following aspects of

OSHA's decisions regarding their workplace: (1) the amount of time **(abatement period)** given an employer to correct a hazardous condition that has been cited, and (2) an employer's request for an extension of an abatement period. Such appeals must be filed within ten working days of a posting. Although opportunities for formal appeals by employees are unlimited, employees may request an informal conference with OSHA officials to discuss any issue relating to the findings and results of a workplace inspection.

Employer Appeals

Employers may appeal a citation, an abatement period, or the amount of a proposed penalty. Before actually filing an appeal, however, an employer may ask for an informal meeting with OSHA's area director. The area director is empowered to revise citations, abatement periods, and penalties in order to settle disputed claims. If the situation is not resolved through this step, an employer may formalize the appeal. Formal appeals are of two types: (1) a petition for modification of abatement, or (2) a notice of contest. The specifics of both are explained in the following paragraphs.

Petition for Modification of Abatement (PMA)

The **PMA** is available to employers who intend to correct the situation for which a citation was issued, but who need more time. As a first step, the employer must make a good-faith effort to correct the problem within the prescribed timeframe. Having done so, the employer may file a petition for modification of abatement. The petition must contain the following information:

- Descriptions of steps taken so far to comply
- How much additional time is needed for compliance and why
- Descriptions of the steps being taken to protect employees during the interim
- Verification that the PMA has been posted for employee information and that the employee representative has been given a copy

Notice of Contest

An employer who does not wish to comply may contest a citation, an abatement period, and/or a penalty. The first step is to notify OSHA's area director in writing. This is known as filing a **notice of contest.** It must be done within 15 working days of receipt of a citation or penalty notice. The notice of contest must clearly describe the basis for the employer's challenge and contain all of the information about what is being challenged (i.e., amount of proposed penalty or abatement period, and so on).

Once OSHA receives a notice of contest, the area director forwards it and all pertinent materials to the Occupational Safety and Health Review Commission **(OSHRC).** OSHRC is an independent agency that is associated with neither OSHA nor the Department of Labor. The Department of Labor describes how OSHRC handles an employer's claim:

> The commission assigns the case to an administrative law judge. The judge may disallow the contest if it is found to be legally invalid, or a hearing may be scheduled for a public place

near the employer's workplace. The employer and the employees have the right to participate in the hearing; the OSHRC does not require that they be represented by attorneys. Once the administrative law judge has ruled, any party to the case may request further review by OSHRC. Any of the three OSHRC commissioners also may, at his or her own motion, bring a case before the Commission for review. Commission rulings may be appealed to the appropriate U.S. Court of Appeals.[19]

Employer appeals are common. In fact, each issue of *Occupational Hazards* carries a special section on "Contested Cases." A review of examples of contested cases can be instructive in both process and outcomes. Two such cases follow.

STATE-LEVEL OSHA PROGRAMS[24]

States are allowed to develop their own safety and health programs. In fact, the OSHAct encourages it. As an incentive, OSHA will fund up to 50 percent of the cost of operating a state program for states with approved plans. States may develop comprehensive plans covering public and private employers or limit their plans to coverage of public employers only. In such cases, OSHA covers employers not included in the state plan.

To develop an OSHA-approved safety and health plan, a state must have adequate legislative authority and must demonstrate the ability to develop standards-setting, enforcement, and appeals procedures within three years; public employee protection; a sufficient number of qualified enforcement personnel; and education, training, and technical assistance programs. When a state satisfies all of these requirements and accomplishes all developmental steps,

---

CASE STUDY

Lorenzo Textile Mill, Inc.

In response to an employee complaint, an OSHA compliance officer inspected Lorenzo Textile Mill, Inc., and cited the company for failure to post a copy of the OSHA cotton dust standard. Lorenzo contested the citation on the following grounds:

- Because the standard is 133 pages in length it does not lend itself to posting.
- Because of the standard's length, a copy was placed on a table in the employees' lunchroom rather than being posted.
- Each new employee was given a copy of the standard along with an explanation of its contents.[20]

OSHA referred the case to an administrative judge. According to *Occupational Hazards,* the judge "concluded that Lorenzo had done all that could be reasonably expected to keep the standard posted, furthermore, since each employee was provided with a copy, the purpose of the posting was accomplished."[21]

CASE STUDY

IGC Contracting, Inc.

Following an order from his supervisor, a carpenter working for IGC reluctantly removed all of the guardrails from stairwell openings, floor openings, and open-side stairs at a Berkley Heights, New Jersey, job site. Before doing so, he telephoned his union representative, who complained to OSHA. The complaint resulted in an investigation, and IGC was cited for willful violation of three standards. The proposed penalty for each violation was $5,000, for a total penalty of $15,000.[22]

IGC contested the citation, arguing that the foreman who ordered the guardrails removed had done so on his own initiative and that no management official knew of his order. IGC claimed that its superintendent would have rescinded the order if he had known about it. According to *Occupational Hazards,*

> Judge Knight found that the foreman was responsible for the day-to-day work and safety of IGC workers, and that he had exposed IGC and other subcontractors' workers to the danger of death or serious physical harm—in direct violation of OSHA safety standards—by ordering the guard-rails removed.[23]

All three of the proposed penalties were upheld. However, the amount of each was reduced by $2,500 for a total penalty of $7,500. The administrative judge was following the guidelines that require consideration of the following factors in assessing penalties: gravity of the violation, company size, and the company's history of good faith. Because IGC is a small company, its penalties were reduced.

OSHA then certifies that a state has the legal, administrative, and enforcement means necessary to operate effectively. This action renders no judgement on how well or poorly a state is actually operating its program, but merely attests to the structural completeness of its program. After this certification, there is a period of at least one year to determine if a state is effectively providing safety and health protection.[25]

Figure 4–3 lists the states that currently have OSHA-approved safety and health plans. Connecticut and New York cover the public sector only. The Virgin Islands and Puerto Rico account for two of the 23 plans.

SERVICES AVAILABLE FROM OSHA

In addition to setting standards and inspecting for compliance, OSHA provides services to help employers meet the latest safety and health standards. The services, typically offered at no cost, are intended for smaller companies, particularly those with especially hazardous processes and/or materials. Three categories of services are available from OSHA: consultation, voluntary protection programs, and training and education services.

Alaska
P. O. Box 21149
Juneau, AL 99802-1149
907-465-2700

Arizona
800 W. Washington
Phoenix, AZ 85007
602-542-5795

California
395 Oyster Pt. Blvd., 3rd Flr.
  Wing C
San Francisco, CA 94080
415-737-2960

Connecticut
200 Folly Brook Blvd.
Wethersfield, CT 06109
203-566-5123

Hawaii
830 Punchbowl St.
Honolulu, HI 96813
808-548-3150

Indiana
State Office Bldg. 1013
100 N. Senate Ave.
Indianapolis, IN 46204-2287
317-232-2665

Iowa
1000 E. Grand Ave.
Des Moines, IA 50319
515-281-3447

Kentucky
U.S. Highway 127 South
Frankfort, KY 40601
502-564-3070

Maryland
501 St. Paul Place, 15th Floor
Baltimore, MD 21202
301-333-4179

Michigan (Labor)
309 N. Washington Square
P. O. Box 30015
Lansing, MI 48909
517-335-8022

Michigan (Public Health)
3423 N. Logan St., Box 30195
Lansing, MI 48909
517-335-8022

Minnesota
443 Lafayette Road
St. Paul, MN 55155
612-296-2342

Nevada
1370 S. Curry St.
Carson City, NV 89710
702-885-5240

New Mexico
1190 St. Francis Drive, N2200
Santa Fe, NM 87503-0968
702-885-5240

New York
One Main St.
Brooklyn, NY 11201
518-457-3518

North Carolina
4 West Edenton St.
Raleigh, NC 27601
919-733-7166

Oregon
21 Labor & Ind. Bldg.
Salem, OR 97310
503-378-3304

Puerto Rico
Prudencio Rivera Martinez
  Bldg.
505 Munoz Rivera Ave.
Hato Rey, PR 00918
809-654-2119-22

Virginia
P. O. Box 12064
Richmond, VA 23241-0064
804-786-2376

South Carolina
P. O. Box 11329
Columbia, SC 29211-1329
803-734-9594

Vermont
120 State St.
Montpelier, VT 05602
802-828-2765

Virgin Islands
Box 890, Christiansted
St. Croix, VI 00820
809-773-1994

Washington
Gen. Admin. Bldg.
Room 334-AX-31
Olympia, WA 98504-0631
206-753-6307

Figure 4–3
States with approved safety and health plans.

SAFETY FACT

Maine OSHA's Top 200 Program

The OSHA of Maine developed one of the most innovative state-level safety programs. Maine's *Top 200 Program* departs from the traditional OSHA approach of *inspect and fine*—an approach that many safety professionals believe encourages businesses to cover up problems instead of identifying and dealing with them. Instead, Maine's OSHA identifies the 200 businesses in the state that record the most injuries each year and makes them an innovative offer. Any company on the list that agrees to identify its own safety problems and work with OSHA officials to correct them is exempt from wall-to-wall inspections. As a result, Maine's business community is identifying and solving an unprecedented number of safety problems. In the eight years prior to implementing the Top 200 Program, a total of 37,000 safety problems had been reported to Maine's OSHA. After just two years of the program, this number had increased to 174,331, and 118,051 of these had been corrected. Along the way, OSHA became a partner to the business community instead of an intrusive arm of big government.

In recognition of the innovative spirit and positive results produced by this initiative, Maine's Top 200 Program received the *Innovations in American Government Award* from the Ford Foundation and the John F. Kennedy School of Government at Harvard University.

Consultation Services

Consultation services provided by OSHA include assistance in (1) identifying hazardous conditions; (2) correcting identified hazards; and (3) developing and implementing programs to prevent injuries and illnesses. To arrange consultation services, employers contact the consultation provider in their state (see Figure 4–4).

The actual services are provided by professional safety and health consultants, who are not OSHA employees. They typically work for state agencies or universities and provide consultation services on a contract basis; OSHA provides the funding. OSHA publication 3047, entitled *Consultation Services for the Employer,* may be obtained from the nearest OSHA office.

Voluntary Protection Programs

OSHA's **Voluntary Protection Programs** (VPPs) serve the following three basic purposes:

■ To recognize companies that have incorporated safety and health programs into their overall management system

State	Telephone	State	Telephone
Alabama	205-348-3033	Nebraska	401-471-4717
Alaska	907-264-2599	Nevada	701-789-0546
Arizona	602-255-5795	New Hampshire	603-271-3170
Arkansas	501-682-4522	New Jersey	609-984-3517
California	415-557-2870	New Mexico	505-827-2885
Colorado	303-491-6151	New York	518-457-5468
Connecticut	203-566-4550	North Carolina	919-733-3949
Delaware	302-571-3908	North Dakota	701-224-2348
District of Columbia	202-576-6339	Ohio	614-644-2631
Florida	904-488-3044	Oklahoma	405-528-1500
Georgia	404-894-8274	Oregon	503-378-3272
Guam	9-011 671-646-9246	Pennsylvania	800-381-1241 (Toll-free) 412-357-2561
Hawaii	808-548-7510	Puerto Rico	809-754-2134-2171
Idaho	208-385-3283	Rhode Island	401-277-2438
Illinois	312-917-2339	South Carolina	803-734-9579
Indiana	317-232-2688	South Dakota	605-688-4101
Iowa	515-281-5352	Tennessee	615-741-7036
Kansas	913-296-4386	Texas	512-458-7254
Kentucky	502-564-6895	Utah	801-530-6868
Louisiana	504-342-9601	Vermont	801-828-2765
Maine	207-289-6460	Virginia	804-367-1986
Maryland	301-333-4219	Virgin Islands	809-772-1315
Massachusetts	616-727-3463	Washington	206-586-0961
Michigan	517-335-8250 (Health) 517-322-1814 (Safety)	West Virginia	304-348-7890
Minnesota	612-297-2393	Wisconsin	608-266-8579 (Health) 414-512-5063 (Safety)
Mississippi	601-987-3961	Wyoming	307-777-7786
Missouri	314-751-3403		
Montana	406-444-6401		

Figure 4–4
State consultation project directory.

- To motivate companies to incorporate health and safety programs into their overall management system
- To promote positive, cooperative relationships among employers, employees, and OSHA

OSHA currently operates three programs under the VPP umbrella. These programs are discussed in the following paragraphs.

Star Program

The **Star Program** recognizes companies that have incorporated safety and health into their regular management system so successfully that their injury rates are below the national average for their industry. This is OSHA's most strenuous program. To be part of the Star Program, a company must demonstrate

- Management commitment
- Employee participation
- An excellent work-site analysis program
- A hazard prevention and control program
- A comprehensive safety and health training program[26]

Merit Program

The **Merit Program** is less strenuous than the Star Program. It is seen as a stepping-stone to recognize companies that have made a good start toward Star Program recognition. OSHA works with such companies to help them take the next step and achieve Star Program recognition.

Demonstration Program

The Department of Labor describes the **Demonstration Program** as follows: "for companies that provide Star-quality worker protection in industries where certain Star requirements can be changed to include these companies as Star participants."[27]

Companies participating in any of the VPPs are exempt from regular programmed OSHA inspections. However, employee complaints, accidents that result in serious injury, or major chemical releases will be "handled according to routine enforcement procedures."[28]

Training and Education Services

Training and education services available from OSHA take several forms. OSHA operates a training institute in Des Plaines, Illinois, that offers a wide variety of services to safety and health personnel from the public and private sectors. The institute has a full range of facilities including classrooms and laboratories in which it offers more than 60 courses.

To promote training and education in locations other than the institute, OSHA awards grants to nonprofit organizations. Colleges, universities, and other nonprofit organizations apply for funding to cover the costs of providing workshops, seminars, or short courses on safety and health topics currently high on OSHA's list of priorities. Grant funds must be used to plan, develop, and present instruction. Grants are awarded annually and require a match of at least 20 percent of the total grant amount.

EMPLOYER RIGHTS AND RESPONSIBILITIES

OSHA is very specific in delineating the rights and responsibilities of employers regarding safety and health. These rights and responsibilities, as set forth in OSHA publication 2056, are summarized in this section.

Employer Rights

The following list of **employer rights** under the OSHAct is adapted from OSHA 2056, 1991 (Revised). Employers have the right to do the following:

- Seek advice and consultation as needed by contacting or visiting the nearest OSHA office.
- Request proper identification of the OSHA compliance officer prior to an inspection.
- Be advised by the compliance officer of the reason for an inspection.
- Have an opening and closing conference with the compliance officer in conjunction with an inspection.
- Accompany the compliance officer on the inspection.
- File a notice of contest with the OSHA area director within 15 working days of receipt of a notice of citation and proposed penalty.
- Apply for a temporary variance from a standard if unable to comply because the materials, equipment, or personnel needed to make necessary changes within the required time are not available.
- Apply for a permanent variance from a standard if able to furnish proof that the facilities or methods of operation provide employee protection at least as effective as that required by the standard.
- Take an active role in developing safety and health standards through participation in OSHA Standards Advisory Committees, through nationally recognized standards-setting organizations, and through evidence and views presented in writing or at hearings.
- Be assured of the confidentiality of any trade secrets observed by an OSHA compliance officer during an inspection.
- Ask NIOSH for information concerning whether any substance in the workplace has potentially toxic effects.[29]

Employer Responsibilities[30]

In addition to the rights set forth in the previous subsection, employers have prescribed responsibilities. The following list of **employer responsibilities** under the OSHAct is adapted from OSHA 2056, 1991 (Revised). Employers must do the following:

- Meet the general duty responsibility to provide a workplace free from hazards that are causing or are likely to cause death or serious physical harm to employees, and comply with standards, rules, and regulations issued under the OSHAct.
- Be knowledgeable of mandatory standards and make copies available to employees for review upon request.
- Keep employees informed about OSHA.
- Continually examine workplace conditions to ensure that they conform to standards.
- Minimize or reduce hazards.
- Make sure employees have and use safe tools and equipment (including appropriate personal protective equipment) that is properly maintained.
- Use color codes, posters, labels, or signs as appropriate to warn employees of potential hazards.
- Establish or update operating procedures and communicate them so that employees follow safety and health requirements.
- Provide medical examinations when required by OSHA standards.
- Provide the training required by OSHA standards.
- Report to the nearest OSHA office within 48 hours any fatal accident or one that results in the hospitalization of five or more employees.
- Keep OSHA-required records of injuries and illnesses and post a copy of the totals from the last page of OSHA Form 200/300 during the entire month of February each year. (This applies to employers with 11 or more employees.)
- At a prominent location within the workplace, post **OSHA Poster 2203** informing employees of their rights and responsibilities.
- Provide employees, former employees, and their representatives access to the Log and Summary of Occupational Injuries and Illnesses (OSHA Form 200/300) at a reasonable time and in a reasonable manner.
- Give employees access to medical and exposure records.
- Give the OSHA compliance officer the names of authorized employee representatives who may be asked to accompany the compliance officer during an inspection.
- Not discriminate against employees who properly exercise their rights under the act.
- Post OSHA citations at or near the work site involved. Each citation or copy must remain posted until the violation has been abated or for three working days, whichever is longer.
- Abate cited violations within the prescribed period.

EMPLOYEE RIGHTS AND RESPONSIBILITIES

Employee Rights

Section 11(c) of the OSHAct delineates **employee rights.** These rights are actually protection against punishment for employees who exercise their right to pursue any of the following courses of action:

- Complain to an employer, union, OSHA, or any other government agency about job safety and health hazards.
- File safety or health grievances.
- Participate in a workplace safety and health committee or in union activities concerning job safety and health.
- Participate in OSHA inspections, conferences, hearings, or other OSHA-related activities.[31]

Employees who feel they are being treated unfairly because of actions they have taken in the interest of safety and health have 30 days in which to contact the nearest OSHA office. Upon receipt of a complaint, OSHA conducts an investigation and makes recommendations based on its findings. If an employer refuses to comply, OSHA is empowered to pursue legal remedies at no cost to the employee who filed the original complaint.

In addition to those just set forth, employees have a number of other rights. Employees may

- Expect employers to make review copies available of OSHA standards and requirements.
- Ask employers for information about hazards that may be present in the workplace.
- Ask employers for information on emergency procedures.
- Receive safety and health training.
- Be kept informed about safety and health issues.
- Anonymously ask OSHA to conduct an investigation of hazardous conditions at the work site.
- Be informed of actions taken by OSHA as a result of a complaint.
- Observe during an OSHA inspection and respond to the questions asked by a compliance officer.
- See records of hazardous materials in the workplace.
- See their medical record.
- Review the annual Log and Summary of Occupational Injuries (OSHA Form 200/300).
- Have an exit briefing with the OSHA compliance officer following an OSHA inspection.
- Anonymously ask NIOSH to provide information about toxicity levels of substances used in the workplace.
- Challenge the abatement period given employers to correct hazards discovered in an OSHA inspection.

- Participate in hearings conducted by the Occupational Safety and Health Review Commission.
- Be advised when an employer requests a variance to a citation or any OSHA standard.
- Testify at variance hearings.
- Appeal decisions handed down at OSHA variance hearings.
- Give OSHA input concerning the development, implementation, modification, and/or revocation of standards.[32]

Employee Responsibilities

Employees have a number of specific responsibilities. The following list of **employee responsibilities** is adapted from OSHA 2056, 1991 (Revised). Employees must

- Read the OSHA poster at the job site and be familiar with its contents.
- Comply with all applicable OSHA standards.
- Follow safety and health rules and regulations prescribed by the employer and promptly use personal protective equipment while engaged in work.
- Report hazardous conditions to the supervisor.
- Report any job-related injury or illness to the employer and seek treatment promptly.
- Cooperate with the OSHA compliance officer conducting an inspection.
- Exercise their rights under the OSHAct in a responsible manner.[33]

KEEPING UP TO DATE ON OSHA

OSHA's standards, rules, and regulations are always subject to change. The development, modification, and revocation of standards is an ongoing process. It is important for prospective and practicing safety and health professionals to stay up to date with the latest actions and activities of OSHA. Following is an annotated list of strategies that can be used to keep current:

- Establish contact with the nearest regional or area OSHA office and periodically request copies of new publications or contact the OSHA Publications Office at the following address:

    OSHA Publications Office
    200 Constitution Avenue, N.W.
    Room N–3101
    Washington, D.C. 20210
    http://www.osha-slc.gov/oshdoc/interp_std_toc

- Review professional literature in the safety and health field. Numerous periodicals carry OSHA updates that are helpful.
- Establish and maintain relationships with other safety and health professionals for the purpose of sharing information, and do so frequently.
- Join professional organizations, review their literature, and attend their conferences.

PROBLEMS WITH OSHA

Federal agencies are seldom without their detractors. Resentment of the federal bureaucracy is intrinsic in the American mind-set. Consequently, complaints about OSHA are common. Even supporters occasionally join the ranks of the critics. Often, the criticisms leveled against OSHA are valid.

Criticisms of OSHA take many different forms. Some characterize OSHA as an overbearing bureaucracy with little or no sensitivity to the needs of employers who are struggling to survive in a competitive marketplace. At the same time, others label OSHA as timid and claim it does not do enough. At different times and in different cases, both points of view have probably been at least partially accurate.

Most criticism of OSHA comes in the aftermath of major accidents or a workplace disaster. Such events typically attract a great deal of media attention, which, in turn, draws the attention of politicians. In such cases, the criticism tends to focus on the question, "Why didn't OSHA prevent this disaster?" At congressional hearings, detractors will typically answer this question by claiming that OSHA spends too much time and too many resources dealing with matters of little consequence while ignoring real problems. Supporters of OSHA will typically answer the question by claiming that a lack of resources prevents the agency from being everywhere at once. There is a measure of validity in both answers.

An example of a disaster that precipitated this type of dialogue is the chemical explosion that occurred in May 1991 at a fertilizer plant in Sterlington, Louisiana. This tragic incident claimed the life of eight employees. An additional 128 were injured. Two weeks later, a subcommittee of the House of Representatives convened to hear testimony relating to the explosion. It turned out that OSHA had inspected the plant six times and had issued eight different citations. However, there had been no follow-up inspection of the plant in almost ten years.[34] Charges of poor follow-up were leveled at OSHA in this case. OSHA officials responded that they follow up to the extent that their resources allow.

In fact, there is evidence that OSHA has made a positive difference since the inception of the OSHAct. In the first 20 years of OSHA's existence, occupational fatalities dropped by 25 percent from 12,000 annually to 9,000 annually.[35] However, a recurrent complaint, and one that is often valid, is a lack of follow-up on OSHA's part.

A study conducted in 1991 by the General Accounting Office (GAO), the investigative arm of Congress, found that OSHA often has to depend on the companies that it cites to volunteer information about whether they have corrected the violations.[36] According to *Occupational Hazards,*

> GAO disclosed that only 6 percent of the more than 50,000 inspections conducted by OSHA in fiscal 1989 were follow-ups to previous inspections that had uncovered violations. In 24 percent of the follow-up inspections, the employer had failed to correct cited conditions.[37]

A study conducted by the National Safe Workplace Institute, a private organization located in Chicago, Illinois, criticizes the federal government for not doing enough to protect the American worker. In a report released in January 1992, the institute questioned the government's priorities, claiming the nation has six fish and game inspectors for every one safety inspector.[38] OSHA officials responded to the criticism by saying the government does not need more inspectors. Rather, it needs to find ways to make employers take more responsibility for the safety and health of their employees.[39]

On the one hand, statistics cited in this section show that OSHA has made a difference in the condition of the workplace in this country. On the other hand, large, centralized bureaucratic agencies rarely achieve a high level of efficiency. This is compounded in OSHA's case by the fact that the agency is subject to the ebb and flow of congressional support, particularly in the area of funding. Consequently, OSHA is likely to continue to be an imperfect organization subject to ongoing criticism.

OTHER FEDERAL AGENCIES AND ORGANIZATIONS

Although OSHA is the most widely known safety and health organization in the federal government, it is not the only one. Figure 4–5 lists government agencies and organizations (including OSHA) with safety and health as part of their mission.

Of those listed, the most important to modern safety and health professionals are NIOSH and OSHRC. The missions of these organizations are summarized in the following paragraphs.

NIOSH

The National Institute for Occupational Safety and Health **(NIOSH)** is part of the Department of Health and Human Services (HHS). (Recall that OSHA is part of the Department of Labor.) NIOSH has two broad functions: research and education. The main focus of the agency's research is on toxicity levels and human tolerance levels of hazardous substances. NIOSH prepares recommendations along these lines for OSHA standards dealing with hazardous substances. NIOSH studies are also published and made available to employers. Each year, NIOSH publishes updated lists of toxic materials and recommended tolerance levels. These publications represent the educational component of NIOSH's mission.

The Department of Health and Human Services describes NIOSH as follows:

> In 1973, NIOSH became a part of the Centers for Disease Control (CDC), an arm of the Public Health Service in the Department of Health and Human Services. NIOSH is unique among federal research institutions because it has the authority to conduct research in the workplace, and to respond to requests for assistance from employers and employees.

NIOSH also consults with the Department of Labor (DOL) and other federal, state, and local government agencies to promote occupational safety and health and makes recommendations to DOL about worker exposure limits.

NIOSH estimates that over 7,000 people are killed at work each year, and nearly 12 million nonfatal injuries occur in the workplace. In addition to death and injury, it is estimated that over 10 million men and women are exposed to hazardous substances in their jobs that can eventually cause fatal or debilitating diseases. To help establish priori-

Figure 4–5
Agencies dealing with safety
and health.

American Public Health Association
1015 Fifteenth Street, N.W.
Washington, D.C. 20005

Bureau of Labor Statistics
U.S. Department of Labor
Washington, D.C. 20212

Bureau of National Affairs, Inc.
Occupational Safety and Health Reporter
1231 25th Street, N.W.
Washington, D.C. 20037

Commerce Clearing House
Employee Safety and Health Guide
4205 W. Peterson Avenue
Chicago, IL 60646

Environmental Protection Agency
401 M Street, S.W.
Washington, D.C. 20001

National Institute for Occupational Safety and
   Health (NIOSH)
4676 Columbia Parkway
Cincinnati, OH 45226

Occupational Safety and Health Administration
U.S. Department of Labor
200 Constitution Avenue
Washington, D.C. 20210

Occupational Safety and Health Review
   Committee (OSHRC)
Washington, D.C. 20210

U.S. Consumer Product Safety Commission
Washington, D.C. 20207

ties in developing research and control of these hazards, NIOSH developed a list of the
ten leading work-related diseases and injuries:

1. Occupational lung diseases
2. Musculoskeletal injuries
3. Occupational cancers
4. Occupational cardiovascular disease
5. Severe occupational traumatic injuries
6. Disorders of reproduction
7. Neurotoxic disorders
8. Noise-induced hearing loss

9. Dermatological conditions
10. Psychological disorders[40]

Figure 4–6 is an organizational chart showing the major divisions in NIOSH. The four principal divisions are described in the following paragraphs.

Division of Biomedical and Behavioral Science[41]

The Division of Biomedical and Behavioral Science **(DBBS)** conducts research in the areas of toxicology, behavioral science, ergonomics, and the health consequences of various physical agents. DBBS investigates problems created by new technologies and develops biological monitoring and diagnostic aids to ensure that the workplace is not responsible for diminished health, functional capacity, or life expectancy of workers.

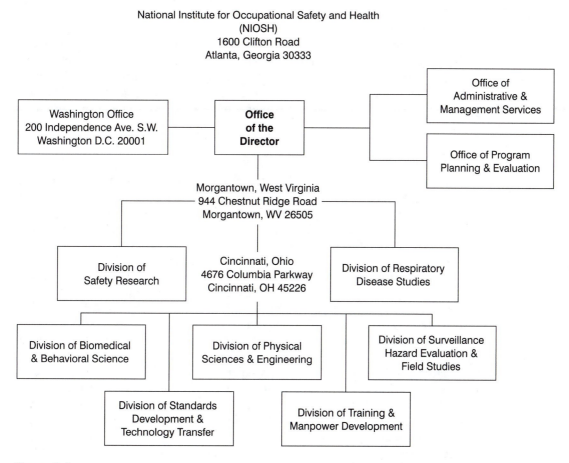

Figure 4–6
The major divisions of NIOSH.

Consultation and data basic to the development of criteria for standards for workplace exposure are furnished.

DBBS conducts laboratory and work-site research into the psychological, behavioral, physiological, and motivational factors relating to job stress as well as those induced by chemical and physical agents. The division assesses physical work capacity and tolerance for environmental conditions as influenced by age, gender, body type, and physical fitness. Interventions and control procedures using various approaches are developed and tested for their ability to reduce undue job stress.

The DBBS toxicology research involves dose-response and methods development studies to evaluate the effects of toxic agents and to develop techniques that utilize biomarkers as indicators of toxic exposures and early indicators of pathological change. The studies include applications of cellular biology, immunochemistry, pharmacokinetics, and neurophysiology. Through laboratory analysis of biological samples from animals exposed experimentally and humans exposed occupationally to workplace hazards, DBBS provides clinical and biochemical consultations for ascertaining the extent of exposure and for diagnosing occupational diseases.

DBBS conducts laboratory and work-site research on hazards from physical agents such as noise, vibration, and nonionizing energy sources. Studies seek to identify exposure factors that are significant to the health and well-being of the work force. Investigations, including instrumentation and methods development, for characterizing and relating exposure factors to biological and performance changes in animal and human populations are also undertaken.

Division of Respiratory Disease Studies[42]

The Division of Respiratory Disease Studies **(DRDS)** is the focal point for the clinical and epidemiological research that NIOSH conducts on occupational respiratory diseases. The division provides legislatively mandated medical and autopsy services and conducts medical research to fulfill NIOSH's responsibilities under the Federal Mine Safety Health Amendments Act of 1977.

The division conducts field studies of occupational respiratory diseases, in addition to designing and interpreting cross-sectional and prospective morbidity and mortality studies relating to occupational respiratory disease. Field studies are conducted at mines, mills, and other industrial plants where occupational respiratory diseases occur among workers. The division uses epidemiological techniques, including studies of morbidity and mortality, to detect common characteristics related to occupational respiratory diseases.

To formulate and implement programs that will identify factors involved in the early detection and differential rates of susceptibility to occupational respiratory diseases, DRDS conducts cell biology research to determine the role of microorganisms and environmental exposure in these diseases. The division also provides autopsy evaluations and a pathology research program. Research is conducted on immunological mechanisms and cell physiology to determine the effects of environmental exposure as it relates to occupational respiratory diseases.

DRDS provides for planning, coordinating, and processing medical examinations mandated under the Federal Mine Safety and Health Amendments Act of 1977 and operates a certification program for medical facilities and physicians who participate in the examination program. DRDS also evaluates and approves employer programs for the examination of employees in accordance with published regulations as well as arranging for the examination of employees who work at locations that lack an approved examination program. The division conducts the National Coal Workers' Autopsy Program and performs research into the postmortem identification and quantification of occupational respiratory exposures.

Division of Surveillance, Hazard Evaluations, and Field Studies[43]

The Division of Surveillance, Hazard Evaluations, and Field Studies **(DSHEFS)** conducts surveillance of the nation's workforce and workplaces to assess the magnitude and extent of job-related illnesses, exposures, and hazardous agents. DSHEFS conducts legislatively mandated health hazard evaluations and industry-wide epidemiological research programs, including longitudinal studies of records and clinical/environmental field studies and surveys. DSHEFS also provides technical assistance on occupational safety and health problems to other federal, state, and local agencies; other technical groups; unions; employers; and employees.

Surveillance efforts are designed for the early detection and continuous assessment of the magnitude and extent of occupational illnesses, disabilities, deaths, and exposures to hazardous agents, using new and existing data sources from federal, state, and local agencies; labor; industry; tumor registries; physicians; and medical centers. DSHEFS also conducts evaluation and validation studies of reporting systems covering occupational illnesses, with the intent of improving methods for measuring the magnitude of occupational health problems nationwide.

Division of Training and Manpower Development[44]

The Division of Training and Manpower Development **(DTMD)** implements Section 21 of the OSHAct, which sets forth training and education requirements. DTMD develops programs to increase the numbers and competence of safety and health professionals. This continuing education program provides short-term technical training courses including seminars, independent study packages, and specialized workshops to federal, state, and local government; private industry; labor unions; and other organizations in the field of safety and health. The curriculum development component develops courseware and other training materials for NIOSH-sponsored training programs, including those represented by in-house faculty as well as those conducted by universities and other outside training organizations.

The educational resource development program continually assesses manpower needs for safety and health practitioners and researchers on a nationwide basis. To help meet the demand, DTMD administers a major training grant program to foster the development of academically based training programs for occupational physicians, occupational health nurses, industrial hygienists, toxicologists, epidemiologists, and safety professionals. DTMD also develops specific criteria for the selection of qualified organizations to conduct research training, graduate education, continuing education, and outreach programs to expand the network of knowledgeable professionals in occupational safety and health.

DTMD initiates special-emphasis projects targeted to physicians, engineers, managers, vocational education students, and science teachers to facilitate the inclusion of occupational safety and health knowledge in their formal program of study. The division establishes a collaborating relationship with the many professional societies and accrediting bodies to formalize the process of long-term commitment through professional networking.

OSHRC

The Occupational Safety and Health Review Committee (OSHRC) is not a government agency. Rather, it is an independent board whose members are appointed by the president and given quasi-judicial authority to handle contested OSHA citations. When a citation, proposed penalty, or abatement period issued by an OSHA area director is contested by an employer, OSHRC hears the case. OSHRC is empowered to review the evidence, approve the recommendations of the OSHA area director, reject those recommendations, or revise them by assigning substitute values. For example, if an employer contests the amount of a proposed penalty, OSHRC is empowered to accept the proposed amount, reject it completely, or change it.

Mining Enforcement and Safety Administration

The mining industry is exempt from OSHA regulations. Instead, mining is regulated by the Metal and Non-Metallic Mine Safety Act. OSHA does regulate those aspects of the industry that are not directly involved in actual mining work. There is a formal memorandum of understanding between OSHA and the Mining Enforcement and Safety Administration (MESA), the agency that enforces the Metal and Non-Metallic Mine Safety Act.

In 1977, Congress passed the Mine Safety and Health Amendment, which established the Mine Safety and Health Administration (MSHA) as a functional unit within the U.S. Department of Labor. MSHA works with MESA to ensure that the two agencies do not become embroiled in jurisdictional disputes.

Federal Railroad Administration

Railroads, for the most part, fall under the jurisdiction of OSHA. The Federal Railroad Administration (FRA) exercises limited jurisdiction over railroads in situations involving working conditions. Beyond this, railroads must adhere to the standards for *General Industry* in C.F.R. Part 1910. OSHA and FRA personnel coordinate to ensure that jurisdictional disputes do not arise.

DISCUSSION CASE

What Is Your Opinion?

"The OSHAct is a nightmare! All it has accomplished is the creation of a department full of governmental bureaucrats who bully private industry." This was the opening line in a debate on government safety regulations sponsored by the Industrial Technology Department of Pomona State University. The OSHA advocate in the debate responded as follows: "The OSHAct is a model of government regulation as it should be. Had private industry been responsive to the safety and health concerns of employees, there would have been no need for government regulation." These are two widely divergent viewpoints. What is your opinion in this matter?

OSHA'S GENERAL INDUSTRY STANDARDS

The most widely applicable OSHA standards are the *General Industry Standards*. These standards are found in 29 C.F.R. 1910. Part 1910 consists of 21 subparts, each carrying an uppercase-letter designation. Subparts A and B contain no compliance requirements. The remaining subparts are described in the following subsections.

Subpart C: General Safety and Health Provisions

The only compliance standard in Subpart C is *Access to Employee Exposure and Medical Records*. Employers that are required to keep medical and exposure records must do the following: (1) maintain the records for the duration of employment plus 30 years, and (2) give employees access to their individual personal records.

Subpart D: Walking–Working Surfaces

Subpart D contains the standards for all surfaces on which employees walk or work. Specific sections of Subpart D are as follows:

1910.21	Definitions
1910.22	General requirements
1910.23	Guarding floor and wall openings and holes
1910.24	Fixed industrial stairs
1910.25	Portable wood ladders
1910.26	Portable metal ladders
1910.27	Fixed ladders
1910.28	Safety requirements for scaffolding

1910.29	Manually propelled mobile ladder stands and scaffolds (towers)
1910.30	Other working surfaces
1910.31	Sources of standards
1910.32	Standards organizations

Subpart E: Means of Egress

Subpart E requires employers to ensure that employees have a safe, accessible, and efficient means of escaping a building under emergency circumstances. Specific sections of Subpart E are as follows:

1910.35	Definitions
1910.36	General requirements
1910.37	Means of egress, general
1910.38	Employee emergency plans and fire prevention plan
1910.39	Source of standards
1910.40	Standards organizations

Emergency circumstances might be caused by fire, explosions, hurricanes, tornadoes, flooding, terrorist acts, earthquakes, nuclear radiation, or other acts of nature not listed here.

Subpart F: Powered Platforms

Subpart F applies to powered platforms, mechanical lifts, and vehicle-mounted work platforms. The requirements of this subpart apply only to employers who use this type of equipment in facility maintenance operations. Specific sections of Subpart F are as follows:

1910.66	Powered platforms for building maintenance
1910.67	Vehicle-mounted elevating and rotating work platforms
1910.68	Manlifts
1910.69	Sources of standards
1910.70	Standards organizations

Subpart G: Health and Environmental Controls

The most widely applicable standard in Subpart G is 1910.95 (occupational noise exposure). Other standards in this subpart pertain to situations where ionizing and/or non-ionizing radiation are present. Specific sections of Subpart G are as follows:

| 1910.94 | Ventilation |
| 1910.95 | Occupational noise exposure |

1910.96	Ionizing radiation
1910.97	Nonionizing radiation
1910.98	Effective dates
1910.99	Sources of standards
1910.100	Standards organizations

Subpart H: Hazardous Materials

Four of the standards in Subpart H are widely applicable. Section 1901.106 is an extensive standard covering the use, handling, and storage of flammable and combustible liquids. Of particular concern are fire and explosions. Section 1910.107 applies to indoor spray-painting processes and processes in which paint is applied in powder form (e.g., electrostatic powder spray).

Section 1910.119 applies to the management of processes involving specifically named chemicals and flammable liquids and gases. Section 1910.120 contains requirements relating to emergency response operations and hazardous waste. All of the standards contained in Subpart H are as follows:

1910.101	Compressed gases (general requirements)
1910.102	Acetylene
1910.103	Hydrogen
1910.104	Oxygen
1910.105	Nitrous oxide
1910.106	Flammable and combustible liquids
1910.107	Spray finishing using flammable and combustible materials
1910.108	Dip tanks containing flammable and combustible materials
1910.109	Explosive and blasting agents
1910.110	Storage and handling of liquefied petroleum gases
1910.111	Storage and handling of anhydrous ammonia
1910.114	Effective dates
1910.115	Sources of standards
1910.116	Standards organizations
1910.119	Process safety management of highly hazardous chemicals
1910.120	Hazardous waste operations and emergency response

Subpart I: Personal Protective Equipment

Subpart I contains three of the most widely applicable standards: 1910.132 General Requirements; 1910.133 Eye and Face Protection; and 1910.134 Respiratory Protection.

The most frequently cited OSHA violations relate to these and the other personal-protective equipment standards. All of the standards in this subpart are as follows:

1910.132	General requirements
1910.133	Eye and face protection
1910.134	Respiratory protection
1910.135	Occupational head protection
1910.136	Occupational foot protection
1910.137	Electrical protective devices
1910.138	Effective dates
1910.139	Sources of standards
1910.140	Standards organizations

Subpart J: General Environment Controls

This subpart contains standards that are widely applicable because they pertain to general housekeeping requirements. An especially important standard contained in this subpart is 1910.146: Permit-Required Confined Spaces. A confined space is one that meets any or all of the following criteria:

- Large enough and so configured that a person can enter it and perform assigned work tasks therein
- Continuous employee occupancy is not intended

The *lockout/tagout* standard is also contained in this subpart. All of the standards in this subpart are as follows:

1910.141	Sanitation
1910.142	Temporary labor camps
1910.144	Safety color code for marking physical hazards
1910.145	Accident prevention signs and tags
1910.146	Permit-required confined space
1910.147	Control of hazardous energy (lockout/tagout)
1910.148	Standards organizations
1910.149	Effective dates
1910.150	Sources of standards

Subpart K: Medical and First Aid

This is a short subpart, the most important section of which pertains to eye-flushing. If employees are exposed to *injurious corrosive materials,* equipment must be provided for quickly flushing the eyes and showering the body. The standard also requires medical personnel to be readily available. "Readily available" can mean that there is a clinic or hospital nearby. If such a facility is not located nearby, employers must have a person on hand who has had first-aid training. The standards in this subpart are as follows:

1910.151	Medical seminars and first aid
1910.153	Sources of standards

Subpart L: Fire Protection

This subpart contains the bulk of OSHA's fire protection standard. These standards detail the employer's responsibilities concerning fire brigades, portable fire-suppression equipment, fixed fire-suppression equipment, and fire-alarm systems. Employers are not required to form fire brigades, but if they choose to, employers must adhere to the standard set forth in 1910.156. The standards in this subpart are as follows:

Fire Protection

1910.155	Scope, application, and definitions applicable to this subpart
1910.156	Fire brigades

Portable Fire-Suppression Equipment

1910.157	Portable fire extinguishers
1910.158	Standpipe and hose systems

Fixed Fire-Suppression Equipment

1910.159	Automatic sprinkler systems
1910.160	Fixed extinguishing systems, general
1910.161	Fixed extinguishing systems, dry chemical
1910.162	Fixed extinguishing systems, gaseous agent
1910.163	Fixed extinguishing systems, water spray and foam

Other Fire Protection Systems

1910.164	Fire detection systems
1910.165	Employee alarm systems

Subpart M: Compressed Gas/Air

This subpart contains just three sections and only one standard, 1910.169. This standard applies to compressed-air equipment that is used in drilling, cleaning, chipping, and hoisting. There are many other uses of compressed air in work settings, but 1910.169 applies only to these applications. The sections in this subpart are as follows:

1910.169	Air receivers

| 1910.170 | Sources of standards |
| 1910.171 | Standards organizations |

Subpart N: Materials Handling and Storage

This subpart contains one broad standard—1910.176—and several more specific standards relating to 1910.176. Subpart N is actually limited in scope. It applies only to the handling and storage of materials, changing rim wheels on large vehicles, and the proper use of specific equipment identified in the standards' titles. All of the standards in this subpart are as follows:

1910.176	Handling materials—general
1910.177	Servicing multi-piece and single-piece rim wheels
1910.178	Powered industrial trucks
1910.179	Overhead and gantry cranes
1910.180	Crawler locomotive and truck cranes
1910.181	Derricks
1910.182	Effective dates
1910.183	Helicopters
1910.184	Slings
1910.189	Sources of standards
1910.190	Standards organizations

Subpart O: Machinery and Machine Guarding

This subpart contains standards relating to specific types of machines. The types of machines covered are identified in the titles of the standards contained in Subpart O. These standards are as follows:

1910.211	Definitions
1910.212	General requirements for all machines
1910.213	Woodworking machinery requirements
1910.214	Cooperage machinery
1910.215	Abrasive wheel machinery
1910.216	Mills and calendars in the rubber and plastics industries
1910.217	Mechanical power presses
1910.218	Forging machines
1910.219	Mechanical power-transmission apparatus
1910.220	Effective dates
1910.221	Sources of standards

| 1910.222 | Standards organizations |

Subpart P: Hand Tools/Portable Power Tools

This subpart contains standards relating to the use of hand tools, portable power tools, and compressed-air-powered tools. The types of tools covered in this subpart, in addition to typical hand tools, include jacks, saws, drills, sanders, grinders, planers, power lawn-mowers, and other tools. The standards contained in this subpart are as follows:

1910.241	Definitions
1910.242	Hand- and portable-powered tools and equipment, general
1910.243	Guarding of portable tools and equipment
1910.244	Other portable tools and equipment
1910.245	Effective dates
1910.246	Sources of standards
1910.247	Standards organizations

Subpart Q: Welding, Cutting, and Brazing

Welding, cutting, and brazing are widely used processes. This subpart contains the standards relating to these processes in all of their various forms. The primary safety and health concerns are fire protection, employee personal protection, and ventilation. The standards contained in this subpart are as follows:

1910.251	Definitions
1910.252	General requirements
1910.253	Oxygen-fuel gas welding and cutting
1910.254	Arc welding and cutting
1910.255	Resistance welding
1910.256	Sources of standards
1910.257	Standards organizations

Subpart R: Special Industries

This subpart is different from others in Part 1910. Whereas other subparts deal with specific processes, machines, and materials, Subpart R deals with specific industries. Each separate standard relates to a different category of industry. The standards contained in this subpart are as follows:

1910.261	Pulp, paper, and paperboard mills
1910.262	Textiles
1910.263	Bakery equipment

1910.264	Laundry machinery and operations
1910.265	Sawmills
1910.266	Pulpwood logging
1910.268	Telecommunications
1910.272	Grain handling facilities
1910.274	Sources of standards
1910.275	Standards organizations

Subpart S: Electrical

This subpart contains standards divided into the following two categories: (1) design of electrical systems, and (2) safety-related work practices. These standards are excerpted directly from the National Electrical Code. Those included in Subpart S are as follows:

1910.301	Introduction
1910.302	Electric utilization systems
1910.303	General requirements
1910.304	Wiring design and protection
1910.305	Wiring methods, components, and equipment for general use
1910.306	Specific-purpose equipment and installations
1910.307	Hazardous (classified) locations
1910.308	Special systems
1910.331	Scope
1910.332	Training
1910.333	Selection and use of work practices
1910.334	Use of equipment
1910.335	Safeguards for personnel protection
1910.399	Definitions applicable to this subpart

Subpart T: Commercial Diving Operations

This subpart applies only to commercial diving enterprises. The standards contained in Subpart T are divided into six categories: (1) general, (2) personnel requirements, (3) general operations and procedures, (4) specific operations and procedures, (5) equipment procedures and requirements, and (6) record keeping. These standards are as follows:

1910.401	Scope and application
1910.402	Definitions
1910.410	Qualifications of dive teams

1910.420	Safe practices manual
1910.421	Pre-dive procedure
1910.422	Procedures during dive
1910.423	Post-dive procedures
1910.424	SCUBA diving
1910.425	Surface-supplied-air diving
1910.426	Mixed-gas diving
1910.427	Liveboating
1910.430	Equipment
1910.440	Record keeping
1910.441	Effective dates

Subpart Z: Toxic and Hazardous Substances

This is an extensive subpart containing the standards that establish *permissible exposure limits* (PELs) for numerous toxic and hazardous substances. All such substances have an assigned PEL which is the amount of a given airborne substance to which employees can be exposed during a specified period of time.

The standards relating to these specific toxic and hazardous substances are contained in 1910.1000 through 1910.1500 and are as follows:

1910.1000	Air contaminants
1910.1001	Asbestos
1910.1002	Coal tar pitch volatiles; interpretation of term
1910.1003	4-Nitrobiphenyl
1910.1004	Alpha-Nephthylamine
1910.1006	Methyl chloromethyl ether
1910.1007	3,3¢–Dichlorobenzidine (and its salts)
1910.1008	Bis-Chloromethyl ether
1910.1009	Beta-Naphthylamine
1910.1010	Benzedrine
1910.1011	4-Aminodiphenyl
1910.1012	Ethyleneimine
1910.1013	Beta-Propiolactone
1910.1014	2-Acetylaminofluorene
1910.1015	4-Dimethylaminoazobenzene
1910.1016	N-Nitrosodimethylamine
1910.1017	Vinyl chloride

1910.1018	Inorganic arsenic
1910.1025	Lead
1910.1027	Cadmium
1910.1028	Benzene
1910.1029	Coke oven emissions
1910.1030	Bloodborne pathogens
1910.1043	Cotton dust
1910.1044	1,2-dibromo-3-chloropropane
1910.1045	Acrylonitrile
1910.1047	Ethylene oxide
1910.1050	Methylenedianiline
1910.1200	Hazard communication
1910.1450	Occupational exposure to hazardous chemicals in laboratories
1910.1499	Sources of standards
1910.1500	Standards organizations

OSHA's General Industry Standards were covered in some depth in this section because they have the broadest application for students of workplace safety. The C.F.R. also contains standards for the maritime industry. These standards are described in the sections that follow, but in less detail than was devoted to OSHA's General Standards, because they are not as widely applicable.

OSHA'S MARITIME STANDARDS

OSHA's Maritime Standards apply to shipbuilding, ship-repairing, and ship-breaking operations not already covered by U.S. Coast Guard regulations. It is important to note that Coast Guard regulations do take precedence over OSHA Maritime Standards and supersede those standards in cases of overlap or conflict. Part 1915 of 29 C.F.R. contains the standards relating to shipbuilding, ship repairing, and ship breaking. Part 1917 contains the standards for marine terminals, and Part 1918 contains longshoring standards. Part 1919 contains the gear-certification standards.

Part 1915: Shipyard Employment

Part 1915 is divided into seven subparts covering the various aspects of shipyard operations. Those subparts are as follows:

1915.1	Purpose and authority
1915.2	Scope and application
1915.3	Responsibility
1915.4	Definitions

1915.5	Reference specifications, standards, and codes
1915.6	Commercial diving operations
1915.7	Competent person

Part 1917: Marine Terminals

Part 1917 is divided into seven subparts covering the various operations associated with marine terminals. Those subparts are as follows:

Subpart A	Scope and definitions
Subpart B	Marine terminal operations
Subpart C	Cargo handling gear and equipment
Subpart D	Specialized terminals
Subpart E	Personal protection
Subpart F	Terminal facilities
Subpart G	Related terminal operations and equipment

Part 1918: Longshoring

Longshoring involves loading and unloading maritime vessels. Part 1918 contains the standards relating to longshoring and is subdivided into the following subparts:

Subpart A	General provisions
Subpart B	Gangways and gear certification
Subpart C	Means of access
Subpart D	Working surfaces
Subpart E	Opening and closing hatches
Subpart F	Ship's cargo-handling gear
Subpart G	Cargo-handling gear and equipment other than ship's gear
Subpart H	Handling cargo
Subpart I	General working conditions
Subpart J	Personal protective equipment

Gear Certification

Cargo gear- and material-handling devices must be certified as being safe for use in the handling, loading, unloading, and transport of materials. Persons who certify the gear and devices must be properly accredited. Part 1919 contains the following standards regulating the accreditation process:

| Subpart C | General safety and health provisions |

Subpart D	Occupational health and environment controls
Subpart E	Personal protective and life-saving equipment
Subpart F	Fire protection and prevention
Subpart G	Signs, signals, and barricades
Subpart H	Materials handling, storage, use, and disposal
Subpart I	Tools (hand and power)
Subpart J	Welding and cutting
Subpart K	Electrical
Subpart L	Scaffolding
Subpart M	Floor and wall openings
Subpart N	Cranes, derricks, hoists, elevators, and conveyors
Subpart O	Motor vehicles, mechanized equipment, and marine vessels
Subpart P	Excavations
Subpart Q	Concrete and masonry construction
Subpart S	Underground construction, caissons, and cofferdams.
Subpart T	Demolition
Subpart U	Blasting and the use of explosives
Subpart V	Power transmission and distribution
Subpart W	Rollover protective structures; overhead protection
Subpart X	Stairways and ladders
Subpart Y	Commercial diving operations
Subpart Z	Toxic and hazardous substances

OSHA'S CONSTRUCTION STANDARDS

These standards apply to employers involved in construction, alteration, and/or repair activities. To further identify the scope of the applicability of its construction standards, OSHA took the terms *construction, alteration,* and *repair* directly from the Davis-Bacon Act. This act provides minimum wage protection for employees working on construction projects. The implication is that if the Davis-Bacon Act applies to an employer, OSHA's Construction Standards also apply.

These standards are contained in Part 1926 of the C.F.R. Subparts A–Z. OSHA does not base citations on material contained in Subparts A and B. Consequently, those subparts have no relevance here. The remaining subparts are as follows:

Subpart C	General safety and health provisions
Subpart D	Occupational health and environmental controls
Subpart E	Personal protective and life-saving equipment

Subpart F	Fire protection and prevention
Subpart G	Signs, signals, and barricades
Subpart H	Materials handling, storage, use, and disposal
Subpart I	Tools—hand and power
Subpart J	Welding and cutting
Subpart K	Electrical
Subpart L	Scaffolding
Subpart M	Floor and wall openings
Subpart N	Cranes, derricks, hoists, elevators, and conveyors
Subpart O	Motor vehicles, mechanized equipment, and marine operations
Subpart P	Excavations
Subpart Q	Concrete and masonry construction
Subpart R	Steel erection
Subpart S	Underground construction, caissons, and cofferdams
Subpart T	Demolition
Subpart U	Blasting and the use of explosives
Subpart V	Power transmission and distribution
Subpart W	Rollover protective structures; overhead protection
Subpart X	Stairways and ladders
Subpart Y	Commercial diving operations
Subpart Z	Toxic and hazardous substances

STANDARDS AND CODES

A **standard** is an operational principle, criterion, or requirement—or a combination of these. A **code** is a set of standards, rules, or regulations relating to a specific area. Standards and codes play an important role in modern safety and health management and engineering. These written procedures detail the safe and healthy way to perform job tasks and, consequently, to make the workplace safer and healthier.

Having written standards and codes that employees carefully follow can also decrease a company's exposure to costly litigation. Courts tend to hand down harsher rulings to companies that fail to develop or adapt, implement, and enforce appropriate standards and codes. Consequently, safety and health professionals should be familiar with the standards and codes relating to their company.

Numerous organizations develop standards for different industries. These organizations can be categorized broadly as follows: the government, professional organizations, and technical/trade associations.

Organizations that fall within these broad categories develop standards and codes in a wide variety of areas including, but not limited to, the following: dust hazards, electricity, emergency electricity systems, fire protection, first aid, hazardous chemicals, instrumentation, insulation, lighting, lubrication, materials, noise/vibration, paint, power, wiring, pressure relief, product storage and handling, piping materials, piping systems, radiation exposure, safety equipment, shutdown systems, and ventilation.

Figure 4–7 is a list of organizations that develop and publish standards and codes covering a wide variety of fields and areas of concern. Figure 4–8 is a reference for applicable standards and codes providers in various areas of concern (see Figure 4–7 to decode the abbreviations).

LAWS AND LIABILITY

The body of law pertaining to workplace safety and health grows continually as a result of a steady stream of liability litigation. Often a company's safety and health profession-

Figure 4–7
Organizations that develop standards and codes.

Organization	Abbreviation
U.S. Government	
Bureau of Mines	BM
Department of Transportation	DOT
U.S. Coast Guard	USCG
Hazardous Materials Regulation Board	HMRB
Federal Aviation Administration	FAA
Environmental Protection Agency	EPA
National Institute for Standards and Technology	NIST
Occupational Safety and Health Administration	OSHA
Professional Associations	
American Conf. of Governmental Industrial Hygienists	ACGIH
American Industrial Hygiene Association	AIHA
American Institute of Chemical Engineers	AIChE
American Society of Mechanical Engineers	ASME
American Soc. of Heating, Refrig. and Air Conditioning Engineers	ASHRAE
Illumination Engineers Society	IES
Institute of Electrical and Electronic Engineers	IEEE
Instrument Society of America	ISA
Technical and Trade Associations	
American Water Works Association	AWWA
Air Conditioning and Refrigeration Institute	ARI
Air Moving and Conditioning Association	AMCA
American Association of Railroads	AAR
American Gas Association	AGA

als are key players in litigation alleging negligence on the part of the company when an accident or health problem has occurred. Because health and safety litigation has become so prevalent today, professionals in the field need to be familiar with certain fundamental legal principles relating to such litigation. These principles are explained in the following section.

Fundamental Legal Principles

The body of law that governs safety and health litigation evolves continually. However, even though cases that set new precedents and establish new principles continue to occur, a number of fundamental legal principles surface frequently in the court. The most important of these—as well as several related legal terms—are summarized in the following paragraphs.

Organization	Abbreviation
Technical and Trade Associations, *continued*	
American Petroleum Institute	API
Chlorine Institute	CI
Compressed Gas Association	CGA
Cooling Tower Institute	CTI
Manufacturing Chemists Association	MCA
Manufacturers Standardization Society	MSS
National Electrical Manufacturers Association	NEMA
National Fluid Power Association	NFPA
Pipe Fabrication Institute	PFI
Scientific Apparatus Makers Association	SAMA
Society and Plastics Industry	SPI
Steel Structures Painting Council	SSPC
Tubular Exchanger Manufacturers Association	TEMA
Testing Organizations	
American National Standards Institute	ANSI
American Society for Testing and Materials	ASTM
National Fire Protection Association	NFPA
Underwriters Laboratories, Inc.	UL
National Safety Council	NSC
Insurance Organizations	
American Insurance Association	ATA
Factory Insurance Association	FIA
Factory Mutual Systems	FMS
Oil Insurance Association	OIA

Area of Concern	Organizations That Develop and Publish Standards*
Dust hazards	ANSI, NFPA, UL, NSC, ACGIH, AIHA, BM, USCG
Electricity	ANSI, NFPA, NSC, ADA, FIA, FM, OIA, API, USCG, OSHA, IEEE, NEMA
Emergency electricity	NFPA, AIA, FM, IEEE, AGA, NEMA, USCG
Fire protection	ANSI, NFPA, UL, NSC, AIA, FIA, OIA, AWWA, API, CGA, MCA, NEMA, BM, USCG, OSHA
First aid	ANSI, NFPA, NSC, AIA, ACGIH, AIHA, CI, MCA, DOT, USCG
Flammability of substances	ANSI, NFPA, UL, NSC, FIA, FM, MCA, SPI, DOT, USCG, NIST
Hazardous chemicals	ANSI, NFPA, UL, NSC, AIA, FIA, FM, ACGIH, AIHA, AIChE, CI, CGA, MCA, DOT, USCG, OSHA
Instrumentation	ANSI, ASTM, NFPA, UL, AIA, FIA, FM, OIA, IEEE, ISA, AWWA, ARI, API, CGA, SAMA, NIST
Insulation	ANSI, ASTM, UL, AIA, FM, OIA, ASHRAE, USCG
Lighting	ASTM, NFPA, UL, AIA, FM, OIA, ASHRAE, USCG
Lubrication	ANSI, NFPA, ASME, AMCA
Materials	ANSI, ASTM, NFPA, UL, NSC, AIA, FM, OIA, ISA, AWWA, CI, CGA, CTI, MCA, TEMA, USCG, HMR
Noise—vibration	ANSI, ASTM, NFPA, UL, NSC, ACGIH, AIHA, ASHRAE, ISA, ARI, AMCA, AGA, NFPA, EPA
Paint/coatings	ANSI, ASTM, UL, AIChE, AWWA, SSPC, HMRB
Power writing	ANSI, NFPA, UL, FIA, FM, OIA, IEEE, API, NEMA, USCG, OSHA
Pressure relief	NFPA, FIA, FM, OIA, API, CI, CGA, USCG, HMRB, OSHA
Product storage/handling	ANSI, NFPA, AIA, FIA, FM, OIA, AIChE, AAR, API, CI, CGA, MCA, USCG, OSHA
Piping materials	ANSI, ASTM, NFPA, UL, AIA, FIA, FM, ASME, ASHRAE, AWWA, ARI, AGA, API, CI, CGA, MSS
Piping systems	ANSI, ASTM, NFPA, UL, AIA, FIA, FM, ASME, ASHRAE, AWWA, ARI, AGA, API, CI, CGA, MSS
Radiation exposure	ANSI, NFPA, NSC, AIA, ACGIH, AIHA, ASME, ISA, DOT
Safety equipment	ANSI, UL, NSC, FM, ACGIH, AIHA, CI, CGA, MCA, BM, USCG, OSHA
Shutdown systems	NFPA, UL, AIA, FIA, OIA, API, USCG
Ventilation	ANSI, NFPA, NSC, AIA, FIA, FM, ACGIH, AIHA, ASHRAE, ISA, CI, DOT, HMRB

Figure 4–8

Reference table of standards providers.

*Refer to Figure 4–7 for an explanation of abbreviations.

Negligence

Negligence means failure to take reasonable care or failure to perform duties in ways that prevent harm to humans or damage to property. The concept of *gross negligence* means failure to exercise even slight care or intentional failure to perform duties properly, regardless of the potential consequences. *Contributory negligence* means that an injured party contributed in some way to his or her own injury. In the past, this concept was used to protect defendants against negligence charges because the courts awarded no damages to plaintiffs who had contributed in any way to their own injury. Modern court cases have rendered this approach outdated with the introduction of *comparative negligence*. This concept distributes the negligence assigned to each party involved in litigation according to the findings of the court.

Liability

Liability is a duty to compensate as a result of being held responsible for an act or omission. A newer, related concept is *strict liability*. This means that a company is liable for damages caused by a product that it produces, regardless of negligence or fault.

Care

Several related concepts fall under the heading of care. *Reasonable care* is the amount that would be taken by a prudent person in exercising his or her legal obligations toward others. *Great care* means the amount of care that would be taken by an extraordinarily prudent person in exercising his or her legal obligations toward others. *Slight care* represents the other extreme: a measure of care less than what a prudent person would take. A final concept in this category is the *exercise of due care*. This means that all people have a legal obligation to exercise the amount of care necessary to avoid, to the extent possible, bringing harm to others or damage to their property.

Ability to Pay

The concept of **ability to pay** applies when there are a number of defendants in a case, but not all have the ability to pay financial damages. It allows the court to assess all damages against the defendant or defendants who have the ability to pay. For this reason, it is sometimes referred to as the "deep pockets" principle.

Damages

Damages are financial awards assigned to injured parties in a lawsuit. *Compensatory damages* are awarded to compensate for injuries suffered and for those that will be suffered. *Punitive damages* are awarded to ensure that a guilty party will be disinclined to engage in negligent behavior in the future.

Proximate Cause

Proximate cause is the cause of an injury or damage to property. It is that action or lack of action that ties one person's injuries to another's lack of reasonable care.

Willful/Reckless Conduct

Behavior that is even worse than gross negligence is **willful/reckless conduct.** It involves intentionally neglecting one's responsibility to exercise reasonable care.

Tort

A **tort** is an action involving a failure to exercise reasonable care that may, as a result, lead to civil litigation.

Foreseeability

The concept of **foreseeability** holds that a person can be held liable for actions that result in damages or injury only when risks could have been reasonably foreseen.

The types of questions around which safety and health litigation often revolve are these: Does the company keep employees informed of rules and regulations? Does the company enforce its rules and regulations? Does the company provide its employees with the necessary training? The concepts set forth in this section come into play as both sides in the litigation deal with these questions from their respective points of view.

Health and safety professionals can serve their companies best by (1) making sure that a policy and corresponding rules and regulations are in place; (2) keeping employees informed about rules and regulations; (3) encouraging proper enforcement practices; and (4) ensuring that employees get the education and training they need to perform their jobs safely.

SUMMARY

1. The impetus for passing the Occupational Safety and Health Act, or OSHAct, was that workplace accidents were causing an average of 14,000 deaths every year in the United States. Each year, 2.5 million workers were disabled in workplace accidents, and approximately 300,000 new cases of occupational diseases were reported annually.

2. The mission of OSHA is to ensure to the extent possible that every working person in the United States has a safe and healthy working environment so that valuable human resources are preserved and protected. The Department of Labor breaks down this mission statement further into the following specific purposes: (a) encourage employers and employees to reduce workplace hazards; (b) implement new safety and health programs; (c) improve existing safety and health programs; (d) encourage research that will lead to innovative ways of dealing with workplace safety and health problems; (e) establish the rights of employers and employees regarding the improvement of workplace safety and health; (f) monitor job-related illnesses and injuries through a system of reporting and record keeping; (g) establish training programs to increase the number of safety and health professionals and to continually improve their competence; (h) establish mandatory workplace safety and health standards and enforce those standards; (i) provide for the development and approval of state-level workplace safety and health programs; and (j) monitor, analyze, and evaluate state-level safety and health programs.

3. The OSHAct covers all employers and all 50 states, the District of Columbia, Puerto Rico, and all other territories that fall under the jurisdiction of the U.S. government with the following exceptions: (a) persons who are self-employed; (b) family farms that employ only immediate members of the family, (c) federal agencies covered by other federal statutes, and (d) state and local governments.

4. OSHA developed standards based on its perception of need at the request of other federal agencies, state and local governments, other standards-setting agencies, labor organizations, or even individual private citizens. OSHA uses the committee approach for developing standards. OSHA's standing committees are the National Advisory Committee on Occupational Safety and Health and the Advisory Committee on Construction Safety and Health.

5. OSHA can take three different types of action on standards: a standard may be adopted, amended, or revoked. Before any of these actions can be undertaken, OSHA must publish its intentions in the *Federal Register*. OSHA has two options for meeting this requirement: a notice of proposed rule making and an advance notice of proposed rule making.

6. Once the standard has been passed, it becomes effective on the date prescribed. However, any person who is opposed to a standard may file an appeal in the court of appeals serving the geographic region in which the complainant lives or does business. Appeal paperwork must be initiated within 60 days of a standard's approval.

7. When an employer is unable to comply with a new standard but may be able if given time, a temporary variance may be requested. OSHA will grant such a variance up to a maximum of one year. Employers must demonstrate that they are making a concerted effort to comply and must take the steps necessary to protect employees while working toward compliance.

8. Employers who feel that their workplace already exceeds the requirements of a new standard may request a permanent variance and must present their evidence to OSHA for inspection. Employees must be informed of the application for a variance and notified of their right to request a hearing.

9. OSHA provides for the centralization and systematization of record-keeping and reporting requirements of the OSHAct to employers of 11 or more workers. Both exempt and nonexempt employers must report the following types of accidents within 48 hours: (a) those that result in deaths, and (b) those that result in the hospitalization of five or more employees.

10. All occupational illnesses and injuries must be reported if they result in one or more of the following: (a) death to one or more workers; (b) one or more days away from work for the employee; (c) restricted motion or restrictions to the work an employee can do; (d) loss of consciousness to one or more workers; (e) transfer of an employee to another job; (f) medical treatment needed beyond in-house first aid; and (g) appear in Appendix B of the OSHAct.

11. All records required by OSHA can be maintained using the following forms: OSHA Form 300, OSHA Form 301, and OSHA Form 200S.

12. Employers are required to post the following material at locations where employee information is normally displayed: OSHA Poster 2203, summaries of variance

requests of all types, copies of all OSHA citations received for failure to meet standards, and the summary page of OSHA Form 200/300.

13. OSHA compliance officers are authorized to take the following action with regard to workplace inspections; (a) enter at reasonable times any site, location, or facility where work is taking place; (b) inspect at reasonable times any condition, facility, machine, equipment, materials, and so on; and (c) question in private any employee or other person formally associated with the company.

14. OSHA is empowered to issue citations and/or set penalties. Citations are for (a) other than serious violations, (b) willful violations, (c) repeat violations, and (d) failure to correct prior violations.

15. Employees may appeal the following aspects of OSHA's decisions regarding their workplace: (a) the amount of time (abatement period) given an employer to correct a hazardous condition that has been cited, and (b) an employer's request for an extension of an abatement period.

16. Employers may petition for modification of abatement or contest a citation, an abatement period, and/or a penalty.

17. States are allowed to develop their own safety and health programs. As an incentive, OSHA will fund up to 50 percent of the cost of operating a state program for states with approved plans. States may develop comprehensive plans covering public and private sector employers or limit their plans to coverage of public employers only.

18. In addition to setting standards and inspecting for compliance, OSHA provides services to help employers meet the latest safety and health standards. Services are typically offered at no cost and are intended for smaller companies, particularly those with especially hazardous processes or materials. Services available from OSHA include consultation, voluntary inspection programs, and training/education. OSHA is not without its detractors. Criticisms of OSHA take many forms, depending on the perspective of the critic. Some characterize OSHA as an overbearing bureaucracy with little or no sensitivity to the needs of employers who are struggling to survive in a competitive marketplace. Others label OSHA as timid and claim it does not do enough. At different times and different places, both points of view have probably been at least partially accurate. Other federal agencies/organizations that play important roles with regard to workplace safety and health are the National Institute for Occupational Safety and Health, which is part of the Department of Health and Human Services, and the Occupational Safety and Health Review Committee, which is an independent board consisting of members appointed by the president and given quasi-judicial authority to handle contested OSHA citations.

19. A standard is an operational principle, criterion, or requirement—or a combination of these. A code is a set of standards, rules, or regulations relating to the specific area. Standards and codes play an important role in modern safety and health management and engineering. These written procedures detail the safe and

healthy way to perform jobs, which, consequently, makes for a safer and healthier workplace.

20. Fundamental legal principles with which safety and health professionals should be familiar are (a) negligence, (b) liability, (c) care, (d) ability to pay, (e) damages, (f) proximate cause, (g) willful/reckless conduct, (h) tort, and (i) foreseeability.

KEY TERMS AND CONCEPTS

Abatement period	Opening conference
Ability to pay	OSHA
Appeals process	OSHA Form 301
Closing conference	OSHA Form 300
Code	OSHA Form 200S
Consultation services	OSHA Poster 2203
Damages	OSHAct
DBBS	OSHRC
Demonstration Program	Other-than-serious violation
DRDS	Permanent variance
DSHEFS	PMA
DTMD	Proposed penalty
Employee responsibilities	Proximate cause
Employee rights	Record keeping
Employer appeals	Repeat violation
Employer responsibilities	Reporting
Employer rights	Standard
Foreseeability	Star Program
Inspection tour	State-level OSHA program
Liability	Temporary emergency standard
Merit Program	Temporary variance
NACOSH	Tort
Negligence	Voluntary Protection Programs
NIOSH	Willful/reckless conduct
Notice of contest	Willful violation
Notice of proposed rule making	Workplace accident
Occupational disease	Workplace inspection

REVIEW QUESTIONS

1. Briefly explain the rationale for the OSHAct.
2. What is OSHA's mission/purpose?
3. List those who are exempted from coverage by OSHA.
4. Explain how the following processes relating to OSHA standards are accomplished: passage of a new standard; request for a temporary variance; appealing a standard.
5. Briefly describe OSHA's latest record-keeping requirements.
6. What are OSHA's reporting requirements?
7. Explain what employers are required to do in order to keep employees informed.
8. Describe how a hypothetical OSHA workplace inspection would proceed from the first step to the last.
9. List and explain three different types of OSHA citations and the typical penalties that accompany them.
10. Describe the process for appealing an OSHA citation.
11. List and briefly explain OSHA's voluntary protection programs.
12. List five employer responsibilities.
13. List five employee rights.
14. Describe the purpose and organization of NIOSH.
15. Define the following legal terms as they relate to workplace safety: *negligence; liability; ability to pay; tort.*

ENDNOTES

1. U.S. Department of Labor. *All About OSHA,* OSHA 2056, 1991 (Revised), p. 1.
2. Ibid.
3. Ibid., p. 2.
4. Ibid., p. 5.
5. Ibid., p. 7.
6. Ibid., p. 9.
7. Hans, M. E. "OSHA Closes in on New Confined Space Standard," *Safety and Health,* June 1991, Vol. 143, No. 6, pp. 59–60.
8. Moretz, S. "Industry Prepares for OSHA's Hazardous Waste Role," *Occupational Hazards,* November 1989, pp. 39–42.
9. Moretz, S. "General Requirements of OSHA's Final Hazardous Waste Standard," *Occupational Hazards,* November 1989, p. 41.
10. U.S. Department of Labor, *All About OSHA,* p. 12.
11. Ibid.
12. Ibid., pp. 12–13.
13. Ibid., p. 17.
14. Ibid., pp. 24–25.
15. Tyson, P. R. "Record-High OSHA Penalties," *Safety and Health,* March 1991, Vol. 143, No. 3, p. 17.
16. Ibid.

17. "OSHA Speeds Up the Decision Process," *Safety and Health,* October 1991, Vol. 144, No. 4, pp. 29–31.
18. Ibid., p. 30.
19. U.S. Department of Labor, *All About OSHA,* p. 27.
20. Davis, S. "Contested Cases, Missing or Misplaced," *Occupational Hazards,* March 1991, p. 47.
21. Ibid.
22. Ibid., p. 28.
23. Ibid.
24. U.S. Department of Labor, *All About OSHA,* p. 28.
25. Ibid.
26. Ibid.
27. Ibid., p. 29.
28. Ibid., p. 32.
29. Ibid., pp. 34–35.
30. Ibid., pp. 35–36.
31. Ibid., p. 37.
32. Ibid., pp. 39–40.
33. Ibid., p. 37.
34. "Fatal Explosion Fuels Process Safety Concerns," *Occupational Hazards,* July 1991, p. 15.
35. "NIOSH Chief 'Jarred' by Talk of Reform," *Occupational Hazards,* July 1991, p. 15.
36. "Failure to Follow-Up," *Occupational Hazards,* July 1991, p. 15.
37. Ibid., p. 25.
38. Associated Press. "Safety Not 1st in Arkansas, Study Shows," *Northwest Florida Daily News* (Fort Walton Beach, FL), January 1, 1992, p. C-2.
39. Ibid.
40. U.S. Department of Health and Human Services. *Occupational Safety & Health for Fiscal Year 1988 Under Public Law 91–596,* Public Health Service, Centers for Disease Control, National Institute for Occupational Safety and Health.
41. U.S. Department of Health and Human Services. "Public Health Service, Centers for Disease Control, National Institute for Occupational Safety and Health." Handout, p. 1.
42. Ibid., p. 2.
43. Ibid., p. 3.
44. Ibid., p. 4.

Workers' Compensation

OVERVIEW OF WORKERS' COMPENSATION

The concept of **workers' compensation** developed as a way to allow injured employees to be compensated appropriately without having to take their employer to court. The underlying rationale for workers' compensation had two aspects: (1) fairness to injured employees, especially those without the resources to undertake legal actions that are often long, drawn out, and expensive; and (2) reduction of costs to employers associated with workplace injuries (e.g., legal, image, and morale costs). Workers' compensation is intended to be a no-fault approach to resolving workplace accidents by rehabilitating injured employees and minimizing their personal losses because of their reduced ability to perform and compete in the labor market.[1] Since its inception as a concept, workers' compensation has evolved into a system that pays out approximately $70 million in benefits and medical costs annually. The national average net cost of workers' compensation in the manufacturing sector is almost $6 per $100 of payroll.

Workers' compensation represents a compromise between the needs of employees and the needs of employers. Employees give up their right to seek unlimited compensation for pain and suffering through legal action. Employers award the prescribed compensation (typically through insurance premiums) regardless of the employee's negligence. The theory is that in the long run both employees and employers will benefit more than either would through legal action. As you will see later in this chapter, although workers' compensation has reduced the amount of legal action arising out of workplace accidents, it has not completely eliminated legal actions.

Objectives of Workers' Compensation

Workers' compensation laws are not uniform from state to state. In fact, there are extreme variations. However, regardless of the language contained in the enabling legislation in a specific state, workers' compensation as a concept has several widely accepted objectives:

1. Replacement of income
2. Rehabilitation of the injured employee
3. Prevention of accidents
4. Cost allocation[2]

The basic premises underlying these objectives are described in the following paragraphs.

Replacement of Income

Employees injured on the job lose income if they are unable to work. For this reason, workers' compensation is intended to replace the lost income adequately and promptly. Adequate **income replacement** is viewed as replacement of current and future income (minus taxes) at a ratio of two-thirds (in most states). Workers' compensation benefits are required to continue even if the employer goes out of business.

Rehabilitation of the Injured Employee

A basic premise of workers' compensation is that the injured worker will return to work in every case possible, although not necessarily in the same job or career field. For this reason, a major objective of workers' compensation is to rehabilitate the injured employee. The **rehabilitation** program is to provide the needed medical care at no cost to the injured employee until he or she is pronounced fit to return to work. The program also provides vocational training or retraining as needed. Both components seek to motivate the employee to return to the labor force as soon as possible.

Accident Prevention

Preventing future accidents is a major objective of workers' compensation. The theory underlying this objective is that employers will invest in **accident prevention** programs in order to hold down compensation costs. The payoff to employers comes in the form of lower insurance premiums that result from fewer accidents.

Cost Allocation

The potential risks associated with different occupations vary. For example, working as a miner is generally considered more hazardous than working as an architect. The underlying principle of **cost allocation** is to spread the cost of workers' compensation appropriately and proportionately among industries ranging from the most to the least hazardous. The costs of accidents should be allocated in accordance with the accident history of the industry so that high-risk industries pay higher workers' compensation insurance premiums than do low-risk industries.[3]

Who Is Covered by Workers' Compensation?

Workers' compensation laws are written at the state level, and there are many variations among these laws. As a result, it is difficult to make generalizations. Complicating the issue further is the fact that workers' compensation laws are constantly being amended, revised, and rewritten. Additionally, some states make participation in a workers' compensation program voluntary; others excuse employers with fewer than a specified number of employees.

In spite of the differences among workers' compensation laws in the various states, approximately 80 percent of the employees in the United States are covered by workers' compensation. Those employees who are not covered or whose coverage is limited vary as the laws vary. However, they can be categorized in general terms as follows:

1. Agricultural employees
2. Domestic employees
3. Casual employees
4. Hazardous work employees
5. Charitable or religious employees
6. Employees of small organizations
7. Railroad and maritime employees
8. Contractors and subcontractors
9. Minors
10. Extraterritoriality employees[4]

According to Hammer, coverage in these types of employment, to the extent there is coverage, varies from state to state as follows:[5]

■ *Agricultural employees* have limited coverage in 38 states, Puerto Rico, and the Virgin Islands. In 15 states, workers' compensation coverage for agricultural employees is voluntary. In these states, employers are allowed to provide coverage if they wish but are not required to do so.

- *Domestic employees* have coverage available in all 50 states and Puerto Rico. However, coverage tends to be limited and subject to minimum requirements regarding hours worked and earnings.

- *Casual employees* are employed in positions in which the work is occasional, incidental, and scattered at irregular intervals. Such employees are not typically afforded workers' compensation coverage.

- *Hazardous employment* is the only type afforded workers' compensation coverage in some states. To qualify, a particular type of employment must be on an approved list of hazardous or especially hazardous jobs. However, the trend in these states is to broaden the list of approved jobs.

- *Charitable or religious employees* are not afforded workers' compensation in most states when this work is irregular, temporary, or for short-term periods.

- *Small organizations* that employ fewer than a stipulated number of employees do not fall under the umbrella of workers' compensation in 26 states.

- *Railroad and maritime workers* are not typically covered by workers' compensation. However, in most cases, they are covered by the Federal Employer's Liability Act. This act disallows the use of common law defenses by employers if sued by an employee for negligence.

- *Contractors and subcontractors* are those who agree to perform a job or service for an agreed amount of money in a nondirected, nonsupervised format. In essence, contract and subcontract employees are viewed as being self-employed. For this reason, they are not covered by workers' compensation.

- *Minors* are afforded regular workers' compensation coverage as long as they are legally employed. In some states, coverage is significantly higher for minors who are working illegally.

- **Extraterritorial employees** are those who work in one state but live in another. In these cases, the employee is usually on temporary duty. Such employees are typically afforded the workers' compensation coverage of their home state.

HISTORICAL PERSPECTIVE

Before workers' compensation laws were enacted in the United States, injured employees had no way to obtain compensation for their injuries except to take their employer to court. Although common law did require employers to provide a safe and healthy work environment, injured employees bore the burden of proof that negligence in the form of unsafe conditions contributed to these injuries. According to the Society of Manufacturing Engineers, prior to passage of workers' compensation, employees often had to sue their employer to receive compensation for injuries resulting from a workplace accident or occupational disease, even when the following circumstances prevailed:

- The employee was disabled or died as the result of a **workplace accident** or **occupational disease.**

- The injury might have been expected to occur when the risks and hazards of the job were considered.

■ **Worker negligence** on the part of a fellow worker or the injured employee clearly caused the injury.[6]

Proving that an injury was the result of employee negligence was typically too costly, too difficult, and too time consuming to be a realistic avenue of redress for most injured employees. According to Somers and Somers, a New York commission determined that it took from six months to six years for an injured worker's case to work its way through the legal system.[7] Typically, injured workers, having lost their ability to generate income, could barely afford to get by, much less pay medical expenses, legal fees, and court costs. Another inhibitor was the **fear factor.** Injured employees who hoped to return to work after recovering were often afraid to file suit because they feared retribution by their employer. Employers not only might refuse to give them their jobs back but also might **blackball** them with other employers. Add to this that fellow employees were often afraid to testify to the negligence of the employer, and it can be seen why few injured workers elected to take their employers to court.

Even with all of these inhibitors, some injured employees still chose to seek redress through the courts in the days before workers' compensation. Those who did faced a difficult challenge because the laws at that time made it easy for employers to defend themselves successfully. All an employer had to do to win a decision denying the injured plaintiff compensation was show that at least one of the following conditions existed at the time of the accident:

1. *Contributory negligence was a factor in the accident.* **Contributory negligence** meant that the injured worker's own negligence contributed to the accident. Even if the employee's negligence was a very minor factor, it was usually enough to deny compensation in the days before workers' compensation.

2. *There was negligence on the part of a fellow worker.* As with contributory negligence, negligence by a fellow employee, no matter how minor a contributing factor it was, could be sufficient to deny compensation.

3. *There was assumption of risk on the part of the injured employee.* If an employee knew that the job involved risk, he or she could not expect to be compensated when the risks resulted in accidents and injuries.[8]

Since the majority of workplace accidents involve at least some degree of negligence on the part of the injured worker or fellow employees, employers typically won these cases. Because it required little more than a verbal warning by the employer to establish grounds for assumption of risk, the odds against an injured employee being awarded compensation become clear.

In his book *American Social Science,* Gagilardo gives an example of a case that illustrates how difficult it was to win compensation in the days before workers' compensation.[9] He relates the example of an employee who contracted tuberculosis while working under clearly hazardous conditions for a candy-making company. She worked in a wet, drafty basement that admitted no sunlight. Dead rats floated in the overflow of a septic tank that covered the basement floor, and a powerful stench permeated the workplace. Clearly, these were conditions that could contribute to the employee contracting tuberculosis. However, she lost the case and was denied compensation. The ruling judge justified the verdict as follows:

> We think that the plaintiff, as a matter of law, assumed the risk attendant upon her remaining in the employment (*Wager* v. *White Star Candy Company*, 217 N.Y. Supp. 173).

Situations such as this eventually led to the enactment of workers' compensation laws in the United States.

WORKERS' COMPENSATION LEGISLATION

Today, all 50 states, the District of Columbia, Guam, and Puerto Rico have workers' compensation laws. However, these laws did not exist prior to 1948. Considering that Prussia passed a workers' compensation law in 1838, the United States was obviously slow to adopt the concept. In fact, the first workers' compensation law enacted in the United States did not pass until 1908, and it applied only to federal employees working in especially hazardous jobs. The driving force behind passage of this law was President Theodore Roosevelt, who as governor of New York had seen the results of workplace accidents firsthand. Montana was the first state to pass a compulsory workers' compensation law. However, it was short-lived. Ruling that the law was unconstitutional, the Montana courts overturned it.

In 1911, the New York Court of Appeals dealt proponents of workers' compensation a serious blow. The New York state legislature had passed a compulsory workers' compensation law in 1910. However, in the case of *Ives* v. *South Buffalo Railway Company* (201 N.Y. 271, 1911) the New York Court of Appeals declared the law unconstitutional based on the contention that it violated the due process clause in the Fourteenth Amendment to the U.S. Constitution.[10]

This ruling had a far-reaching impact. According to Hammer, "The prestige of the New York court influenced legislation in many of the other states to believe that any compulsory law also would be held unconstitutional."[11] However, even with such precedent-setting cases as *Ives* on the books, pressure for adequate workers' compensation grew as unsafe working conditions continued to result in injuries, diseases, and deaths. In fact, shortly after the New York Court of Appeals released its due process ruling, tragedy struck in a New York City textile factory.

On March 25, 1911, the building that housed the Triangle Shirtwaist Factory on its eighth floor caught fire and burned.[12] As a result of the fire, 149 of the company's 600 workers were dead and another 70 were injured. Although the cause of the accident could not be determined, it was clear to investigators and survivors alike that unsafe conditions created by the management of the company prevented those who died or were injured from escaping the fire.

Exit passageways on each floor of the building were unusually narrow (20 inches wide), which made it difficult for employees to carry out bolts of material. A wider exit on each floor was kept locked to force employees to use the narrow exit. The two elevators were slow and able to accommodate only small groups at a time.

As the fire quickly spread, employees jammed into the narrow passageways, crushing each other against the walls and underfoot. With all exits blocked, panic-stricken employees began to jump out of windows and down the elevator shafts. When the pandemonium subsided and the fire was finally brought under control, the harsh realization of why so many had been trapped by the deadly smoke and flames quickly set in.

The owners were brought into court on charges of manslaughter. Although they were not convicted, the tragedy did focus nationwide attention on the need for a safe workplace and adequate workers' compensation. As a result, new, stricter fire codes were adopted in New York, and in spite of the state court's ruling in *Ives,* the state legislature passed a workers' compensation law.

The next several years saw a flurry of legislation in other states relating to workers' compensation. In response to demands from workers and the general public, several states passed limited or noncompulsory workers' compensation laws. Many such states held back out of fear of being overturned by the courts. Others, particularly Washington, publicly disagreed with the New York Court of Appeals and passed compulsory laws. The constitutionality debate continued until 1917 when the U.S. Supreme Court ruled that workers' compensation laws were acceptable.

MODERN WORKERS' COMPENSATION

Since 1948, all states have had workers' compensation laws. However, the controversy surrounding workers' compensation has not died. As medical costs and insurance premiums have skyrocketed, many small businesses have found it difficult to pay the premiums. Unrealistic workers' compensation rates are being cited more and more frequently as contributing to the demise of small business in America.

The problem has even developed into an economic development issue. Business and industrial firms are closing their doors in those states with the highest workers' compensation rates and moving to states with lower rates. States with lower rates are using this as part of their recruiting package to attract new businesses and industry. Where low-rate states border high-rate states, businesses are beginning to move their offices across the border to the low-rate state while still doing business in the high-rate state.

Critics are now saying that workers' compensation has gotten out of hand and is no longer fulfilling its intended purpose. To understand whether this is the case, one must begin with an examination of the purpose of workers' compensation. According to Hammer, the U.S. Chamber of Commerce identified the following six basic objectives of workers' compensation.

1. To provide an appropriate level of income and medical benefits to injured workers or to provide income to the worker's dependents regardless of fault.
2. To provide a vehicle for reducing the amount of personal injury litigation in the court system.
3. To relieve public and private charities of the financial strain created by workplace injuries that go uncompensated.
4. To eliminate time-consuming and expensive trials and appeals.

5. To promote employer interest and involvement in maintaining a safe work environment through the application of an experience-rating system.
6. To prevent accidents by encouraging frank, objective, and open investigations of the causes of accidents.[13]

Early proponents of workers' compensation envisioned a system in which both injured workers and their employers would win. Injured workers would receive prompt compensation, adequate medical benefits, and appropriate rehabilitation to allow them to reenter the workforce and be productive again. Employers would avoid time-consuming, expensive trials and appeals and would improve relations with employees and the public in general.

What proponents of workers' compensation did not anticipate were the following factors: (1) employees who would see workers' compensation as a way to ensure themselves a lifelong income without the necessity of work; (2) enormous increases in the costs of medical care with corresponding increases in workers' compensation insurance premiums; and (3) the radical differences among workers' compensation laws passed by the various states.

Not all employees abide by the spirit of workers' compensation (i.e., rehabilitation in a reasonable amount of time). Attempted abuse of the system was perhaps inevitable. Unfortunately, such attempts result in a return to what workers' compensation was enacted to eliminate: time-consuming, drawn-out, expensive legal battles and the inevitable appeals.

Proponents of workers' compensation reform can cite a long list of cases that illustrate their point. The city of Pittsfield, Massachusetts, was being overwhelmed by workers' compensation claims in 1989. One of the more remarkable cases concerned a city worker who was receiving workers' compensation benefits as the result of a back injury. While collecting benefits, he was a star player for a local softball team.[14] According to Bellow, he eventually agreed to waive his right to compensation for a lump sum settlement of $12,000 plus $3,000 for his lawyer, and city officials considered themselves cheaply rid of him.[15]

Another Pittsfield city employee in the Department of Public Works was injured and began collecting workers' compensation at a rate of $295.50 per week. In addition to his job with the city, this worker owned a small diesel oil company. When his workers' compensation benefits were called into question because he owned a business that produced an income, the injured employee sold the business to his son. According to Bellow, "Some city officials speculated that [he] sold his business to save his pension, because people who are employed are no longer entitled to a pension."[16]

One of the favorite examples of opponents of workers' compensation is that of the "fat deputy sheriff." According to Brydoff,

> The deputy sheriff was already despondent over the breakup of the extramarital affair he was having with a married colleague, when his supervisor made an expensive mistake. He rated the already overstressed sheriff's job performance as substandard and told him he was too fat. The sheriff promptly filed a claim for workers' compensation benefits to cover job-related pressures stemming from his performance evaluation that contributed to his mental injury.[17]

Medical costs have skyrocketed in the United States since the 1960s. There are many reasons for this. During this same period, the costs associated with other basic human needs including food, clothing, transportation, shelter, and education have also increased markedly. Increases in medical costs can be explained, at least partially, as the normal cost of living increases experienced in other sectors of the economy. However, the costs associated with medical care have increased much faster and much more than the costs in these other areas. The unprecedented increases can be attributed to two factors: (1) technological developments that have resulted in extraordinary but costly advances in medical care; and (2) a proliferation of **litigation** that has driven the cost of malpractice insurance steadily up. Each of these factors has contributed to higher medical costs. For example, X-ray machines that cost thousands of dollars have been replaced by magnetic resonance imaging (MRI) systems that may cost millions. **Malpractice** suits that once might not even have gone to court now result in multimillion-dollar settlements. Such costs are, of course, passed on to whoever pays the medical bill—in this case, employers who must carry workers' compensation insurance. California's workers' compensation system, for example, costs employers $7 billion annually in insurance premiums. Between 1985 and 1987, the cost of workers' compensation premiums increased by 60 percent. This trend continues.

In addition to contributing individually to increased medical costs, technology and litigation have interacted in such a way as to increase costs even further. This interaction occurs as follows. First, an expensive new technology is developed that enhances the predictive and/or prescriptive capabilities of the medical profession. Second, expensive malpractice suits force medical practitioners to be increasingly cautious and, accordingly, to order even more tests than a patient's symptoms might suggest. Finally, the tests involve expensive new technologies, adding even more to the cost of medical care.

Early supporters of the concept did not anticipate the radical differences among workers' compensation laws in the various states. The laws themselves differ, as do their interpretations. The differences are primarily in the areas of benefits, penalties, and who is covered. These differences translate into differences in the rates charged for workers' compensation insurance. As a result, the same injury incurred under the same circumstances but in different states can yield radically different benefits for the employee.

The potential for abuse, steadily increasing medical costs that lead to higher insurance premiums, and differences among workers' compensation laws all contribute to the controversy that still surrounds this issue. As business and industry continue to protest that workers' compensation has gotten out of hand, it will continue to be a heated issue in state legislatures. As states try to strike the proper balance between meeting the needs of the workforce while simultaneously maintaining a positive environment for doing business, workers' compensation will be an issue with which they will have to deal.

WORKERS' COMPENSATION INSURANCE

The costs associated with workers' compensation must be borne by employers as part of their overhead. In addition, employers must also ensure that the costs will be paid even if they go out of business. The answer for most employers is workers' compensation insurance.

SAFETY FACT

Back Injuries Cost More with Workers' Compensation

One of the most common workplace injuries is the back injury. Almost one-fourth of all workers' compensation claims involve back injuries. According to the California Workers' Compensation Institute (CWCI), medical care costs are 43 percent higher when part of a workers' compensation claim than when part of a group medical plan. Back injuries treated under workers' compensation use medical services at a higher rate than those treated under a group medical plan. Workers' compensation claimants have more total visits for medical treatment, more total procedures, more procedures per week, and more visits per week.

In most states, workers' compensation insurance is compulsory. Exceptions to this are New Jersey, South Carolina, Texas, and Wyoming. New Jersey allows ten or more employers to form a group and self-insure. Texas requires workers' compensation only for *carriers,* as defined in Title 25, Article 911–A, Section II, Texas state statutes. Wyoming requires workers' compensation only for employers involved in specifically identified *extrahazardous occupations.*

A common thread woven through all of the various compensation laws is the requirement that employers carry workers' compensation insurance. There are three types: **state funds, private insurance,** and **self-insurance.** Figure 5–1 summarizes the methods of insurance coverage allowed in a representative sample of states. Regardless of the method of coverage chosen, rates can vary greatly from company to company and state to state. Rates are affected by a number of different factors including the following:

- Number of employees
- Types of work performed (risk involved)
- Accident experience of the employer
- Potential future losses
- Overhead and profits of the employer
- Quality of the employer's safety program
- Estimates by actuaries

According to Hammer, insurance companies use one of the following six methods in determining the premium rates of employers:[18]

1. *Schedule rating.* Insurance companies establish baseline safety conditions and evaluate the employer's conditions against the baselines. Credits are awarded for conditions that are better than the baseline, and debits are assessed for conditions that are worse. Insurance rates are adjusted accordingly.

2. *Manual rating.* A manual of rates is developed that establishes rates for various occupations. Each occupation may have a different rate based on its perceived level of

State	State Fund	Private Insurer	Individual Employer Self-Insurance	Group of Employer's Self-Insurance
Alabama	No	Yes	Yes	Yes
Arkansas	No	Yes	Yes	Yes
California	Competitive	Yes	Yes	No
Florida	No	Yes	Yes	Yes
Indiana	No	Yes	Yes	No
Kansas	No	Yes	Yes	Yes
Maryland	Competitive	Yes	Yes	Yes
Montana	Competitive	Yes	Yes	Yes
Nebraska	No	Yes	Yes	No
New Jersey	No	Yes	Yes	No
New Mexico	No	Yes	Yes	No
Texas	No	Yes	No	No
Washington	Exclusive	No	Yes	Yes
Wisconsin	No	Yes	Yes	No

Figure 5–1
Workers' compensation coverage methods allowed for selected states.

hazard. The overall rate for the employer is a pro-rata combination of all the individual rates.

3. *Experience rating.* Employers are classified by type. Premium rates are assigned based on predictions of average losses for a given type of employer. Rates are then adjusted either up or down according to the employer's actual experience over the past three years.

4. *Retrospective rating.* Employees pay an established rate for a set period. At the end of the period, the actual experience is assessed, and an appropriate monetary adjustment is made.

5. *Premium discounting.* Large employers receive discounts on their premiums based on their size. The theory behind this method is that it takes the same amount of time to service a small company's account as it does a large company's, but the large company produces significantly more income for the insurer. Premium discounts reward the larger company for its size.

6. *Combination method.* The insurer combines two or more of the other methods to arrive at premium rates.

The trend nationwide for the past decade has been for premiums to increase markedly. For example, over the past ten years, some states experienced increases of over 60 percent. This trend will ensure that workers' compensation remains a controversial issue in the state legislatures.

RESOLUTION OF WORKERS' COMPENSATION DISPUTES

One of the fundamental objectives of workers' compensation is to avoid costly, time-consuming litigation. Whether this objective is being accomplished is questionable. When an injured employee and the employer's insurance company disagree on some aspect of the compensation owed (e.g., weekly pay, length of benefits, degree of disability), the disagreement must be resolved. Most states have an arbitration board for this purpose. Neither the insurance company nor the injured employee is required to hire an attorney. However, many employees do. There are a number of reasons for this. Some don't feel they can adequately represent themselves. Others are fearful of the "big business running over the little guy" syndrome. In any case, workers' compensation litigation is still very common and expensive.

Allowable attorney fees are set by statute, administrative rule, or policy in most states. In some states, attorney fees can be added to the injured employee's award. In others, the fee is a percentage of the award. Here are some examples of how attorney fees are handled in several states:

- Alaska establishes attorney fees by statute at the following rates: 25 percent on the first $1,000 and 10 percent on the balance. It is unlawful for attorneys to accept fees that are not statutorily approved.
- Delaware establishes attorney fees by statute at the following rate: 30 percent on the total award or a flat fee of $2,250, whichever is smaller.
- Arizona establishes attorney fees by statute at a flat rate of 25 percent. However, there are no statutory provisions preventing attorneys from charging additional fees not set forth in the law.
- Georgia establishes attorney fees by administrative rule at the following rates: 25 percent to 33 percent. These fees can be added to the employee's award in certain cases.
- Indiana establishes attorney fees by administrative rule at the following rates: 20 percent on the first $5,000; 15 percent on the next $5,000; and 10 percent on the balance. In certain cases, attorney fees can be added to the employee's award.

INJURIES AND WORKERS' COMPENSATION

The original workers' compensation concept envisioned compensation for workers who were injured in on-the-job accidents. What constituted an accident varied from state to state. However, all original definitions had in common the characteristics of being *sud-*

den and *unexpected*. Over the years, the definition of an accident has undergone continual change. The major change has been a trend toward the elimination of the *sudden* characteristic. In many states the gradual onset of an injury or disease as a result of prolonged exposure to harmful substances or a harmful environment can now be considered an accident.

A **harmful environment** does not have to be limited to its physical components. Psychological factors (such as stress) can also be considered. In fact, the highest rate of growth in workers' compensation claims over the past two decades has been in the area of stress-related injuries. For example, ten years ago, there were 1,282 stress-related claims filed by employees in California. Today this number has increased to over 7,000 claims. This increase in California is indicative of a nationwide trend.

The National Safety Council maintains statistical records of the numbers and types of injuries suffered in various industries in the United States. Industries are divided into the following categories: agriculture, mining, construction, manufacturing, transportation/public utilities, trade, services, and public sector. Injuries in these industrial sectors are classified according to the type of accident that caused them. Accident types include overexertion, being struck by or against an object, falls, bodily reactions, caught in or between objects, motor vehicle accident, coming in contact with radiation or other caustics, being rubbed or abraded, and coming in contact with temperature extremes.

According to the National Safety Council, over 30 percent of all disabling work injuries are the result of overexertion when all industry categories are viewed in composite. The next most frequent cause of injuries is struck by/struck against objects at 24 percent. Falls account for just over 17 percent. The remainder are fairly evenly distributed among the other accident types just listed.[19]

AOE and COE Injuries

Workers' compensation benefits are owed only when the injury arises out of employment **(AOE)** or occurs in the course of employment **(COE).** When employees are injured undertaking work prescribed in their job description, work assigned by a supervisor, or work normally expected of employees, they fall into the AOE category. Sometimes, however, different circumstances determine whether the same type of accident is considered to be AOE. For example, say, a soldering technician burns her hand while repairing a printed circuit board that had been rejected by a quality control inspector. This injury would be classified as AOE. Now suppose the same technician brings a damaged printed circuit board from her home stereo to work and burns her hand while trying to repair it. This injury would not be covered because the accident did not arise from her employment. Determining whether an injury should be classified as AOE or COE is often a point of contention in workers' compensation litigation.

Who Is an Employee?

Another point of contention in workers' compensation cases is the definition of the term *employee*. This is an important definition because it is used to determine AOE and COE. A person who is on the company's payroll, receives benefits, and has a supervisor is clearly an **employee.** However, a person who accepts a service contract to perform a specific task or set of tasks and is not directly supervised by the company is not considered an employee. Although definitions vary from state to state, there are common characteristics. In all definitions, the workers must receive some form of remuneration for work done, and the employer must benefit from this work. Also, the employer must supervise and direct the work, both process and result. These factors—supervision and direction—are what set **independent contractors** apart from employees and exclude them from coverage. Employers who use independent contractors sometimes require the contractors to show proof of having their own workers' compensation insurance.

Unless an employer provides transportation, employees are not generally covered by workers' compensation when traveling to and from work. However, in some circumstances, they can be covered. Hammer gives the following example:

> John normally finishes work early and makes his way home over freeways that have little traffic at that time. However, his employer calls an unexpected conference which delays John's return until the freeways are extremely crowded. As a result, John is injured in an accident. The exposure to abnormal hazard makes him eligible for workers' compensation.[20]

Hammer gives another example of a worker in New York who became so elated about his job he danced a jig. During the process, he fell down and broke his leg. As a result, he filed for workers' compensation. In *Bletter v. Harcourt,* 250 N.E. 2nd 572 (1969), the court of appeals granted the compensation, stating that the joy felt by the employee was work related and was, therefore, compensable.[21]

DISABILITIES AND WORKERS' COMPENSATION

Injuries that are compensable typically fall into one of four categories: (1) temporary partial disability, (2) temporary total disability, (3) permanent partial disability, and (4) permanent total disability (Figure 5–2). Determining the extent of disability is often a contentious issue. In fact, it accounts for more workers' compensation litigation than any

---

SAFETY MYTH

Workers' Compensation Claims and Character

A leopard doesn't change its spots. A dishonest employee will always abuse the workers' compensation system with fraudulent claims. Right? Maybe not. A company in Oklahoma reduced its workers' compensation costs in just two years from $486,000 to $47,000. How did this company accomplish a reduction of this magnitude? With character training based on biblical principles. The program focuses on such character traits as honesty, attentiveness, gratitude, and dependability. At monthly training sessions, employees discuss one character trait and how it applies on the job.

Figure 5–2
Types of disabilities.

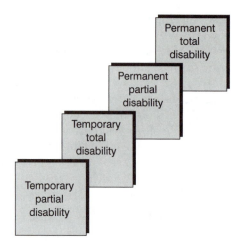

other issue. Further, when a disability question is litigated, the case tends to be complicated because the evidence is typically subjective, and it requires hearing officers, judges, or juries to determine the future.

Temporary Disability

Temporary disability is the state that exists when it is probable that an injured worker, who is currently unable to work, will be able to resume gainful employment with no or only partial disability. Temporary disability assumes that the employee's condition will substantially improve. Determining whether an employee is temporarily disabled is not normally difficult. Competent professionals can usually determine the extent of the employee's injuries, prescribe the appropriate treatment, and establish a timeline for recovery. They can then determine if the employee will be able to return to work and when the return might take place.

There is an important point to remember when considering a temporary disability case. The ability to return to work relates only to work with the company that employed the worker at the time of the accident.

Temporary disability can be classified as either *temporary total disability* or *temporary partial disability*. A **temporary total disability** classification means the injured worker is incapable of any work for a period of time but is expected to recover fully. Most workers' compensation cases fall in this classification. A **temporary partial disability** means the injured worker is capable of light or part-time duties. Depending on the extent of the injury, temporary partial disabilities sometimes go unreported. This practice is allowable in some states. It helps employers hold down the cost of their workers' compensation premium. This is similar to not reporting a minor fender bender to your automobile insurance agent.

Most states prescribe in law the benefits owed in temporary total disability cases. Factors prescribed typically include a set percentage of an employee's wage that must be paid and a maximum period during which benefits can be collected. Figure 5–3 shows this information for a geographically distributed selection of states. Since workers' compensation legislation changes continually, this figure is provided only as an illustration of how benefits are prescribed in the laws of the various states. Actual rates are subject to change.

Permanent Partial Disability

Permanent partial disability is the condition that exists when an injured employee is not expected to recover fully. In such cases, the employee will be able to work again but not at full capacity. Often employees who are partially disabled must be retrained for another occupation.

Permanent partial disabilities can be classified as *schedule* or *nonschedule* disabilities. **Schedule disabilities** are typically the result of nonambiguous injuries such as the loss of a critical but duplicated body part (e.g., arm, ear, hand, finger, or toe). Since such injuries are relatively straightforward, the amount of compensation that they generate and the period of time that it will be paid can be set forth in a standard schedule. A compilation of information from such schedules for a geographically distributed list of states is shown in Figure 5–4. Since workers' compensation legislation changes continually, this figure is provided only as an example. Actual rates are subject to change continually.

Figure 5–3
Temporary total disability benefits for selected states.
Source: U.S. Department of Labor.

State	Percent of Employee's Wage	Maximum Period
Alabama	66⅔	Duration of disability
Arkansas	66⅔	450 weeks
California	66⅔	Duration of disability
Florida	66⅔	350 weeks
Indiana	66⅔	500 weeks
Kansas	66⅔	Duration of disability
Maryland	66⅔	Duration of disability
Montana	66⅔	Duration of disability
Nebraska	66⅔	Duration of disability
New Jersey	70	400 weeks
New Mexico	66⅔	700 weeks
Texas	66⅔	401 weeks
Washington	60–75	Duration of disability
Wisconsin	60–75	Duration of disability

Figure 5–4
Permanent partial disability
benefits for selected states.
*Based on actual wages lost
and not subject to minimums.
Source: U.S. Department of
Labor.

State	Percent of Employee's Wage	Maximum Period
Alabama	66⅔	300 weeks
Arkansas	66⅔	450 weeks
California	66⅔	619.25 weeks
Florida	*	525 weeks
Indiana	60⅔	500 weeks
Kansas	66⅔	415 weeks
Maryland	66⅔	Duration of disability
Montana	66⅔	500 weeks
Nebraska	66⅔	300 weeks
New Jersey	70	600 weeks
New Mexico	66⅔	500 weeks
Texas	66⅔	300 weeks
Washington	—	—
Wisconsin	66⅔	1,000 weeks

Nonschedule injuries are less straightforward and must be dealt with on a case-by-case basis. Disabilities in this category tend to be the result of head injuries, the effects of which can be more difficult to determine. The amount of compensation awarded and the period over which it is awarded must be determined by studying the evidence. Awards are typically made based on a determination of percent disability. For example, if it is determined that an employee has a 25 percent disability, the employee might be entitled to 25 percent of the income he or she could have earned before the injury with normal career progression factored in.

Four approaches to handling permanent partial disability cases have evolved. Three are based on specific theories, and the fourth is based on a combination of two or more of these theories. The three theories are (1) whole-person theory, (2) wage-loss theory, and (3) loss of wage-earning capacity theory.

Whole-Person Theory

The **whole-person theory** is the simplest and most straightforward of the theories for dealing with permanent partial disability cases. Once it has been determined that an injured worker's capabilities have been permanently impaired to some extent, this theory is applied like a subtraction problem. What the worker can do after recuperating from the injury is determined and subtracted from what he or she could do before the accident. Factors such as age, education, and occupation are not considered.

Wage-Loss Theory

The **wage-loss theory** requires a determination of how much the employee could have earned had the injury not occurred. The wages actually being earned are subtracted from what could have been earned, and the employee is awarded a percentage of the difference. No consideration is given to the extent or degree of disability. The only consideration is loss of actual wages.

Loss of Wage-Earning Capacity Theory

The most complex of the theories for handling permanent partial disability cases is the **loss of wage-earning capacity theory,** because it is based not just on what the employee earned at the time of the accident, but also on what he or she might have earned in the future. Making such a determination is obviously a subjective undertaking. Factors considered include past job performance, education, age, gender, advancement potential at the time of the accident, among others. Once future earning capacity has been determined, the extent to which it has been impaired is estimated, and the employee is awarded a percentage of the difference. Some states prescribe maximum amounts of compensation and maximum periods within which it can be collected. For example, in Figure 5–4, Alabama sets the maximum period for collecting on a nonschedule injury as 300 weeks. Maryland, on the other hand, awards compensation for the duration of the disability. Scheduled disabilities are typically compensated for the duration in all states.

The use of schedules has reduced the amount of litigation and controversy surrounding permanent partial disability cases. This is the good news aspect of schedules. The bad news aspect is that they may be inherently unfair. For example, a surgeon who loses his hand would receive the same compensation as a laborer with the same injury if the loss of a hand is scheduled.

Permanent Total Disability

A **permanent total disability** exists when an injured employee's disability is such that he or she cannot compete in the job market. This does not necessarily mean that the employee is helpless. Rather, it means an inability to compete reasonably. Handling permanent total disability cases is similar to handling permanent partial disability cases except that certain injuries simplify the process. In most states, permanent total disability can be assumed if certain specified injuries have been sustained (i.e., loss of both eyes or both arms). In some states, compensation is awarded for life. In others, a time period is specified. Figure 5–5 shows the maximum period that compensation can be collected for a geographically distributed list of states. Notice in this figure that the time periods range from 401 weeks (Texas) to life (California and Wisconsin).

MONETARY BENEFITS OF WORKERS' COMPENSATION

The **monetary benefits** accruing from workers' compensation vary markedly from state to state. The actual amounts are of less importance than the differences among

Figure 5–5

Duration of permanent total disability benefits for selected states.

Source: U.S. Department of Labor.

State	Maximum Paid
Alabama	Duration of disability
Arkansas	Duration of disability
California	Life
Florida	Duration of disability
Indiana	500 weeks
Kansas	Duration of disability
Maryland	Duration of disability
Montana	Duration of disability
Nebraska	Duration of disability
New Jersey	450 weeks (life in some cases)
New Mexico	700 weeks
Texas	401 weeks (life in some cases)
Washington	Duration of disability
Wisconsin	Life

them. Of course, the amounts set forth in schedules change frequently. However, for the purpose of comparison, consider that at one time the loss of a hand in Pennsylvania resulted in an award of $116,245. The same injury in Colorado brought only $8,736.

When trying to determine a scheduled award for a specific injury, it is best to locate the latest schedule for the state in question. One way to do this is to contact the following agency:

U.S. Department of Labor
Employment Standards Administration
Office of State Liaison and Legislative Analysis
Division of State Workers' Compensation Programs
200 Constitution Avenue, N.W.
Washington, D.C. 20210

Death and Burial Benefits

Workers' compensation benefits accrue to the families and dependents of workers who are fatally injured. Typically, the remaining spouse receives benefits for life or until remarriage. However, in some cases, a time period is specified. Dependents typically receive benefits until they reach the legal age of maturity unless they have a condition or circumstances that make them unable to support themselves even after attaining that age. Figure 5–6 contains the death benefits accruing to surviving spouses and children for a geographically distributed list of states. Since the actual amounts of benefits are subject to change, these are provided for the purpose of illustration and comparison only.

Further expenses are provided in addition to death benefits in all states except Oklahoma. As is the case with all types of workers' compensation, the amount of burial benefits varies from state to state and is subject to change. Figure 5–7 contains the maximum burial benefits for a geographically distributed list of states.

State	Percent of Employee's Wage		Maximum Period
	(Spouse Only)	(Spouse and Children)	
Alabama	50	66⅔	500 weeks
Arkansas	35	66⅔	Widow/widowerhood; children until 18 or married
California	—	—	—
Florida	50	66⅔	Widow/widowerhood; children until 18 or married
Indiana	66⅔	66⅔	500 weeks
Kansas	66⅔	66⅔	Widow/widowerhood; children until 18
Maryland	66⅔	66⅔	Widow/widowerhood; children until 18
Montana	66⅔	66⅔	Surviving spouse— 500 weeks; children until 18
Nebraska	66⅔	75	Widow/widowerhood; children until 18
New Jersey	—	70	Widow/widowerhood; children until 18
New Mexico	66⅔	66⅔	700 weeks
Texas	66⅔	66⅔	Widow/widowerhood; children until 18
Washington	60	70	Widow/widowerhood; children until 18
Wisconsin	66⅔	—	300 weeks

Figure 5–6
Death benefits for surviving spouses and children for selected states.
Source: U.S. Department of Labor.

Figure 5–7
Maximum burial allowances for selected states.*
*Figures are provided for the purpose of comparison and illustration only. They are subject to change.
Source: U.S. Department of Labor.

State	Maximum Allowance
Alabama	$1,000
Arkansas	3,000
California	2,000
Florida	2,500
Indiana	2,000
Kansas	3,200
Maryland	1,200
Montana	1,400
Nebraska	2,000
New Jersey	2,000
New Mexico	3,000
Texas	2,500
Washington	2,000
Wisconsin	1,500

MEDICAL TREATMENT AND REHABILITATION

All workers' compensation laws provide for payment of the medical costs associated with injuries. Most states provide full coverage, but some limit the amount and duration of coverage. For example, in Arkansas, employer liability ceases six months after an injury occurs in those cases in which the employee is able to continue working or six months after he or she returns to work in cases where there is a period of recuperation. In either case, the employer's maximum financial liability is $10,000.[22] In Ohio, medical benefits for silicosis, asbestosis, and coal miner's pneumoconiosis are paid only in the cases of temporary total or permanent total disability.[23]

The laws also specify who is allowed or required to select a physician for the injured employee. The options can be summarized as follows:

■ *Employee selects physician of choice.* This option is available in Alaska, Arizona, Delaware, Hawaii, Illinois, Kentucky, Louisiana, Maine, Massachusetts, Mississippi, Nebraska, New Hampshire, North Dakota, Ohio, Oklahoma, Oregon, Rhode Island, Texas, the Virgin Islands, Washington, West Virginia, Wisconsin, and Wyoming.

■ *Employee selects physician from a list provided by the state agency.* This option applies in Connecticut, Nevada, New York, and the District of Columbia.

■ *Employee selects physician from a list provided by the employer.* This option applies in Georgia, Tennessee, and Virginia.

■ *Employer selects the physician.* This option applies in Alabama, Florida, Idaho, Indiana, Iowa, Maryland, Montana, New Jersey, New Mexico, North Carolina, South Carolina, and South Dakota.

■ *Employer selects the physician, but the selection may be changed by the state agency.* This option applies in Arkansas, Colorado, Kansas, Minnesota, Missouri, Utah, and Vermont.

■ *Employer selects the physician, but after a specified period of time, the employee may choose another.* This option applies only in Puerto Rico.

Rehabilitation and Workers' Compensation

Occasionally an injured worker will need rehabilitation before he or she can return to work. There are two types of rehabilitation: medical and vocational. Both are available to workers whose ability to make a living is inhibited by physical and/or mental work-related problems.

Medical rehabilitation consists of providing whatever treatment is required to restore to the extent possible any lost ability to function normally. This might include such services as physical therapy or the provision of prosthetic devices. **Vocational rehabilitation** involves providing the education and training needed to prepare the worker for a new occupation. Whether the rehabilitation services are medical or vocational in nature or both, the goal is to restore the injured worker's capabilities to the level that existed before the accident.

ADMINISTRATION AND CASE MANAGEMENT

It is not uncommon for minor injuries to go unreported. The employee might be given the rest of the day off or treated with first aid and returned to work. This is done to avoid time-consuming paperwork and to hold down the cost of workers' compensation insurance. However, if an accident results in a serious injury, several agencies must be notified. What constitutes a serious injury, like many workers' compensation issues, can differ from state to state. However, as a rule, an injury is serious if it requires over 24 hours of active medical treatment (this does not include passive treatment such as observation). Of course, a fatality, a major disfigurement, or the loss of a limb or digit is also considered serious and must be reported.

At a minimum, the company's insurer, the state agency, and the state's federal counterpart must be notified. Individual states may require that additional agencies be notified. All establish a time frame within which notification must be made. Once the notice of injury has been filed, there is typically a short period before the victim or dependents can begin to receive compensation unless in-patient hospital care is required. However, when payments do begin to flow, they are typically retroactive to the date of the injury.

State statutes also provide a maximum time period that can elapse before a compensation claim is filed. The notice of injury does not satisfy the requirement of filing a **claim notice.** The two are separate processes. The statute of limitations on claim notices varies from state to state. However, most limit the period to no more than a year except in cases of work-related diseases in which the exact date of onset cannot be determined.

All such activities—filing injury notices, filing claim notices, arriving at settlements, and handling disputes—fall under the collective heading of administration and case management. Most states have a designated agency that is responsible for administration and case management. In addition, some states have independent boards that conduct hearings and/or hear appeals when disputes arise.

Once a workers' compensation claim is filed, an appropriate settlement must be reached. Three approaches can be used to settle a claim: (1) direct settlement, (2) agreement settlement, and (3) public hearing. The first two are used in uncontested cases, the third in contested cases.

1. ***Direct settlement.*** The employer or its insurance company begins making what it thinks are the prescribed payments. The insurer also sets the period over which payments will be made. Both factors are subject to review by the designated state agency. This approach is used in Arkansas, Michigan, Mississippi, New Hampshire, Wisconsin, and the District of Columbia.

2. *Agreement settlement.* The injured employee and the employer or its insurance company work out an agreement on how much compensation will be paid and for how long. Such an agreement must be reached before compensation payments begin. Typically, the agreement is reviewed by the designated state administrative agency. In cases where the agency disapproves the agreement, the worker continues to collect compensation at the agreed-on rate until a new agreement can be reached. If this is not possible, the case becomes a contested case.

3. *Public hearing.* If an injured worker feels he or she has been inadequately compensated or unfairly treated, a hearing can be requested. Such cases are known as *contested cases.* The hearing commission reviews the facts surrounding the case and renders a judgment concerning the amount and duration of compensation. Should the employee disagree with the decision rendered, civil action through the courts is an option.

COST ALLOCATION

Workers' compensation is a costly concept. From the outset, one of the basic principles has been cost allocation. Cost allocation is the process of spreading the cost of workers' compensation across an industry so that no individual company is overly burdened. The cost of workers' compensation includes the costs of premiums, benefits, and administration. These costs have risen steadily over the years.

When workers' compensation costs by industry are examined, there are significant differences. For example, the cost of workers' compensation for a bank is less than one-half of one percent of gross payroll. For a ceramics manufacturer, the percentage might be as high as three or four percent.

Cost allocation is based on the experience rating of the industry. In addition to being the fairest method (theoretically) of allocating costs, this approach is also supposed to give employers an incentive to initiate safety programs. Opinions vary as to the fairness and effectiveness of this approach. Arguments against it include the following: (1) small firms do not have a sufficient number of employees to produce a reliable and accurate picture of the experience rating; (2) firms that are too small to produce an experience rating are rated by class of industry, thereby negating the incentive figure; (3) premium rates are more directly sensitive to experience levels in larger firms but are less so in smaller companies; and (4) in order to hold down experience ratings, employers may put their efforts into fighting claims rather than preventing accidents. Not much hard research has been conducted to determine the real effects of cost allocation. Such research is badly needed to determine if the theoretical construct of workers' compensation is, in reality, valid.

PROBLEMS WITH WORKERS' COMPENSATION

There are serious problems with workers' compensation in the United States. On the one hand, there is evidence of abuse of the system. On the other hand, many injured workers who are legitimately collecting benefits suffer a substantial loss of income. Complaints about workers' compensation are common from all parties involved with the system (employers, employees, and insurance companies).

Earlier in this chapter, the example of the overweight deputy sheriff who applied for benefits because he was distraught over the breakup of his extramarital love affair and a poor performance evaluation was cited as an example of abuse. This individual is just one of thousands who are claiming that job stress has disabled them to the point that workers' compensation is justified. In 1980, there were so few stress claims they were not even recorded as a separate category. Today, they represent a major and costly category.

Stress claims are more burdensome than physical claims because they are typically reviewed in an adversarial environment. This leads to the involvement of expert medical witnesses and attorneys. As a result, even though the benefits awarded for stress-related injuries are typically less than those awarded for physical injuries, the cost of stress claims is often higher because of the litigation.

In addition to abuse, a steadily increasing caseload is also a problem. For example, California saw its caseload increase by 40,000 in one year.[24] As this trend continues, the cost of workers' compensation will increase on a parallel track.

Although the cost of workers' compensation is increasing steadily, the amount of compensation going to injured workers is often disturbingly low. In a given year, if workers' compensation payments in the United States amount to around $27 billion (which is typical), $17 billion of this goes to benefits. Almost $10 billion is taken up by medical costs.[25] The amount of wages paid to injured workers in most states is 66 percent. This phenomenon is not new. According to Seigel,

> In the neighborhood of one-half million families a year are now left without income due to disabling occupational accidents, and those who are compensated receive far less in relation to their income than was the case 40 years ago.
>
> Because of an information system which leans heavily upon dramatic and publicized disasters to stimulate governmental progress, workman's compensation in America continues to be a national disgrace.[26]

The most fundamental problem with workers' compensation is that it is not fulfilling its objectives. Lost income is not being adequately replaced, the number of accidents has not decreased, and the effectiveness of cost allocation is questionable. Clearly, the final chapter on workers' compensation has not yet been written.

SPOTTING WORKERS' COMPENSATION FRAUD/ABUSE

There is evidence of waste, fraud, and abuse of the system in all states that have passed workers' compensation laws. However, the public outcry against fraudulent claims is making states much less tolerant of, and much more attentive to, abuse. For example, the Ohio legislature passed a statute that allows criminal charges to be brought against employees, physicians, and lawyers who give false information in a workers' compensation case. This is a positive trend. However, even these measures will not completely eliminate abuse.

For this reason, it is important for organizations to know how to spot employees who are trying to abuse the system by filing fraudulent workers' compensation claims. Following are some factors that should cause employers to view claims with suspicion. However, just because one or more of these factors is present does not mean that an employee is attempting to abuse the system. Rather, these are simply factors that should raise cautionary flags:[27]

1. The person filing the claim is never home or available by telephone or has an unlisted telephone number.
2. The injury in question coincides with a layoff, termination, or plant closing.
3. The person filing the claim is active in sports.
4. The person filing the claim has another job.
5. The person filing the claim is in line for early retirement.
6. The rehabilitation report contains evidence that the person filing the claim is maintaining an active lifestyle.
7. No organic basis exists for disability. The person filing the claim appears to have made a full recovery.
8. The person filing the claim receives all mail at a post office box and will not divulge a home address.
9. The person filing the claim is known to have skills such as carpentry, plumbing, or electrical that could be used to work on a cash basis while feigning a disability.
10. There are no witnesses to the accident in question.
11. The person filing the claim has relocated out of state or out of the country.
12. Demands for compensation are excessive.
13. The person filing the claim has a history of filing.
14. Doctor's reports are contradictory in nature.
15. A soft-tissue injury is claimed to have produced a long-term disability.
16. The injury in question occurred during hunting season.

These factors can help organizations spot employees who may be trying to abuse the workers' compensation system. It is important to do so because legitimate users of the system are hurt, as are their employers, by abusers of the system. If one or more of these factors is present, employers should investigate the claim carefully before proceeding.

THE FUTURE OF WORKERS' COMPENSATION

The future of workers' compensation can be summarized in one word: **reform.** The key to reforming workers' compensation is finding a way to allocate more of the cost to benefits and medical treatment and less to administration and litigation. In 1989, California, which is the bellwether state for workers' compensation, took the lead in the reform

---

SAFETY FACT

Workers' Compensation Scams in Florida

Hidden Payroll
The owner of a Citrus County roofing and painting business was arrested on charges of evading $832,000 in workers' compensation premiums. These premiums are based on payroll, and investigators charged that the roofer disguised about $1.5 million in payroll costs as subcontractor expenses.

Forklifted Foot
A Jacksonville warehouse manager was arrested on workers' compensation fraud charges after collecting $166,836 from a total disability claim. Although he claimed a forklift ran over his foot causing him to be unable to walk without assistance, investigators obtained a video tape showing him walking his daughter down the aisle at her wedding.

Slip 'n Fall
A Gainesville woman was arrested on two counts of grand theft and two counts of filing false and fraudulent insurance claims. The charges stem from a series of 11 *slip and fall* accidents during a four-year period. Investigators allege the woman staged accidents in grocery and department stores to file claims for various back, head, arm, and leg injuries.

Drugstore Deal
The owner of a North Florida drugstore was arrested following a 62-count federal indictment charging that he submitted phony prescription claims for more than $1 million in Medicaid and state insurance reimbursements.

Insurance Mill
A suspended surgeon who owned an orthopedic clinic in Sunrise was arrested based on allegations of involvement in billing insurers for treatments not rendered, altering medical tests, and soliciting patients through runners.

Chiropractic Claims
A Pompano Beach chiropractor was arrested in an alleged scheme to bill insurers for patient visits and treatments never provided.

Source: Florida Department of Labor, Division of Workers' Compensation movement with its Workers' Compensation Improvement Act. Key elements of the act included the following:

- Stabilizing workers' compensation costs over the long term
- Streamlining administration of the system
- Reducing the costs associated with the resolution of medical issues
- Limiting stress-related claims
- Limiting vocational rehabilitation benefits
- Increasing benefits paid for temporary and permanent disabilities
- Reducing the amount that insurers may charge for overhead
- Providing more public input into the setting of rates

At the time the act was proposed, California's workers' compensation costs had a history of increasing at a rate of over 11 percent per year. California paid out more in claims every year than any other state, but ranked among the lowest nationwide in benefits paid to injured workers. California's problems with workers' compensation paint a picture in microcosm of the problems facing all states.

In 1990, Colorado's governor made workers' compensation reform a top priority. According to the *Rocky Mountain News,*

> The high cost of insurance is only a symptom of the problem. . . . Colorado operates under a no-fault system, which means there's no attempt to assign blame to employer or employee if a worker is injured on the job. As a result, workers aren't motivated to hold claims down and in fact may be prompted to attempt scams.[28]

The reform package proposed in Colorado has the following key points:

- Taking litigation out of the process to the extent possible
- Improving the management of medical treatment and overall case management
- Streamlining the claim notification and processing system
- Requiring more sufficient justification from insurance carriers concerning their rates.[29]

The examples of California and Colorado are indicative of what is taking place nationwide. The states are moving toward the enactment of reforms that will lower overall costs but improve benefits. This is a trend that is likely to continue for the next two decades.

COST REDUCTION STRATEGIES

Safety and health professionals are responsible for helping their organizations hold down workers' compensation costs. Of course, the best way to accomplish this goal is to maintain a safe and healthy workplace, thereby preventing the injuries that drive the costs up. This section presents numerous other strategies that have proven effective in reducing workers' compensation costs after injuries have occurred, which happen in even the safest environments.

DISCUSSION CASE

What Is Your Opinion?

Dr. Maxine McGuire, Professor of Safety and Health at Orange Beach Community College, made the following statement to her class: "Workers compensation laws were passed to get the courts and the lawyers out of the process when employees are injured. The laws have not worked. There are more fraud, abuse, and lawsuits than ever." Is workers' compensation working as a concept? What is your opinion?

General Strategies

Regardless of the type of organization, there are several rules of thumb that can help reduce workers' compensation claims. These general strategies are as follows:

1. *Stay in touch with the injured employee.* Let injured employees know that they have not been forgotten and that they are not isolated. Answer all of their questions and try to maintain their loyalty to the organization.

2. *Have a return-to-work program and use it.* The sooner an injured employee returns to work, even with a reduced workload, the lower workers' compensation costs will be. Reduced costs can, in turn, lower the organization's insurance premium. When using the return-to-work strategy, be cautious. Communicate with the employee and his or her medical treatment team. Have a clear understanding of the tasks that can be done and those that should be avoided, such as how much weight the employee can safely lift.

3. *Determine the cause of the accident.* The key to preventing future accidents and incidents is determining the cause of the accident in question. And the key to holding down workers' compensation costs is preventing accidents. Eliminating the root cause of every accident is fundamental to any cost-containment effort.

Specific Strategies

In addition to the general strategies just presented, there are numerous specific cost-containment strategies that have proven to be effective. These specific strategies are presented in this section.

1. *Cultivate job satisfaction.* According to Jon Gice, increasing job satisfaction is just as important as eliminating physical hazards in the workplace.[30] High stress, aggression, alienation, and social maladjustment—all factors associated with, and aggravated by, a lack of job satisfaction—can make employees less attentive while working. An inattentive employee is an accident waiting to happen. Gice recommends the following strategies for improving job satisfaction:[31]

■ Recognize and reward employees.

- Communicate frequently and openly with employees about job-related problems.
- Give employees as much control over their work as possible.
- Encourage employees to talk freely among themselves.
- Practice conflict management.
- Provide adequate staffing and expense budgets.
- Encourage employees to use employee assistance programs.

2. *Make safety part of the culture.* The Ohio Division of Safety and Hygiene recommends the following steps for making safety part of the organizational culture as a way to reduce workers' compensation costs:[32]

- Ensure visible, active leadership, involvement, and commitment from senior management.
- Involve employees at all levels in the safety program and recognize them for their efforts.
- Provide comprehensive medical care, a part of which is a return-to-work program.
- Ensure effective communication throughout the organization.
- Coordinate all safety and health processes.
- Provide orientation and training for all employees.
- Have written safe work practices and procedures.
- Have a comprehensive written safety policy.
- Keep comprehensive safety records and analyze the data contained in those records.

3. *Have a systematic cost-reduction program.* To reduce costs, an organization should have a systematic program that can be applied continually and consistently. Alpha Meat Packing Company of South Gate, California, has had success using the following strategies:[33]

- Insert safety notes and reminders in employees' paycheck envelopes.
- Call injured employees at home to reassure them that they will have a job when they return.
- Keep supervisors trained on all applicable safety and health issues, procedures, rules, and so on.
- Hold monthly meetings to review safety procedures, strategies, and techniques.
- Reward employees who give suggestions for making the workplace safer.

4. *Use integrated managed care.* Managed care is credited by many with reducing workers' compensation costs nationwide. Others claim the reduction in cost is because managed care dangerously restricts the types and amount of health care provided to injured employees. GE Aircraft Engines is an advocate of managed care that is fully integrated as follows:[34]

- A comprehensive safety and hazard-prevention program to keep employees safe and healthy.

- On-site medical clinics.
- Plant compensation teams that include nurses, rehabilitation technicians, and third-party adjudicators.
- Claims management.
- Company-wide safety and health database.
- Return-to-work program.

SUMMARY

1. Workers' compensation was developed to allow injured employees to be compensated without the need for litigation. It has four main objectives: replacement of income, rehabilitation, accident prevention, and cost allocation.
2. Prior to the enactment of workers' compensation laws, employees' only recourse when injured was through the courts and the prevailing laws favored employers.
3. Early workers' compensation laws were ruled unconstitutional. The constitutional debate continued until 1917 when the U.S. Supreme Court ruled that workers' compensation laws were acceptable.
4. All 50 states have workers' compensation laws, but they vary markedly. All laws are enacted to provide benefits, pay medical costs, provide for rehabilitation when necessary, decrease litigation, and encourage accident prevention.
5. There are three types of workers' compensation insurance: state funds, private insurers, and self-insurance. Six methods are used for determining insurance premium rates for employers: schedule rating, manual rating, experience rating, retrospective rating, premium discounting, and a combination of these.
6. Although an often-stated objective of workers' compensation is the reduction of costly litigation, many cases still go to court. This is particularly true in stress-related cases.
7. Workers' compensation applies when an injury can be categorized as arising out of employment (AOE) or occurring in the course of employment (COE).
8. The definition of an employee can vary from state to state. However, a key concept in distinguishing between an employee and an independent contractor is direction (supervision). Employees are provided direction by the employer; contractors are not.
9. Injuries that are compensable through workers' compensation fall into one of four categories: (1) temporary partial disability, (2) temporary total disability, (3) permanent partial disability, and (4) permanent total disability.
10. Three theoretical approaches to handling permanent partial disability cases are (1) the whole-person theory, (2) the wage-loss theory, and (3) the loss of wage-earning capacity theory.
11. Workers' compensation benefits accrue to the families and dependents of workers who are fatally injured. Typically, the remaining spouse receives benefits for life or until remarriage. Dependents typically receive benefits until they reach the legal age of maturity.

12. All workers' compensation laws provide for payment of medical expenses within specified time periods, but there are differences on how a physician may be selected. The options are (a) employee selects physician of choice; (b) employee selects physician from a list provided by the state agency; (c) employee selects physician from a list provided by the employer; (d) employer selects the physician; (e) employer selects the physician, but the selection may be changed by the state agency; or (f) employer selects the physician but after a specified period of time the employee may choose another.

13. The definition of a serious injury can vary from state to state. However, as a rule an injury is serious if it requires over 24 hours of active medical treatment (this does not include passive treatment such as observation).

14. Workers' compensation claims can be settled in one of three ways. The first two are uncontested; the third is for contested claims. These approaches are (a) direct settlement, (b) agreement settlement, or (c) public hearing.

15. Cost allocation is the process of attempting to spread the cost of workers' compensation across an industry so that no individual company is overly burdened. The cost of workers' compensation includes the cost of premiums, benefits, and administration.

16. The problems being experienced with workers' compensation can be summarized as follows: Workers' compensation is not achieving its intended objectives. It has not succeeded in taking litigation out of the process. The cost of the system rises steadily, but the benefits to injured workers have decreased in real terms. There is evidence of abuse. Therefore, the future of workers' compensation can be summarized in one word—reform.

17. Cost reduction strategies include staying in touch with injured employees and determining the causes of accidents.

═══════ KEY TERMS AND CONCEPTS ═══════

Accident prevention	Experience rating
Agreement settlement	Extraterritorial employees
AOE	Fear factor
Assumption of risk	Harmful environment
Blackball	Income replacement
Claim notice	Independent contractor
COE	Litigation
Contributory negligence	Loss of wage-earning capacity theory
Cost allocation	Malpractice
Direct settlement	Manual rating
Employee	Medical rehabilitation

Monetary benefits

Occupational disease

Permanent partial disability

Permanent total disability

Premium discounting

Private insurance

Public hearing

Reform

Rehabilitation

Retrospective rating

Schedule disabilities

Schedule rating

Self-insurance

State funds

Stress claims

Temporary partial disability

Temporary total disability

Vocational rehabilitation

Wage-loss theory

Whole-person theory

Worker negligence

Workers' compensation

Workplace accident

REVIEW QUESTIONS

1. Explain the underlying rationale of workers' compensation as a concept.
2. List four objectives of workers' compensation.
3. List five types of employees who may not be covered by workers' compensation.
4. What is meant by the term *contributory negligence?*
5. What is meant by the term *assumption of risk?*
6. Explain the reasons for the unprecedented increases in medical costs in the United States.
7. What are the three types of workers' compensation insurance?
8. Insurance companies use one of six methods for determining the premium rates of employers. Select three and explain them.
9. How can one determine if an injury should be considered serious?
10. Explain the concepts of AOE and COE.
11. Distinguish between an employee and an independent contractor.
12. Define the following terms: *temporary disability* and *permanent disability.*
13. Explain the following theories of handling permanent partial disability cases: whole-person, wage-loss, loss of wage-earning capacity.
14. Distinguish between medical and vocational rehabilitation.
15. What are the three approaches for settling workers' compensation claims?
16. Explain the theory of cost allocation.
17. Summarize briefly the problems most widely associated with workers' compensation.
18. What types of actions are workers' compensation reform movements likely to recommend in the future?
19. Explain the most common workers' compensation cost reduction strategies.

ENDNOTES

1. Society of Manufacturing Engineers. *Manufacturing Management* (Dearborn, MI: Society of Manufacturing Engineers, 1988), pp. 12–24.
2. Ibid.
3. Ibid.
4. Hammer, W. *Occupational Safety Management and Engineering,* 4th ed. (Upper Saddle River, NJ: Prentice Hall, 1989), p. 36.
5. Ibid.
6. Society of Manufacturing Engineers. *Manufacturing Management,* pp. 12–24.
7. Somers, H. M., and Somers, A. R. *Workmen's Compensation* (New York: John Wiley & Sons, 1945), p. 29.
8. Ibid.
9. Gagilardo, D. *American Social Science* (New York: Harper & Row, 1949), p. 149.
10. Hammer, W. *Occupational Safety Management and Engineering,* p. 32.
11. Ibid.
12. Ibid.
13. Ibid.
14. Bellow, Daniel O. "Workers' Compensation Abuses Dismay, Alarm Pittsfield Officials," *Berkshire Eagle,* Pittsfield, MA, May 2, 1989, p. 84.
15. Ibid.
16. Ibid.
17. Brydoff, Carol, "Workman's Compensation Needs Reform Now—Consider the Case of the Fat Deputy Sheriff," *Tribune,* Oakland, CA, August 1, 1988, p. B9.
18. Hammer, W. *Occupational Safety Management and Engineering,* p. 39.
19. Society of Manufacturing Engineers. *Accident Facts,* 1990 ed. (Chicago: Society of Manufacturing Engineers), p. 36.
20. Hammer, W. *Occupational Safety Management and Engineering,* p. 42.
21. Ibid.
22. U.S. Department of Labor, *State Workers' Compensation Laws,* January 1996.
23. Ibid.
24. National Safety Council, *Accident Facts,* 1996 ed. (Chicago: National Safety Council), p. 43.
25. Ibid.
26. Seigel, N. "The Fellow-Employee Suit: A Study in Legal Futility and Misguided Legislation," *Risk Management Today* (New York: American Management Association, 1960), p. 74.
27. Kizorek, B. "Video Surveillance and Workers' Comp Fraud," *Occupational Hazards,* February 1994, p. 28.
28. "Romer Pushes Workers' Comp Reform," *Rocky Mountain News,* November 13, 1990, p. A2.
29. Ibid.

30. Gice, Jon. A report entitled "The Relationship between Job Satisfaction and Workers Compensation Claims," CPCU Society, 720 Providence Road, Malvern, PA 19355-0709.

31. Ibid.

32. Ohio Division of Safety and Hygiene. "Ohio Prompts 10 Steps to Reduced Comp Costs," *Occupational Hazards,* August 1996, p. 59.

33. Workers' Comp Update. "Meatpacking Industry Cuts Comp Claims," *Occupational Hazards,* May 1996, p. 103.

34. Goforth, Candace. "Workers' Comp: Is Managed Care the Answer?," *Occupational Hazards,* October 1996, p. 126

CHAPTER SIX

Ergonomic Hazards and Repetitive Strain Injuries

Seeking to improve productivity, & quality by reducing workplace stressors, reducing the risk of injuries & illnesses. and increasing efficiency.

MAJOR TOPICS

- Ergonomics Defined
- Factors Associated with Physical Stress
- OSHA's Ergonomics Guidelines
- Worksite Analysis Program for Ergonomics
- Hazard Prevention and Control
- Medical Management Program
- Training and Education
- Common Indicators of Problems
- Identifying Specific Ergonomic Problems
- Ergonomic Problem-Solving Strategies
- Economics of Ergonomics
- Repetitive Strain Injury Defined
- Classifications of RSIs
- Preventing RSI

The history of workplace development in the Western world is characterized by jobs and technologies designed to improve processes and productivity. All too often in the past, little or no concern was given to the impact of the job process or technology on workers. As a result, work processes and machines have sometimes been unnecessarily dangerous. Another result has been that new technologies have sometimes failed to live up to expectations. This is because, even in the age of high technology, human involvement in work processes is still the key to the most significant and enduring productivity improvements. If a machine or system is uncomfortable, difficult, overly complicated, or dangerous with which to interact, human workers will not be able to derive the full benefit of its use.

The proliferation of uncomfortable and dangerous workplace conditions, whether created by job design or unfriendly technologies, is now widely recognized as harmful to

it involves:
*a. using special design & evaluation techniques to make
tasks, objects & environment more compatible with
human abilities.*

Ergonomic Hazards and Repetitive Strain Injuries   145

productivity, quality, and worker safety and health. The advent of the science of ergonomics is making the workplace more physically friendly. This, in turn, is making the workplace a safer and healthier place.

ERGONOMICS DEFINED

Minimizing the amount of physical stress in the workplace requires continuous study of the ways in which people and technology interact. The insight learned from this study must then be used to improve the interaction. This is a description of the science of **ergonomics**. For the purpose of this book, ergonomics is defined as follows:

> *multidisciplinary*   *Physiological aspects to*
> Ergonomics is the science of conforming the workplace and all of its elements to
> the worker.

The word *ergonomics* is derived from the Greek language. *Ergon* is Greek for *work;* *nomos* means *laws.* Therefore, in a literal sense, ergonomics means work laws. In practice, it consists of the scientific principles (laws) applied in minimizing the physical stress associated with the workplace (work). Figure 6–1 summarizes some of the widely accepted benefits of ergonomics.

The *Occupational Health & Safety Letter* carried the following news brief, which illustrates both the health and financial liabilities that can grow out of ergonomic problems:

> OSHA cited Crane and Company, Inc., of Dalton, Massachusetts, for 52 instances of alleged willful violations involving excessive ergonomic stress in its paper mills and one repeat record-keeping violation last month. Crane, employing 1,128 workers, makes currency paper, stationery, and other paper products. The alleged violations occurred in three departments (bordering, finishing, and papeterie). The operations performed by workers in these departments were videotaped by OSHA inspectors and then analyzed by an agency Health Response team. Company records showed that between 1986 and 1989 a total of 71 women worked in three departments, OSHA said. Of those, 35 reported 60 cases of ergonomic-related disorders, the agency alleged. There were 30 surgical procedures performed on the women in attempts

Figure 6–1
Benefits of ergonomics.

- Improved health and safety for workers
- Higher morale throughout the workplace
- Improved quality
- Improved productivity
- Improved competitiveness
- Decreased absenteeism and turnover
- Fewer workplace injuries/health problems

to relieve their problems, and seven of the women have had two or more surgical procedures, officials added.[1]

This example illustrates a key point. There are benefits to be derived from ergonomics (as listed in Figure 6–1). There are also problems, both financial and health-related, that can result from giving too little attention to ergonomics. The matter is complicated further by the fact that health problems tend to multiply a company's financial problems. Consequently, modern safety and health professionals need to be well-versed in ergonomics.

FACTORS ASSOCIATED WITH PHYSICAL STRESS[2]

In a book translated from work originally produced by the Swedish Work Environment Fund, the National Safety Council lists eight variables that can influence the amount of **physical stress** experienced on the job:

1. Sitting   Standing
2. Stationary   Moveable/mobile
3. Large demand for strength/power   Small demand for strength/power
4. Good horizontal work area   Bad horizontal work area
5. Good vertical work area   Bad vertical work area
6. Nonrepetitive motion   Repetitive motion
7. Low surface   High surface
8. No negative environmental factors   Negative environmental factors

The following paragraphs summarize the extent of this influence and the forms that it can take.

Sitting vs. Standing

Generally speaking, sitting is less stressful than standing. Standing for extended periods, particularly in one place, can produce unsafe levels of stress on the back, legs, and feet.

SAFETY FACT

High Cost of Pain

When employees are in pain, they respond by taking sick leave from work. Sick leave due to unspecified pain costs employers more than $3 billion dollars in lost work days. Approximately 17 million employees in the United States take an average of three days of sick leave per year to deal with unspecified pain (e.g., headaches, neck pain, back pain, and menstrual pain). Often the pain is not work-related.

Although less so than standing, sitting can be stressful unless the appropriate precautions are taken. These precautions include proper posture, a supportive back rest, and frequent standing/stretching movement.

Stationary vs. Mobile

Stationary jobs are those done primarily at one workstation. Of course, even these jobs involve movement at the primary workstation and occasional movement to other areas. Mobile jobs, on the other hand, require continual movement from one station to another. The potential for physical stress increases with stationary jobs when workers fail to take such precautions as periodically standing/stretching/moving. The potential for physical stress increases with mobile jobs when workers carry materials as they move from station to station.

Large vs. Small Demand for Strength/Power

In classifying jobs by these two criteria, it is important to understand that repeatedly moving small amounts of weight over a period of time can have a cumulative effect equal to the amount of stress generated by moving a few heavy weights. Regardless of whether the stress results from lifting a few heavy objects or repeated lifting of lighter objects, jobs that demand larger amounts of strength/power are generally more stressful than those requiring less.

Good vs. Bad Horizontal Work Area

A good **horizontal work area** is one that is designed and positioned so that it does not require the worker to bend forward or to twist the body from side to side. Horizontal work areas that do require these movements are bad. Bad horizontal work surfaces increase the likelihood of physical stress.

Good vs. Bad Vertical Work Area

Good **vertical work areas** are designed and positioned so that workers are not required to lift their hands above their shoulders or bend down in order to perform any task. Vertical work areas that do require these movements are bad. Bad vertical work areas increase the likelihood of physical stress.

Nonrepetitive vs. Repetitive Motion

Jobs that are **repetitive** involve short-cycle motion that is repeated continually. Nonrepetitive jobs involve a variety of tasks that are not, or only infrequently, repeated. Repetition can lead to monotony and boredom. When this happens, the potential for physical stress increases.

Low vs. High Surface Contact

Surface stress can result from contact with hard surfaces such as tools, machines, and equipment. High surface contact jobs tend to be more stressful in a physical sense than are low surface contact jobs.

Presence vs. Absence of Environmental Factors

Generally, the more **environmental factors** with which a worker has to contend on the job, the more stressful the job. For example, personal protective equipment, although conducive to reducing environmental hazards, can increase the amount of physical stress associated with the job.

OSHA'S ERGONOMICS GUIDELINES

In January 1989, OSHA published guidelines for general safety and health program management.[3] **OSHA's ergonomic guidelines** are voluntary and are designed to provide employers with the information and guidance needed to meet their obligations under the OSHAct regarding ergonomics.

OSHA followed these guidelines with another set designed specifically for the meat-packing industry.[4] These specific guidelines represent a model for guidelines that are likely to be developed for other specific industries. Meat packing was singled out for attention because of the high incidence of **cumulative trauma disorders (CTDs)** associated with meat packing. CTDs are injuries that result from an accumulation of repetitive motion stress. For example, using scissors continually over time can cause a CTD in the hand and wrist.

OSHA explains the selection of the meat-packing industry as a model as follows:

> Why meat packing? Most importantly, CTDs are particularly prevalent in the meat packing industry. Although ergonomic hazards are by no means confined to meat packing, the incidence and severity of CTDs and other workplace injuries and illnesses in this industry demand that effective programs be implemented to protect workers from these hazards. These should be part of the employer's overall safety and health management programs.[5]

Figure 6–2 is a checklist of the program elements of OSHA's ergonomic guidelines for the meat-packing industry. These same elements are likely to be contained in any future industry-specific guidelines developed by OSHA. These elements are explained in the following sections. The major program elements recommended by OSHA are (1) worksite analysis, (2) hazard prevention and control, (3) medical management, and (4) training and education.

WORKSITE ANALYSIS PROGRAM FOR ERGONOMICS*

Although complex analyses are best performed by a professional ergonomist, the "ergonomic team"—or any qualified person—can use this program to conduct a **work-**

Figure 6–2
Checklist of program elements of OSHA's ergonomic guidelines for the meat-packing industry.

Application of standard :
doesn't apply to construction,
routine operations, agri,
apply to :
stent handling jobs (nurse)
shipping receiving & delivery
Baggage handlers
warehouse work
Beverage & water handling
garbage & trash collecting
assembly line working
meat, poultry
& fish processing

Worksite Analysis
Hazard Prevention and Control

- Engineering controls
- Work practice controls
- Personal protective equipment
- Administrative controls

Medical Management
Training and Education

- General training
- Job-specific training
- Training for supervisors
- Training for managers
- Training for engineers and maintenance personnel

SAFETY MYTH

Monday Morning Syndrome Debunked

More workplace injury claims are filed on Monday than on any other day of the week. More than 23 percent of all workers' compensation claims are filed on Monday. This has led to a belief that employees are injuring themselves during non-work-related activities over the weekend and filing workers' compensation claims on Monday. The popular belief is that employees who don't carry personal medical insurance use this ploy to gain medical coverage under workers' compensation. However, a study conducted by Brian P. McCall and David Card under the auspices of the National Bureau of Economic Research raises questions about this belief. According to this study, the high incidence of workers' compensation claims filed on Monday can be attributed to such factors as the following:

- Effect on the body of weekend activities
- Lack of activity over the weekend, coupled with a sudden return to physical work on Monday
- Effect of cold weather and age on muscles (Monday morning strains and sprains).

Source: McCall, David P., and Card, David. "Is Workers' Compensation Covering Uninsured Medical Costs?" Working Paper No. 5058, National Bureau of Economic Research, 1996, p. 18.

site analysis and identify stressors in the workplace. The purpose of the following information is to give a starting point for finding and eliminating those tools, techniques, and conditions that may be the source of ergonomic problems.

In addition to analyzing current workplace conditions, planned changes to existing and new facilities, processes, materials, and equipment should be analyzed to ensure that changes made to enhance production will also reduce or eliminate ergonomic risk factors. As emphasized before, this program should be adapted to each individual workplace.

The discussion of the recommended program for worksite analysis is divided into four main parts: (1) gathering information from available sources; (2) conducting baseline screening surveys to determine which jobs need closer analysis; (3) performing ergonomic job hazard analyses of those workstations with identified risk factors; and after implementing control measures, (4) conducting periodic surveys and follow-up studies to evaluate changes.

Information Sources

Records Analysis and Tracking

The essential first step in worksite analysis is **records analysis and tracking** to develop the information necessary to identify ergonomic hazards in the workplace. Existing medical, safety, and insurance records, including OSHA-200/300 logs, should be analyzed for evidence of injuries or disorders associated with CTDs. Health-care providers should participate in this process to ensure confidentiality of patient records.

Incidence Rates

Incidence rates for upper extremity disorders and/or back injuries should be calculated by counting the incidence of CTDs and reporting the number for each 100 full-time workers per year by facility:

$$\text{Incidence rate} = \frac{(\text{number of new cases/yr}) \times (200{,}000 \text{ work hrs}) \text{ per facility*}}{\text{number of hours worked/facility/yr}}$$

Screening Surveys

The second step in worksite analysis is to conduct baseline screening surveys. Detailed baseline screening surveys identify jobs that put employees at risk of developing CTDs. If the job places employees at risk, an effective program will then require the ergonomic job hazard analysis.*

---

*Adapted from OSHA 3123, 1991 (reprinted).
*Adapted from OSHA 3123.
*The same method should be applied to departments, production lines, or job types within the facility.

Checklist

The survey is performed with an ergonomic checklist. This checklist should include components such as posture, materials handling, and upper extremity factors. (The checklist should be tailored to the specific needs and conditions of the workplace.)

Ergonomic Risk Factors

Identification of ergonomic hazards is based on **ergonomic risk factors:** conditions of a job process, workstation, or work method that contribute to the risk of developing CTDs. Not all of these risk factors will be present in every CTD-producing job, nor is the existence of one of these factors necessarily sufficient to cause a CTD.

CTD Risk Factors

Some of the risk factors for CTDs of the upper extremities include the following:

- Repetitive and/or prolonged activities
- Forceful exertions, usually with the hands (including pinch grips)
- Prolonged static postures
- Awkward postures of the upper body, including reaching above the shoulders or behind the back, and twisting the wrists and other joints to perform tasks
- Continued physical contact with work surfaces (e.g., contact with edges)
- Excessive vibration from power tools
- Cold temperatures
- Inappropriate or inadequate hand tools

Back Disorder Risk Factors

Risk factors for back disorders include items such as the following:

- Bad body mechanics such as continued bending over at the waist, continued lifting from below the knees or above the shoulders, and twisting at the waist, especially while lifting
- Lifting or moving objects of excessive weight or asymmetric size
- Prolonged sitting, especially with poor posture
- Lack of adjustable chairs, footrests, body supports, and work surfaces at workstations
- Poor grips on handles
- Slippery footing

Multiple Risk Factors

Jobs, operations, or workstations that have **multiple risk factors** have a higher probability of causing CTDs. The combined effect of several risk factors in the development of CTDs is sometimes referred to as *multiple causation.*

Ergonomic Job Hazard Analyses

At this point, the employer has identified—through the information sources and screening surveys just discussed—jobs that place employees at risk of developing CTDs. As an essential third step in the worksite analysis, an effective ergonomics program requires a job hazard analysis for each job so identified.

The job hazard analysis should be routinely performed by a qualified person for jobs that put workers at risk of developing CTDs. This type of analysis helps to verify lower risk factors at light duty or restricted activity work positions and to determine if risk factors for a work position have been reduced or eliminated to the extent feasible.

Workstation Analysis

An adequate **workstation analysis** would be expected to identify all risk factors present in each studied job or workstation. For upper extremities, three measures of repetitiveness are the total hand manipulations per cycle, the cycle time, and the total manipulations or cycles per work shift. Force measurements may be noted as an estimated average effort and a peak force. They may be recorded as light, moderate, or heavy.

Tools should be checked for excessive vibration. The tools, personal protective equipment, and dimensions and adjustability of the workstation should be noted for each job hazard analysis. Finally, hand, arm, and shoulder postures and movements should be assessed for levels of risk.

Lifting Hazards

For manual materials handling, the maximum weight-lifting values should be calculated.

Videotape Method

The use of videotape, where feasible, is suggested as a method for analysis of the work process. Slow-motion videotape or equivalent visual records of workers performing their routine job tasks should be analyzed to determine the demands of the task on the worker and how each worker actually performs each task.

Periodic Ergonomic Surveys

The fourth step in worksite analysis is to conduct periodic review. Periodic surveys should be conducted to identify previously unnoticed factors or failures or deficiencies in work practices or engineering controls. The periodic review process should also include feedback, follow-up, and trend analysis.

Feedback and Follow-up

A reliable system should be provided for employees to notify management about conditions that appear to be hazardous and to utilize their insight and experience to determine work practice and engineering controls. This might be initiated by an ergonomic questionnaire and maintained through an active safety and health committee or by employee participation with the ergonomic team. Reports of ergonomic hazards or signs and symptoms of potential CTDs should be investigated by ergonomic screening surveys and appropriate ergonomic hazard analysis in order to identify risk factors and controls.

Trend Analysis

Trends of injuries and illnesses related to actual or potential CTDs should be calculated, using several years of data where possible. Trends should be calculated for several departments, process units, job titles, or workstations. These trends may also be used to determine which work positions are most hazardous and need to be analyzed by the qualified person.

Using standardized job descriptions, incidence rates may be calculated for work positions in successive years to identify trends. Using trend information can help to determine the priority of screening surveys and/or ergonomic hazard analyses.

HAZARD PREVENTION AND CONTROL

Engineering solutions, where feasible, are the preferred method for ergonomic **hazard prevention and control**. The focus of an ergonomics program is to make the job fit the person—not to make the person fit the job. This is accomplished by redesigning the workstation, work methods, or tool to reduce the demands of the job, including high force, repetitive motion, and awkward postures. A program with this goal requires research into currently available controls and technology. It should also include provisions for utilizing new technologies as they become available and for in-house research and testing. Following are some examples of engineering controls that have proven to be effective and achievable.

Workstation Design

Workstations should be designed to accommodate the persons who actually use them; it is not sufficient to design for the average or typical worker. Workstations should be easily adjustable and should be either designed or selected to fit a specific task, so that they are comfortable for the workers who use them. The work space should be large enough to allow for the full range of required movements, especially where knives, saws, hooks, and similar tools are used.

Design of Work Methods

Traditional work method analysis considers static postures and repetition rates. This should be supplemented by addressing the force levels and the hand and arm postures involved. The tasks should be altered to reduce these and the other stresses associated with CTDs. The results of such analyses should be shared with the health-care providers to assist in compiling lists of light-duty and high-risk jobs.

Keys to Successful Ergonomic Program:
→ commitment on the part of top management
→ written program
→ employee involvement
→ continuous monitoring of the program
→ adjusting as necessary based on results of monitoring

Tool Design and Handles

Tools should be selected and designed to minimize the risks of upper extremity CTDs and back injuries. In any tool design, a variety of sizes should be available. Examples of criteria for selecting tools include the following:

- Designing tools to be used by either hand, or providing tools for both left- and right-handed workers.
- Using tools with triggers that depress easily and are activated by two or more fingers.
- Using handles and grips that distribute the pressure over the fleshy part of the palm, so that the tool does not dig into the palm.
- Designing/selecting tools for minimum weight; counterbalancing tools heavier than one or two pounds.
- Selecting pneumatic and power tools that exhibit minimal vibration and maintaining them in accordance with manufacturer's specifications or with an adequate vibration-monitoring program. Wrapping handles and grips with insulation material (other than wraps provided by the manufacturer for this purpose) is normally *not* recommended, as it may interfere with a proper grip and increase stress.

MEDICAL MANAGEMENT PROGRAM[*]

An effective **medical management program** for cumulative trauma disorders is essential to the success of an employer's ergonomic program in industries with a high incidence of CTDs. It is not the purpose of these guidelines to dictate medical practice for an employer's **health-care providers**. Rather, they describe the elements of a medical management program for CTDs to ensure early identification, evaluation, and treatment of signs and symptoms; to prevent their recurrence; and to aid in their prevention. Medical management of CTDs is a developing field, and health-care providers should monitor developments on the subject. These guidelines represent the best information currently available.

A physician or occupational health nurse (OHN) with training in the prevention and treatment of CTDs should supervise the program. Each work shift should have access to health-care providers in order to facilitate treatment, surveillance activities, and recording of information. Where such personnel are not employed full-time, the part-time employment of appropriately trained health-care providers is recommended.

In an effective ergonomics program, health-care providers should be part of the ergonomics team interacting and exchanging information routinely to prevent and treat CTDs properly. The major components of a medical management program for the prevention and treatment of CTDs are trained first-level health-care providers, health surveillance, employee training and education, early reporting of symptoms, appropriate medical care, accurate record keeping, and quantitative evaluation of CTD trends throughout the plant.

---

[*] Adapted from OSHA 3123, 1991 (reprinted).

Trained and Available Health-Care Providers

Appropriately trained health-care providers should be available at all times and on an ongoing basis as part of the ergonomic program. In an effective medical management program, first-level health-care providers should be knowledgeable in the prevention, early recognition, evaluation, treatment, and rehabilitation of CTDs, as well as in the principles of ergonomics, physical assessment of employees, and OSHA record-keeping requirements.

Periodic Workplace Walk-Through

In an effective program, health-care providers should conduct periodic, systematic workplace walk-throughs to remain knowledgeable of operations and work practices, to identify potential light duty jobs, and to maintain close contact with employees. Health-care providers should also be involved in identifying risk factors for CTDs in the workplace as part of the ergonomic team.

These walk-through surveys should be conducted every month or whenever a particular job task changes. A record should be kept documenting the date of the walk-through, area(s) visited, risk factors recognized, and action initiated to correct identified problems. Follow-up should be initiated to correct problems identified and should be documented to ensure that corrective action is taken when indicated.

Symptoms Survey

Those responsible for the medical management program should develop a standardized measurement to determine the extent of work-related disorder symptoms in each area of the plant. This measurement will help to determine which jobs are exhibiting problems and to measure progress of the ergonomic program.

Institute a Survey

A **symptoms survey** of employees should be conducted to measure employee awareness of work-related disorders and to report the location, frequency, and duration of discomfort. Body diagrams should be used to facilitate gathering this information. Surveys normally should not include employee's personal identifiers to encourage employee participation.

The survey is one method for identifying areas or jobs where potential CTD problems exist. The major strength of the survey approach is in collecting data on the number of workers who may be experiencing some form of CTD. Reported pain symptoms by several workers on a specific job would indicate the need for further investigation of that job.

Conduct the Survey Annually

Conducting the survey annually should help detect any major change in the prevalence, incidence, and/or location of reported symptoms.

Keep List of Light-Duty Jobs

The ergonomist or other qualified person should analyze the physical procedures used in the performance of each job, including lifting requirements, postures, hand grips, and frequency of repetitive motion.

The ergonomist and health-care providers should develop a list of jobs with the lowest ergonomic risk. For such jobs, the ergonomic risk should be described. This information will assist health-care providers in recommending assignments to light- or restricted-duty jobs. The light-duty job should, therefore, not increase ergonomic stress on the same muscle-tendon groups. Health-care providers should likewise develop a list of known high-risk jobs. Supervisors should periodically review and update the lists.

Health Surveillance

Baseline

The purpose of baseline health surveillance is to establish a base against which changes in health-care status can be evaluated, not to prevent people from performing work. Prior to assignment, all new and transferred workers who are to be assigned to positions involving exposure of a particular body part to ergonomic stress should receive baseline health surveillance.

Conditioning Period Follow-Up

New and transferred employees should be given the opportunity during a four- to six-week break-in period to condition their muscle-tendon groups prior to working at full capacity. Health-care providers should perform a follow-up assessment of these workers after the break-in period (or after one month, if the break-in period is longer than a month) to determine if conditioning of the muscle-tendon groups has been successful; whether any reported soreness or stiffness is transient and consistent with normal adaptation to the job or whether it indicates the onset of a CTD; and if problems are identified, what appropriate action and further follow-up are required.

Periodic Health Surveillance

Periodic health surveillance—every two to three years—should be conducted on all workers who are assigned to positions involving exposure of a particular body part to ergonomic stress. The content of this assessment should be similar to that outlined for the baseline. The worker's medical and occupational history should be updated.

Employee Training and Education

Health-care providers should participate in the training and education of all employees, including supervisors and other plant management personnel, on the different types of CTDs and means of prevention, causes, early symptoms, and treatment of CTDs. This information should be reinforced during workplace walk-throughs and the individual health surveillance appointments. All new employees should be given such education during orientation. This demonstration of concern and the distribution of information should facilitate the early recognition of CTDs prior to the development of more severe and disabling conditions and increase the likelihood of compliance with prevention and treatment.

Encourage Early Report of Symptoms

Employees should be encouraged by health-care providers and supervisors to report early signs and symptoms of CTDs to the in-plant health facility. This allows for timely and appropriate evaluation and treatment without fear of discrimination or reprisal by employers. It is important to avoid any potential disincentives for employee reporting, such as limits on the number of times that an employee may visit the health unit.

Protocols for Health-Care Providers

Health-care providers should use written protocols for health surveillance and the evaluation, treatment, and follow-up of workers with signs or symptoms of CTDs. A qualified health-care provider should prepare the protocols. These protocols should be available in the plant health facility. Additionally, the protocols should be reviewed and updated annually and/or as state-of-the-art evaluation and treatment of these conditions changes.

Evaluation, Treatment, and Follow-Up of CTDs

If CTDs are recognized and treated appropriately early in their development, a more serious condition can likely be prevented. Therefore, a good medical management program that seeks to identify and treat these disorders early is important.

OSHA Record-Keeping Forms

The Occupational Safety and Health Act and record-keeping regulations in Title 29, Code of Federal Regulations (C.F.R.) 1904 provide specific recording requirements that comprise the framework of the occupational safety and health recording system. The Bureau of Labor Statistics (BLS) has issued guidelines that provide official agency interpretations concerning the record-keeping and reporting of occupational injuries and illnesses. These guidelines—U.S. Department of Labor, BLS: *Record-Keeping Guidelines for Occupational Injuries and Illnesses*, September 1986 (or later editions as published)—provide supplemental instructions for the OSHA record-keeping forms (OSHA Forms 200/300, 101, and 200-S) and should be available in every plant health-care facility. Since health-care providers often provide information for OSHA logs, they should be aware of record-keeping requirements and participate in fulfilling them.

Monitor Trends

Health-care providers should periodically (e.g., quarterly) review health-care facility sign-in logs, OSHA-200/300 forms, and individual employee medical records to monitor trends for CTDs in the plant. This ongoing analysis should be made in addition to the symptoms survey to monitor trends continuously and to substantiate the information obtained in the annual symptoms survey. The analysis should be done by department, job title, work area, and so on.

The information gathered from the annual symptoms survey will help to identify areas or jobs where potential CTD problems exist. This information may be shared with anyone in the plant, since employees' personal identifiers are not solicited. The analysis of medical records (e.g., sign-in logs and individual employee medical records) may reveal areas or jobs of concern, but it may also identify individual workers who require further follow-up. The information gathered while analyzing medical records is confidential; thus, care must be exercised to protect the individual employee's privacy. The information gained from the CTD trend analysis and symptoms survey will help determine the effectiveness of the various programs initiated to decrease CTDs in the plant.

TRAINING AND EDUCATION*

The fourth major program element for an effective ergonomics program is training and education. The purpose of training and education is to ensure that employees are sufficiently informed about the ergonomic hazards to which they may be exposed and thus able to participate actively in their own protection.

Training and education allow managers, supervisors, and employees to understand the hazards associated with a job or process, their prevention and control, and their medical consequences. A training program should include all affected employees, engineers and maintenance personnel, supervisors, and health-care providers.

The program should be designed and implemented by qualified persons. Appropriate special training should be provided for personnel responsible for administering the program. The program should be presented in language and at a level of understanding appropriate for the individuals being trained. It should provide an overview of the potential risk of illnesses and injuries, their causes and early symptoms, the means of prevention, and treatment.

The program should also include a means for adequately evaluating its effectiveness. This might be achieved by using employee interviews, testing, and observing work practices, to determine if those who received the training understand the material and the work practices to be followed.

COMMON INDICATORS OF PROBLEMS

Does my company have ergonomic problems? Are injuries/illnesses occurring because too little attention is paid to ergonomic factors? These are questions that modern safety and health professionals should ask themselves. But how does one answer such questions? According to the National Safety Council, the factors discussed in the following paragraphs can be examined to determine if ergonomic problems exist in a given company.[6]

---

* Adapted from OSHA 3123, 1991 (reprinted).

Apparent Trends in Accidents and Injuries

By examining accident reports, record-keeping documents such as OSHA Form 200 (Log and Summary of Occupational Injuries and Illnesses), first aid logs, insurance forms, and other available records of illnesses or injuries, safety and health professionals can identify trends if they exist. A pattern or a high incidence rate of a specific type of injury typically indicates that an ergonomic problem exists.

Incidence of Cumulative Trauma Disorders

Factors associated with CTDs include a high level of repetitive work, greater than normal levels of hand force, awkward posture, high levels of vibration, high levels of mechanical stress, extreme temperatures, and repeated hand-grasping/pinch-gripping. By observing the workplace and people at work, safety and health professionals can determine the amount of exposure that employees have to these factors and the potential for ergonomics-related problems.

Absenteeism and High Turnover Rates

High absentee rates and high turnover rates can be indicators of ergonomic problems. People who are uncomfortable on the job to the point of physical stress are more likely to miss work and/or leave for less stressful conditions.

Employee Complaints

A high incidence of **employee complaints** about physical stress or poor workplace design can indicate the presence of ergonomic problems.

Employee-Generated Changes

Employees tend to adapt the workplace to their needs. The presence of many workplace adaptations, particularly those intended to decrease physical stress, can indicate the presence of ergonomic problems. Have employees added padding, modified personal protective equipment, brought in extra lighting, or made other modifications? Such **employee-generated changes** may be evidence of ergonomic problems.

Poor Quality

Poor quality, although not necessarily caused by ergonomic problems, can be the result of such problems. Poor quality is at least an indicator that there may be ergonomic problems. Certainly, poor quality is an indicator of a need for closer inspection.

Manual Material Handling

The incidence of musculoskeletal injuries is typically higher in situations that involve a lot of **manual material handling.** Musculoskeletal injuries increase significantly when the job involves one or more of the following: lifting large objects, lifting bulky objects, lifting objects from the floor, and lifting frequently. When such conditions exist, the company has ergonomic problems.

IDENTIFYING SPECIFIC ERGONOMIC PROBLEMS

Specific ergonomic problems are identified by conducting a task analysis of the job in question. Figure 6–3 lists the types of problems that can be identified by a thorough task analysis.[7] The National Safety Council recommends using one or more of the following approaches for conducting a task analysis.

General Observation

General observation of a worker or workers performing the task(s) in question can be an effective task analysis technique. The effectiveness is usually enhanced if the workers are not aware that they are being observed. When observing employees at work, be especially attentive to tasks requiring manual material handling and repetitive movements.

Questionnaires and Interviews

Questionnaires and **interviews** can be used for identifying ergonomic problems. Questionnaires are easier to distribute, tabulate, and analyze, but interviews generally provide more in-depth information.

Videotaping and Photography

Videotaping technology has simplified the process of task analysis considerably. Videotaping records the work being observed as it is done, it is silent so it is not intrusive, and such capabilities as freeze and playback enhance the observer's analysis capabilities sig-

Figure 6-3
Problems that can be pin-pointed by a task analysis.

- Tasks that involve potentially hazardous movements
- Tasks that involve frequent manual lifting
- Tasks that involve excessive wasted motion or energy
- Tasks that are part of a poor operations flow
- Tasks that require unnatural or uncomfortable posture
- Tasks with high potential for psychological stress
- Tasks with a high fatigue factor
- Tasks that could/should be automated
- Tasks that involve or lead to quality control problems

nificantly. Photography can also enhance the observer's analysis capabilities by recording each motion or step involved in performing a task. If photography is used, be aware that flashes can be disruptive. High-speed film will allow you to make photographs without using a flash.

Drawing or Sketching

Making a neat sketch of a workstation or a drawing showing workflow can help identify problems. Before using a drawing or sketch as part of a task analysis, make sure that it is accurate. Ask an employee who is familiar with the area or process sketched to check the drawing.

Measuring the Work Environment

Measurements can help identify specific ergonomic problems. How far must a worker carry the material manually? How high does a worker have to lift an object? How much does an object weigh? How often is a given motion repeated? Answers to these and similar questions can enhance the effectiveness of the analysis process.

Understanding the Ergonomics of Aging

When identifying specific ergonomic problems in the workplace, don't overlook the special challenges presented by aging workers. A good ergonomics program adapts the job to the person. Since nearly 30 percent of the workforce is 45 years of age or older, organizations must be prepared to adapt workstations to employees whose physical needs are different from those of their younger counterparts.

In adapting workstations and processes for employees who are 45 or older, keep the following rules of thumb in mind:[8]

- Nerve conduction velocity, hand-grip strength, muscle mass, range of motion, and flexibility all begin to diminish about age 45.
- Weight and mass tend to increase through about the early fifties.
- Height begins to diminish beginning around age 30.
- Lower back pain is more common in people 45 years of age and older.
- Visual acuity at close range diminishes with age.

These rules of thumb mean that safety and health professionals cannot take a "one-size-fits-all" approach to ergonomics. Adaptations for older workers must be individualized and should take aging factors into account.

ERGONOMIC PROBLEM-SOLVING STRATEGIES

The factors that influence stress were explained earlier in this chapter. These factors can be combined in different ways, and the ways in which they are combined determine the type and amount of stress experienced. For the purpose of recommending ergonomic problem-solving strategies, the National Safety Council suggests using the following combinations:

- Seated repetitive work with light parts
- Seated work with larger parts
- Seated control work
- Standing work
- Standing for heavy lifting and/or carrying work in one place or in motion
- Work with hands above chest height
- Work with hand tools
- Work with VDTs[9]

Seated Repetitive Work with Light Parts[10]

This type of work can produce more physical stress than one might suspect. Back, neck, shoulder, and lower leg pain are commonly associated with this type of work. Some of the problems associated with seated repetitive work are the result of the nature of the job. The fixed work position and repetitive motion can contribute to ergonomic problems. To solve these problems, it may be necessary to modify both the job and the workstation. Improvement strategies include the following:

- Include other work tasks to break the monotony of repetition.
- Use job rotation, with workers rotating from one or more different jobs.
- Adjust the height of the work surface and/or position.
- Use an adjustable chair equipped with hand, wrist, or arm supports as appropriate.
- Make sure that there is sufficient leg room (height, width, and depth).
- Use ergonomic devices to adjust the height and angle of work (see Figure 6–4).

Seated Work with Larger Parts[11]

This type of work, which involves interacting with objects that may be too large to manipulate manually, is associated with assembly and welding jobs. Problems associated with this type of work are typically related to posture, illumination, reach, and lifting. Ergonomic strategies for improving work conditions include the following:

- Use technology to lift and position the work for easy access that does not require bending, twisting, and reaching.
- Use supplemental lighting at the worksite.
- Use adjustable chairs and work surfaces as appropriate (see Figure 6–5).

Mazda confronted the problems associated with this type of work when it built a new facility in Flat Rock, Michigan. Because of the ergonomic hazards commonly associated with traditional automobile assembly lines, Mazda designed its Flat Rock facility with a tilt-up line. According to LaBar,

Figure 6–4

Adjust the angle of work to reduce stress.

Source: From a drawing by Robin J. Miller.

Mazda installed a tilt line in the trim and final shop to allow employees to work underneath cars without having to bend down, crawl into pits, or stretch their arms. To further reduce awkward bending and reaching motions, Mazda's overhead conveyor line was designed to move at different heights along the assembly process, depending on which part of the car is being worked on.[12]

Seated Control Work[13]

This type of work involves sitting in one location and using wheels, levers, knobs, handles, and buttons to control a process, system, or piece of equipment. The physical stress associated with this type of work is typically the result of excessive vibration or bending and twisting to achieve better visibility. Ergonomic strategies for improving work conditions include the following:

- Use an adjustable swivel chair with inflatable back and seat support.
- Provide comfortable and convenient locations for control devices.
- Use control devices that meet the following standards: finger control systems that do not require more than five newtons (1.1 pounds); hand levers that do not exceed 20 newtons (4.5 pounds).
- Position the control seat so that a clear line of sight exists between the work and the person controlling it.
- Provide a ladder if the workstation is more than 14 inches above ground.

Figure 6–5
Use adjustable work surfaces
to reduce stress.
Source: From a drawing by
Robin J. Miller.

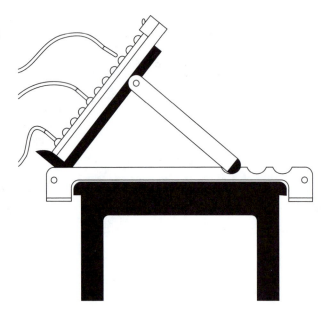

Standing Work[14]

This category includes most jobs that are performed while standing. Such jobs do not involve a great deal of repetitive motion but do involve handling medium to heavy materials. An example is a machine operator's job (lathe, mill, drill, punch, saw, and so on). Physical stress associated with this type of work includes leg, arm, and back strains. Occasionally, side strains occur when bending and twisting are necessary. Ergonomic strategies for improving work conditions include the following:

■ Use adjustable machines and work surfaces to ensure the optimum height and position. When the machine height cannot be adjusted, portable platforms can serve the same purpose (Figure 6–6).

■ When purchasing new machines, make sure there is a recess at the bottom for feet. This will allow the operator to stand close to the machine without having to bend over. Also, look for machines that have easily accessible controls that fall within a comfortable reach zone for operators.

■ Provide ample free space around machines for moving material in and out and to allow for ease of movement in servicing machines.

Standing for Heavy Lifting and Carrying[15]

This type of work involves heavy lifting and moving material while standing. Lifting and moving may be a relatively small part of the job but are required somewhat regularly. The physical stress most commonly associated with this type of work is back and muscle strains resulting from improper lifting. Falls can also be a problem. Ergonomic strategies for improving work conditions include the following:

- Eliminate manual lifting to the extent possible using various lifting and hoisting technologies.
- Where manual lifting is necessary, train workers in proper lifting techniques.
- Provide sufficient room around all objects to allow lifting without twisting.
- Supply the appropriate personal protection equipment such as sure-grip shoes and gloves.
- Keep floors around materials to be lifted clean and dry to prevent slips.
- Do not allow manual carrying of heavy objects upstairs. Stairs increase the physical stress of carrying and, in turn, the potential for injury.

Work with Hands Above the Chest[16]

This type of work can be done in either a standing or sitting position. It might involve material handling, but does not have to. Physical stress associated with this type of work includes neck, upper body, and heart strain. Of these, the most potentially dangerous is heart strain. Prolonged work with the arms above the shoulder level requires the heart to work harder to pump blood to the elevated areas. Ergonomic strategies for improving work conditions include the following:

- Eliminate manual lifting to the extent possible by raising the work floor using lifts and various other technologies.

Figure 6–6
Adjust machines or work surfaces to reduce stress.
Source: From a drawing by Robin J. Miller.

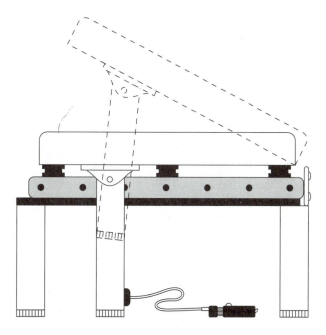

- Use extension arms or poles when the work floor cannot be raised.
- When purchasing new machines, look for machines with controls that are easily accessible below the horizontal plane of a worker's shoulders.

Work with Hand Tools[17]

All of the types of work presented in this section might involve the use of hand tools to some extent. However, because hand tools introduce a variety of potential hazards that are indigenous to their use, they are best examined as a separate work type. Physical stress associated most commonly with the use of hand tools includes carpal tunnel syndrome (CTS) and muscle strains of the lower arm, hands, and wrist. Ergonomic strategies for improving the work conditions focus primarily on improving hand positions during the use of tools, enhancing the worker's grip on tools, and minimizing the amount of twisting involved. Following are some of these strategies:

- Select tools that are designed to keep hands in the rest position (palm down, gently curved, thumb outstretched, and knuckle of the index finger higher than that of the little finger).
- Reduce stress on the hand by selecting tools that have thick, rather than thin, handles (a good range for the diameter is 0.8 to 1.2 inches).
- Select tools that have enhanced gripping surfaces on handles such as knurling, filing, or other enhancements.
- To the extent possible, eliminate twisting by selecting tools designed so that the direction of movement/function is the same as the direction in which force is applied or by using technology (e.g., power screwdriver).
- For tools that do not involve twisting, select handles that have an oval-shaped cross section.
- Select tools with handles made of hard, nonpermeable materials that will not absorb toxic liquids that could be harmful to the skin.

Work with VDTs

The video display terminal, primarily because of the all-pervasive integration of personal computers in the workplace, is now the most widely used piece of office equipment. This fact coupled with the ergonomic hazards associated with VDTs have created a whole new range of concerns for safety and health professionals. Using ergonomics to design a workspace will make it easier, safer, more comfortable, and more efficient to use. Following are some strategies that can be used to reduce the hazards associated with VDTs:[18]

- *Arrange the keyboard properly.* It should be located in front of the user, not to the side. Body posture and the angle formed by the arms are critical factors (see Figure 6-7).
- *Adjust the height of the desk.* Taller employees often have trouble working at *average* height desks. Raising the desk with wooden blocks can solve this problem.

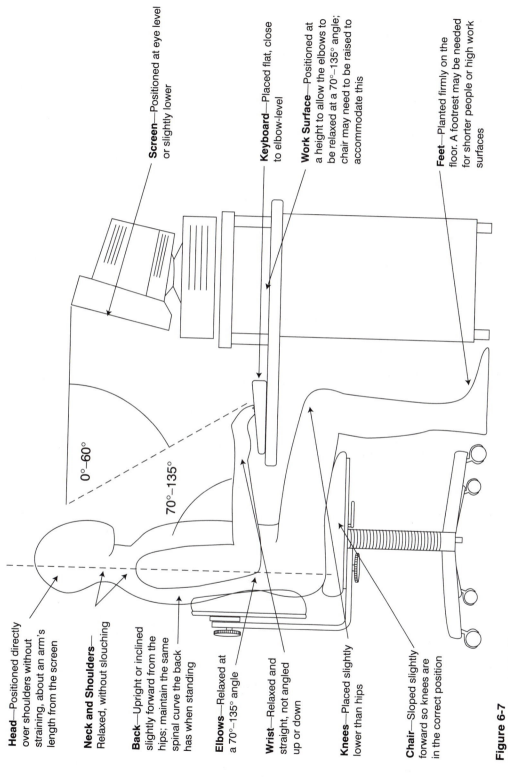

Head—Positioned directly over shoulders without straining, about an arm's length from the screen

Neck and Shoulders—Relaxed, without slouching

Back—Upright or inclined slightly forward from the hips; maintain the same spinal curve the back has when standing

Elbows—Relaxed at a 70°–135° angle

Wrist—Relaxed and straight, not angled up or down

Knees—Placed slightly lower than hips

Chair—Sloped slightly forward so knees are in the correct position

Screen—Positioned at eye level or slightly lower

Keyboard—Placed flat, close to elbow-level

Work Surface—Positioned at a height to allow the elbows to be relaxed at a 70°–135° angle; chair may need to be raised to accommodate this

Feet—Planted firmly on the floor. A footrest may be needed for shorter people or high work surfaces

0°–60°

70°–135°

Figure 6-7
Ergonomics of VDTs. The diagram on the left highlights optimal postures and positions for the computer user.

167

- *Adjust the tilt of the keyboard.* The rear portion of the keyboard should be lower than the front.
- *Encourage employees to use a soft touch on the keyboard and when clicking a mouse.* A hard touch increases the likelihood of injury.
- *Encourage employees to avoid wrist resting.* Resting the wrist on any type of edge can increase pressure on the wrist.
- *Place the mouse within easy reach.* Extending the arm to its full reach increases the likelihood of injury.
- *Remove dust from the mouse ball cavity.* Dust can collect, making it difficult to move the mouse. Blowing out accumulated dust once a week will keep the mouse easy to manipulate.
- *Locate the VDT at a proper height and distance.* The VDT's height should be such that the top line on the screen is slightly below eye level. The optimum distance between the VDT and user will vary from employee to employee, but it will usually be between 16 and 32 inches.
- *Minimize glare.* Glare from a VDT can cause employees to adopt harmful postures. Changing the location of the VDT, using a screen hood, and closing or adjusting blinds and shades can minimize glare.
- *Reduce lighting levels.* Reducing the lighting level in the area immediately around the VDT can eliminate vision strain.
- *Dust the VDT screen.* VDT screens are magnets to dust. Built-up dust can make the screen difficult to read, contributing to eye strain.
- *Eliminate telephone cradling.* Cradling a telephone receiver between an uplifted shoulder and the neck while typing can cause a painful disorder called cervical radiculopathy (compression of the cervical vertebrae in the neck). Employees who need to talk on the telephone while typing should wear a headphone.
- *Require typing breaks.* Continuous typing for extended periods should be avoided. Repetitive strain injuries are cumulative. Breaking up the repetitive motion in question (typing and clicking) can help prevent the accumulation of strain.

DISCUSSION CASE

What Is Your Opinion?

"We need to bring in an ergonomics consultant. I'm concerned that all of a sudden we are going to begin seeing CTD and back injuries resulting from some of our new processes and workstations. What do you think?" Mark Patron, Safety Director for Manufacturing Technologies Corporation, thought the question over before answering. "We don't need a consultant. The risk factors for CTD and back problems are well known in the business. Let me check a couple of reference books and get back to you." Patron sees ergonomic issues as a normal part of his job. His boss thinks an ergonomics expert is needed. What is your opinion?

ECONOMICS OF ERGONOMICS

Perhaps the most underresearched subject relating to ergonomics is the cost effectiveness of safety and health. There are two main reasons for this: (1) such research is often more complex and extensive than the safety and health measures that have been undertaken; and (2) many decision makers think such studies are irrelevant because promoting safety and health is the right thing to do, regardless of costs. As with safety and health in general, there are no in-depth studies available that conclusively pin down the cost benefits of specific ergonomic measures.

According to Bedal,

> Does the application of ergonomic principles make good business sense? The good news is most industry experts believe ergonomics does make good business sense. But the bad news is that if you're looking for studies to prove ergonomics is worth the investment, you'll be hard pressed to find them. Very few true cost-benefit analyses on applying ergonomic principles in an industrial setting have been done.[19]

There is disagreement among well-informed professionals as to whether ergonomic improvements should be expected to meet the test of cost-benefit analysis. However, there is growing support for research that will produce reliable data. This point of view is beginning to characterize the outlook of safety and health professionals in government and academe.[20] Consequently, it is important to understand the problems in attempting to undertake or participate in cost-benefit studies.

Inhibitors of Cost-Benefit Studies

A number of factors inhibit hard research into the economics of ergonomics. Bedal summarizes the most problematic of these as follows:

- ■ Record-keeping systems in industry are not sufficient to support such studies. As a result, a base of comparison from which to work has not been established.
- ■ Industry does not track injuries and illnesses in ways that provide the controls necessary for true hard research. There are no control groups against which to compare groups of injured workers.
- ■ It is difficult and sometimes impossible to determine what improvements can be attributed directly to specific ergonomic strategies and what improvements should be attributed to other factors.
- ■ Undertaking hard research studies requires a commitment of both time and money. A longitudinal study of the effects of a given ergonomic improvement might take three to five years to produce reliable data.
- ■ Follow-up evaluations of injuries that summarize the direct and indirect costs do not exist to the extent necessary to contribute to hard research.[21]

These inhibitors cannot be overcome unless industrial firms are willing to invest enormous sums of money and time. Although the data produced by such investments would be valuable, it is generally not perceived as adding to a company's bottom line. Therefore, industry is likely to continue to rely on academe and governmental agencies for research. In today's intensely competitive international marketplace, industry is likely to invest all available funds in efforts that are perceived as improving productivity and quality. Consequently, hard research into the economics of ergonomics may remain a low priority for some time to come.

REPETITIVE STRAIN INJURY (RSI) DEFINED

The personal computer has become an all-pervasive and universal work tool. Jobs from the shop floor to the executive office now involve frequent, repetitive computer use. This means that people in the workplace are typing and clicking at an unprecedented pace. Frequent and, for some, constant computer use have led to an explosion of injuries heretofore seen mostly in the meat-packing industry. Collectively, these injuries are known as **repetitive strain injuries**, or RSIs.

Definition

RSI is an umbrella term that covers a number of cumulative trauma disorders (CTDs) caused by forceful or awkward hand movements repeated frequently over time. Other aggravating factors include poor posture, an improperly designed workstation, and job stress. RSIs occur to the muscles, nerves, and tendons of the hands, arms, shoulders, and neck.

CLASSIFICATIONS OF RSIs[22]

For years, RSI has been incorrectly referred to as *carpal tunnel syndrome,* which, in reality, is actually one type of RSI. This is like referring to all trees as oaks. Figure 6-8 is a checklist of the most common RSIs organized into four broad classifications.

Muscle and Tendon Disorders

Tendons connect muscles to bones. They can accommodate very little in the way of stretching and are prone to injury if overused. Overworking a tendon can cause small tears in it. These tears can become inflamed and cause intense pain. This condition is known as **tendinitis.**

Myofacial muscle damage can also be caused by overexertion. It manifests itself in soreness that persists even when resting. Muscles may burn and be sensitive to the touch. When sore muscles become inflamed and swell, the symptoms are aggravated even further by nerve compression. Tendons that curve around bones are encased in protective coverings called *sheathes.* Sheathes contain a lubricating substance known as synovial fluid. When tendons rub against the sheath too frequently, friction is produced. The body responds by producing additional synovial fluid. Excess build-up of this fluid

Figure 6-8
RSI checklist—types of injury
by classification.

Muscle and Tendon Disorders
- ✓ Tendinitis
- ✓ Muscle damage (myofacial)
- ✓ Tenosynovitis
- ✓ Stenosing tenosynovitis
  - DeQuervain's disease
  - Trigger finger (flexor tenosynovitis)
- ✓ Shoulder tendinitis
- ✓ Bicipital tendinitis
- ✓ Rotator cuff tendinitis
- ✓ Forearm tendinitis
  - Flexor carpi radialis tendinitis
  - Extensor tendinitis
  - Flexor tendinitis
- ✓ Epicondylitis
- ✓ Ganglion cysts

Cervical Radiculopathy

Tunnel Syndromes
- ✓ Carpal tunnel syndrome
- ✓ Radial tunnel syndrome
- ✓ Sulcus ulnaris syndrome
- ✓ Cubital tunnel syndrome
- ✓ Guyons canal syndrome

Nerve and Circulation Disorders
- ✓ Thoracic outlet syndrome
- ✓ Raynaud's disease

can cause swelling which, in turn, causes pressure on the surrounding nerves, causing a condition known as **tenosynovitis.**

Chronic tenosynovitis is known as *stenosing tenosynovitis* of which there are two types: DeQuervain's disease and flexor tenosynovitis (trigger finger). DeQuervain's disease affects the tendon at the junction of the wrist and thumb. It causes pain when the thumb is moved or when the wrist is twisted. Flexor tenosynovitis involves the locking of a digit in a bent position, hence the term *trigger finger*. However, it can occur in any finger.

Shoulder tendinitis is of two types: bicipital and rotator cuff tendinitis. Bicipital tendinitis occurs at the shoulder joint where the bicep muscle attaches. The rotator cuff is a group of muscles and tendons in the shoulder that move the arm away from the body and turn it in and out. Pitchers in baseball and quarterbacks in football often experience rotator cuff tendinitis.

Forearm tendinitis is of three types: flexor carpi radialis tendinitis, extensor tendinitis, and flexor tendinitis. Flexor carpi radialis tendinitis causes pain in the wrist at the base of the thumb. Extensor tendinitis causes pain in the muscles in the top of the hand, making it difficult to straighten the hands. Flexor tendinitis causes pain in the fingers, making them difficult to bend.

Epicondylitis and *ganglion cysts* are two muscle and tendon disorders. Epicondylitis (lateral) affects the outside of the elbow, whereas epicondylitis (medial) affects the inside. The common term for this disorder is tennis "elbow." Ganglion cysts grow on the tendon, tendon sheath, or synovial lining, typically on top of the hand, on the nail bed, above the wrist, or on the inside of the wrist.

Cervical Radiculopathy

This disorder is most commonly associated with holding a telephone receiver on an upraised shoulder while typing. This widely practiced act can cause compression of the cervical discs in the neck, making it painful to turn the head. Putting the body in an unnatural posture while using the hands is always dangerous.

Tunnel Syndromes

Tunnels are conduits for nerves that are formed by ligaments and other soft tissues. Damage to the soft tissues can cause swelling that compresses the nerves that pass through the tunnel. These nerves are the median, radial, and ulnar nerves that pass through a tunnel in the forearm and wrist. Pain experienced with tunnel injuries can be constant and intense. In addition to pain, people with a tunnel injury might experience numbness, tingling, and a loss of gripping power. The most common **tunnel syndromes** are carpal tunnel syndrome, radial tunnel syndrome, sulcus ulnaris syndrome, cubital tunnel syndrome, and guyon's canal syndrome.

Nerve and Circulation Disorders

When friction or inflammation cause swelling, both nerves and arteries can be compressed, restricting the flow of blood to muscles. This can cause a disorder known as *thoracic outlet syndrome*. The symptoms of this disorder are pain in the entire arm, numbness, coldness, and weakness in the arm, hand, and fingers.

If the blood vessels in the hands are constricted, *Raynaud's disease* can result. Symptoms include painful sensitivity, tingling, numbness, coldness, and paleness in the fingers. It can affect one or both hands. This disorder is also known as *vibration syndrome* because it is associated with vibrating tools.

PREVENTING RSI[23]

The best way to prevent RSI is to make employees aware of the hazards that can cause it. These hazards include poor posture at the workstation, inappropriate positioning of the hands and arms, a heavy hand on a keyboard or mouse, and any other act that repeatedly puts the body in an unnatural posture while using the hands. Ergonomically sound workstations can help prevent RSI, especially when they can be modified to fit the individual employee. However, even the best ergonomic design cannot prevent a heavy hand on the keyboard or mouse. Consequently, ergonomics is only part of the answer. Following are some prevention strategies that can be applied in any organization:

1. *Teach employees the warning signs.* RSI occurs cumulatively over time. It sneaks up on people. Employees should be aware of the following warning signs: weakness in the hands or forearms, tingling, numbness, heaviness in the hands, clumsiness, stiffness, lack of control over the fingers, cold hands, and tenderness to the touch.

2. *Teach employees how to stretch.* Employees whose jobs involve repetitive motion work such as typing can help prevent RSI by using stretching exercises. Limbering up the hands and forearms each day before starting work and again after long breaks such as the lunch hour will help eliminate the stress on muscles and tendons that can lead to RSI.

3. *Teach employees to start slowly.* Long-distance runners typically start slowly, letting their bodies adjust and their breathing find its rhythm. They pick up the pace steadily, until eventually settling in at a competitive pace. This approach is an excellent example of how employees in jobs that are RSI prone should work. Teach employees to limber up, then begin slowly and increase their pace gradually.

4. *Avoid the use of wrist splints.* Teach employees to position their hands properly without using wrist splints. Splints can cause the muscles that they support to atrophy, thereby actually increasing the likelihood of problems.

5. *Start an exercise group.* Exercises that strengthen the hands and forearms coupled with exercises that gently stretch hand and forearm muscles can be an excellent preventive measure. Exercises that strengthen the back can help improve posture, and good posture helps prevent RSI.

6. *Select tools wisely.* RSI and other cumulative trauma disorders (CTDs) are most frequently associated with the repetitive use of VDTs and hand tools. Selecting and using hand tools properly can help prevent RSI and other CTDs. Figure 6-9 is a checklist for the proper selection and use of hand tools.

Figure 6-9
Checklist for safe selection
and use of hand tools.

Use Anthropometric Data

Anthropometric data has to do with human body dimensions.
Such data can be used to determine the proper handle length,
grip span, tool weight, and trigger length when selecting tools.

Reduce Repetition

Repetition is a hazard that can and should be reduced using
such strategies as the following:

- Limit overtime.
- Change the process.
- Provide mechanical assists.
- Require breaks.
- Encourage stretching and strengthening exercises.
- Automate where possible.
- Rotate employees regularly.
- Distribute work among more employees.

Reduce the Force Required

The more force required, the more potential for damage to soft tissue. Required force can be reduced using the following strategies:

- Use power tools wherever possible.
- Use the power grip instead of the pinch grip.
- Spread the force over the widest possible area.
- Eliminate slippery, hard, and sharp gripping surfaces.
- Use jigs and fixtures to eliminate the pinch grip.

Minimize Awkward Postures

Awkward postures contribute to CTDs. The following strategies
can reduce posture hazards:

- Keep the wrist in a neutral position.
- Keep elbows close to the body (90°–110° where bent).
- Avoid work that requires overhead reaching.
- Minimize forearm rotation.

SUMMARY

1. Ergonomics is the science of conforming the workplace and all of its elements to
   the worker.
2. The word *ergonomics* is derived from the Greek language. *Ergon* means work and
   nomos means laws.

3. Common factors that can influence the amount of physical stress associated with a given job include the following: sitting, standing, demand for strength/power, work area, type of motion, amount of surface contact, and environmental factors.

4. OSHA's ergonomic guidelines have the following program elements: worksite analysis, hazard prevention and control, medical management, and training/ education.

5. Common indicators of the existence of ergonomic problems include the following: trends in accidents and injuries, incidence of cumulative trauma disorders (CTDs), absenteeism, high turnover rates, employee complaints, employee-generated changes, poor quality, and high incidence of manual material handling.

6. Procedures for identifying specific ergonomic problems include the following: general observation, questionnaires, interviews, videotaping, photography, drawing, sketching, and measuring.

7. In devising ergonomic problem-solving strategies, work might be divided into the following categories: seated repetitive work with light parts, seated work with larger parts, seated control work, standing work, standing work with heavy lifting or carrying work from place to place, work with hands above chest height, and work with hand tools.

8. There are two reasons why little hard research has been conducted concerning the economics of ergonomics: (1) the cost of such research is often more than the ergonomic improvements that were made; and (2) the feeling that ergonomic improvements should not have to be justified on the basis of cost effectiveness.

KEY TERMS AND CONCEPTS

Cervical radiculopathy

Cumulative trauma disorders (CTDs)

Employee complaints

Employee-generated changes

Environmental factors

Ergonomic risk factors

Ergonomics

General observation

Hazard prevention and control

Health-care providers

Horizontal work area

Incidence rates

Interviews

Manual material handling

Medical management program

Multiple risk factors

OSHA's ergonomic guidelines

Physical stress

Questionnaires

Records analysis and tracking

Repetitive motion

Repetitive strain injury

Symptoms survey

Tendinitis

Tenosynovitis

Tunnel syndromes

Vertical work area

Worksite analysis

Workstation analysis

REVIEW QUESTIONS

1. Define the term *ergonomics*. Explain its origins.
2. Explain how the following opposing factors can influence the amount of physical stress associated with a job: sitting vs. standing, large vs. small demand for strength/power, nonrepetitive vs. repetitive motion.
3. Name the four program elements of OSHA's ergonomic guidelines.
4. What are the four main parts of OSHA's recommended worksite analysis program for ergonomics?
5. List five risk factors associated with CTDs.
6. Briefly explain the steps in conducting an ergonomic job hazard analysis.
7. Briefly explain the components of a hazard prevention and control program.
8. Who should participate in an ergonomic training program?
9. List and briefly explain three common indicators of the existence of ergonomic problems.
10. Describe three approaches that can be used to pinpoint specific ergonomic problems.
11. Describe an ergonomic problem-solving strategy for each of the following types of work: seated repetitive work with light parts, work with hands above the chest height, and work with hand tools.
12. Explain why so little hard research has been done concerning the economics of ergonomics.
13. Define the term *RSI*. Describe the most common types.

ENDNOTES

1. Morris, B. K. (ed.). "Ergonomic Problems at Paper Mill Prompt Fine," *Occupational Health & Safety Letter*, August 8, 1990, Vol. 20, No. 16, p. 127.
2. National Safety Council. *Making the Job Easier: An Ergonomics Idea Book* (Chicago: National Safety Council, 1988), pp. 1–3.
3. *Federal Register*, January 26, 1989, Vol. 54, No. 16, pp. 3904–3916.
4. OSHA 3123, 1991 (reprinted), U.S. Department of Labor, p. 1.
5. Ibid.
6. National Safety Council. *Ergonomics: A Practical Guide* (Chicago: National Safety Council, 1988), pp. 2–1 through 2–5.
7. Ibid., p. 31.
8. LaBur, Gregg. "The Age(ing) of Ergonomics," *Occupational Hazards*, April 1996, pp. 32–33.
9. National Safety Council. *Making the Job Easier: An Ergonomics Idea Book*, pp. 25–32.
10. Ibid., pp. 5–6.
11. Ibid., pp. 13–24.
12. LaBar, G. "Ergonomics: The Mazda Way," *Occupational Hazards*, April 1990, p. 44.
13. National Safety Council. *Making the Job Easier: An Ergonomics Idea Book*, pp. 25–32.

14.   Ibid., pp. 33–58.
15.   Ibid., pp. 59–78.
16.   Ibid., pp. 79–84.
17.   Ibid. pp. 85–102.
18.   Carson, Roberta. "Ergonomic Innovations: Free to a Good Company," *Occupational Hazards,* January 1996, pp. 61–64.
19.   Bedal, T. "The Economics of Ergonomics: What Are the Paybacks?" *Safety & Health*, October 1990, Vol. 142, No. 4, p. 34.
20.   Ibid., p. 38.
21.   Ibid., p. 39.
22.   Pascarelli, Emil, and Quilter, Deborah. *Repetitive Strain Injury* (New York: John Wiley & Sons, Inc., 1994), pp. 49–62.
23.   Carson, Roberta. "Ergonomically Designed Tools," *Occupational Hazards,"* September 1995. p. 50.

CHAPTER SEVEN

Stress and Safety

WORKPLACE STRESS DEFINED

Stress is a pathological, and therefore generally undesirable, **human reaction** to psychological, social, occupational, or environmental stimuli.[1] Stress has been defined as the reaction of the human organism to a threatening situation.[2] The stressor is an external stimuli, and **stress** is the response of the human body to this stimuli.

This is a clinical definition of stress. Other definitions relate more directly to **workplace stress.** Finding a universally acceptable definition should be a priority. According to Jefferson Singer of Connecticut College in New London, employers and employees disagree on how stress should be defined and what should be done to reduce it.[3]

Corporations tend to see stress as an individually based problem that is rooted in an employee's lifestyle, psychological makeup, and personality. Unions view stress as the result of excessive demands, poor supervision, or conflicting demands.[4] According to Singer, "The best definition probably includes both sets of factors."[5]

However it is defined, stress is a serious problem in the modern workplace. Stress-related medical bills and the corresponding absentee rates cost employers over $150 billion annually. Almost 15 percent of all occupational disease claims are stress-related.

Workplace stress involves the emotional state resulting from a perceived difference between the level of occupational demand and a person's ability to cope with this demand. Since preparations and emotions are involved, workplace stress is considered a subjective state. An environment that a worker finds to be stressful may generate feelings of tension, anger, fatigue, confusion, and/or anxiety.

Workplace stress is primarily a matter of person–workload fit. The status of the person–workload fit can influence the acceptance of the work and the level of acceptable performance of that work. The perception of workload may be affected by the worker's needs and his or her level of job satisfaction. Workplace stress is further influenced by the relation between job demands and the worker's ability to meet those demands. Because workplace stress may be felt differently by different people in similar situations, it must be concluded that there are many causes of workplace stress.

SOURCES OF WORKPLACE STRESS

The sources of on-the-job stress may involve physical working conditions, work overload, **role ambiguity**, lack of **feedback**, personality, personal and family problems, and/or role conflict. Other sources of workplace stress are discussed in the following paragraphs.

- **Task complexity** relates to the number of different demands made on the worker. A job perceived as being too complex may cause feelings of inadequacy and result in emotional stress. Repetitive and monotonous work may lack complexity so that the worker becomes bored and dissatisfied with the job and possibly experiences some stress associated with the boredom.

- **Control** over the job assignment can also be a source of workplace stress. Most workers experience less stress when they participate in determining the work routine, including schedule and selection of tasks. Several studies have indicated that workers prefer to take control of their job assignment and experience less workload stress if given this opportunity.[6] A related source of stress that has been introduced in the age of high technology is from electronic monitoring. According to a study conducted at the University of Wisconsin at Madison, "Video display terminal workers who are electronically monitored suffered greater health problems than those who are not."[7]

- A **feeling of responsibility** for the welfare or safety of family members may produce on-the-job stress. Being responsible for the welfare of his or her family may cause a worker to feel that options to take employment risks are limited. A worker may then perceive that he or she is "trapped in the job." Overly constrained employment options may lead to anxiety and stress. The feeling of being responsible for the safety of the general public has also been shown to be a stressor. Air traffic controllers are known to experience intense stress when their responsibility for public safety is tested by a near-accident event. A feeling of great responsibility associated with a job can transform a routine activity into a stress-inducing task.

- **Job security** involves the risk of unemployment. A worker who believes that his or her job is in jeopardy will experience anxiety and stress. The ready availability of other rewarding employment and a feeling that one's professional skills are needed reduce the stress associated with job security issues.

- **Workload demands** can stimulate stress when they are perceived as being overwhelming. These demands may involve time constraints and cognitive constraints such as speed of decision making and mandates for attention. Workload demands may also be physically overwhelming if the worker is poorly matched to the physical requirements of the job or is fatigued. Whenever the worker believes the workload to be too demanding, stress can result.

- **Psychological support** from managers and co-workers gives a feeling of acceptance and belonging and helps defuse stress. A lack of such support may increase the perception of a burdensome workload and result in stress.

- The lack of **environmental safety** can also be a cause of stress. Feeling that one is in danger can be a stressor. Workers need to feel safe from environmental hazards such as extreme temperatures, pressure, electricity, fire, explosives, toxic materials, ionizing radiation, noises, and dangerous machinery. To reduce the potential for stress due to environmental hazards, workers should feel that their managers are committed to safety and that their company has an effective safety program.

A poll for Northwestern National Life Insurance Company,[8] in which workers across the country were interviewed, shows an epidemic of workplace stress. Job stress sources found by this study include the following:

- The company was recently purchased by another company.
- Downsizing or layoffs have occurred in the past year.
- Employee benefits were significantly cut recently.
- Mandatory overtime is frequently required.
- Employees have little control over how they do their work.
- The consequences of making a mistake on the job are severe.
- Workloads vary greatly.
- Most work is machine-paced or fast-paced.
- Workers must react quickly and accurately to changing conditions.
- Personal conflicts on the job are common.
- Few opportunities for advancement are available.
- Workers cope with a great deal of bureaucracy in getting work done.
- Staffing, money, or technology is inadequate.
- Pay is below the going rate.
- Employees are rotated among shifts.

The study showed that stress may lead to decreased productivity, higher absenteeism and job turnover, poor morale, and greater numbers of stress-related illnesses.

Technological developments have also introduced new sources of stress into the workplace. The proliferation of computers is increasing stress levels in ways that were not anticipated. According to Caroline Dow and Douglas Covert of the University of

Evansville, "Workplaces using computers and video display terminals (VDTs) are unwittingly breeding stress, especially among women."[9]

Dow and Covert conducted an experiment in which they observed the effect of a 16-kilohertz pure tone sound on the stress levels of college-age women. They found that anxiety and irritation were the first responses to the sound. The loudness of the sound was not a major factor; even quiet tones caused noticeable stress.[10]

HUMAN REACTIONS TO WORKPLACE STRESS

Human reactions to workplace stress may be grouped into the following categories: subjective or emotional (anxiety, aggression, guilt); behavioral (being prone to accidents, trembling); cognitive (inability to concentrate or make decisions); physiological (increased heart rate and blood pressure); and organizational (absenteeism and poor productivity). Continual or persistent stress has been linked to many physiological problems.[11] Initially, the effects may be psychosomatic, but with continued stress, the symptoms show up as actual organic dysfunction. The most common forms of stress-related diseases are gastrointestinal, particularly gastric or duodenal ulcers. Research has linked some autoimmune diseases with increased long-term workplace stress.[12]

The human response to workplace stress can be compared to a rubber band being stretched. As the stress continues to be applied, the rubber band stretches until a limit is reached when the rubber band breaks. For humans, various physical and psychological changes are observed with the repetitive stimuli of stress. Until the limit is reached, the harmful effects can be reversed. With an increase in intensity or duration of the stress beyond the individual's limit, the effects on the human become pathological.

M. Selye identified three stages of the human stress response: (1) alarm, (2) resistance, and (3) exhaustion.[13] The alarm reaction occurs when the stress of a threat is sensed. The **stage of alarm** is characterized by pallor, sweating, and an increased heart rate. This stage is usually short. It prepares the body for whatever action is necessary.

When the stress is maintained, the **stage of resistance** initiates a greater physical response. The alarm symptoms dissipate, and the body develops an adaptation to the stress. The capacity for adaptation during this stage is limited.

Eventually, with sustained stress, the **stage of exhaustion** is reached. This stage is demonstrated by the body's failure to adapt to the continued stress. Psychosomatic diseases such as gastric ulcers, colitis, rashes, and autoimmune disorders may begin during this stage. The tendency to develop a specific stress-related disease may be partially predetermined by heredity, personal habits such as smoking, and personality.

From an evolutionary viewpoint, the adverse effects of stress on health may be considered to be a maladaptation of humans to stress. What does this tell us? Either we (1) learn to do away with all stress (unlikely); (2) avoid all stressful situations (equally unlikely); (3) learn to adapt to being sick because of stress (undesirable); or (4) learn to adapt to workplace stress (the optimal choice). The first step in learning to adapt to stress is understanding the amount of stress to which we are subjected.

SAFETY FACT

Workplace Stress and Health

The physical and mental health of working people is affected by their *psychological environment*. Autocratic, insensitive managers and supervisors can produce enough stress in employees to cause physical illness. When managers and supervisors show an interest in employees, empower them to participate, and provide positive reinforcement, fewer illnesses or cases of work-related depression are reported.

MEASUREMENT OF WORKPLACE STRESS

Workplace stress can be seen as an individual's psychological reaction to the work environment. Although psychological response cannot be directly measured in physical terms, one method commonly employed uses a measurement of mental workload. Mental workload can be measured in one of three ways:

1. With **subjective ratings**, the workers are asked to rate their perceived level of workload. The perceived workload is then viewed as a direct reflection of workplace stress. The workers may be asked to rate their mood in relation to the work situation. The data gathered by this method is obviously subjective and state-dependent. **State-dependent** data are directly related to the circumstances or state under which they are collected and, therefore, have a built-in state bias.

2. **Behavioral time-sharing** techniques require the simultaneous performance of two tasks. One of the tasks is considered the primary or most important; the other is of secondary importance. The decrease in performance efficiency of the secondary task is considered an index of workload for behavioral time-sharing or human multitasking. Workplace stress is thought to increase as behavioral time-sharing increases.

3. **Psychophysiological techniques** require simultaneous measurement of heart rate and brain waves, which are then interpreted as indexes of mental workload and workplace stress.

Behavioral time-sharing and psychophysiological techniques are related to theoretical models, making data easier to interpret. These two techniques also require sophisticated equipment and data collection methods.

Subjective ratings may be collected using questionnaires or survey instruments. These instruments may ask about the physical working conditions, the individual's health and mental well-being, and perceived overall satisfaction with the job. The data may then be compared to standardized scales developed by various researchers.[14]

Psychosocial questionnaires evaluate workers' emotions about their jobs. Workers may be asked about job satisfaction, workload, pace, opportunities for advancement, management style, and organizational climate. Psychosocial questionnaires are another form of subjective ratings and are also subject to state-dependent bias in the data.

Regardless of the measurement method, since workplace stress is dependent on personal awareness, no direct means of measuring workplace stress are now available.

SHIFT WORK, STRESS, AND SAFETY

Shift work can require some employees to work when the majority of people are resting. In some cases, shift work requires rotating between two or three different starting times, which may vary by eight hours or more. Shift work has traditionally been required by the medical community, the transportation industry, utilities, security, and, increasingly, by retail sales.

Basic physiological functions are scheduled by the biological clock called the **circadian rhythm**. Most children in the United States grow up on the day shift, going to school during the day and sleeping at night. After a life of being on the day shift, the body perceives a change in work shift as being stressful. If the person takes a job starting at midnight, his or her body will still expect to be sleeping at night and active during the day.

Many physical and psychological functions are affected by circadian rhythm. Blood pressure, heart rate, body temperature, and urine production are measurably slower at night. These same functions are normally faster during the day (active time).

Behavioral patterns also follow the circadian slower-at-night and more-active-during-the-day pattern. Sleep demand and sleep capacity for people age 14 or older is greatest at night and least during daylight hours. Alertness has been determined to be decreased at night.[15]

Workers surveyed have consistently reported lower job satisfaction with rotating shifts.[16] Day-shift workers with the same task definitions report higher job satisfaction and less stress than their second- or third-shift counterparts.[17] Rotating shifts over several weeks can result in desensitization to the circadian rhythms. With this desensitization comes a measurable loss in productivity, increased numbers of accidents, and reported subjective discomfort.[18] After returning to a predictable shift, workers regained their biological clock and circadian rhythm.

Not working the normal day-shift hours results in an increase in workplace stress, with rotating shifts being the most stressful. From a safety viewpoint, shift workers are subjected to more workplace stress in terms of weariness, irritability, depression, and a lack of interest in work. Shift work increases workplace stress and may lead to a less safe worker.

IMPROVING SAFETY BY REDUCING WORKPLACE STRESS

Not all sources of stress on the job can be eliminated, and employment screening is unlikely to identify all those who are sensitive to stress. People can learn to adapt to stress, however. **Training** can help people recognize and deal with stress effectively. Employees need to know what is expected of them at any given time and to receive recognition when it is deserved. Managers can reduce role ambiguity and stress caused by lack of feedback by providing frequent feedback.

DISCUSSION CASE

What Is Your Opinion?

"We are beginning to see more and more stress-related problems. I'm afraid that if we don't deal with the issue, stress is going to cause even more serious injuries." "Nonsense. We all have stress. There has always been stress on the job. It goes with the territory. A few employees go to a seminar on workplace stress and, all of a sudden, everyone is complaining about stress." This discussion took place between the Safety Director and the CEO of Gulf Coast Electric Company. Is stress really a legitimate workplace hazard? What is your opinion?

Stress can result from low participation or lack of **job autonomy**. A manager can help employees realize their full potential by helping them match their career goals with the company's goals and giving them more control over their jobs.

Managers can help design jobs in ways that lead to worker satisfaction, thereby lessening work stress. **Physical stress** can be reduced by improving the work environment and establishing a sound safety and health program. Managers can also assist in the effort to provide varied and independent work with good possibilities for contact and collaboration with fellow workers and for personal development.

Organizational approaches to coping with work stress include avoiding a monotonous, mechanically **controlled pace**, standardized motion patterns, and constant repetition of **short-cycle operations**. Other stress-inducing work design features to avoid include jobs that do not make use of a worker's knowledge and initiative, that lack human contact, and that have authoritarian-type supervision.

There are also several individual approaches to coping with stress. One of the most important factors in dealing with stress is learning to recognize its symptoms and taking them seriously. Handling stress effectively should be a lifelong activity that gets easier with practice. Keeping a positive mental attitude can help defuse some otherwise stressful situations.

People can analyze stress-producing situations and decide what is worth worrying about. Individuals can effectively respond to a stressful workload by **delegating responsibility** instead of carrying the entire load. Relaxation techniques can also help reduce the effects of stress. Some common relaxation methods include meditation, biofeedback, music, and exercise.

The Northwestern National Life Insurance Company recommends the following strategies for workplace stress reduction:[19]

- Management recognizes workplace stress and takes steps regularly to reduce this stress.
- Mental health benefits are provided in the employee's health insurance coverage.
- The employer has a formal **employee communications** program.

- Employees are given information on how to cope with stress.
- Workers have current, accurate, and clear **job descriptions**.
- Management and employees talk openly with one another.
- Employees are free to talk with each other during work.
- Employers offer exercise and other stress reduction classes.
- Employees are recognized and rewarded for their contributions.
- Work rules are published and are the same for everyone.
- Child-care programs are available.
- Employees can work **flexible hours**.
- Perks are granted fairly based on a person's level in the organization.
- Workers have the training and **technology access** that they need.
- Employers encourage work and personal support groups.
- Workers have a place and time to relax during the work day.
- Elder-care programs are available.
- Employees' work spaces are not crowded.
- Workers can put up personal items in their work areas.
- Management appreciates humor in the workplace.

In addition to the strategies recommended by the Northwestern National Life Insurance Company, several others have been identified. Writing for *Occupational Hazards*, S. L. Smith recommends the following ways to reduce stress in the workplace:[20]

- Match workload and pace to the training and abilities of employees.
- Make an effort to match work schedules with the personal lives of employees.
- Clearly define work roles.
- Before giving employees additional duties beyond their normal work roles, make sure they receive the necessary training.
- Promote teamwork among employees and encourage it throughout the organization.
- Involve employees in making decisions that affect them.
- Inform employees in a timely manner of organizational changes that might affect them.

There is no one clear answer to workplace stress. The suggestions given here are a good starting place for management and employees to begin the process of being aware of and dealing effectively with workplace stress.

STRESS IN SAFETY MANAGERS

Safety and health management can be a stressful profession. Robert Scherer of Wright State University identified the following four conditions that frequently trigger stress among modern safety and health professionals:[21]

1. Role overload

2. Coping with regulatory breakdown
3. Communication breakdown
4. Competing loyalties

According to Scherer, safety and health professionals are sometimes overloaded when corporate *downsizing* results and they are delegated more and more responsibilities. Trying to keep up with the ever-changing multitude of regulations is a stress-inducing challenge. Communication relating to safety and health is always a challenge. However, when economic forces focus an organization's attention on other matters, it can be even more difficult than usual to get the safety and health message across. This increased difficulty can lead to increased stress. Line managers who are more concerned with meeting production quotas than with the safety of their employees sometimes try to influence safety and health managers—their colleagues and sometimes their friends—to look the other way. This subjects safety and health managers to the pressures of competing loyalties.[22]

Scherer recommends that safety and health managers cope with these four common triggers of stress by applying the following strategies: (1) prioritize activities by focusing on those that present the most risk to the organization; (2) work closely with the organization's legal staff and subscribe to an on-line CD/ROM updating service; (3) formalize communication and hold regularly scheduled safety and health meetings for all operating employees; and (4) focus on the risks to the organization and refuse to take sides.[23]

STRESS AND WORKERS' COMPENSATION

There are serious problems with workers' compensation in the United States. On the one hand, there is evidence of abuse of the system. On the other hand, many injured workers who are legitimately collecting benefits suffer a substantial loss of income. Complaints

---

SAFETY MYTH

Managers Cannot Reduce Employee Stress

Of course, the workplace is stressful! This fact cannot be helped. Some people can handle stress, and some cannot. There is nothing managers can do. Right? Wrong. There is a lot that managers can do. Managers can provide training to help employees learn to recognize stress and deal with it. Managers can empower employees, giving them as much control as possible over their jobs. Managers can improve communication with employees. Managers can provide clear, understandable job descriptions that eliminate ambiguity. Managers can provide childcare or childcare assistance. Managers can provide exercise and stress reduction programs. There are many strategies that managers can employ to reduce stress on the job.

about workers' compensation are common from all parties involved with the system (employers, employees, and insurance companies).

In Chapter 5, the example of the overweight deputy sheriff who applied for benefits because he was distraught over the breakup of his extramarital love affair and a poor performance evaluation was cited as an example of abuse. This individual is just one of thousands who are claiming that job stress has disabled them to the point that workers' compensation is justified. In 1980, there were so few stress claims that they were not even recorded as a separate category. By 1990, they represented a major and costly category.

Stress claims are more burdensome than physical claims because they are typically reviewed in an adversarial environment. This leads to the involvement of expert medical witnesses and attorneys. As a result, even though the benefits awarded for stress-related injuries are typically less than those awarded for physical injuries, the cost of stress claims is often higher because of litigation.

SUMMARY

1. Stress is a pathological human response to psychological, social, occupational, or environmental stimuli.
2. Workplace stress involves a worker's feelings resulting from perceived differences between the demands of the job and the person's capacity to cope with these demands.
3. Sources of workplace stress include environmental conditions, work overload, role ambiguity, lack of feedback, personality, personal and family problems, and role conflict. Other sources of workplace stress are task complexity, lack of control over the job, public safety responsibility, job security, lack of psychological support, and environmental safety concerns.
4. A life insurance company poll of workers found that workplace stress could be caused by company reorganizations or buyout, layoffs, mandatory overtime, varying workloads, work pace, lack of opportunity for advancement, bureaucracy, shortages of staff money or technology, low pay, or rotating shifts.
5. Human reaction to workplace stress may be grouped into five categories: (a) subjective, (b) behavioral, (c) cognitive, (d) physiological, and (e) organizational.
6. Psychosomatic reaction to stress may eventually lead to autoimmune disease.
7. Until an individual's limit is reached, the effects of stress may be reversed. After that limit, with continuing stress, the effects can become pathological.
8. Research has shown three stages of human reaction to stress: (a) alarm, (b) resistance, and (c) exhaustion.
9. The best policy regarding stress is to learn to adapt to it. Efforts to rid the workplace of all sources of stress are unlikely to succeed.
10. Mental workload can be measured in three ways: (a) subjective ratings that are state-dependent, (b) behavioral time-sharing, and (c) psychophysiological techniques.
11. Psychosocial questionnaires study how workers feel about their jobs.
12. Shift work occurs when the majority of people are at leisure. The circadian rhythm, or biological clock within the body, determines when a person will be comfortable

either working or sleeping. Workers surveyed have reported lower job satisfaction with rotating shifts. Shift work may result in loss of productivity and an increased number of accidents.

13. All sources of stress on the job cannot be eliminated.

14. Managers can reduce workplace stress by reducing role ambiguity and increasing feedback and job autonomy. Managers can also lessen workers' stress by reducing exposure to physical hazards, varying the work pace, and eliminating monotonous and/or short-cycle operations.

15. Workplace stress reduction, according to a life insurance study, can be accomplished by providing employee mental health insurance benefits, improving employee–management communications, providing workers with information about how to deal with stress, providing job descriptions, talking with employees regularly, recognizing and rewarding contributions, having published work rules, and offering child-care and elder-care programs. Other stress reducers include permitting flexible work hours, granting perks fairly, giving adequate training and technology access, providing a place and time to relax, having uncrowded workplaces with space to put up personal items, and having a management that keeps a sense of humor.

16. Individuals can effectively respond to a stressful workload by delegating responsibility and learning how to relax. Relaxation methods include meditation, biofeedback, music, and exercise.

17. One of the most important factors in dealing with stress is learning to recognize its symptoms and taking the symptoms seriously.

===== KEY TERMS AND CONCEPTS =====

Behavioral time-sharing	Physical stress
Circadian rhythm	Psychological support
Control	Psychophysiological techniques
Controlled pace	Psychosocial questionnaires
Delegate responsibility	Role ambiguity
Employee communications	Shift work
Environmental safety	Short-cycle operations
Feedback	Stage of alarm
Feeling of responsibility	Stage of exhaustion
Flexible hours	Stage of resistance
Human reaction	State-dependent
Job autonomy	Stress
Job description	Subjective rating
Job security	Task complexity

Technology access   Workload demands

Training   Workplace stress

========== REVIEW QUESTIONS ==========

1. Define *stress*.
2. How is workplace stress different from general stress?
3. List five sources of workplace stress and give an on-the-job example for each source.
4. Explain why lack of job autonomy may cause workplace stress.
5. Give five categories of human reaction to workplace stress.
6. How are psychosomatic reactions to stress and actual physiological illness related?
7. How are some autoimmune diseases and workplace stress related?
8. Explain three stages of human reaction to stress.
9. Explain three ways in which mental workload can be measured.
10. Discuss efforts to rid the workplace of all causes of workplace stress.
11. What type of data do psychosocial questionnaires provide? Discuss the bias in this type of data.
12. Discuss how shift work causes workplace stress. Give suggestions for minimizing workplace stress from shift work.
13. Give specific steps that can be taken by managers to help reduce workplace stress.
14. Discuss at least five methods to reduce workplace stress, according to the life insurance company research.
15. Explain how individuals can reduce workplace stress.

========== ENDNOTES ==========

1. Fraser, T. M. *The Worker at Work* (New York: Taylor & Francis, 1989), p. 103.
2. Grandjean, E. *Fitting the Task to the Man, A Textbook of Occupational Ergonomics*, 4th ed. (New York: Taylor & Francis, 1988), pp. 175–176.
3. "Causes and Cures for Stress Still Misunderstood," *Occupational Health & Safety Letter*, November 28, 1990, Vol. 20, No. 24, p. 196.
4. Ibid.
5. Ibid.
6. Grandjean, E. *Fitting the Task to the Man*, pp. 176–177.
7. "Electronic Monitoring Causes Worker Stress," *Occupational Health & Safety Letter*, October 17, 1990, Vol. 20, No. 21, p. 168.
8. Cope, L. "Quiz Developed to Determine Workplace Stress," *Tallahassee Democrat*, June 2, 1991, pp. 1E, 3E.
9. "Computer Sound in Workplace Linked to Stress," *Occupational Health & Safety Letter*, August 22, 1990, Vol. 20, No. 17, p. 141.
10. Ibid.
11. Fraser, T. M. *The Worker at Work*, pp. 104, 110–111.

12. Konz, S. *Work Design: Industrial Ergonomics* (Worthington, OH: Publishing Horizons, 1990), pp. 347–390.
13. Fraser, T. M. *The Worker at Work*, pp. 111–115.
14. Grandjean, E. *Fitting the Task to the Man*, pp. 177–178.
15. Fraser, T. M. *The Worker at Work,* pp. 116–117.
16. Ibid.
17. Ibid.
18. Ibid.
19. Cope, L. "Quiz Developed to Determine Workplace Stress."
20. Smith, S. L. "Combating Stress," *Occupational Hazards*, March 1994, p. 57.
21. Scherer, R. As presented in the article, "Stress Homes in on Our Safety's Ranks," *Occupational Hazards*, January 1994, pp. 156–157.
22. Ibid., p. 157.
23. Ibid.

CHAPTER EIGHT

Mechanical Hazards and Machine Safeguarding

Mechanical hazards are those associated with power-driven machines, whether automated or manually operated. Concerns about **mechanical hazards** date back to the Industrial Revolution and the earliest days of mechanization. Machines driven by steam, hydraulic, and/or electric power introduced new hazards into the workplace. In spite of advances in safeguarding technologies and techniques, mechanical hazards are still a major concern today. In addition, automated machines have introduced new concerns.

COMMON MECHANICAL INJURIES

In an industrial setting, people interact with machines that are designed to drill, cut, shear, punch, chip, staple, stitch, abrade, shape, stamp, and slit such materials as metals, composites, plastics, and elastomers. If appropriate safeguards are not in place or if workers fail to follow safety precautions, these machines can apply the same procedures to humans. When this happens, the types of **mechanical injuries** that result are typically the result of cutting, tearing, shearing, crushing, breaking, straining, or puncturing (see Figure 8–1). Information about each of these hazards is provided in the following paragraphs.

Cutting and Tearing

A cut occurs when a body part comes in contact with a sharp edge. The human body's outer layer is arranged from the outside in as follows: *epidermis,* which is the tough outer covering of the skin; *dermis,* which constitutes the greatest part of the skin's thickness; *capillaries,* which are tiny blood vessels that branch off the small arteries and veins in the dermis; *veins,* which are blood vessels that collect blood from the capillaries and return it to the heart; and *arteries,* which are the larger vessels that carry blood from the heart to the capillaries in the skin. The seriousness of **cutting** or **tearing** the skin depends on how much damage is done to the skin, veins, arteries, muscles, and even bones.

Shearing

To understand what **shearing** is, think of a paper cutter. It shears the paper. Power-driven shears for severing paper, metal, plastic, elastomers, and composite materials are widely used in manufacturing. In times past, such machines often amputated fingers and hands. Such tragedies would typically occur when operators reached under the shearing blade to make an adjustment or to place materials there and activate the blade before fully removing their hand. Safeguards against shearing accidents are explained later in this chapter.

Figure 8-1
Some common mechanical hazards.

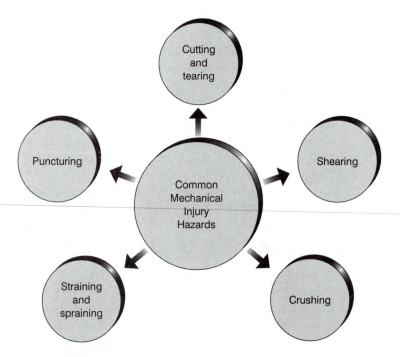

Crushing

Injuries from **crushing** can be particularly debilitating, painful, and difficult to heal. They occur when a part of the body is caught between two hard surfaces that progressively move together, thereby crushing anything between them. Crushing hazards can be divided into two categories: *squeeze-point* types and *run-in points*.

Squeeze-point hazards exist where two hard surfaces, at least one of which must be in motion, push close enough together to crush any object that may be between them. The process can be slow, as in a manually operated vice, or fast, as with a metal-stamping machine.

Run-in point hazards exist where two objects, at least one of which is rotating, come progressively closer together. Any gap between them need not become completely closed. It need only be smaller than the object or body part lodged in it. Meshing gears and belt pulleys are examples of run-in point hazards.

Body parts can also be crushed in other ways, for example, a heavy object falling on a foot or a hammer hitting a finger. However, these are impact hazards, which are covered in Chapter 9.

Breaking

Machines used to deform engineering materials in a variety of ways can also cause broken bones. A **break** in a bone is known as a fracture. Fractures are classified as simple, compound, complete, and incomplete.

A simple fracture is a break in a bone that does not pierce the skin. A compound fracture is a break that has broken through the surrounding tissue and skin. A complete fracture divides the affected bone into two or more separate pieces. An incomplete fracture leaves the affected bone in one piece but cracked.

Fractures are also classified as transverse, oblique, and comminuted. A transverse fracture is a break straight across the bone. An oblique fracture is diagonal. A comminuted fracture exists when the bone is broken into a number of small pieces at the point of fracture.

Straining and Spraining

There are numerous situations in an industrial setting when **straining** of muscles or **spraining** of ligaments is possible. A strain results when muscles are overstretched or torn. A sprain is the result of torn ligaments in a joint. Strains and sprains can cause swelling and intense pain.

Puncturing

Punching machines that have sharp tools can puncture a body part if safety precautions are not observed or if appropriate safeguards are not in place. **Puncturing** results when an object penetrates straight into the body and pulls straight out, creating a wound in the shape of the penetrating object. The greatest hazard with puncture wounds is the potential for damage to internal organs.

SAFEGUARDING DEFINED

All of the hazards explained in the previous section can be reduced by the application of appropriate safeguards. C.F.R. 1910 Subpart O contains the OSHA standards for machinery and machine guarding (1910.211–1910.222). The National Safety Council defines **safeguarding** as follows:

> . . . machine safeguarding is to minimize the risk of accidents of machine–operator contact. The contact can be:
>
> 1. An individual making the contact with the machine—usually the moving part—because of inattention caused by fatigue, distraction, curiosity, or deliberate chance taking;
> 2. From the machine via flying metal chips, chemical and hot metal splashes, and circular saw kickbacks, to name a few;
> 3. Caused by the direct result of a machine malfunction, including mechanical and electrical failure.[1]

Safeguards can be broadly categorized as point-of-operation guards, point-of-operation devices, and feeding/ejection methods. The various types of safeguards in these categories are explained later in this chapter.

RISK ASSESSMENT[2]

Risk assessment is the process of quantifying the level of risk associated with the operation of a given machine. It should be a structured and systematic process that answers the following four specific questions:

1. How *severe* are potential injuries?
2. How *frequently* are employees exposed to the potential hazards?
3. What is the *possibility* of avoiding the hazard if it does occur?
4. What is the *likelihood* of an injury should a safety control system fail?

The most widely used risk assessment technique is the decision tree, coupled with codes representing these four questions and defined levels of risk. Figure 8-2 is an example of a risk assessment decision tree. In this example, the codes and their associated levels of risk are as follows:

S = Severity

Question 1: Severity of potential injuries

S1	Slight injury (bruise, abrasion)
S2	Severe injury (amputation or death)

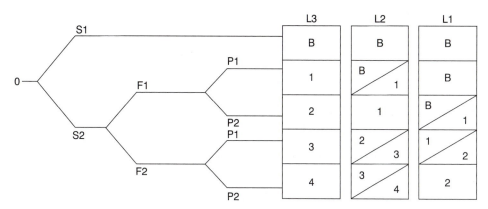

Figure 8-2
Risk-assessment decision tree.

F = Frequency

Question 2: Frequency of exposure to potential hazards
F1   Infrequent exposure
F2   From frequent to continuous exposure

P = Possibility

Question 3: Possibility of avoiding the hazard if it does occur
P1   Possible
P2   Less possible to not impossible

L = Likelihood

Question 4: Likelihood that the hazard will occur
L1   Highly unlikely
L2   Unlikely
L3   Highly likely

Risk Levels

Associated risk factors ranging from lowest (B) to highest (4)
By applying the decision tree in Figure 8-2 or a similar device, the risk associated with the operation of a given machine can be quantified. This allows safety personnel to assign logical priorities for machine safeguarding and hazard prevention.

REQUIREMENTS FOR ALL SAFEGUARDS

The various machine motions present in modern industry involve mechanisms that rotate, reciprocate, or do both. This equipment includes tools, bits, chucks, blades, spokes, screws, gears, shafts, belts, and a variety of different types of stock. Safeguards can be devised to protect workers from harmful contact with such mechanisms while at the same time allowing work to progress at a productive rate. The National Safety Council has established the following requirements for safeguards.[3]

1. *Prevent contact.* Safeguards should prevent human contact with any potentially harmful machine part. The prevention extends to machine operators and any other person who might come in contact with the hazard.

2. *Be secure and durable.* Safeguards should be attached so that they are secure. This means that workers cannot render them ineffective by tampering with or disabling them. This is critical because removing safeguards in an attempt to speed production is a common practice. Safeguards must also be durable enough to withstand the rigors of the workplace. Worn-out safeguards won't protect workers properly.

3. *Protect against falling objects.* Objects falling onto moving machine mechanisms increase the risk of accidents, property damage, and injury. Objects that fall on a moving part can be quickly hurled out, creating a dangerous projectile. Therefore, safeguards must do more than just prevent human contact. They must also shield the moving parts of machines from falling objects.

4. *Create no new hazard.* Safeguards should overcome the hazards in question without creating new ones. For example, a safeguard with a sharp edge, unfinished surface, or protruding bolts introduces new hazards while protecting against the old.

5. *Create no interference.* Safeguards can interfere with the progress of work if they are not properly designed. Such safeguards are likely to be disregarded or disabled by workers feeling the pressure of production deadlines.

6. *Allow safe maintenance.* Safeguards should be designed to allow the more frequently performed maintenance tasks (e.g., lubrication) to be accomplished without the removal of guards. For example, locating the oil reservoir outside the guard with a line running to the lubrication point will allow for daily maintenance without removing the guard.

Design and construction of safeguards are highly specialized activities requiring a strong working knowledge of machines, production techniques, and safety. However, it is critical that all of the factors explained in this section be considered and accommodated during the design process.

POINT-OF-OPERATION GUARDS[4]

Guards are most effective when used at the point of operation, which is where hazards to humans exist. Point-of-operation hazards are those caused by the shearing, cutting, or bending motions of a machine. Pinch-point hazards result from guiding material into a machine or transferring motion (e.g., from gears, pressure rollers, or chains and sprockets). Single-purpose safeguards, since they guard against only one hazard, typically are permanently fixed and nonadjustable. Multiple-purpose safeguards, since they guard against more than one hazard, typically are adjustable.

Point-of-operation guards are of three types, each with its own advantages and limitations: fixed, interlocked, and adjustable (Figure 8–3).

- **Fixed guards** provide a permanent barrier between workers and the point of operation. They offer the following advantages: They are suitable for many specific applications, can be constructed in-plant, require little maintenance and are suitable for high-production, repetitive operations. Limitations include the following: They sometimes limit visibility, are often limited to specific operations, and sometimes inhibit normal cleaning and maintenance.

- **Interlocked guards** shut down the machine when the guard is not securely in place or is disengaged. The main advantage of this type of guard is that it allows safe access to the machine for removing jams or conducting routine maintenance without the need for taking off the guard. There are also limitations. Interlocked guards require careful adjustment and maintenance and, in some cases, can be easily disengaged.

- **Adjustable guards** provide a barrier against a variety of different hazards associated with different production operations. They have the advantage of flexibility. However, they do not provide as dependable a barrier as other guards do, and they require frequent maintenance and careful adjustment.

Figures 8–4 and 8–5 show point-of-operation guards used in modern manufacturing settings.

POINT-OF-OPERATION DEVICES[5]

A number of different **point-of-operation devices** can be used to protect workers. The most widely used are explained in the following paragraphs.

Figure 8–3
Point-of-operation guards.

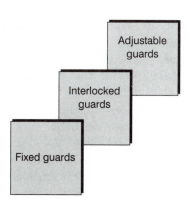

Figure 8–4
Series 12 PRO-TECH-TOR
GATE GUARD used on an
open-back power press.
Courtesy of PROTECH SYSTEMS.

- **Photoelectric devices** are optical devices that shut down the machine whenever the light field is broken. These devices allow operators relatively free movement. They do have limitations, including the following: They do not protect against mechanical failure, they require frequent calibration, and they can be used only with machines that can be stopped.

- **Radio-frequency devices** are capacitance devices that brake the machine if the capacitance field is interrupted by a worker's body or another object. These devices have the same limitations as photoelectric devices.

- **Electromechanical devices** are contact bars that allow only a specified amount of movement between the worker and the hazard. If the worker moves the contact bar beyond the specified point, the machine will not cycle. These devices have the limitation of requiring frequent maintenance and careful adjustment.

- **Pullback devices** pull the operator's hands out of the danger zone when the machine starts to cycle. These devices eliminate the need for auxiliary barriers. However, they also have limitations. They limit operator movement, must be adjusted for each individual operator, and require close supervision to ensure proper use.

- **Restraint devices** hold the operator back from the danger zone. They work well, with little risk of mechanical failure. However, they do limit the operator's movement, must be adjusted for each individual operator, and require close supervision to ensure proper use.

■ **Safety trip devices** include trip wires, trip rods, and body bars. All of these devices stop the machine when tripped. They have the advantage of simplicity. However, they are limited in that all controls must be activated manually. They protect only the operator and may require the machine to be fitted with special fixtures for holding work.

■ **Two-hand controls** require the operator to use both hands concurrently to activate the machine (e.g., a paper-cutter or metal-shearing machine). This ensures that hands cannot stray into the danger zone. Although these controls do an excellent job of protecting the operator, they do not protect onlookers or passers-by. In addition, some two-hand controls can be tampered with and made operable using only one hand.

■ **Gates** provide a barrier between the danger zone and workers. Although they are effective at protecting operators from machine hazards, they can obscure the work, making it difficult for the operator to see.

Figure 8–5
Series 17 CHECKMATE
RIVET GUARD used on a
foot-operated riveting
machine.
Courtesy of PROTECH SYSTEMS.

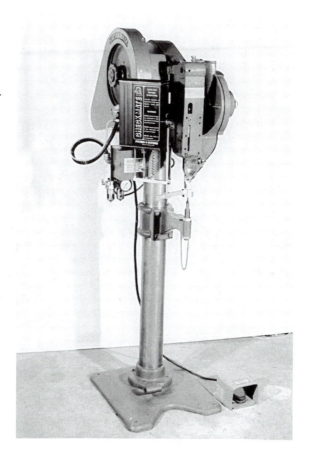

Figure 8–6 is a point-of-operation device that shuts down the machine when an operator's hand breaks a light beam. How quickly will the machine shut down? The stopping distance equation is as follows:

$$SD = \text{Stoptime (seconds} \times \text{hand speed constant (63''/second)}$$

FEEDING AND EJECTION SYSTEMS[6]

Feeding and ejection systems can be effective safeguards if properly designed and used. The various types of feeding and ejection systems available for use with modern industrial machines are summarized in the following paragraphs.

- **Automatic feed** systems feed stock to the machine from rolls. Automatic feeds eliminate the need for operators to enter the danger zone. Such systems are limited in the types and variations of stock that they can feed. They also typically require an auxiliary barrier guard and frequent maintenance.
- **Semiautomatic feed** systems use a variety of approaches for feeding stock to the machine. Prominent among these are chutes, moveable dies, dial feeds, plungers,

Figure 8–6
Series 25 EAGLE EYE INFRA-RED LIGHT BARRIER. A point-of-operation guarding system on a roller press machine.
Courtesy of PROTECH SYSTEMS.

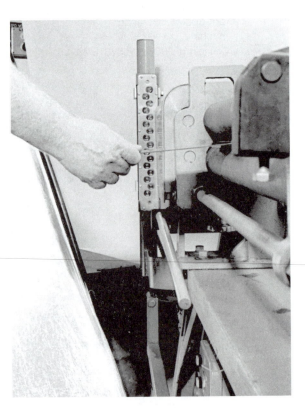

and sliding bolsters. They have the same advantages and limitations as automatic feed systems.

- ■ **Automatic ejection** systems eject the work pneumatically or mechanically. The advantage of either approach is that operators don't have to reach into the danger zone to retrieve workpieces. However, these systems are restricted to use with relatively small stock. Potential hazards include blown chips or debris and noise. Pneumatic ejectors can be quite loud.
- ■ **Semiautomatic ejection** systems eject the work using mechanisms that are activated by the operator. Consequently, the operator does not have to reach into the danger zone to retrieve workpieces. These systems do require auxiliary barriers and can be used with a limited variety of stock.

ROBOT SAFEGUARDS

Robots have become commonplace in modern industry. The safety and health concerns relating to robots are covered in Chapter 22. Only the guarding aspects of robot safety are covered in this section. The main hazards associated with robots are (1) entrapment of a worker between a robot and a solid surface, (2) impact with a moving robot arm, and (3) impact with objects ejected or dropped by the robot.

The best guard against these hazards is to erect a physical barrier around the entire perimeter of a robot's **work envelope** (the three-dimensional area established by the robot's full range of motion). This physical barrier should be able to withstand the force of the heaviest object that a robot could eject.

Various types of shutdown guards can also be used. A guard containing a sensing device that automatically shuts down the robot if any person or object enters its work envelope can be effective. Another approach is to put sensitized doors or gates in the perimeter barrier that automatically shut down the robot as soon as they are opened.

These types of safeguards are especially important because robots can be deceptive. A robot that is not moving at the moment may simply be at a stage between cycles. Without warning, it might make sudden and rapid movements that could endanger any person inside the work envelope.

LOCKOUT/TAGOUT SYSTEMS

One of the most effective safeguarding approaches in use today is the **lockout/tagout system**. This method was especially designed to protect against the unexpected startup of a machine that is supposed to be turned off. This is important because OSHA statistics show that six percent of all workplace fatalities are caused by the unexpected activation of machines while they are being serviced, cleaned, or otherwise maintained.[7] OSHA's lockout/tagout standard is C.F.R. 1910.147 and 1910.331.

In a lockout system, a padlock is placed through a gate covering the activating mechanism or is applied in some other manner to prevent a machine from being turned on until the lock is removed. The lock usually has a label that gives the name, depart-

---

DISCUSSION CASE

What Is Your Opinion?

John Martin, Director of Manufacturing, is not happy with his colleague Pete Chang who is director of safety and health at Robbins Engineering Corporation. Chang has ordered two machines in the Manufacturing Department *tagged out* until point-of-operation guards are replaced. The machining supervisor, in an attempt to increase output, had his machinists remove the guards. As a result, there have been several minor injuries and a couple of more serious near misses. The issue is short-term productivity vs. the safety and health of employees. What is your opinion on this issue?

---

ment, and telephone extension of the person who put it on. It may also carry a message such as the following: *"This lock is to be removed only by _____ ."* The name in the blank is the person who applied the lock (see Figure 8–7).

A tagout system is exactly like a lockout system except a tag is substituted for the lock. Tags should be used only in cases where a lock is not feasible. Figure 8–8 is an example of a tag that might be used in a modern industrial setting. Sometimes, tags and locks may be used together.

Figure 8–7
Lockout system.

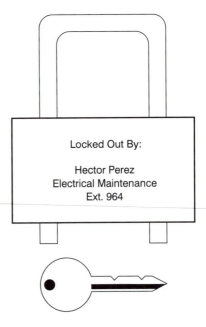

Locked Out By:

Hector Perez
Electrical Maintenance
Ext. 964

Figure 8–8
Tagout system.

The following examples demonstrate why lockout/tagout systems are so important: (1) An employee is cleaning the guarded side of an operating granite saw when he is pulled into it and killed; (2) an employee is cleaning scrap from under a shear when another employee hits the activation button and decapitates him; and (3) an employee is cleaning paper from a waste crusher when he falls into it and is crushed.[8]

Accidents such as these were the driving force behind the development of OSHA Standard 29 C.F.R. 1910.147 and its follow-on Instruction STD 1–7.3. The standard covers all general industry in the United States and requires the use of lockout or tagout systems to protect against the unexpected activation of machines and equipment. Instruction STD 1–7.3 clarifies the standard to require the use of lockout systems wherever possible. Tagout systems are to be used only in those instances where lockout systems are not feasible. "Instruction STD 1–7.3 states that when an employer uses tags instead of locks, they must demonstrate that the tagout program will provide a level of safety equivalent to a lockout program."[9] It is estimated by OSHA that full compliance with the lockout/tagout standard will prevent 120 accidental deaths, 29,000 serious injuries, and 32,000 minor injuries every year.[10]

OSHA identified the following four factors that are essential for effective lockout/tagout systems:

1. *Attention to detail.* It is important to have procedures covering all steps and allowing for all contingencies.
2. *Extensive training.* Employers must teach workers about lockout/tagout procedures and why it is important to use them.
3. *Reinforcement of training.* Employers must reinforce the training provided by stressing the importance of safety principles and practices.
4. *Disciplinary action.* Violators of lockout/tagout procedures and other general safety principles and practices must be disciplined.[11]

GENERAL PRECAUTIONS

The types of safeguards explained in this chapter are critical. In addition to these specific safeguards, there are also a number of general precautions that apply across the board in settings where machines are used. Some of the more important general precautions are as follows:

- All operators should be trained in the safe operation and maintenance of their machines.
- All machine operators should be trained in the emergency procedures to take when accidents occur.
- All employees should know how to activate emergency shutdown controls. This means knowing where the controls are and how to activate them.
- Inspection, maintenance, adjustment, repair, and calibration of safeguards should be carried out regularly.
- Supervisors should ensure that safeguards are properly in place when machines are in use. Employees who disable or remove safeguards should be disciplined appropriately.
- Operator teams (two or more operators) of the same system should be trained in coordination techniques and proper use of devices that prevent premature activation by a team member.
- Operators should be trained and supervised to ensure that they dress properly for the job. Long hair, loose clothing, neckties, rings, watches, necklaces, chains, and earrings can become caught in equipment and, in turn, pull the employee into the hazard zone.
- Shortcuts that violate safety principles and practices should be avoided. The pressures of deadlines should never be the cause of unsafe work practices.
- Other employees who work around machines but do not operate them should be made aware of the emergency procedures to take when an accident occurs.

BASIC PROGRAM CONTENT

Machine safeguarding should be organized, systematic, and comprehensive. A company's safeguarding program should have at least the following elements:

- Safeguarding policy that is part of a broader company-wide safety and health policy.
- Machine/hazard analysis
- Lockout/tagout (materials and procedures)
- Employee training
- Comprehensive documentation
- Periodic safeguarding audits (at least annually)

Problem	Action
Machine operating without the safety guard.	Stop machine immediately and activate the safety guard.
Maintenance worker cleaning a machine that is operating.	Stop machine immediately and lock or tag it out.
Visitor to the shop is wearing a necktie as he observes a lathe in operation.	Immediately pull the visitor back and have him remove the tie.
An operator is observed disabling a guard.	Stop the operator, secure the guard, and take disciplinary action.
A robot is operating without a protective barrier.	Stop the robot and erect a barrier immediately.
A machine guard has a sharp, ragged edge.	Stop the machine and eliminate the sharp edge and ragged burrs by rounding it off.

Figure 8–9
Selected examples of problems and corresponding actions.

TAKING CORRECTIVE ACTION

What should be done when a mechanical hazard is observed? The only acceptable answer to this question is, take *immediate corrective action*. The specific action indicated will depend on what the problem is. Figure 8–9 shows selected examples of problems and corresponding corrective actions.

These are only a few of the many different types of problems that require corresponding corrective action. Regardless of the type of problem, the key to responding is immediacy. As shown in the examples given earlier in this chapter, waiting to take corrective action can be fatal.

=========== SUMMARY ===========

1. The most common mechanical injuries are cutting and tearing, shearing, crushing, breaking, straining, spraining, and puncturing.
2. Safeguarding involves devices or methods that minimize the risk of accidents resulting from machine–operator contact. The contact might result from an individual making contact with a machine, flying metal chips, chemical or hot metal splashes, stock kickbacks, or mechanical malfunction.
3. All safeguards should have the following characteristics: prevent contact, be secure and durable, protect against falling objects, create no new hazards, create no interference, and allow safe maintenance.

4. Point-of-operation devices come in a variety of different types including the following: photoelectric, radio frequency, electromechanical, pullback, restraint, safety trip, two-hand controls, trips, and gates.

5. Feeding and ejection systems can be effective safeguards if properly designed and used. These systems come in two types: automatic and semiautomatic.

6. The main mechanical hazards associated with robots are as follows: (1) entrapment of a worker between a robot and a solid surface, (2) impact with a moving robot arm, and (3) impact with objects ejected or dropped by the robot. The best safeguard for a robot is a barrier around the perimeter of its work envelope. Sensitized doors or gates in the barrier can also decrease the hazard potential.

7. A lockout system is a padlock placed to prevent a machine from being accidentally turned on. A tagout system uses a tag to tell workers that a machine should not be activated until the tag is removed by the person who put it on. OSHA Standard 29 C.F.R. 1910.147 requires the use of lockout/tagout systems. Instruction STD 1–7.3 clarifies the standard.

8. OSHA identified the following four factors as essential in order for lockout/tagout systems to be effective: (1) attention to detail, (2) extensive training, (3) reinforcement of training, and (4) disciplinary action.

9. When hazards or hazardous behavior is observed, corrective action should be taken immediately. Waiting to act can be fatal.

KEY TERMS AND CONCEPTS

Adjustable guards

Automatic ejection

Automatic feed

Breaking

Crushing

Cutting

Electromechanical devices

Fixed guards

Gates

Interlocked guards

Lockout/Tagout systems

Mechanical hazards

Mechanical injuries

Photoelectric devices

Point-of-operation devices

Point-of-operation guards

Pullback devices

Puncturing

Radio-frequency devices

Restraint devices

Risk assessment

Safeguarding

Safety trip devices

Semiautomatic ejection

Semiautomatic feed

Shearing

Spraining

Straining

Tearing

Two-hand controls

Work envelope

REVIEW QUESTIONS

1. List and briefly explain the common types of mechanical injury hazards.
2. Explain the concept of safeguarding.
3. What are the requirements all safeguards should meet?
4. Describe the three types of point-of-operation guards.
5. Describe four types of point-of-operation devices.
6. What are the relative advantages and disadvantages of feeding and ejection systems?
7. Describe the primary hazards associated with robots.
8. Explain how to guard against the hazards associated with robots.
9. What is a lockout system?
10. What is a tagout system?
11. What impact might a lockout/tagout system have if carefully followed nationwide?
12. List the four factors OSHA feels are essential in order for lockout/tagout systems to be effective.
13. Explain the concept of risk assessment.

ENDNOTES

1. National Safety Council. *Guards: Safeguarding Concepts Illustrated*, 5th ed. (Chicago: National Safety Council, 1987), p. 1.
2. EN 954, Part I, "Safety of Machinery—Principles of Safety Related to Control Systems." *European Union,* 1997.
3. National Safety Council. *Guards: Safeguarding Concepts Illustrated*, pp. 2–3.
4. Ibid., p. 36.
5. Ibid., pp. 38–39.
6. Ibid., p. 44.
7. Caruey, A. "Lock Out the Chance for Injury," *Safety & Health,* May 1991, Vol. 143, No. 5, p. 46.
8. Ibid.
9. Ibid.
10. Ibid.
11. Ibid.

CHAPTER NINE

Falling, Impact, Acceleration, and Lifting Hazards

MAJOR TOPICS

- Causes of Falls
- Kinds of Falls
- Walking and Slipping
- Slip and Fall Prevention Programs
- OSHA Fall Protection Standards
- Ladder Safety
- Impact and Acceleration Hazards
- Lifting Hazards
- Standing Hazards
- Forklift Safety

Some of the most common accidents in the workplace happen as the result of slipping, falling, and improper lifting. Impact from a falling object is also a common cause of accidents. This chapter provides the information needed by modern safety and health professionals to prevent such accidents.

CAUSES OF FALLS

In a typical year, more than 10,000 workers will lose their lives in falls. More than 16 percent of all disabling work-related injuries are the result of falls.[1] Clearly, falls are a major concern of safety and health professionals. According to Kohr, the primary causes of falls are

1. A foreign object on the walking surface
2. A design flaw in the walking surface
3. Slippery surfaces
4. An individual's impaired physical condition[2]

A **foreign object** is any object that is out of place or in a position to trip someone or to cause a slip. There is an almost limitless number of **design flaws** that might cause a fall. A poorly designed floor covering, a ladder that does not seat properly, or a catwalk that gives way are all examples of design flaws that might cause falls. **Slippery surfaces** are particularly prevalent in industrial plants where numerous different lubricants and cleaning solvents are used.

Automobile accidents are often caused when a driver's attention is temporarily drawn away from the road by a visual distraction. This is also true in the workplace. Anything that distracts workers visually can cause a fall. When a person's physical condition is impaired for any reason, the potential for falls increases. This is a particularly common problem among aging workers. Understanding these causes is the first step in developing fall prevention techniques.

KINDS OF FALLS

Falling from ladders and other elevated situations is covered later in this chapter. This section deals with the more common surface falls. According to Miller, such falls can be divided into the following four categories:

- **Trip and fall** accidents occur when workers encounter an unseen foreign object in their path. When the employees' foot strikes the object, he or she trips and falls.
- **Stump and fall** accidents occur when a worker's foot suddenly meets a sticky surface or a defect in the walking surface. Expecting to continue at the established pace, the worker falls when his or her foot is unable to respond properly.
- **Step and fall** accidents occur when a person's foot encounters an unexpected step down (e.g., a hole in the floor or a floorboard that gives way). This can also happen when an employee thinks he or she has reached the bottom of the stairs when, in reality, there is one more step.
- **Slip and fall** accidents occur when the worker's center of gravity is suddenly thrown out of balance (e.g., an oily spot causes a foot to shoot out from under the worker).[3]

Miller summarizes his finding relating to surface falls:

> The most common is the slip and fall accident. Foot contact is broken, and the individual attempts to right himself. A recovery of equilibrium is reflexive and not under conscious control in most cases. If the pedestrian strikes the walking surface with a fleshy part of the body, the injuries are likely to be minimal. If, on the other hand, the victim strikes a bony body part in the fall, the injuries are likely to be more severe.[4]

WALKING AND SLIPPING

Judging by the number of injuries that occur each year as the result of slipping, it is clear that walking can be hazardous to a worker's health. This is, in fact, the case when walking on an unstable platform. Goldsmith defines a stable platform for walking as "a walking surface with a high degree of traction, free from obstructions and therefore safe."[5] It follows that an unstable platform is one lacking traction, or one on which there are obstructions, or both.

Measuring Surface Traction

In order to understand **surface traction**, one must have a basis for comparison. An effective way for comparing the relative traction of a given surface is to use the **coefficient of friction**. Goldsmith describes the coefficient of friction as "a numerical correlation of the resistance of one surface (for example, a shoe) against another surface (the floor)."[6]

Figure 9–1 is a continuum showing coefficient of friction ratings from very slippery to good traction.[7] Surfaces with a coefficient of friction of 0.2 or less are very slippery and very hazardous. At the other end of the continuum, surfaces with a coefficient of friction of 0.4 or higher have good traction.

Goldsmith lists a number of different common surfaces along with their respective coefficients of friction with leather-soled shoes. To gain a feel for what different coefficients actually mean, consider the following: (1) ice has a coefficient of friction of 0.10; (2) concrete has a coefficient of 0.43; (3) linoleum has a coefficient of 0.33; and (4) waxed white oak has a coefficient of 0.24. These numbers show that ice is very slippery and very hazardous, concrete has good traction, linoleum is slippery but not dangerously so, and waxed white oak is slippery and hazardous.[8]

Factors That Decrease Traction

The coefficients of friction presented in the previous section all decrease, meaning that the hazard potential increases when the surface in question is wet. For example, the coefficient of concrete drops from 0.43 to 0.37, which moves it down one category on the traction continuum.[9]

Consequently, **good housekeeping** can be a major factor in reducing slip and fall hazards. Water, oil, soap, coolant, and cleaning solvents left on a floor can decrease traction and turn an otherwise safe surface into a danger zone. Rubber-soled shoes can decrease slipping hazards somewhat, but changing the type of shoe is not enough to ensure safety. Additional precautions are needed.

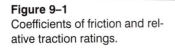

Figure 9–1
Coefficients of friction and relative traction ratings.

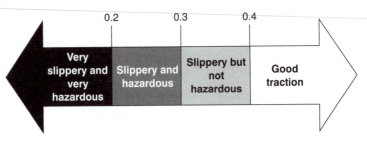

Strategies for Preventing Slips

Modern safety and health professionals are concerned with preventing slips and falls. Slip prevention should be a part of the company's larger safety and health program. Here are some strategies that can be used to help prevent slipping:

1. *Choose the right material from the outset.* Where the walking surface is to be newly constructed or an existing surface is to be replaced, safety and health professionals should encourage the selection of surface materials that have the highest possible coefficient of friction. Getting it right from the start is the best way to prevent slipping accidents.

2. *Retrofit an existing surface.* If it is too disruptive or too expensive to replace a slippery surface completely, **retrofit** it with friction enhancement devices or materials. Such devices/materials include runners, skid strips, carpet, grooves, abrasive coatings, grills, and textured coverings.

3. *Practice good housekeeping.* Regardless of the type of surface, keep it clean and dry. Spilled water, grease, oil, solvents, and other liquids should be removed immediately. When the surface is wet intentionally, as when cleaning or mopping, rope off the area and erect warning signs.

4. *Require nonskid footwear.* Employees who work in areas where slipping is likely to be a problem should be required to wear shoes with special nonskid soles. This is no different from requiring steel-toed boots to protect against falling objects. **Nonskid footwear** should be a normal part of a worker's personal protective equipment.

5. *Inspect surfaces frequently.* Employees who are working to meet production deadlines may be so distracted that they don't notice a wet surface, or they may notice it but feel too rushed to do anything about it. Consequently, safety and health professionals should conduct frequent inspections and act immediately when a hazard is identified.[10]

SLIP AND FALL PREVENTION PROGRAMS[11]

A company's overall safety and health program should include a slip and fall prevention component. According to Kohr, such a component should have the following elements:

1. *A policy statement/commitment.* Statement to convey management's commitment. Areas that should be included in the policy statement are management's intent, scope of activity, responsibility, accountability, the safety professional's role, authority, and standards.

2. *Review and acceptance of walkways.* Contain the criteria that will be used for reviewing all walking surfaces and determining if they are acceptable. For example, a criterion might be a minimum coefficient of friction value. Regardless of the criteria, the methodology that will be used for applying them to the **review and acceptance** of walkways should also be explained.

3. *Reconditioning and retrofitting.* Include recommendations and timetables for **reconditioning** or retrofitting existing walking surfaces that do not meet review and acceptance criteria.

4. *Maintenance standards and procedures.* State the maintenance standards for walking surfaces (e.g., How often should surfaces be cleaned, resurfaced, replaced, etc.?). In addition, this section should contain procedures for meeting the standards.

5. *Inspection, audits, tests, and records.* Provide a comprehensive list of inspections, audits, and tests (including the types of tests) that will be done, how frequently, and where. Give records of the results.

6. *Employee footwear program.* Specify the type of footwear required of employees who work on different types of walking surfaces.

7. *Defense methods for legal claims.* Outline the company's legal defenses so that aggressive action can be taken immediately should a lawsuit be filed against the company. In such cases, it is important to be able to show that the company has not been negligent (e.g., the company has a slip and fall prevention program that is in effect).

8. *Measurement of results.* Contain the following two parts: (1) an explanation of how the program will be evaluated and how often (e.g., comparison of yearly, quarterly, or monthly slip and fall data); (2) records of the results of these evaluations.

OSHA FALL PROTECTION STANDARDS

The OSHAct mentions fall protection in several places. Although the General Industry Standards are silent on fall protection, the problem is covered in the following subparts:

Subpart D	Walking/working surfaces
Subpart F	Powered platforms, manlifts, and vehicle-mounted work platforms
Subpart R	Special industries

In addition to these OSHA standards, the American National Standards Institute (ANSI) publishes a Fall Protection Standard (ANSI Z359.1: *Safety Requirements for Personal Fall Arrest Systems, Subsystems, and Components*). The most comprehensive and most controversial fall protection standard is **OSHA's Fall Protection Standard** for the construction industry (Subpart M of 29 C.F.R. 1926).

OSHA's Fall Protection Standard for Construction

OSHA's current Fall Protection Standard sets the *trigger height* at six feet. This means that any construction employee working higher than six feet off the ground must use a fall protection device such as a safety harness and line.

This trigger height means that virtually every small residential builder and roofing contractor is subject to the standard. Because most residential builders and roofing contractors are small and fall protection equipment is expensive, Subpart M of 29 C.F.R. 1926 is a source of much controversy.

OSHA officials argue that the six-foot trigger height saves up to 80 lives per year and prevents more than 56,000 injuries. The rationale is that six percent of all lost-time fall injuries in the construction industry are caused by falls from less than 10 feet. Opponents counter that the cost of complying with the standard is almost $300 million annually. Commercial contractors, whose employees typically work much higher than the 6- to 16-foot range are not concerned about the height controversy. Consequently, OSHA is under intense pressure to waive the six-foot trigger height for residential builders and roofers.

LADDER SAFETY

Jobs that involve the use of ladders introduce their own set of safety problems, one of which is an increased potential for falls. The National Safety Council recommends that ladders be inspected before every use and that employees who use them follow a set of standard rules.[12]

Inspecting Ladders

Taking a few moments to look over a ladder carefully before using it can prevent a fall. The National Safety Council recommends the following when inspecting a ladder:

- See if the ladder has the manufacturer's instruction label on it.
- Determine whether the ladder is strong enough.
- Read the label specifications about weight capacity and applications.
- Look for the following conditions: cracks on side rails; loose rungs, rails, or braces; damaged connections between rungs and rails.
- Check for heat damage and corrosion.
- Check wooden ladders for moisture that might cause them to conduct electricity.
- Check metal ladders for burrs and sharp edges.
- Check fiberglass ladders for signs of **blooming** deterioration of exposed fiberglass).[13]

Dos and Don'ts of Ladder Use

Many accidents involving ladders result from improper use. Following a simple set of rules for the proper use of ladders can reduce the risk of falls and other ladder-related accidents. The National Safety Council recommends the following dos and don'ts of ladder use:

- Check for slipperiness on shoes and ladder rungs.
- Limit a ladder to one person at a time.
- Secure the ladder firmly at the top and bottom.
- Set the ladder's base on a firm, level surface.
- Apply the **four-to-one ratio** (base one foot away from the wall for every four feet between the base and the support point).

- Face the ladder when climbing up or down.
- Barricade the base of the ladder when working near an entrance.
- Don't lean a ladder against a fragile, slippery, or unstable surface.
- Don't lean too far to either side while working (stop and move the ladder).
- Don't rig a makeshift ladder; use the real thing.
- Don't allow more than one person at a time on a ladder.
- Don't allow your waist to go any higher than the last rung when reaching upward on a ladder.
- Don't carry tools in your hands while climbing a ladder.
- Don't place a ladder on a box, table, or bench to make it reach higher.[14]

OSHA standards for walking and working surfaces and **ladder safety** are set forth in 29 C.F.R. Part 1910 (Subpart D). The standards contained in Subpart D are as follows:

1910.21	Definitions
1910.22	General requirements
1910.23	Guarding floor and wall openings and holes
1910.24	Fixed industrial stairs
1910.25	Portable wood ladders
1910.26	Portable metal ladders
1910.27	Fixed ladders
1910.28	Safety requirements for scaffolding
1910.29	Manually propelled mobile ladder stands and scaffolds (towers)
1910.30	Other working surfaces
1910.31	Sources of standards
1910.32	Standards organizations

IMPACT AND ACCELERATION HAZARDS

An employee working on a catwalk drops a wrench. The falling wrench accelerates over the 20-foot drop and strikes an employee below. Had the victim not been wearing a hard hat he might have sustained serious injuries from the impact. A robot loses its grip on a part, slinging it across the plant and striking an employee. The impact from the part breaks one of the employee's ribs. These are examples of accidents involving **acceleration** and **impact**. So is any type of fall since, having fallen, a person's rate of fall accelerates until striking a surface (impact). Motor vehicle accidents are also acceleration/impact instances.

Since falls were covered in the previous section, this section will focus on hazards relating to the acceleration and impact of objects. Approximately 25 percent of the workplace accidents that occur each year as the result of acceleration and impact involve objects that become projectiles.

Protection from Falling/Accelerating Objects

Objects that fall, are slung from a machine, or otherwise become projectiles pose a serious hazard to the heads, faces, feet, and eyes of workers. Consequently, protecting workers from projectiles requires the use of appropriate personal protective equipment and strict adherence to safety rules by all employees.

Head Protection

Approximately 120,000 people sustain head injuries on the job each year.[15] Falling objects are involved in many of these accidents. These injuries occur in spite of the fact that many of the victims were wearing hard hats. Such statistics have been the driving force behind the development of tougher, more durable hard hats.

According to Bross, "For nearly 70 years, the conventional hard hat—hard outer-shell and absorbing suspension system—has succeeded in helping prevent injuries at worksites across the nation and around the world."[16] Originally introduced in 1919, the hard hats first used for **head protection** in an industrial setting were inspired by the helmets worn by soldiers in World War I. Such early versions were made of varnished resin-impregnated canvas. As material technology evolved, hard hats were made of vulcanized fiber, then aluminum, then fiberglass. Today's hard hats are typically made from the thermoplastic material polyethylene, using the injection-molding process.[17] Basic hard hat design has not changed radically since before World War II. They are designed to provide limited protection from impact primarily to the top of the head, and thereby reduce the amount of impact transmitted to the head, neck, and spine.[18]

The American National Standards Institute (ANSI) standard for hard hats is Z89–1986. OSHA subsequently adopted this standard as its hard hat standard (29 C.F.R. 1010.135). According to Bross,

> This standard calls for testing hard hats for impact attenuation and penetration resistance as well as electrical insulation. Specifically, hard hats are tested to withstand a 40-foot-pound impact, which is equivalent to a two-pound hammer falling about 20 feet. Hard hats are also designed to limit penetration of sharp objects that may hit the top of the hard-hat shell and to provide some lateral penetration protection.[19]

Hard hats can help reduce the risk associated with falling or projected objects, but only if they are worn. The use of hard hats in industrial settings that might have falling objects has been mandated by federal law since 1971.[20] In addition to making the use of hard hats mandatory when appropriate and supervising to ensure compliance, Feuerstein recommends the use of incentives.[21] According to Feuerstein,

> It would seem that the sweetest offer a head-injury prevention program makes is a work environment free of injuries from falling objects. But sometimes this ultimate reward is too abstract to excite employees. They need to be led into safety for its own sake by concrete incentives, such as intra-department competition, monetary rewards for good suggestions, points toward prizes, and peer recognition for the most improved behavior.[22]

Resources expended promoting the use of hard hats are resources wisely invested. "Work accidents resulting in head injuries cost employers and workers an estimated $2.5

billion per year in workers' compensation insurance, medical expenses and accident investigation as well as associated costs due to lost time on the job and substitute workers. That is an average cost of $22,500 for each worker who received a head injury."[23]

Eye and Face Protection

Eye and face protection typically consists of safety goggles or face shields. The ANSI standard for face and eye protective devices is Z87.1–1989. OSHA has also adopted this standard. It requires that nonprescription eye and face protective devices pass two impact tests: a high mass, low-speed test and a low mass, high-speed test. Figure 9–2 summarizes the purpose of the tests and their individual requirements concerning impact and penetration. Figures 9–3, 9–4, and 9–5 are examples of the types of devices available for **eye and face protection**.

The high-mass impact test determines the level of protection provided by face and eye protective devices from relatively heavy, pointed objects that are moving at low speeds. The high-velocity impact test determines the level of protection provided from low-mass objects moving at high velocity.

Foot Protection

The OSHA regulations for **foot protection** are found in 29 C.F.R. 1910.132 and 126. Foot and toe injuries account for almost 20 percent of all disabling workplace injuries in the United States.[24] There are over 100,000 foot and toe injuries in the workplace each year.[25] According to Kelly, the major kinds of injuries to the foot and toes are from

- Falls/impact from sharp and/or heavy objects (this type accounts for 60 percent of all injuries)
- Compression when rolled over by or pressed between heavy objects
- Punctures through the sole of the foot

SAFETY FACT

Lateral Protection and Hard Hats

Conventional hard hats are designed to deflect a downward vertical blow. However, every year, there are head injuries on the job from lateral blows. Consequently, the American National Standards Institute (ANSI) has added a new type of hard hat to its protective headgear standard (ANSI Z89.1). Under this standard, conventional hard hats are classified as Type I. Hard hats that provide lateral protection are classified as Type II. Expect lateral protection to become the workplace standard eventually because 70 percent of the employees injured each year while wearing hard hats received blows to the unprotected areas of the head.

DISCUSSION CASE

What Is Your Opinion?

Two safety and health professionals are debating an issue over lunch. "Hard hats don't really do any good because employees won't wear them. It's like dieting. The best diet in the world won't work if you don't follow it." "I don't agree," said the second safety professional. "Hard hats have prevented thousands of head injuries. If employees don't wear them, it's because managers don't enforce their own rules." Join this debate. What is your opinion?

- Conductivity of electricity or heat
- Electrocution from contact with an energized, conducting material
- Slips on unstable walking surfaces
- Hot liquid or metal splashed into shoes or boots
- Temperature extremes[26]

The key to protecting workers' feet and toes is to match the protective measure with the hazard. This involves the following steps: (1) identify the various types of hazards present in the workplace, (2) identify the types of footwear available to counter the hazards, and (3) require that proper footwear be worn. Shoes selected should meet all

---

SAFETY FACT

Employers and Employees Share Responsibility for Eye Protection

Workers incur more than 80,000 eye injuries on the job each year. The regulations for eye protection are found in OSHA 1910.132-133. They require employers to provide appropriate safety equipment, first aid facilities, and training. The training must include the following:

- Conditions requiring eye protection
- Types of eye protection
- Proper use of personal protection devices (goggles, face shields, and so on) such as how to put on, adjust, and take off
- Limitations of the various types of personal protection equipment
- Proper care of personal protection devices

Employers must comply with OSHA regulations, but employees must do their part as well. They must take the training seriously and apply it on their jobs.

High Mass Impact Test—Purpose

This test is intended to ensure the level of mechanical integrity of a protective device and a level of protection from relatively heavy, pointed objects traveling at low speeds. Frames shall be capable of resisting impact from a 500 gram (17.6 ounces) missile with a 30-degree conical heat-treated tip and a 1-mm (.039 inches) radius dropped form a height of 130 cm (51.2 inches). No parts or fragments shall be ejected from the protector that could contact an eye.

High Velocity Impact Test—Purpose

This test is intended to ensure a level of protection from high-velocity, low-mass projectiles. Frames shall be capable of resisting impact from 6.35-mm (¼ inch) steel balls weighing 1.06 grams (.04 ounce) at 150 feet per second (fps) from 0 degrees to 90 degrees for frames; 250 fps for goggles, 300 fps for face shields.

Impact Test—Drop Ball

A 25.4-mm (1-inch) steel ball, weighing 68 grams (24 ounces), free fall from 127 cm (50 inches).

Lens Thickness

Thickness is 3.0 mm (.118 inch) except lenses that withstand high velocity impact, then 2.0-mm (.079-inch) thickness is acceptable.

Impact Test—Penetration

Lens shall be capable of resisting penetration from a Singer needle on a holder weighing 1.56 ounces dropped freely from 50 inches.

Figure 9–2
ANSI Standard Z87.1–1989.

applicable ANSI standards and have a corresponding ANSI rating. For example, "A typical ANSI rating is Z41PT83M1–75C–25. This rating means that the footwear meets the 1983 ANSI standard and the steel toe cap will withstand 75 foot pounds of impact and 2,500 pounds of compression."[27]

Modern safety boots are available that provide comprehensive foot and toe protection. The best safety boots will provide all of the following types of protection:

- *Steel toe* for impact protection
- *Rubber or vinyl* for chemical protection
- *Puncture-resistant soles* for protection against sharp objects
- *Slip-resistant soles* for protection against slippery surfaces
- *Electricity-resistant material* for protection from electric shock.

Employers are not required to provide footwear for employees, but they are required (29 C.F.R. 1910.132 and 136) to provide training on foot protection. The training must cover the following topics as a minimum:

- Conditions when protective footwear should be worn

- Type of footwear needed in a given situation
- Limitations of protective footwear
- Proper use of protective footwear

LIFTING HAZARDS

Back injuries that result from improper lifting are among the most common in an industrial setting. In fact, back injuries account for approximately $12 billion in workers' compensation costs annually. Putnam relates the following statistics concerning workplace back injuries:

- Lower back injuries account for 20 to 25 percent of all workers' compensation claims.
- Thirty-three to 40 percent of all worker's compensation costs are related to lower back injuries.
- Each year there are approximately 46,000 back injuries in the workplace.
- Back injuries cause 100 million lost workdays each year.
- Approximately 80 percent of the population will experience lower back pain at some point in their lives.[28]

Back injuries in the workplace are typically caused by improper lifting, reaching, sitting, and bending. **Lifting hazards** such as poor posture, ergonomic factors, and personal lifestyles also contribute to back problems. Consequently, a company's overall safety and health program should have a **back safety/lifting** component.

Back Safety/Lifting Program[29]

Prevention is critical in back safety. Consequently, safety and health professionals need to know how to establish back safety programs that overcome the hazards of lifting and other activities. Dr. Alex Kaliokin recommends the following six-step program:

Figure 9–3
Safety glasses that wrap around for lateral protection.
Courtesy of ELVEX Corporation, Bethel, CT.

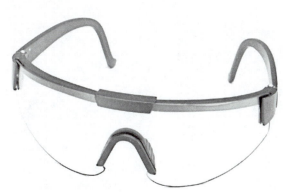

Figure 9–4
Eye safety shield combined
with ear and head protection.
Courtesy of ELVEX Corporation,
Bethel, CT.

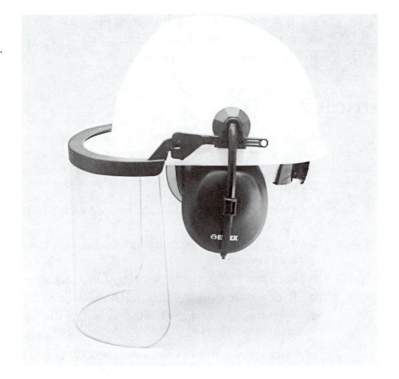

Figure 9–5
Faceshield (Huntsman®
Model K Facesaver® with
8154L window attached).
Courtesy of Kedman Company,
Huntsman Products Division.

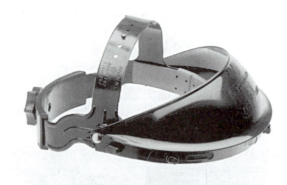

SAFETY FACT

NIOSH Backbelt Controversy

For years, employees whose jobs involve lifting have worn backbelts to protect the lower back from injury. However, the National Institute for Occupational Safety and Health (NIOSH) challenges the benefits of backbelts. According to NIOSH officials, the facts concerning backbelts are as follows:

- The claimed benefits of backbelts are unproven.
- Backbelts should not be considered personal protective equipment.
- Backbelts do not prevent injuries associated with lifting, pushing, pulling, twisting, or bending.
- Backbelts can put a temporary strain on the cardiovascular system.

Advocates counter that even if the benefits of backbelts have not been scientifically proven, there is anecdotal evidence of their value. According to advocates, backbelts are an expression of management's concern for employees, and they serve as a reminder to use proper lifting techniques.

Another argument made against using backbelts is the *Hercules Syndrome* in which employees wearing belts think they can pick up even more weight than usual. Because they are wearing a backbelt, employees may be less attentive to proper lifting techniques. This is a controversy that is likely to continue for a long time.

---

1. *Display poster illustrations.* Posters that illustrate proper lifting, reaching, sitting, and bending techniques should be displayed strategically throughout the workplace. This is as important in offices as in the plant. Clerical and office personnel actually sustain a higher proportion of back injuries than employees in general. Sitting too long without standing, stretching, and walking can put as much pressure on the back as lifting.

2. *Preemployment screening.* **Preemployment screening** can identify people who already have back problems when they apply. This is important because more than 40 percent of back injuries occur in the first year of employment and the majority of these injuries are back related.

3. *Regular safety inspections.* Periodic inspections of the workplace can identify potential problem areas so that corrective action can be taken immediately. Occasionally bringing a workers' compensation consultant in to assist with an inspection can help identify hazards that company personnel might miss.

4. *Education and training.* Education and training designed to help employees understand how to lift, bend, reach, stand, walk, and sit safely can be the most effective preventive measure undertaken. Companies that provide back safety training report a significant decrease in back injuries.

5. *Use external services.* A variety of external health-care agencies can help companies extend their programs. Identify local health-care-providing agencies and organi-

zations, what services they can provide, and a contact person in each. Maintaining a positive relationship with these **external service** contact people can increase the services available to employers.

6. *Map out the prevention program.* The first five steps should be written down and incorporated in the company's overall safety and health program. The written plan should be reviewed periodically and updated as needed.

In spite of a company's best efforts, back injuries will still occur. Consequently, modern safety and health managers should be familiar with the treatment and therapy that injured employees are likely to receive. According to Putnam, "Aggressive treatment for **reconditioning** addresses five goals: restoring function, reducing pain, minimizing deficits in strength, reducing lost time, and returning the body to pre-injury fitness levels."[30]

A concept that is gaining acceptance in bridging the gap between treatment/therapy and a safe return to work is known as **work hardening**.[31] Putnam describes work hardening as follows:

In specifically designed facilities known as "work centers" a broad range of work stations, exercise equipment, and aggressive protocols can focus on work-reconditioning. The objectives are:

- A return to maximum physical abilities;
- Improvement of general body fitness;
- Training to limit the possibility of re-injury;
- Work simulation that duplicates demands.[32]

The work centers referred to by Putnam replicate in as much detail as possible the injured employee's actual work environment. In addition to undergoing carefully controlled and monitored therapy in the work center, the employee is encouraged to use exercise equipment. Employees who undergo work center therapy should have already completed a program of acute physical therapy and pain management and they should be medically stable.[33]

Health and safety managers can help facilitate the fastest possible safe resumption of duties by injured employees by identifying local health-care providers that use the work-hardening approach. Such services and local providers of them should be made known to higher management so that the company can take advantage of them.

Proper Lifting Techniques

One of the most effective ways to prevent back injuries is to teach employees proper lifting techniques. Following are lifting techniques that should be taught as part of an organization's safety program.

Plan Ahead

- Determine if you can lift the load. Is it too heavy or too awkward?
- Decide if you need assistance.
- Check your route to see whether it has obstructions and slippery surfaces.

Lift with Your Legs, Not Your Back

- Bend at your knees, keeping your back straight.
- Position your feet close to the object.
- Center your body over the load.
- Lift straight up smoothly; don't jerk.
- Keep your torso straight; don't twist while lifting or after the load is lifted.
- Set the load down slowly and smoothly with a straight back and bent knees; don't let go until the object is on the floor.

Push, Don't Pull

- Pushing puts less strain on your back; don't pull objects.
- Use rollers under the object whenever possible.

NIOSH Lifting Guidelines

The National Institute for Occupational Safety and Health (NIOSH) originally developed guidelines for lifting and lowering in 1981. The guidelines include a formula for calculating the recommended weight limit (RWL) for a given lifting job. The 1981 formula was simple and easy to use because it considered only a few factors that affect a lifting task. In 1993, the NIOSH guidelines were revised. The new formula takes nonsymmetrical lifting and lifting of items that don't have handles into account. Another important change is the addition of a new **multitask analysis strategy.** This strategy gives safety professionals a method for considering a variety of related lifting variables and how they interact. This is a much more complicated method than the original, but it is also much more accurate and realistic.

The multitask analysis strategy is particularly useful when dealing with tasks in which the lifting variables change throughout the task. David B. Mahone explains such tasks using the example of a stacking job in which "each item is placed at a different vertical location in the stack."[34] The ergonomics of the task change with each successive item added to the stack (e.g., reach span required, height of lift) as do the corresponding hazards.

Because of its complexity, the new NIOSH *Lifting Equation* is more difficult to use than the original version. Safety professionals are not likely to get involved in actually solving the various formulas that make up the equation unless they are also mathematicians. However, software that allows safety professionals to simply plug selected values into the formulas is becoming readily available and is recommended. These values may be easily collected using nothing more than a stopwatch and a tape measure.

To apply the new lifting equation, safety professionals need to understand the types of information that they must collect and either turn it over to mathematicians or plug it into a computer program. This information is as follows:

LC: Load constant (always use 51 pounds or 23 kilograms)

HM:	Horizontal line measured from the midpoint between the ankles forward to the midpoint between the hands, at both the origin and destination of lift
VM:	Vertical line from the floor to the hands (also measured at the origin and destination of the lift)
DM:	Vertical distance between the origin and destination of the lift
AM:	Turning or twisting angle of asymmetry
FM:	Average frequency rate of lifting measured in lifts per minute
CM:	Coupling value (Does the item to be lifted have a good, fair, or poor grasping mechanism?)

STANDING HAZARDS

Consider this statement by Roberta Carson, a certified professional ergonomist: "Prolonged standing or walking is common in industry and can be very painful. Low back pain, sore feet, varicose veins, swelling in the legs, general muscular fatigue, and other health problems have been associated with prolonged standing or walking."[35] Carson recommends the following precautions for minimizing standing hazards:[36]

Anti-Fatigue Mats

Anti-fatigue mats provide cushioning between the feet and hard working surfaces such as concrete floors. This cushioning effect can reduce muscle fatigue and lower back pain. However, too much cushioning can be just as bad as too little. Consequently, it is important to test mats on a trial basis before buying a large quantity. Mats that become slippery when wet should be avoided. In areas where chemicals are used, be sure to select mats that will hold up to chemicals.

Shoe Inserts

When anti-fatigue mats are not feasible because employees must move from area to area and, correspondingly, from surface to surface, shoe inserts may be the answer. Such inserts are worn inside the shoe and provide the same type of cushioning the mats provide. Shoe inserts can help reduce lower back, foot, and leg pain. It is important to ensure proper fit. If inserts make an employee's shoes too tight, they will do more harm than good. In such cases, employees may need to wear a slightly larger shoe size.

Foot Rails

Foot rails added to work stations can help relieve the hazards of prolonged standing. Foot rails allow employees to elevate one foot at a time four or five inches. The elevated foot rounds out the lower back, thereby relieving some of the pressure on the spinal column.

Placement of a rail is important. It should not be placed in a position that inhibits movement or becomes a tripping hazard.

Workplace Design

A well-designed workstation can help relieve the hazards of prolonged standing. The key is to design workstations so that employees can move about while they work and can adjust the height of the workstation to match their physical needs.

Sit/Stand Chairs

Sit/stand chairs are higher-than-normal chairs that allow employees who typically stand while working to take quick mini-breaks and return to work without the hazards associated with getting out of lower chairs. They have the advantage of giving the employee's feet, legs, and back an occasional rest without introducing the hazards associated with lower chairs.

Proper Footwear

Proper footwear is critical for employees who stand for prolonged periods. Well-fitting, comfortable shoes that grip the worksurface and allow free movement of the toes are best.

FORKLIFT SAFETY

OSHA issues its standards for forklift safety in 29 C.F.R. 1910.118 under the heading "Powered Industrial Truck." These regulations apply to forklifts, platform lift trucks, and motorized hand trucks. The latest edition of OSHA's forklift standard is based primarily on a standard produced by the American Society of Mechanical Engineers entitled, "Safety for Low Lift and High Lift Trucks" (ASME B56.1).

Forklifts are different from cars and trucks in several ways. Employees who drive forklifts should understand how they are different. The primary differences are as follows:

- Forklifts are typically steered by the rear wheels.
- An empty forklift can be more difficult to steer than one with a load.
- Forklifts are frequently driven in reverse.
- Forklifts have three-point suspension so that the center of gravity can move from the rear of the vehicle closer to the front when it is loaded.

Because of these differences, it is important to ensure that only properly trained employees drive forklifts and that these employees follow some basic rules of accident prevention. The rules fall into four categories: (1) general, (2) lifting, (3) traveling, and (4) placing.

General Rules

The rules in this section are general and apply to all phases of forklift operation:

- Keep arms, hands, and legs inside the vehicle at all times.
- Face in the direction of travel at all times.
- If the load blocks your view, drive backward.
- Allow plenty of room for braking; at least three vehicle lengths.
- Make sure there is sufficient overhead clearance before moving a load.

Rules for Picking Up a Load

- Make sure the load is within the capacity of the forklift.
- Make sure forks are positioned properly.
- Make sure the load is properly balanced.
- Make sure the load is secure.
- Raise the load to the proper height.
- Run the forks all the way into the pallet, and tilt the mast back to stabilize the load before moving.
- Back out and stop completely, before lowering the load.

Rules for Traveling with a Load

- Always give pedestrians the right-of-way.
- Never allow passengers on the forklift.
- Keep the forks low while moving.
- Keep the load tilted back slightly while moving.
- Drive slowly; a forklift is not a car.
- Slow down at all intersections; stop and sound the horn at blind intersections.
- Drive up and back down ramps and inclines.
- Never lift or lower the load when traveling.
- Keep to the right just as you do when driving a car.
- Watch for oil, grease, and wet spots, which could inhibit traction.
- Cross railroad tracks at a skewed angle, never at a right angle.
- Watch for edges on loading docks and other changes in elevation.

Rules for Placing a Load

- Stop the forklift completely before raising or lowering the load.
- Move slowly and cautiously with the load raised.
- Never walk or stand under a raised load or allow anyone else to do so.
- Be certain the forks have cleared the pallet before turning and before changing height.

- Stack the load square and straight.
- Check behind and on both sides before backing up.

SUMMARY

1. The primary causes of falls are a foreign object on the walking surface, a design flaw in the walking surface, a slippery surface, and a person's impaired physical condition.
2. Most falls fit into one of four categories: trip and fall, stump and fall, step and fall, and slip and fall.
3. The coefficient of friction between surfaces is an effective method for comparing the traction of a walking surface. A coefficient of friction of 0.20 or less means the surface is very slippery and very hazardous. A coefficient of 0.40 or higher means there is good traction. Coefficients of friction decrease when a surface is wet.
4. Strategies for preventing slips include the following: choose the right material from the outset, retrofit an existing surface, practice good housekeeping, require nonskid footwear, and inspect surfaces frequently.
5. A slip and fall prevention program should have the following components: policy statement/commitment; review and acceptance of walkways; reconditioning and retrofitting; maintenance standards and procedures; inspection, audits, tests, and records; employee footwear; legal defenses; and measurement of results.
6. Dos and don'ts of ladder safety include checking for slipperiness, allowing only one person on the ladder at a time, securing the base and top on a level surface, applying the four-to-one ratio, facing the ladder when climbing, avoiding leaning, and always holding on with one hand.
7. Protection from impact injuries from falling or projected objects includes personal protective equipment to protect the head (hard hats), eyes and face (goggles or shields), and feet (footwear).
8. Back/lifting safety programs should have the following components: poster illustrations, preemployment screening, regular safety inspections, education and training, external services, and a written map of the program.
9. Standing hazards can be minimized by using the following: anti-fatigue mats, shoe inserts, foot rails, improved workplace design, and sit/stand chairs.

KEY TERMS AND CONCEPTS

Acceleration	Eye and face protection
Back safety/lifting	Foot protection
Blooming	Foreign object
Coefficient of friction	Four-to-one ratio
Design flaw	Good housekeeping
External services	Head protection

Impact

Ladder safety

Lifting hazards

Multitask analysis strategy

Nonskid footwear

OSHA Fall Protection Standard

Preemployment screening

Reconditioning

Retrofit

Review and acceptance

Slip and fall

Slippery surface

Step and fall

Stump and fall

Surface traction

Trip and fall

Work hardening

REVIEW QUESTIONS

1. List the primary causes of falls.
2. Explain briefly the most common kinds of falls.
3. Explain how surface traction is measured.
4. List and briefly explain five strategies for preventing slips.
5. Describe the various components of a slip and fall prevention program.
6. Explain the "trigger height" controversy over OSHA's Fall Protection Standard for construction.
7. What should an employee look for when inspecting a ladder?
8. Briefly summarize the evolution of hard hats in this country.
9. List six major kinds of injuries to the foot and toes that occur in the workplace each year.
10. What are the typical causes of back injuries in the United States?
11. Describe the six-step back safety/lifting program.
12. List four ways to minimize standing hazards.
13. Explain the strategies for proper lifting that should be taught as part of the safety program.

ENDNOTES

1. Kohr, R. L. "Slip Slidin' Away," *Safety & Health*, November 1989, Vol. 140, No. 5, p. 52.
2. Ibid.
3. Miller, B. C. "Falls: A Cast of Thousands Cost of Millions," *Safety & Health*, February 1988, Vol. 137, No. 2, p. 24.
4. Ibid.
5. Goldsmith, A. "Natural Walking, Unnatural Falls," *Safety & Health,* December 1988, Vol. 138, No. 6, p. 44.
6. Ibid., p. 46.
7. Ibid.
8. Ibid.

9. Ibid.
10. Ibid.
11. Kohr, R. L. "Slip Slidin' Away," p. 52.
12. National Safety Council. "Ladder Safety Is No Accident," *Today's Supervisor*, June 1991, pp. 8–9.
13. Ibid., p. 8.
14. Ibid., p. 9.
15. Bross, M. S. "Advances Lead to Tougher, More Durable Hard Hat," *Occupational Health & Safety*, February 1991, Vol. 60, No. 2, p. 22.
16. Ibid.
17. Ibid.
18. Ibid., p. 23.
19. Ibid.
20. Feuerstein, P. "Head Protection Looks Up," *Safety & Health*, September 1991, Vol. 144, No. 3, p. 38.
21. Ibid., p. 39.
22. Ibid.
23. Bross, M. S. "Advances Lead to Tougher, More Durable Hard Hat," p. 23.
24. Dutton, C. "Make Foot Protection a Hit," *Safety & Health*, November 1988, Vol. 138, No. 5, p. 30.
25. Ibid.
26. Kelly, S. M. "Start Out with the Right Footwear," *Safety & Health*, February 1990, Vol. 141, No. 2, pp. 48–49.
27. Dutton, C. "Make Foot Protection a Hit," pp. 31–32.
28. Putnam, A. "How to Reduce the Cost of Back Injuries," *Safety & Health*, October 1988, Vol. 138, No. 4, pp. 48–49.
29. Kaliokin, A. "Six Steps Can Help Prevent Back Injuries and Reduce Compensation Costs," *Safety & Health*, October 1988, Vol. 138, No. 4, p. 50.
30. Putnam, A. "How to Reduce the Cost of Back Injuries," p. 50.
31. Ibid.
32. Ibid.
33. Ibid.
34. Mahone, D. B. "Job Re-Design, Not 'Quick Fixes,' Thwarts Many Back Injury Hazards," *Occupational Health & Safety*, January 1994, p. 52.
35. Carson, R. "Stand By Your Job," *Occupational Health & Safety*, April 1994, p. 38.
36. Ibid., pp. 40–42.
37. *Official OSHA Safety Handbook* (Neenah, WI: J. J. Keller & Associates, Inc., 1996), pp. 69–75.

Heat and Temperature Hazards

Part of providing a safe and healthy workplace is appropriately controlling the temperature, humidity, and air distribution in work areas. A work environment in which the temperature is not properly controlled can be uncomfortable. Extremes of either heat or cold can be more than uncomfortable—they can be dangerous. Heat stress, cold stress, and burns are major concerns of modern safety and health professionals. This chapter provides the information that professionals need to know to overcome the hazards associated with extreme temperatures.

THERMAL COMFORT[1]

Thermal comfort in the workplace is a function of a number of different factors. Temperature, humidity, air distribution, personal preference, and acclimatization are all determinants of comfort in the workplace. However, determining optimum conditions is not a simple process.

To fully understand the hazards posed by temperature extremes, safety and health professionals must be familiar with several basic concepts related to thermal energy. The most important of these are summarized here:

- **Conduction** is the transfer of heat between two bodies that are touching, or from one location to another within a body. For example, if an employee touches a workpiece that has just been welded and is still hot, heat will be conducted from the workpiece to the hand. Of course, the result of this heat transfer is a burn.

■ **Convection** is the transfer of heat from one location to another by way of a moving medium (a gas or a liquid). Convection ovens use this principle to transfer heat from an electrode by way of gases in the air to whatever is being baked.

■ **Metabolic heat** is produced within a body as a result of activity that burns energy. All humans produce metabolic heat. This is why a room that is comfortable when occupied by just a few people may become uncomfortable when it is crowded. Unless the thermostat is lowered to compensate, the metabolic heat of a crowd will cause the temperature of a room to rise to an uncomfortable level.

■ **Environmental heat** is produced by external sources. Gas or electric heating systems produce environmental heat as do sources of electricity and a number of industrial processes.

■ **Radiant heat** is the result of electromagnetic non-ionizing energy that is transmitted through space without the movement of matter within that space.

THE BODY'S RESPONSE TO HEAT

The human body is equipped to maintain an appropriate balance between the metabolic heat that it produces and the environmental heat to which it is exposed. Sweating and the subsequent evaporation of the sweat are the body's way of trying to maintain an acceptable temperature balance. According to Alpaugh, this balance can be expressed as a function of the various factors in the following equation.[2]

$$H = M \pm R \pm C - E$$

where
H = body heat
M = internal heat gain (metabolic)
R = radiant heat gain
C = convection heat gain
E = evaporation (cooling)

The ideal balance when applying this equation is no new heat gain. As long as heat gained from radiation, convection, and metabolic processes does not exceed that lost through the evaporation induced by sweating, the body experiences no stress or hazard. However, when heat gain from any source or sources is more than the body can compensate for by sweating, the result is **heat stress**.

Heat stress can manifest itself in a number of ways depending on the level of stress. The most common types of heat stress are heat stroke, heat exhaustion, heat cramps, heat rash, transient heat fatigue, and chronic heat fatigue. These various types of heat stress can cause a number of undesirable bodily reactions including prickly heat, inadequate venous return to the heart, inadequate blood flow to vital body parts, circulatory shock, cramps, thirst, and fatigue. Different types of heat stress and how to prevent them are covered in the next section.

HEAT STRESS AND ITS PREVENTION

Heat stress is a major concern of modern safety and health professionals. According to North, "Ads on television and in magazines glorify heat. They say: 'Visit tropic lands, bask in the hot sun while wearing the newest tanning lotion, enjoy a home sauna and live in the sun belt.' But in spite of all the messages to the contrary, heat can be harmful—even deadly, under extreme conditions."[3] This section covers various types of heat stress, how it can be prevented, and actions to take when heat stress occurs in spite of preventive measures.

Heat Stroke: Cause, Symptoms, Treatment, and Prevention

Heat stroke is a type of heat stress that occurs as a result of a rapid rise in the body's core temperature. Heat stroke is very dangerous and should be dealt with immediately because it can be fatal. One can recognize heat stroke by observing the symptoms shown in Figure 10–1: (1) hot, dry, mottled skin, (2) confusion and/or convulsions, and (3) loss of consciousness. In addition to these observable symptoms, a victim of heat stroke will have a rectal temperature of 104.5°F or higher that will typically continue to climb.

Several factors can make an individual susceptible to heat stroke. These factors include the following: (1) obesity, (2) poor physical condition, (3) alcohol intake, (4) cardiovascular disease, and (5) prolonged exertion in a hot environment. This last factor can cause heat stroke even in a healthy individual. An employee who has one or more of the first four characteristics is even more susceptible to heat stroke.

In cases of heat stroke, the body's ability to sweat becomes partially impaired or actually breaks down altogether. This sets in motion a situation in which the temperature begins to increase uncontrollably. If this situation is not reversed quickly, heat stroke can be fatal.

If an employee becomes a heat stroke victim, action must be taken immediately to reduce his or her body core temperature. Do not wait until medical help arrives to begin. The victim should be immersed in chilled water if facilities are available. If not, wrap the victim in a wet, thin sheet and fan continuously, adding water periodically to keep the sheet wet.

Prevention strategies include the following: (1) medical screening as part of the employment process to identify applicants who have one or more susceptibility characteristics; (2) gradual **acclimatization** to hot working conditions spread over at least a full week; (3) rotation of workers out of the hot environment at specified intervals during

Figure 10–1
Observable symptoms of heat stroke.

- Skin is hot, dry, and typically red and mottled
- Loss of consciousness
- Confusion and/or convulsions

the work day; (4) use of personal protective clothing that is cooled; and (5) monitoring employees carefully and continually.

An example of using special clothing can be found at the nuclear power plant Three Mile Island in Unit 2. According to Hildebrand, during the summer months, the temperature in Unit 2 can exceed 90°F, making it difficult for employees to work for sustained periods. To counter this problem, Unit 2's safety and health team implemented a program that involves the use of special self-cooling clothing of the following two types: (1) an ice vest that consists of a light fitting vest with 60 small pockets of ice with a total ice capacity of about eight pounds; and (2) a vinyl one-piece coverall with a built-in air distribution system that directs cool air against the body and exhausts warm air. The ice vest is used when mobility is important and the length of exposure is relatively short. The body suit is used when a longer exposure time is necessary.[4]

Heat Exhaustion: Cause, Symptoms, Treatment, and Prevention

Heat exhaustion is a type of heat stress that occurs as a result of water and/or salt depletion. Figure 10–2 summarizes the observable symptoms of heat exhaustion. In addition to these symptoms, a victim of heat exhaustion may have a normal or even lower-than-normal oral temperature, but will typically have a higher-than-usual rectal temperature (i.e., 99.5°F to 101.3°F or 37.5°C to 38.5°C).

Heat exhaustion can be brought on by prolonged exertion in a hot environment and a failure to replace the water and/or salt lost through sweating. It can be compounded by a failure to acclimate employees gradually to working in a hot environment.

In cases of heat exhaustion, the body becomes dehydrated. This, in turn, decreases the volume of blood circulating. The various body parts must then compete for a smaller volume of blood. This causes circulatory strain, which manifests itself in the types of symptoms summarized in Figure 10–2.

SAFETY FACT

Heat Stroke Treatment

An employee whose skin is hot, dry, and red may have heat stroke. If, in addition to these symptoms, the employee seems confused or goes into convulsions, he or she probably has heat stroke. In such cases, waiting for medical professionals might be a fatal mistake. Call them. However, while waiting for medical help to arrive, take immediate action to lower the worker's core body temperature.

Use any method available to reduce the core body temperature quickly. Treat the victim by immersing him or her in cold water, wrapping in a wet sheet, fanning, or rubbing from head to toe continuously with crushed ice. Regardless of the method, remember, the key is to reduce the core body temperature quickly.

Figure 10–2
Observable symptoms of
heat exhaustion.

Fatigue

Nausea and/or vomiting

Headache

Lightheadedness

Clammy, moist skin

Pale or flushed complexion

Fainting when trying to stand

Rapid pulse

A victim of heat exhaustion should be moved to a cool, but not cold, environment and allowed to rest lying down. Fluids should be taken slowly but steadily by mouth until the urine volume indicates that the body's fluid level is once again in balance.

Prevention of heat exhaustion should be handled on the job in the same way as it is at athletic events. Professional baseball and football teams have spring training and a preseason to help get players acclimated to working conditions. Gradual acclimatization over at least a week is also important for employees who will work in the heat. During games, professional athletes replace the fluids that they lose by drinking one of the drinks commercially produced for this purpose. Such drinks are better than water because they contain appropriate amounts of salt, electrolytes, and other important elements lost during sweating. Employees working in the heat should have such fluids readily available and drink them frequently.

Electrolyte imbalance is a problem with heat exhaustion and heat cramps. When people sweat in response to exertion and environmental heat, they lose more than just water. They also lose salt and electrolytes. **Electrolytes** are minerals that are needed for the body to maintain its proper metabolism and for cells to produce energy. Loss of electrolytes causes these functions to break down. For this reason, it is important to use commercially produced drinks that contain water, salt, sugar, potassium, or electrolytes to replace those lost through sweating.

Heat Cramps: Cause, Symptoms, Treatment, and Prevention

Heat cramps are a type of heat stress that occur as a result of salt and potassium depletion. Observable symptoms are primarily muscle spasms that are typically felt in the arms, legs, and abdomen. Heat cramps are caused by salt and potassium depletion from profuse sweating as a result of working in a hot environment. Drinking water without also replacing salt exacerbates the problem.

During heat cramps, salt is lost, water that is taken in dilutes the body's electrolytes, and excess water enters the muscles causing cramping. When heat cramps occur, the appropriate response is to replenish the body's salt and potassium supply orally. This can be done with commercially produced fluids that contain carefully measured amounts of salts, potassium, electrolytes, and other elements.

To prevent heat cramps, acclimate workers to the hot environment gradually over a period of at least a week. Then, ensure that fluid replacement is accomplished with a product that contains the appropriate amount of salt, potassium, and electrolytes.

Heat Rash: Cause, Symptoms, Treatment, and Prevention

Heat rash is a type of heat stress that manifests itself as small, raised bumps or blisters that cover a portion of the body and give off a prickly sensation that can cause discomfort. It is caused by prolonged exposure to hot and humid conditions in which the body is continuously covered with sweat that does not evaporate because of the high humidity. The sweat gland ducts become clogged with retained sweat that does not evaporate. The sweat backs up in the system and causes minor inflammation.

Heat rash is simply and easily treated by removing the victim to a cooler, less humid environment, cleaning the affected area, and changing wet clothes for dry. Special lotions are available to speed the healing process. Heat rash can be prevented by resting in a cool, nonhumid environment and periodically changing into dry clothing.

Heat Fatigue: Cause, Symptoms, Treatment, and Prevention

Transient heat fatigue is a form of heat stress that manifests itself in temporary sluggishness, lethargy, and impaired performance (mental and/or physical). Employees who are not acclimated to working in a hot environment are especially susceptible to transient heat fatigue. The degree and frequency of transient heat fatigue is also a function of physical conditioning.

Well-conditioned employees who are properly acclimated will suffer this form of heat stress less frequently and less severely than poorly conditioned employees will. Consequently, preventing transient heat fatigue involves physical conditioning and acclimatization.

Chronic heat fatigue is similar to transient heat fatigue except that it does not abate after an appropriate rest period. Employees who experience chronic heat fatigue should be moved into jobs that do not involve working in a hot environment. Prolonged chronic heat fatigue, if not relieved, can cause both physiological and psychological stress. The psychological stress can manifest itself in substance abuse and other psychosocially unacceptable behavior.

BURNS AND THEIR EFFECTS

One of the most common hazards associated with heat in the workplace is the burn. Burns can be especially dangerous because they disrupt the normal functioning of the skin, which is the body's largest organ and the most important in terms of protecting other organs. It is necessary first to understand the composition of, and purpose served by, the skin to understand the hazards that burns can represent.

Human Skin

Human skin is the tough, continuous outer covering of the body. It consists of the following two main layers: (1) the outer layer, which is known as the **epidermis**; and (2) the inner layer, which is known as the **dermis, cutis,** or **corium**. The dermis is connected to the underlying subcutaneous tissue.

The skin serves several important purposes including the following: protection of body tissue, sensation, secretion, excretion, and respiration (see Figure 10–3). Protection from fluid loss, water penetration, ultraviolet radiation, and infestation by microorganisms is a major function of the skin. The sensory functions of touching, sensing cold, feeling pain, and sensing heat involve the skin.

The skin helps regulate body heat through the sweating process. It excretes sweat that takes with it electrolytes and certain toxins. This helps keep the body's fluid level in balance. By giving off minute amounts of carbon dioxide and absorbing small amounts of oxygen, the skin also aids slightly in respiration.

What makes burns particularly dangerous is that they can disrupt any or all of these functions depending on their severity. The deeper the penetration, the more severe the burn.

Severity of Burns

The severity of a burn depends on several factors. The most important of these is the depth to which the burn penetrates. Other determining factors include location of the burn, age of the victim, and amount of burned area.

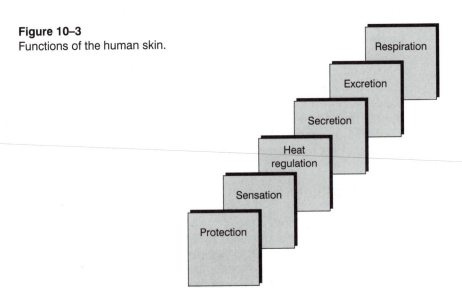

Figure 10–3
Functions of the human skin.

Respiration

Excretion

Secretion

Heat regulation

Sensation

Protection

The most widely used method of classifying burns is by degree (i.e., first-, second-, or third-degree burns). Modern safety and health professionals should be familiar with these classifications and what they mean.

First-degree burns are minor and result only in a mild inflammation of the skin, known as *erythema*. Sunburn is a common form of first-degree burn. It is easily recognizable as a redness of the skin that makes the skin sensitive and moderately painful to the touch.

Second-degree burns are easily recognizable from the blisters that form on the skin. If a second-degree burn is superficial, the skin will heal with little or no scarring. A deeper second-degree burn will form a thin layer of coagulated, dead cells that feels leathery to the touch. A temperature of approximately 210°F can cause a second-degree burn in as little as 15 seconds of contact.

Third-degree burns are very dangerous and can be fatal depending on the amount of body surface affected. A third-degree burn penetrates through both the epidermis and the dermis. A deep third-degree burn will penetrate body tissue. Third-degree burns can be caused by both moist and dry hazards. Moist hazards include steam and hot liquids; these cause burns that appear white. Dry hazards include fire and hot objects or surfaces; these cause burns that appear black and charred.

In addition to the depth of penetration of a burn, the amount of surface area covered is also a critical concern. This amount is expressed as a percentage of **body surface area**, or **BSA**. Figure 10–4 shows how the percentage of BSA can be estimated. Burns covering over 75 percent of BSA are usually fatal.

Using the first-, second-, and third-degree burn classifications in conjunction with BSA percentages, burns can be classified further as minor, moderate, or critical. According to Blocker, these classifications can be summarized as described in the following paragraphs.[5]

Minor Burns

All first-degree burns are considered **minor**. Second-degree burns covering less than 15 percent of the body are considered minor. Third-degree burns can be considered minor provided they cover only 2 percent or less of BSA.

Figure 10–4
Estimating percentage of body surface area (BSA) burned.

Right arm	9% of BSA
Left arm	9% of BSA
Head/neck	9% of BSA
Right leg	18% of BSA
Left leg	18% of BSA
Back	18% of BSA
Chest/stomach	18% of BSA
Perineum	1% of BSA

Moderate Burns

Second-degree burns that penetrate the epidermis and cover 15 percent or more of BSA are considered **moderate**. Second-degree burns that penetrate the dermis and cover from 15 to 30 percent of BSA are considered moderate. Third-degree burns can be considered moderate provided they cover less than 10 percent of BSA and are not on the hands, face, or feet.

Critical Burns

Second-degree burns covering more than 30 percent of BSA or third-degree burns covering over 10 percent of BSA are considered **critical**. Even small-area third-degree burns to the hands, face, or feet are considered critical because of the greater potential for infection to these areas by their nature. In addition, burns that are complicated by other injuries (fractures, soft tissue damage, and so on) are considered critical.

CHEMICAL BURNS

Chemicals are widely used in modern industry even by companies that do not produce them as part of their product base. Many of the chemicals produced, handled, stored, transported and/or otherwise used in industry can cause burns similar to those caused by heat (i.e., first-, second-, and third-degree burns). The hazards of chemical burns are very similar to those of thermal burns.

Chemical burns, like thermal burns, destroy body tissue; the extent of destruction depends on the severity of the burn. However, chemical burns continue to destroy body tissue until the chemicals are washed away completely.

The National Safety Council describes the physiological process in cases of chemical burns:

> Many concentrated chemical solutions have an affinity for water. When they come in contact with body tissue, they withdraw water from it so rapidly that the original chemical composition of the tissue (and hence the tissue itself) is destroyed. In fact a strong caustic may dissolve even dehydrated animal tissue. The more concentrated the solution, the more rapid is the destruction.[6]

The severity of the burn produced by a given chemical depends on the following factors:

- Corrosive capability of the chemical
- Concentration of the chemical
- Temperature of the chemical or the solution in which it is dissolved
- Duration of contact with the chemical[7]

Effects of Chemical Burns

Different chemicals have different effects on the human body. The harmful effects of selected widely used chemicals are summarized in Figure 10–5.[8] These are only a few of the many chemicals widely used in industry today. All serve an important purpose; however, all carry the potential for serious injury.

The primary hazardous effects of chemical burns are infection, loss of body fluids, and shock, and are summarized in the following paragraphs.[9]

Infection

The risk of **infection** is high with chemical burns—as is it with heat-induced burns—because the body's primary defense against infection-causing microorganisms (the skin) is penetrated. This is why it is so important to keep burns clean. Infection in a burn wound can cause *septicemia* (blood poisoning).

Fluid Loss

Body **fluid loss** in second- and third-degree burns can be serious. With second-degree burns, the blisters that form on the skin often fill with fluid that seeps out of damaged tissue under the blister. With third-degree burns, fluids are lost internally and, as a result, can cause the same complications as a hemorrhage. If these fluids are not replaced properly, the burns can be fatal.

Shock

Shock is a depression of the nervous system. It can be caused by physical and/or psychological trauma. In cases of serious burns, it may be caused by the intense pain that can occur when skin is burned away, leaving sensitive nerve endings exposed. Shock from burns can come in the following two forms: (1) primary shock, which is the first

Chemical	Potential Harmful Effect
Acetic Acid	Tissue damage
Liquid bromide	Corrosive effect on the respiratory system and tissue damage
Formaldehyde	Tissue hardening
Lime	Dermatitis and eye burns
Methylbromide	Blisters
Nitric/sulfuric acid mixture	Severe burns and tissue damage
Oxalic acid	Ulceration and tissue damage
White phosphorus	Ignites in air causing thermal burns
Silver nitrate	Corrosive/caustic effect on the skin
Sodium (metal)	Ignites with moisture causing thermal burns
Trichloracetic acid	Tissue damage

Figure 10–5
Harmful effects of selected widely used chemicals.

stage and results from physical pain and/or psychological trauma; and (2) secondary shock, which comes later and is caused by a loss of fluids and plasma proteins as a result of the burns.

First Aid for Chemical Burns

There is a definite course of action that should be taken when chemical burns occur, and the need for immediacy cannot be overemphasized. According to the National Safety Council, the proper response in cases of chemical burns "is to wash off the chemical by flooding the burned areas with copious amounts of water as quickly as possible. This is the only method for limiting the severity of the burn, and the loss of even a few seconds can be vital."[10]

In the case of chemical burns to the eyes, the continuous flooding should continue for at least 15 minutes. The eyelids should be held open to ensure that chemicals are not trapped under them.

According to the National Safety Council,

> The Committee on Industrial Ophthalmology, Council of Industrial Health of the American Medical Association has noted the tremendous saving of eyesight among industrial employees brought about by immediate and thorough flushing of harmful chemicals from the eyes by copious amounts of water. It is the belief of the committee that this is the most effective and practical emergency first aid treatment of eyes injured by chemicals.[11]

Another consideration when an employee comes in contact with a caustic chemical is his or her clothing. If chemicals have saturated the employee's clothes, they must be removed quickly. The best approach is to remove the clothes while flooding the body or the affected area. If necessary for quick removal, clothing should be ripped or cut off.

The critical need to apply water immediately in cases of chemical burns means that water must be readily available. Health and safety professionals should ensure that special eye wash and shower facilities are available wherever employees handle chemicals.

OVERVIEW OF COLD HAZARDS

Temperature hazards are generally thought of as extremes of heat. This is natural because most workplace temperature hazards do relate to heat. However, temperature extremes at the other end of the spectrum—cold—can also be hazardous. Employees who work outdoors in colder climates and employees who work indoors in such jobs as meat packing are subjected to cold hazards.

The major injuries associated with extremes of cold can be classified as being either generalized or localized. A generalized injury from extremes of cold is hypothermia. Localized injuries include frostbite, frostnip, and trenchfoot. According to Alpaugh, "The main factors contributing to cold injury are exposure to humidity and high winds, contact with wetness or metal, inadequate clothing, age, and general health. Physical conditions that worsen the effects of cold include allergies, vascular disease, excessive smoking and drinking, and specific drugs and medicine."[12]

The wind-chill factor increases the level of hazard posed by extremes of cold. Health and safety professionals need to understand this concept and how to make it part of their deliberations when developing strategies to prevent cold stress injuries.

Wind-chill Factor

The human body is able to sense cold. What the body actually senses when it is cold is a combination of temperature and wind velocity. Wind or air movement causes the body to sense coldness beyond what the thermometer actually registers as the temperature. This phenomenon is known as the **wind-chill factor**. It is the cooling effect produced by a combination of temperature, wind velocity, and/or air movement.[13]

When the temperature and the wind-chill factor are considered together, the result is an equivalent temperature that is lower than the thermometer reading. For example, an actual temperature reading of 10°F coupled with a 15-mph wind results in an equivalent temperature of −18°F. Wind increases the hazard of cold stress noticeably.

Frostbite, Frostnip, and Trenchfoot

The less severe disorders that can result from cold stress are frostbite, frostnip, and trenchfoot. Alpaugh describes **frostbite** as follows:

> Frostbite occurs when there is freezing of the fluids around the cells of the body tissues. This freezing is from exposure to extremely low temperatures. The condition results in damage to and loss of tissue. The most vulnerable parts of the body are the nose, cheeks, ears, fingers, and toes. Damage from frostbite can affect either the outer layers of the skin only, or it can include tissue beneath these outer layers. Damage from frostbite can be serious; scarring, tissue death, and amputation are all possibilities, as is permanent loss of movement in the affected parts.[14]

Frostbite is similar to burns in that it has three degrees. With first-degree frostbite, there is freezing but no blistering or peeling. With second-degree frostbite, there is freezing accompanied by blistering and peeling. With third-degree frostbite, there is freezing accompanied by death of skin and/or tissue. The first sign of frostbite is typically a sensation of cold and numbness. These symptoms may be accompanied by tingling, stinging, aching, or cramps. Frostbite of the outer layer of skin results in a whitish, waxy look. Deep frostbite results in tissue that is cold, pale, and solid.[15]

Frostnip is less severe than frostbite. It causes the skin to turn white and typically occurs on the face and other exposed parts of the body. There is no tissue damage with frostnip. However, if the exposed area is not either covered or removed from exposure to the cold, frostnip can become frostbite.

Trenchfoot is a condition that manifests itself as a tingling, itching, swelling, and pain. If these symptoms are not treated, this condition can lead to more serious injury including blistering, death of tissue, and ulceration. Trenchfoot is caused by continuous exposure of the feet simultaneously to a cold, but not freezing, environment and moisture.[16]

Hypothermia

Hypothermia is the condition that results when the body's core temperature drops to dangerously low levels. If this condition is not reversed, the patient literally freezes to death. Figure 10–6 summarizes the various symptoms of hypothermia that can be observed. The number of symptoms and their severity depend on how low the body's core temperature drops.

A person's susceptibility to hypothermia is increased by sedative drugs and alcohol. Sedatives interfere with the transmission of impulses from the nerve endings in the skin to the brain. This can cause a person to miss natural signals that he or she is in danger. Alcohol dilates blood vessels near the skin's surface. This, in turn, increases the amount and rate of heat loss, which results in an even lower body temperature.[17]

PREVENTING COLD STRESS

There are numerous strategies for preventing **cold stress**. Some of them are simple, common-sense strategies that employees should learn and practice. These include wearing appropriate protective clothing, limiting the duration of exposure to cold, replacing fluids (this is just as important in cold environments as it is in hot environments), eating a proper diet to ensure that the body is able to generate metabolic heat, and keeping the feet and all other extremities dry.

Cold Stress Prevention Program

Modern safety and health professionals in settings where cold stress is an issue should establish a cold stress prevention component as part of their overall safety and health program. The National Safety Council recommends that such a program contain the following elements:

Figure 10–6
Observable symptoms of hypothermia.

- Uncontrollable shivering
- Sensation of cold
- Weakened pulse
- Slow or irregular heartbeat
- Slow, slurred speech
- Incoherence and confusion
- Irregular breathing
- Memory lapses
- Fatigue or listlessness
- Exhaustion

DISCUSSION CASE

What Is Your Opinion?

Two safety professionals—one from northern Michigan and one from south Florida—met at a national conference. Both work for companies that require employees to work out-of-doors year round. The safety director from south Florida claims that the most difficult problem he has is preventing heat-related incidents and injuries. Hearing this, the safety director from northern Michigan responded, "If you think heat hazards are bad, try dealing with cold hazards. I'll take heat stroke over hypothermia any day." Is one extreme of temperature more difficult or more dangerous than its opposite? What is your opinion?

---

1. *Medical supervision and screening.* Medical screening involves identifying individuals who are particularly susceptible to cold stress (i.e., applicants who are in poor physical condition, overweight, and/or have cardiovascular problems). Medical supervision involves medical checkups.

2. *Orientation and training.* Employees should learn about the hazards associated with extremes of cold and how to protect themselves and fellow workers, including the use of proper clothing, appropriate work scheduling, proper work practices, and first aid procedures.

3. *Work practices.* Employees should understand and use proper work practices including regularly scheduled fluid replacement and periodic rest breaks in a warm environment.

4. *Engineering and administrative controls.* Responsible company officials should reduce hazard levels as much as possible through the application of appropriate engi-

Figure 10-7
Checklist of engineering and administrative controls for preventing cold stress.

1. Use central or spot heating, warm air jets, contact warm plates, or radiant heaters to provide warmth.
2. Shield the work area from wind.
3. Cover the handles of metal tools with insulating material.
4. Remove metal chairs or cover them with insulating material.
5. Provide heated tents or shelter and require workers to use them periodically for warming breaks.
6. Provide warmed drinks for fluid replacement (nonalcoholic, caffeine-free, properly balanced).
7. Require an acclimatization period of all new workers.
8. Use the buddy system and rotate the employees frequently.
9. Design the job to minimize sitting or standing still.
10. Design the job so that as many tasks as possible can be performed in a warm environment.

neering and administrative controls. Figure 10–7 summarizes engineering and administrative controls that can be used to decrease the potential for cold stress.[18]

In addition to the engineering controls listed in Figure 10–7, organizations should ensure that the air velocity in refrigerated rooms is minimized; power tools and other mechanical devices are used wherever appropriate to reduce the metabolic rate of employees; and all processes that can be modified or redesigned to reduce cold stress hazards are, in fact, changed appropriately. These are engineering controls. Additional administrative controls not listed in Figure 10–7 include the following: scheduling the coldest work for the warmest part of the day; moving any work that can be done in a warmer setting to such a setting; assigning extra workers to the most difficult tasks to cut down on the metabolic rate of each individual employee; providing relief workers so that regular employees can take warming and rest breaks; and providing training for all employees in the basic provisions of cold stress prevention and responding to cold stress emergencies.[19]

SUMMARY

1. Important thermal-related terms include conduction, convection, metabolic heat, environmental heat, and radiant heat.
2. The most common forms of heat stress are heat stroke, heat exhaustion, heat rash, and heat fatigue.
3. The skin serves several important purposes including protection of body tissue, sensation, secretion, excretion, and respiration.
4. The most widely used method of classifying burns is by degree: first-, second-, and third-degree burns. The amount of surface area covered by burns is expressed as a percentage of body surface area (BSA). Burns are also classified as minor, moderate, and critical.
5. The severity of chemical burns depends on the corrosive capability of the chemical, concentration and temperature of the chemical, and the duration of contact.
6. The primary hazards associated with chemical burns beyond the damage to the body tissues are infection, fluid loss, and shock. The most important first aid for chemical burns is immediate and continual flushing with water.
7. Wind or air movement causes the body to sense coldness beyond what the thermometer registers. This phenomenon is known as the wind-chill factor.
8. The most common forms of cold stress are hypothermia, frostbite, frostnip, and trenchfoot.
9. Cold stress can be prevented by applying the following strategies: medical screening/supervision, orientation and training, proper work practices, and engineering/administrative controls.

KEY TERMS AND CONCEPTS

Acclimatization	Cold stress
Body surface area (BSA)	Conduction
Chemical burn	Convection
Chronic heat fatigue	Corium

Critical burn

Cutis

Dermis

Electrolytes

Environmental heat

Epidermis

First-degree burn

Fluid loss

Frostbite

Frostnip

Heat cramps

Heat exhaustion

Heat rash

Heat stress

Heat stroke

Hypothermia

Infection

Metabolic heat

Minor burn

Moderate burn

Radiant heat

Second-degree burn

Shock

Third-degree burn

Transient heat fatigue

Trenchfoot

Wind-chill factor

REVIEW QUESTIONS

1. Define the following thermal comfort-related terms: conduction, convection, and metabolic heat.
2. What is heat stroke? How can it be recognized?
3. How can heat stroke be prevented?
4. What is heat exhaustion? How can it be recognized?
5. How can heat exhaustion be prevented?
6. What are heat cramps?
7. How can heat cramps be prevented?
8. Describe the various purposes served by the skin.
9. Describe and differentiate among first-, second-, and third-degree burns.
10. Describe and differentiate among minor, moderate, and critical burns.
11. List the factors that determine the severity of a chemical burn.
12. Explain the hazards of chemical burns besides tissue damage.
13. What should you do if an employee accidentally splashes a caustic chemical on himself or herself?
14. Explain the concept of wind-chill.
15. Describe the following types of cold stress: frostbite, trenchfoot, and hypothermia.
16. Describe the various components of a cold stress prevention program.

ENDNOTES

1. Alpaugh, E. L. (Revised by T. J. Hogan) *Fundamentals of Industrial Hygiene,* 3rd ed. (Chicago: National Safety Council, 1988) pp. 259–260.
2. Ibid., p. 266.

3. North, C. "Heat Stress," *Safety & Health,* April 1991, Vol. 141, No. 4, p. 55.

4. Hildebrand, J. E. "Radiation: Nuclear Power Industry Poses Unique Problems," *Occupational Health & Safety,* May 1987, Vol. 56, No. 5, p. 28.

5. Blocker, T. G., Jr. *Studies on Burns and Wound Healing* (Austin: University of Texas, 1965), p. 63.

6. National Safety Council, "Chemical Burns," Data Sheet 1–523 Rev. 87, p. 1.

7. Ibid.

8. Ibid., pp. 3–4.

9. Ibid., p. 2.

10. Ibid.

11. Ibid.

12. Alpaugh, E. L. *Fundamentals of Industrial Hygiene,* p. 261.

13. Ibid., p. 262.

14. Ibid., p. 261.

15. Ibid., p. 262.

16. Ibid., p. 261.

17. Ibid.

18. Ibid., p. 264.

19. Ibid.

Pressure Hazards

PRESSURE HAZARDS DEFINED

Pressure is defined in physics as the force exerted against an opposing fluid or thrust distributed over a surface. This may be expressed in force or weight per unit of area, such as psi (pounds per square inch). A **hazard** is a condition with the potential of causing injury to personnel, damage to equipment or structures, loss of material, or lessening of the ability to perform a prescribed function. Thus, a **pressure hazard** is a hazard caused by a dangerous condition involving pressure. Critical injury and damage can occur with relatively little pressure. OSHA defines high-pressure cylinders as those designated with a service pressure of 900 psi or greater.

We perceive pressure in relation to the earth's atmosphere. Approximately 21 percent of the atmosphere is oxygen, with most of the other 79 percent being nitrogen. In addition to oxygen and nitrogen, the atmosphere contains trace amounts of several inert gases: argon, neon, krypton, xenon, and helium.

At sea level, the earth's atmosphere averages 1,013 H 10 N/m^2, or 1.013 millibars, or 760 mm Hg (29.92 inches), or 14.7 psi, depending on the measuring scale used.[1] The international system of measurement utilizes newtons per square meter (N/m²). However, in human physiology studies, the typical unit is millimeters of mercury (mm Hg).

Atmospheric pressure is usually measured using a **barometer.** As the altitude above sea level increases, atmospheric pressure decreases in a nonlinear fashion. For example, at 18,000 feet above sea level, the barometric pressure is equal to 390 mm Hg. Half of this pressure, around 195 mm Hg, can be found at 23,000 feet above sea level.

Boyle's law states that the product of a given pressure and volume is constant with a constant temperature:

$$P_1V_1 = P_2V_2, \text{ when } T \text{ is constant}$$

Air moves in and out of the lungs because of a pressure gradient or difference in pressure. When atmospheric pressure is greater than pressure within the lungs, air flows down this pressure gradient from the outside into the lungs. This is called **inspiration,** inhalation, or breathing in, and occurs with greater lung volume than at rest. When pressure in the lungs is greater than atmospheric pressure, air moves down a pressure gradient outward from the lungs to the outside. **Expiration** occurs when air leaves the lungs and the lung volume is less than the relaxed volume, increasing pressure within the lungs.

Gas exchange occurs between air in the lung alveoli and gas in solution in blood. The pressure gradients causing this gas exchange are called partial pressures. **Dalton's law of partial pressures** states that, in a mixture of theoretically ideal gases, the pressure exerted by the mixture is the sum of the pressures exerted by each component gas of the mixture:

$$P_A = P_O + P_N + P_{else}$$

Air entering the lungs immediately becomes saturated with water vapor. Water vapor, although it is a gas, does not conform to Dalton's law. The partial pressure of water vapor in a mixture of gases is not dependent on its fractional concentration in that mixture. Water vapor partial pressure, instead, is dependent on its temperature. From this exception to Dalton's law comes the fact that at the normal body temperature of 37°C, water vapor maintains a partial pressure of 47 mm Hg as long as that temperature is maintained. With this brief explanation of how pressure is involved in human breathing, we now focus on the various sources of pressure hazards.

SOURCES OF PRESSURE HAZARDS

There are many sources of pressure hazard—some natural, most created by humans. Since the human body is comprised of approximately 85 percent liquid, which is virtually incompressible, increasing pressure does not create problems by itself. Problems can result from air being trapped or expanded within body cavities.

When sinus passages are blocked so that air cannot pass easily from the sinuses to the nose, expansion of the air in these sinuses can lead to problems. The same complications can occur with air trapped in the middle ear's Eustachian tube. As Boyle's law states, gas volume increases as pressure decreases. Expansion of the air in blocked sinus passages or the middle ear occurs with a rapid increase in altitude or rapid ascent underwater. This can cause pain and, if not eventually relieved, disease. Under extreme circumstances of rapid ascent from underwater diving or high-altitude decompression, lungs can rupture.

Nitrogen absorption into the body tissues can become excessive during underwater diving and breathing of nitrogen-enriched air. Nitrogen permeation of tissues occurs in proportion to the partial pressure of nitrogen taken in. If the nitrogen is permeating tissues faster than the person can breathe it out, bubbles of gas may form in the tissues.

Decompression sickness can result from the decompression that accompanies a rapid rise from sea level to at least 18,000 feet or a rapid ascent from around 132 feet to 66 feet underwater. Several factors influence the onset of decompression sickness:

- A *history* of previous decompression sickness increases the probability of another attack.
- *Age* is a component. Being over 30 increases the chances of an attack.
- *Physical fitness* plays a role. People in better condition have a reduced chance of the sickness. Previously broken bones and joint injuries are often the sites of pain.
- *Exercise* during the exposure to decompression increases the likelihood and brings on an earlier onset of symptoms.
- *Low temperature* increases the probability of the sickness.
- *Speed of decompression* also influences the sickness. A rapid rate of decompression increases the possibility and severity of symptoms.
- *Length of exposure* of the person to the pressure is proportionately related to the intensity of symptoms. The longer the exposure, the greater the chances of decompression sickness.

A reduction in partial pressure can result from reduced available oxygen and cause a problem in breathing known as **hypoxia.** Too much oxygen or oxygen breathed under pressure that is too high is called **hyperoxia.** Another partial pressure hazard, **nitrogen narcosis,** results from a higher-than-normal level of nitrogen pressure.

When breathed under pressure, nitrogen causes a reduction of cerebral and neural activity. Breathing nitrogen at great depths underwater can cause a feeling of euphoria and loss of reality. At depths greater than 100 feet (30 meters), nitrogen narcosis can occur even when breathing normal air. The effects may become pathogenic at depths greater than 200 feet, with motor skills threatened at depths greater than 300 feet. Cognitive processes deteriorate quickly after reaching a depth of 325 feet.

BOILERS AND PRESSURE HAZARDS

A boiler is a closed vessel in which water is heated to form steam, hot water, or high-temperature water under pressure.[2] Potential safety hazards associated with boilers and other pressurized vessels include the following:

- Design, construction, and/or installation errors

SAFETY MYTH

Respirators and Employees with Beards

In oxygen-deprived settings, workers must wear respirators. The mask of a respirator must fit snugly against the face and form a good seal. Consequently, an employee with a beard should be required to shave before being allowed into an environment that requires wearing a respirator. Right? Not necessarily. The issue is the seal, not the beard. Different types of beards grow differently on different people. If a beard interferes with the seal, trim it or shave it off. If it doesn't, keep it. Many bearded people are able to achieve a proper seal with no problem.

- Poor or insufficient training of operators
- Human error
- Mechanical breakdown/failure
- Failure or blockage of control and/or safety devices
- Insufficient or improper inspections
- Improper application of equipment
- Insufficient preventive maintenance[3]

Through years of experience, a great deal has been learned about how to prevent accidents associated with boilers. The Traveler's Insurance Company recommends the following daily, weekly, monthly, and yearly accident prevention measures:

1. *Daily check.* Check the water to make sure that it is at the proper level. Vent the furnace thoroughly before starting the fire. Warm up the boiler using a small fire. When the boiler is operating, check it frequently.

2. *Weekly check.* At least once every week, test the low-water automatic shutdown control and record the results of the test on a tag that is clearly visible.

3. *Monthly check.* At least once every month, test the safety valve and record the results of the test on a tag that is clearly visible.

4. *Yearly check.* The low-level automatic shutdown control mechanism should be either replaced or completely overhauled and rebuilt. Arrange to have the vendor or a third-party expert test all combustion safeguards, including fuel pressure switches, limit switches, motor starter interlocks, and shutoff valves.[4]

HIGH-TEMPERATURE WATER HAZARDS[5]

High-temperature water (HTW) is exactly what its name implies—water that has been heated to a very high temperature, but not high enough to produce steam. In some cases, HTW can be used as an economical substitute for steam (e.g., in industrial heating systems). It has the added advantage of releasing less energy (pressure) than steam does.

In spite of this, there are hazards associated with HTW. Human contact with HTW can result in extremely serious burns and even death. The two most prominent sources of hazards associated with HTW are operator error and improper design. Proper training and careful supervision are the best guards against operator error.

Design of HTW systems is a highly specialized process that should be undertaken only by experienced engineers. Mechanical forces such as water hammer, thermal expansion, thermal shock, or faulty materials cause system failures more often than do thermodynamic forces. Therefore, it is important to allow for such causes when designing an HTW system.

The best designs are simple and operator-friendly. Designing too many automatic controls into an HTW system can create more problems than it solves by turning operators into mere attendants who are unable to respond properly to emergencies.

HAZARDS OF UNFIRED PRESSURE VESSELS[6]

Not all pressure vessels are fired. Unfired pressure vessels include compressed air tanks, steam-jacketed kettles, digesters, and vulcanizers, as well as others that can create heat internally by various means rather than by external fire. The various means of creating internal heat include (1) chemical action within the vessel, and (2) application of some heating medium (electricity, steam, hot oil, and so on) to the contents of the vessel. The potential hazards associated with unfired pressure vessels include hazardous interaction between the material of the vessel and the materials that will be processed in it; inability of the filled vessel to carry the weight of its contents and the corresponding internal pressure; inability of the vessel to withstand the pressure introduced into it plus pressure caused by chemical reactions that occur during processing; and inability of the vessel to withstand any vacuum that might be created accidentally or intentionally.

The most effective preventive measure for overcoming these potential hazards is proper design. Specifications for the design and construction of unfired pressure vessels include requirements in the following areas: working pressure range, working temperature range, type of materials to be processed, stress relief, welding/joining measures, and radiography. Designs that meet the specifications set forth for unfired pressure vessels in such codes as the ASME Code (Section VIII) will overcome most predictable hazards.

Beyond proper design, the same types of precautions taken when operating fired pressure vessels can be used when operating unfired vessels. These include continual inspection, proper housekeeping, periodic testing, visual observation (for detecting cracks), and the use of appropriate safety devices.

HAZARDS OF HIGH-PRESSURE SYSTEMS[7]

The hazards most commonly associated with high-pressure systems are leaks, pulsation, vibration, release of high-pressure gases, and whiplash from broken high-pressure pipe, tubing, or hose. Strategies for reducing these hazards include limiting vibration through the use of vibration dampening (use of anchored pipe supports); decreasing the potential for leaks by limiting the number of joints in the system; using pressure gauges; placing shields or barricades around the system; using remote control and monitoring; and restricting access.

PRESSURE DANGERS TO HUMANS

The term *anoxia* refers to the rare case of no oxygen being available. **Hypoxia,** a condition that occurs when the available oxygen is reduced, can occur while ascending to a high altitude or when oxygen in air has been replaced with another gas, which may happen in some industrial situations.

Altitude sickness is a form of hypoxia associated with high altitudes. Ascent to an altitude of 10,000 feet above sea level can result in a feeling of malaise, shortness of breath, and fatigue. A person ascending to 14,000 to 15,000 feet may experience euphoria, along with a reduction in powers of reason, judgment, and memory. Altitude sickness includes a loss of **useful consciousness** at 20,000 to 25,000 feet.[8] After approximately five minutes at this altitude, a person may lose consciousness. The loss of consciousness comes at approximately one minute or less at 30,000 feet. Over 38,000 feet, most people lose consciousness within 30 seconds and may fall into a coma and possibly die.

Hyperoxia, or an increased concentration of oxygen in air, is not a common situation. Hyperbaric chambers or improperly calibrated scuba equipment can create conditions that may lead to convulsions if pure oxygen is breathed for greater than three hours. Breathing air at a depth of around 300 feet can be toxic and is equivalent to breathing pure oxygen at a depth of 66 feet.

At high pressures of oxygen, around 2,000 to 5,000 mm Hg, dangerous cerebral problems such as dizziness, twitching, vision deterioration, and nausea may occur. Continued exposure to these high pressures will result in confusion, convulsion, and eventual death.

Changes in total pressure can induce **trapped gas effects.** With a decrease in pressure, trapped gases will increase in volume (according to Boyle's law). Trapped gases in the body include air pockets in the ears, sinuses, and chest. Divers refer to the trapped gas phenomenon as the *squeeze.* Jet travel causes the most commonly occurring instance of trapped gas effects. Takeoff and landing may cause relatively sudden shifts in pressure, which may lead to discomfort and pain. With very rapid ascent or descent, injury can develop.

Lung rupture can be caused by a swift return to the surface from diving or decompression during high-altitude flight. This event is rare and happens only if the person is holding his or her breath during the decompression.

Evolved gas effects are associated with the absorption of nitrogen into body tissues. When breathed, nitrogen can be absorbed into all body tissues in concentrations proportional to the partial pressure of nitrogen in air. When a person is ascending in altitude, on the ground, in flight, or under water, nitrogen must be exhaled at a rate equal to or exceeding the absorption rate to avoid evolved gas effects.

If the nitrogen in body tissues such as blood is being absorbed faster than it is being exhaled, bubbles of gas may form in the blood and other tissues. Gas bubbles in the tissues may cause decompression sickness, which can be painful and occasionally fatal. Early symptoms of this disorder occur in body bends or joints such as elbows, knees, and shoulders. The common name for decompression sickness is the **bends.**

When the formation of gas bubbles is due to rapid ambient pressure reduction, it is called **dysbarism.**[9] The major causes of dysbarism are (1) the release of gas from the blood, and (2) the attempted expansion of trapped gas in body tissues. The sickness may occur with the decompression associated with rapidly moving from sea level (considered zero) to approximately 20,000 feet above sea level. Dysbarism is most often associated with underwater diving or working in pressurized containers (such as airplanes). Obese and older people seem to be more susceptible to dysbarism and decompression sickness.

Dysbarism manifests itself in a variety of symptoms. The **creeps** are caused by bubble formation in the skin, which causes an itchy, crawling, rashy feeling in the skin. Coughing and choking, resulting from bubbles in the respiratory system, is called the **chokes.** Bubbles occurring in the brain, although rare, may cause tingling and numbing, severe headaches, spasticity of muscles, and in some cases, blindness and paralysis. Dysbarism of the brain is rare. Rapid pressure change may also cause pain in the teeth and sinuses.[10]

Aseptic necrosis of bone is a delayed effect of decompression sickness. Bubbles in the capillaries supplying the bone marrow may become blocked with gas bubbles, which can cause a collection of platelets and blood cells to build up in a bone cavity. The marrow generation of blood cells can be damaged as well as the maintenance of healthy bone cells. Some bone areas may become calcified with severe complications when the bone is involved in a joint.

MEASUREMENT OF PRESSURE HAZARDS

Confirming the point of pressurized gas leakage can be difficult. After a gas has leaked out to a level of equilibrium with its surrounding air, the symptoms of the leak may disappear. There are several methods of detecting pressure hazards:

- *Sounds* can be used to signal a pressurized gas leak. Gas discharge may be indicated by a whistling noise, particularly with highly pressurized gases escaping through small openings. Workers should not use their fingers to probe for gas leaks as highly pressurized gases may cut through tissue, including bone.

- *Cloth streamers* may be tied to the gas vessel to help indicate leaks. Soap solutions may be smeared over the vessel surface so that bubbles are formed when gas escapes. A stream of bubbles indicates gas release.

- *Scents* may be added to gases that do not naturally have an odor. The odor sometimes smelled in homes that cook or heat with natural gas is not the gas but a scent added to it.

- *Leak detectors* that measure pressure, current flow, or radioactivity may be useful for some types of gases.

- *Corrosion* may be the long-term effect of escaping gases. Metal cracking, surface roughening, and general weakening of materials may result from corrosion.

There are many potential causes of gas leaks. The most common of these are as follows:

- *Contamination* by dirt can prevent the proper closing of gas valves, threads, gaskets, and other closures used to control gas flow.
- *Overpressurization* can overstress the gas vessel, permitting gas release. The container closure may distort and separate from gaskets, leading to cracking.
- *Excessive temperatures* applied to dissimilar metals that are joined may cause unequal thermal expansion, loosening the metal-to-metal joint and allowing gas to escape. Materials may crack because of excessive cold, which may also result in gas escape. Thermometers are often used to indicate the possibility of gas release.
- *Operator errors* may lead to hazardous gas release from improper closure of valves, inappropriate opening of valves, or overfilling of vessels. Proper training and supervision can reduce operator errors.

Destructive as well as nondestructive methods may be used to detect pressure leaks and incorrect pressure levels. **Nondestructive testing** methods do not harm the material being tested. Nondestructive methods may include mixing dye penetrants and magnetic or radioactive particles with the gas and then measuring the flow of the gas. Ultrasonic and X-ray waves are another form of nondestructive testing and are often used to characterize materials and detect cracks or other leakage points.

Destructive testing methods destroy the material being checked. Proof pressures generate stresses to the gas container, typically 1.5 to 1.667 times the maximum expected operating pressure for that container. Strain measurements may also be collected to indicate permanent weakening changes to the container material that remain after the pressure is released. **Proof pressure tests** often call for the pressure to be applied for a specified time and released. Stress and strain tests are then applied to the material. Proof pressure tests may or may not result in the destruction of the container being tested.

REDUCING PRESSURE HAZARDS

The reduction of pressure hazards often requires better maintenance and inspection of equipment that measures or uses high-pressure gases. Proper storage of pressurized containers reduces many pressure hazards. Pressurized vessels should be stored in locations away from cold or heat sources, including the sun. Cryogenic compounds (those that have been cooled to unusually low temperatures) may boil and burst the container when not kept at the proper temperatures. The whipping action of pressurized flexible hoses can also be dangerous. Hoses should be firmly clamped at the ends when pressurized.

Gas compression can occur in sealed containers exposed to heat. For this reason, aerosol cans must never be thrown into or exposed to a fire. Aerosol cans may explode violently when exposed to heat, although most commercially available aerosols are contained in low-melting point metals that melt before pressure can build up.

Pressure should be released before working on equipment. Gauges can be checked before any work on the pressurized system is begun. When steam equipment is shut down, liquid may condense within the system. This liquid and/or dirt in the system may become a propellant, which may strike bends in the system, causing loud noises and possible damage.

Water hammer is a series of loud noises caused by liquid flow suddenly stopping.[11] The momentum of the liquid is conducted back upstream in a shock wave. Pipe fittings and valves may be damaged by the shock wave. Reduction of this hazard involves using air chambers in the system and avoiding the use of quick-closing valves.

Negative pressures or **vacuums** are caused by pressures below atmospheric level. Negative pressures may result from hurricanes and tornadoes. Vacuums may cause collapse of closed containers. Building code specifications usually allow for a pressure differential. Vessel wall thickness must be designed to sustain the load imposed by the differential in pressure caused by negative pressure. Figure 11–1 describes several methods to reduce the hazards associated with pressurized containers.

- Install valves so that failure of the valve does not result in a hazard.
- Do not store pressurized containers near heat or sources of ignition.
- Train and test personnel dealing with pressurized vessels. Only tested personnel should be permitted to install, operate, maintain, calibrate, or repair pressurized systems. Personnel working on pressure systems should wear safety face shields or goggles.
- Examine valves periodically to ensure that they are capable of withstanding working pressures.
- Operate pressure systems only under the conditions for which they were designed.
- Relieve all pressure from the system before performing any work.
- Label pressure system components to indicate inspection status as well as acceptable pressures and flow direction.
- Connect pressure relief devices to pressure lines.
- Do not use pressure systems and hoses at pressure exceeding the manufacturer's recommendations.
- Keep pressure systems clean.
- Keep pressurized hoses as short as possible.
- Avoid banging, dropping, or striking pressurized containers.
- Secure pressurized cylinders by a chain to prevent toppling.
- Store acetylene containers upright.
- Examine labels before using pressurized systems to ensure correct matching of gases and uses.

Figure 11–1
Reduction of pressure hazards.

DISCUSSION CASE

What Is Your Opinion?

On a camping trip, Mary Carpenter—safety director for a small manufacturer of pressurized metal containers—saw something that really bothered her. While cleaning up his camp sight, a man threw all of his trash on a large fire. She noticed two aerosol cans being thrown in the fire and quickly warned the man of the danger of explosion. He laughed and shrugged off her warning saying, "There is no danger. The can will melt before it explodes." Who is right in this situation? What is your opinion?

SUMMARY

1. Pressure is defined in physics as the force exerted against an opposing fluid or solid.
2. Pressure is perceived in relation to the earth's atmosphere.
3. Barometers are used to measure atmospheric pressure.
4. Boyle's law states that the product of a given pressure and volume is constant with constant temperature:

$$P_1V_1 = P_2V_2 \text{ when } T \text{ is constant}$$

5. Inspiration is breathing air into the lungs.
6. Expiration occurs when air leaves the lungs.
7. Dalton's law of partial pressures states that in a mixture of ideal gases, the pressure exerted by the mixture is the sum of the pressures exerted by each component gas of the mixture:

$$P_A = P_O + P_N + P_{else}$$

8. Water vapor, although a gas, does not conform to Dalton's law.
9. Increasing pressure on the body does not create problems by itself.
10. Decompression sickness can occur from the decompression involved with a rapid rise from sea level to 18,000 feet or a rapid ascent from around 132 feet to 66 feet underwater.
11. Under extreme circumstances of rapid ascent from underwater diving or high-altitude decompression, lung rupture can occur.
12. Factors involved with decompression sickness include previous exposure history, age, physical fitness, exercise, low temperatures, speed of decompression, and length of exposure.
13. The bends are an example of decompression sickness.
14. Aseptic necrosis of bone can be a delayed effect of decompression sickness.
15. Hypoxia is a reduction of available oxygen.

16. Excessive nitrogen absorption into body tissues can occur from breathing nitrogen-enriched air and is called nitrogen narcosis.
17. Evolved gas effects are associated with the absorption of nitrogen into body tissues.
18. Altitude sickness is a form of hypoxia.
19. Altitude sickness may involve a loss of useful consciousness.
20. Hyperoxia is an increased concentration of oxygen in air and is not common.
21. Trapped gas effects can result from changes in total pressure.
22. Dysbarism is the rapid formation of gas bubbles in the tissue due to rapid ambient pressure reduction.
23. The creeps are caused by bubble formation in the skin.
24. Formation of bubbles in the respiratory tract is called the chokes.
25. Several methods are used to detect pressure hazards: sounds; cloth streamers; soap solutions; scents; leak detectors; visual checks for corrosion or contamination; and overpressurization indicated by a meter, excessive temperatures indicated by a thermometer, or operator errors.
26. Detection of pressure hazards includes destructive and nondestructive testing.
27. Proof pressures can be used to test container strength to contain a pressurized gas.
28. Pressurized cylinders and other vessels should be stored away from cold or heat sources, including the sun.
29. Aerosol cans should not be discarded in fires or by any method using heat.
30. Water hammer is a series of loud noises caused by pressurized liquid flow suddenly stopping.
31. Negative pressures or vacuums are caused by pressures below atmospheric level.

KEY TERMS AND CONCEPTS

Altitude sickness	Hyperoxia
Aseptic necrosis	Hypoxia
Barometer	Inspiration
Bends	Negative pressures
Boyle's law	Nitrogen narcosis
Chokes	Nondestructive testing
Creeps	Pressure
Dalton's law of partial pressures	Pressure hazard
Decompression sickness	Proof pressure
Destructive testing	Trapped gas effects
Dysbarism	Useful consciousness
Evolved gas effects	Vacuum
Expiration	Water hammer
Hazard	

REVIEW QUESTIONS

1. Against which references is pressure measured? How are these references measured?
2. Define inspiration and expiration.
3. Explain Dalton's law of partial pressure.
4. Does water vapor conform to Dalton's law?
5. Briefly discuss decompression sickness.
6. What do length of exposure, the bends, the chokes, and aseptic necrosis of bone have in common?
7. Define hypoxia and hyperoxia.
8. Explain nitrogen narcosis.
9. Discuss altitude sickness.
10. What is the relationship between trapped gas effects and dysbarism?
11. What is the difference between destructive and nondestructive testing?
12. Briefly explain proof pressures.
13. What causes vacuums?

ENDNOTES

1. Fraser, T. M. *The Worker at Work* (New York: Taylor & Francis, 1989), p. 300.
2. National Safety Council. *Accident Prevention Manual for Industrial Operations: Engineering and Technology,* 9th ed. (Chicago: National Safety Council, 1988), p. 485.
3. Ibid., p. 484.
4. Ibid., p. 488.
5. Ibid., pp. 487–89.
6. Ibid., p. 489.
7. Ibid.
8. Fraser, T. M. *The Worker at Work,* p. 315.
9. Hammer, W. *Occupational Safety Management and Engineering* (Upper Saddle River, NJ: Prentice Hall, 1989), p. 336.
10. Ibid., p. 339.
11. Ibid., p. 329.

Electrical Hazards

ELECTRICAL HAZARDS DEFINED

Electricity is the flow of negatively charged particles called **electrons** through an electrically conductive material. Electrons orbit the nucleus of an atom, which is located approximately in the atom's center. The negative charge of the electrons is neutralized by particles called **neutrons,** which act as temporary energy repositories for the interactions between positively charged particles called **protons** and electrons.

Figure 12–1 shows the basic structure of an atom, with the positively charged nucleus in the center. The electrons are shown as energy bands of orbiting negatively charged particles. Each ring of electrons contains a particular quantity of negative charges. The basic characteristics of a material are determined by the number of electron rings and the number of electrons in the outer rings of its atoms. A **positive charge** is present when an atom (or group of atoms) in a material has too many electrons in its outer shell. In all other cases the atom or material carries a **negative charge.**

Electrons that are freed from an atom and are directed by external forces to travel in a specific direction produce **electrical current,** also called *electricity.* **Conductors** are substances that have many free electrons at room temperature and can pass electricity. **Insulators** do not have a large number of free electrons at room temperature and do not conduct electricity. Substances that are neither conductors nor insulators can be called **semiconductors.**

Figure 12–1
Basic structure of an atom.

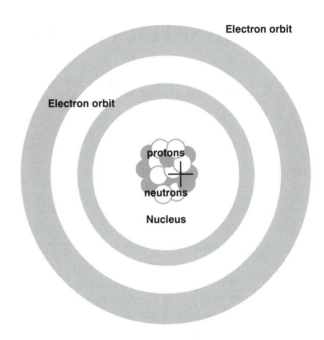

Electrical current passing through the human body causes a shock. The quantity and path of this current determines the level of damage to the body. The path of this flow of electrons is from a negative source to a positive point, since opposite charges attract one another.

When a surplus or deficiency of electrons on the surface of a material exists, **static electricity** is produced. This type of electricity is named "static" because there is no positive material nearby to attract the electrons and cause them to move. Friction is not required to produce static electricity, although it can increase the charge of existing static electricity. When two surfaces of opposite static electricity charges are brought into close range, a discharge, or spark, will occur. The spark from static electricity is often the first clue that such static exists. A common example is the sparks that come from rustling woolen blankets in dry heated indoor air.

The **potential difference** between two points in a circuit is measured by **voltage.** The higher the voltage, the more likely it is that electricity will flow between the negative and positive points.

Pure conductors offer little **resistance** to the flow of electrons. Insulators, on the other hand, have very high resistance to electricity. Semiconductors have a medium-range resistance to electricity. The higher the resistance, the lower the flow of electrons. Resistance is measured in **ohms.**

Electrical current is produced by the flow of electrons. The unit of measurement for current is **amperes** or **amps.** One amp is a current flow of 6.28×10^{18} electrons per second. Current is usually designated by *I*. **Ohm's law** describes the relationship among volts, ohms, and amps. One ohm is the resistance of a conductor that has a current of one amp under the potential of one volt. Ohm's law is stated as

$$V = IR$$

where

V = potential difference in volts
I = current flow in amps
R = resistance to current flow in ohms

Power is measured in wattage, or **watts,** and can be determined from Ohm's law:

$$W = VI \text{ or } W = I^2R$$

where

W = power in watts

Most industrial and domestic use of electricity is supplied by **alternating current**, or **AC** current. In the United States, standard AC circuits cycle 60 times per second. The number of cycles per second is known as **frequency** and is measured in **hertz.** Since voltage cycles in AC current, an **effective current** for AC circuits is computed, which is slightly less than the peak current during a cycle.

A **direct current**, or **DC** current, has been found to generate as much heat as an AC current that has a peak current 41.4 percent higher than the DC. The ratio of effective current to peak current can be determined by

(Effective current)/(Peak current) = (100%)/(100% + 41.4%) = 0.707 or 70.7%

Effective voltages are computed using the same ratios as effective current. A domestic 110-volt circuit has an effective voltage of 110 volts, with peaks of voltage over 150 volts.

The path of electrical current must make a complete loop for the current to flow. This loop includes the source of electrical power, a conductor to act as the path, a device to use the current, called a **load,** and a path to the ground. The earth maintains a relatively stable electrical charge and is a good conductor. The earth is considered to have **zero potential** because of its massive size. Any electrical conductor pushed into the earth is said to have zero potential. The earth is used as a giant common conductor back to the source of power.

Electrical hazards occur when a person makes contact with a conductor carrying a current and simultaneously contacts the ground or another object that includes a conductive path to the ground. This person completes the circuit loop by providing a load for the circuit and thereby enables the current to pass through his or her body. People can be protected from this danger by insulating the conductors, insulating the people, or isolating the danger from the people.

The **National Electrical Code,** or **NEC,** is published by the National Fire Protection Association (NFPA). This code specifies industrial and domestic electrical safety precautions. The NEC categorizes industrial locations and gases relative to their degree of fire hazard and describes in detail the safety requirements for industrial and home wiring. The NEC has been adopted by many jurisdictions as the local electrical code. The **National Board of Fire Underwriters** sponsors **Underwriters Laboratories** (UL). The UL determines whether equipment and materials for electrical systems are safe in the various NEC location categories. The UL provides labels for equipment that it approves as safe within the tested constraints.

Typical 110-volt circuit wiring has a **hot wire** carrying current, a **neutral wire,** and a **ground wire.** The neutral wire may be called a **grounded conductor**, with the ground wire being called a **grounding conductor.** Neutral wires usually have white insulation, hot wires have red or black insulation, and ground wires have green insulation or are bare. Figure 12–2 shows a typical three-wire circuit.

The hot wire carries an effective voltage of 110 volts with respect to the ground. The neutral wire carries nearly zero voltage with respect to the ground. If the hot wire makes contact with an unintended conductor, such as a metal equipment case, the current can bypass the load and go directly to the ground. With the load skipped, the ground wire is a low-resistance path to the earth and carries the highest current possible for that circuit.

A **short circuit** is a circuit in which the load has been removed or bypassed. The ground wire in a standard three-wire circuit provides a direct path to the ground, bypassing the load. Short circuits can be another source of electrical hazard if a human is the conductor to the ground, thereby bypassing the load.

SOURCES OF ELECTRICAL HAZARDS

Short circuits are one of many potential electrical hazards that can cause electrical shock. Another hazard is water, which considerably decreases the resistivity of materials, including humans. The resistance of wet skin can be around 450 ohms, whereas dry skin may have an average resistance of 100,000 ohms. According to Ohm's law, the higher the resistance, the lower the current flow. When the current flow is reduced, the probability of electrical shock is also reduced.

The major causes of electrical shock are

- Contact with a bare wire carrying current. The bare wire may have deteriorated insulation or be normally bare.
- Working with electrical equipment that lacks the UL label for safety inspection.
- Electrical equipment that has not been properly grounded. Failure of the equipment can lead to short circuits.
- Working with electrical equipment on damp floors or other sources of wetness.
- Static electricity discharge.

Figure 12–2
Typical three-wire circuit.

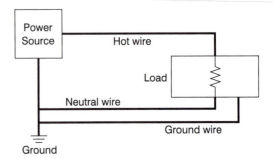

- Using metal ladders to work on electrical equipment. These ladders can provide a direct line from the power source to the ground, again causing a shock.
- Working on electrical equipment without ensuring that the power has been shut off.
- Lightning strikes.

Figure 12–3 depicts some of these electrical shock hazards.

Electrostatic Hazards

Electrostatic hazards may cause minor shocks. Shocks from static electricity may result from a single discharge or multiple discharges of static. Sources of electrostatic discharge include the following:

- Briskly rubbing a nonconductive material over a stationary surface. One common example of this is scuffing shoes across a wool or nylon carpet. Multilayered clothing may also cause static sparks.[1]
- Moving large sheets of plastic, which may discharge sparks.

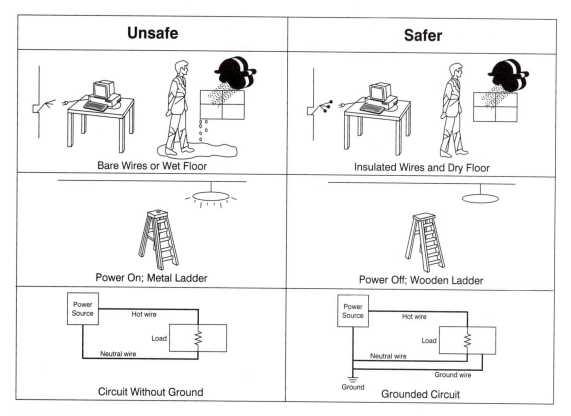

Figure 12–3
Electrical shock hazards.

- The explosion of organic and metallic dusts, which have occurred from static buildup in farm grain silos and mine shafts.
- Conveyor belts. Depending on their constituent material, they can rub the materials being transported and cause static sparks.
- Vehicle tires rolling across a road surface.
- Friction between a flowing liquid and a solid surface.[2]

The rate of discharge of electrical charges increases with lower humidity. Electrostatic sparks are often greater during cold, dry winter days. Adding humidity to the air is not commonly used to combat static discharge, however, because higher humidity may result in an uncomfortable working environment and adversely affect equipment.[3]

Arcs and Sparks Hazards

With close proximity of conductors or contact of conductors to complete a circuit, an electric **arc** can jump the air gap between the conductors and ignite combustible gases or dusts. When the electric arc is a discharge of static electricity, it may be called a **spark.** A spark or arc may involve relatively little or a great deal of power and is usually discharged into a small space.

Combustible and Explosive Materials

High currents through contaminated liquids may cause the contaminants to expand rapidly and explode. This situation is particularly dangerous with contaminated oil-filled circuit breakers or transformers. A poor match between current or polarity and capacitors can cause an explosion. In each of these cases, the conductor is not capable of carrying a current of such high magnitude. Overheating from high currents can also lead to short circuits, which in turn may generate fires and/or explosions.

Lightning Hazards

Lightning is static charges from clouds following the path of least resistance to the earth, involving very high voltage and current. If this path to the earth involves humans, serious disability may result, including electrocution. Lightning may also damage airplanes from intracloud and cloud-to-cloud flashes. Electrical equipment and building structures are commonly subject to lightning hazards. Lightning tends to strike the tallest object on the earth below the clouds. A tree is a common natural path for lightning.

Improper Wiring

Improper wiring permits equipment to operate normally but can result in hazardous conditions. The section of this chapter on detection of electrical hazards discusses tests to identify unsafe wire practices. One common mistake is to **jump the ground wire** to the neutral wire. In this case, the ground wire is actually connected to the neutral wire.

Equipment usually operates in a customary way, but the hazard occurs when low voltages are generated on exposed parts of the equipment, such as the housing. If the neutral circuit becomes corroded or loose, the voltage on the ground wire increases to a dangerous level.

Improper wiring is another common wiring error. When the ground is connected improperly, the situation is referred to as **open ground.** Usually the equipment with this miswiring will operate normally. If a short occurs in the equipment circuitry without proper grounding, anyone touching that equipment may be severely shocked.

With **reversed polarity,** the hot and neutral wires have been reversed. A worker who is not aware that the black lead (hot) and white lead (neutral) have been reversed could be injured or cause further confusion by connecting the circuit to another apparatus. If a short between the on/off switch and the load occurred, the equipment may run indefinitely, regardless of the switch position. In a reversed polarity light bulb socket, the screw threads become conductors.[4]

Temporary wiring installations sometimes remain in place for years until an accident occurs. Flexible wiring should rarely be substituted for fixed wiring in permanent buildings. A loose knot should be tied in a flexible cord when the plug is installed or replaced. The knot can prevent a pull on the cord from being transmitted to electrical connections such as the plug.

Insulation Failure

The degradation of insulation can cause a bare wire and resulting shock to anyone coming in contact with that wire. Most **insulation failure** is caused by environments toxic to insulation. These environments include the following:

- Direct sunlight or other sources of ultraviolet light, which can induce gradual breakdown of plastic insulation material.

- Sparks or arcs from discharging static electricity, which can result in burned-through holes in insulation.

- Repeated exposure to elevated temperatures, which can produce slow but progressive degradation of insulation material.

- Abrasive surfaces, which can result in erosion of the material strength of the insulation.

- Substance incompatibility with the atmosphere around the insulation and the insulation material, which can induce chemical reactions. Such reactions may include oxidation or dehydration of the insulation and eventual breakdown.

- Animals such as rodents or insects chewing or eating the insulation material, leading to exposure of the circuit. Insects can also pack an enclosed area with their bodies so tightly that a short circuit occurs. This is a common occurrence with electrical systems near water, such as pump housings and television satellite dishes.

- Moisture and humidity being absorbed by the insulation material, which may result in the moisture on the insulation carrying a current.

Equipment Failure

There are several ways in which **equipment failure** can cause electrical shocks. Electrical equipment designers attempt to create devices that are explosion-proof, dust-ignition-proof, and spark-proof. Following are some of the more common types of equipment failure:

- Wet insulation can become a conductor and cause an electrical shock.
- Portable tool defects can result in the device's housing carrying an electric current. Workers do not expect tool housings to be charged and may be shocked when they touch a charged tool housing.
- Broken power lines carry great amperage and voltage and can cause severe disability.
- When equipment is not properly grounded or insulated, an unshielded worker may receive a substantial electrical shock.

Hazardous Locations for Electrical Equipment

The NEC classifies **hazardous locations** for electrical equipment. There are three basic classes: Class I for flammable vapors and gases, Class II for combustible dusts, and Class III for ignitable fibers. There are also two divisions of hazard categories. Division I has more stringent requirements for electrical installation than Division II does. Figure 12–4 gives examples for each location category.[5]

		Division	
Class	**Group**	**I**	**II**
I. Flammable vapors and gases	A. Acetylene B. Hydrogen C. Ether D. Hydrocarbon fuels and solvents	Normally explosive; flammable paint spray areas	Not normally in explosive concentration; adjacent to paint spray area
II. Combustible dusts	E. Metal dusts F. Carbon dusts G. Flour, starch, grain, plastic, or chemical dusts	Conductive or ignitable dusts may be present; grain mills or processors	Not normally in ignitable concentration; grain storage areas
III. Ignitable fibers	Textiles, woodworking	Handled or used in manufacturing; cotton gins	Stored or handled in storage, not in manufacturing; excelsior storage

Figure 12–4
Hazardous electrical equipment location categories.

ELECTRICAL HAZARDS TO HUMANS

The greatest danger to humans suffering electrical shock results from current flow. The voltage determines whether a particular person's natural resistance to current flow will be overcome. Skin resistance can vary between 100,000 ohms and 600,000 ohms, depending on skin moisture.[6] Some levels of current **"freeze"** a person to the conductor; the person cannot voluntarily release his or her grasp. **Let-go current** is the highest current level at which a person in contact with the conductor can release the grasp of the conductor. Figure 12–5 shows the relationship between amperage dosage and danger with a typical domestic 60-cycle AC current.

The severity of injury with electrical shock depends on the dosage of current, as shown in Figure 12–5, but also on the path taken through the body by the current. The path is influenced by the resistance of various parts of the body at the time of contact with the conductor. The skin is the major form of resistance to current flow. Current

Dose in Current in Milliamps	Effect on Human Body
Less than 1	No sensation, no perceptible effect.
1	Shock perceptible, reflex action to jump away. No direct danger from shock but sudden motion may cause accident.
More than 3	Painful shock.
6	Let-go current for women.
9	Let-go current for men.
10–15	Local muscle contractions. Freezing to the conductor for 2.5% of the population.
30–50	Local muscle contractions. Freezing to the conductor for 50% of the population.
50–100	Prolonged contact may cause collapse and unconsciousness. Death may occur after 3 minutes of contact due to paralysis of the respiratory muscles.
100–200	Contact of more than a quarter of a second may cause ventricular fibrillation of the heart and death. AC currents continuing for more than one heart cycle may cause fibrillation.
Over 200	Clamps and stops the heart as long as the current flows. Heart beating and circulation may resume when current ceases. High current can produce respiratory paralysis, which can be reversed with immediate resuscitation. Severe burns to the skin and internal organs. May result in irreparable body damage.

Figure 12–5
Current effects on the human body.

paths through the heart, brain, or trunk are generally much more injurious than paths through extremities.

DETECTION OF ELECTRICAL HAZARDS

Several items of test equipment can be used to verify electrical equipment safety. A **circuit tester** is an inexpensive piece of test equipment with two wire leads capped by probes and connected to a small bulb. Most circuit testers test at least a 110- to 220-volt range. This simple tester can ensure that power has been turned off before electrical maintenance begins. The tester may also be used to determine whether housings and other equipment parts are carrying a current. When one of the leads makes contact with a hot wire and the other lead connects to a grounded conductor, the bulb lights.

A **receptacle wiring tester** is a device with two standard plug probes for insertion into an ordinary 110-volt outlet and a probe for the ground. Indicator lights show an improperly wired receptacle (outlet). This tester will not indicate a properly wired outlet, but it will reveal an incorrectly wired one. However, there are several types of miswiring that are not disclosed by using this tester, including the ground wire to neutral wire mistake. Figure 12–6 shows the meaning of lit indicator lights on the receptacle wiring tester.

A **continuity tester** may be used to determine whether a conductor is properly grounded or has a break in the circuit. Continuity is checked on circuits that are disconnected from a power source. Continuity testers often have an alligator clip on one end of a wire and a bulb and probe on the other end of the same wire. One terminal of the tester can be connected to the equipment housing; the other terminal is connected to a known ground. If the bulb does not light, the equipment is shown to be improperly grounded. With a circuit, the bulb will light when a current is capable of passing through the complete circuit. The unlit bulb of a continuity tester will indicate a break in the circuit.

REDUCING ELECTRICAL HAZARDS

Grounding of electrical equipment is the primary method of reducing electrical hazards. The purpose of grounding is to safeguard people from electrical shocks, reduce the prob-

---

SAFETY FACT

Electrical Hazards in the Construction Industry

Construction workers are more likely to die from electrocution than are any other types of workers. An average of 200 construction workers die from electric shock every year, making electrocution the second leading cause of death for construction workers in the United States in most years. Most electrocutions in the construction industry occur during the summer months between 11:00 A.M. and 3:00 P.M., corresponding with the peak months and times of day for construction work.

Figure 12–6
Receptacle wiring tester indicator lights.

Situation	Lights		
	1	2	3
Correct wiring	On	On	Off
Ground jumped to neutral	On	On	Off
Neutral and ground reversed	On	On	Off
Reversed polarity	On	Off	On
Open ground	On	Off	Off

ability of a fire, and protect equipment from damage. Grounding ensures a path to the earth for the flow of excess current. Grounding also eliminates the possibility of a person being shocked by contact with a charged capacitor. The actual mechanism of grounding was discussed at the beginning of this chapter.

Electrical system grounding is achieved when one conductor of the circuit is connected to the earth. Power surges and voltage changes are attenuated and usually eliminated with proper system grounding. **Bonding** is used to connect two pieces of equipment by a conductor. Bonding can reduce potential differences between the equipment and thus reduce the possibility of sparking. Grounding, in contrast, provides a conducting path between the equipment and the earth. Bonding and grounding together are used for entire electrical systems.

Separate equipment grounding involves connecting all metal frames of the equipment in a permanent and continuous manner. If an insulation failure occurs, the current should return to the system ground at the power supply for the circuit. The equipment ground wiring will be the path for the circuit current, enabling circuit breakers and fuses to operate properly. The exposed metal parts of the equipment shown in Figure 12–7 must be grounded or provided with double insulations.[7]

A **ground fault interrupter (GFI),** also called a **ground fault circuit interrupter (GFCI),** can detect the flow of current to the ground and open the circuit, thereby inter-

---

DISCUSSION CASE

What Is Your Opinion?

Two students of occupational safety and health are talking about their recent lesson on electrical hazards. The first student says, "Can you believe that a person can be frozen to an electrically charged wire and be unable to let go of it?" "No way," responded the second student. I've been shocked before. The current knocked my hand away." "It's true," said the first. "You need to take better notes. Professor Gorman talked about the 'freeze' point in his lecture." Which student is right? What is your opinion?

Figure 12-7
Equipment requiring grounding or double insulation.

Portable electric tools such as drills and saws.

Communication receivers and transmitters.

Electrical equipment in damp locations.

Television antenna towers.

Electrical equipment in flammable liquid storage areas.

Electrical equipment operated with over 150 volts.

rupting the flow of current. When the current flow in the hot wire is greater than the current in the neutral wire, a **ground fault** has occurred. The GFI provides a safety measure for a person who becomes part of the ground fault circuit. The GFI cannot interrupt current passing between two circuits or between the hot and neutral wires of a three-wire circuit. To ensure safety, equipment must be grounded as well as protected by a GFI.

There are several options for reducing the hazards associated with static electricity. The primary hazard of static electricity is the transfer of charges to surfaces with lower potential. Bonding and grounding are two means of controlling static discharge. **Humidification** is another mechanism for reducing electrical static; it was discussed in the section on sources of electrical hazards. Raising the humidity above 65 percent reduces charge accumulation.[8]

Anti-static materials have also been used effectively to reduce electrical static hazards. Such materials either increase the surface conductivity of the charged material or absorb moisture, which reduces resistance and the tendency to accumulate charges.

Ionizers and **electrostatic neutralizers** ionize the air surrounding a charged surface to provide a conductive path for the flow of charges. **Radioactive neutralizers** include a radioactive element that emits positive particles to neutralize collected negative electrical charges. Workers need to be safely isolated from the radioactive particle emitter.

Fuses consist of a metal strip or wire that will melt if a current above a specific value is conducted through the metal. Melting the metal causes the circuit to open at the fuse, thereby stopping the flow of current. Some fuses are designed to include a time lag before melting to allow higher currents during startup of the system or as an occasional event.

Magnetic circuit breakers use a solenoid, a type of coil, to surround a metal strip that connects to a tripping device. When the allowable current is exceeded, the magnetic force of the solenoid retracts the metal strip, opening the circuit. **Thermal circuit breakers** rely on excess current to produce heat and bending in a sensitive metal strip. Once bent, the metal strip opens the circuit. Circuit breakers differ from fuses in that they are usually easier to reset after tripping and often provide a lower time lag or none at all before being activated.

Double insulation is another means of increasing electrical equipment safety. Most double insulated tools have plastic nonconductive housings in addition to standard insulation around conductive materials.

There are numerous methods of reducing the risk of electrocution by lightning. Figure 12–8 lists the major precautions to take.[9]

Another means of protecting workers is **isolating the hazard** from the workers or vice versa. **Interlocks** automatically break the circuit when an unsafe situation is detected. Interlocks may be used around high-voltage areas to keep personnel from entering the area. Elevator doors typically have interlocks to ensure that the elevator does not move when the doors are open. **Warning devices** to alert personnel about detected hazards may include lights, colored indicators, on/off blinkers, audible signals, or labels.

It is better to design safety into the equipment and system than to rely on human behavior such as reading and following labels. Figure 12–9 summarizes the many methods of reducing electrical hazards.

OSHA'S ELECTRICAL STANDARDS

OSHA's standards relating to electricity are found in 29 C.F.R. 1910 (Subpart S). They are extracted from the National Electrical Code. This code should be referred to when more detail is needed than appears in OSHA's excerpts. Subpart S is divided into the following two categories of standards: (1) Design of Electrical Systems, and (2) Safety-Related Work Practices. The standards in each of these categories are as follows:

- Place lightning rods so that the upper end is higher than nearby structures.
- Avoid standing in high places or near tall objects. Be aware that trees in an open field may be the tallest object nearby.
- Do not work with flammable liquids or gases during electrical storms.
- Ensure proper grounding of all electrical equipment.
- If inside an automobile, remain inside the automobile.
- If in a small boat, lie down in the bottom of the boat.
- If in a metal building, stay in the building and do not touch the walls of the building.
- Wear rubber clothing if outdoors.
- Do not work touching or near conducting materials, especially those in contact with the earth such as fences.
- Avoid using the telephone during an electrical storm.
- Do not use electrical equipment during the storm.
- Avoid standing near open doors or windows where lightning may enter the building directly.

Figure 12–8
Lightning hazard control.

- Ensure that power has been disconnected from the system before working with it. Test the system for de-energization. Capacitors can store current after power has been shut off.
- Allow only fully authorized and trained people to work on electrical systems.
- Do not wear conductive material such as metal jewelry when working with electricity.
- Screw bulbs securely into their sockets. Ensure that bulbs are matched to the circuit by the correct voltage rating.
- Periodically inspect insulation.
- If working on a hot circuit, use the buddy system and wear protective clothing.
- Do not use a fuse with a greater capacity than was prescribed for the circuit.
- Verify circuit voltages before performing work.
- Do not use water to put out an electrical fire.
- Check the entire length of electrical cord before using it.
- Use only explosion-proof devices and nonsparking switches in flammable liquid storage areas.
- Enclose uninsulated conductors in protective areas.
- Discharge capacitors before working on the equipment.
- Use fuses and circuit breakers for protection against excessive current.
- Provide lightning protection on all structures.
- Train people working with electrical equipment on a routine basis in first aid and cardiopulmonary resuscitation (CPR).

Figure 12–9
Summary of safety precautions for electrical hazards.

Design of Electrical Systems

1910.302	Electric utilization systems
1910.303	General requirements
1910.304	Wiring design and protection
1910.305	Wiring methods, components, and equipment for general use
1910.306	Specific-purpose equipment and installations
1910.307	Hazardous (classified) locations
1910.308	Special systems

Safety-Related Work Practices

1910.331	Scope
1910.332	Training

1910.333	Selection and use of work practices
1910.334	Use of equipment
1910.335	Safeguards for personal protection

===== SUMMARY =====

1. Electricity is the flow of negatively charged particles through an electrically conductive material.
2. Atoms have a centrally located nucleus that consists of protons and neutrons. Electrons orbit the nucleus.
3. Conductors are substances that have many free electrons at room temperature and can pass electricity.
4. Insulators do not have a large number of free electrons at room temperature and do not conduct electricity.
5. When a surplus or deficiency of electrons on the surface of a material exists, static electricity is produced.
6. Resistance is measured in ohms.
7. Current flow is measured in amperes or amps.
8. Ohm's law is $V = IR$, where V = volts, I = amps, and R = ohms.
9. Power is measured in watts.
10. Watts *(W)* are calculated by $W = VI$ or $W = I^2R$.
11. Frequency is measured in hertz.
12. A load is a device that uses electrical current.
13. The NEC specifies industrial and domestic electrical safety precautions.
14. The UL determines whether equipment and materials for electrical systems are safe in the various NEC location categories.
15. Common 110-volt circuits include a hot wire, a neutral wire, and a ground wire.
16. A short circuit is one in which the load has been removed or is bypassed.
17. Sources of electrical hazards include contact with a bare wire, deteriorated insulation, equipment lacking the UL label, improper grounding of equipment, short circuits, dampness, static electricity discharge, metal ladders, power sources remaining on during electrical maintenance, and lightning strikes.
18. Electrostatic hazards may cause minor shocks.
19. A spark or arc involves little power and is discharged into a small space.
20. If the conductor is not capable of carrying a particular amperage of current, the material surrounding the conductor may become overheated and explode or burst into flame.
21. Lightning is a collection of static charges from clouds following the path of least resistance to the earth.
22. Lightning tends to strike the tallest object on the earth.
23. Jumping the ground wire to the neutral wire is unsafe wiring.
24. Open grounds are those with improperly connected ground wires.
25. Reversing the hot and neutral wires results in reversed polarity and an unsafe situation.

26. Flexible wiring should not be substituted for fixed wiring in permanent buildings.
27. Most insulation failure is caused by an environment toxic to insulation.
28. Electrical equipment designers attempt to create devices that are explosion-proof, dust-ignition-proof, and spark-proof.
29. The NEC classifies hazardous locations for electrical equipment.
30. The greatest danger to humans with electrical shock is current flow and the path through the body that the current takes.
31. Above a particular amperage of current, people freeze to conductors and are unable to let go of the conductor.
32. Circuit testers can ensure that power has been turned off.
33. A receptacle wiring tester indicates improperly wired outlets.
34. A continuity tester may be used to check whether a conductor is properly grounded or has a break in the circuit.
35. Grounding ensures a path to the earth for the flow of excess current.
36. Bonding and grounding increase the safety for entire electrical systems.
37. A GFI can detect current flow to the ground and open the circuit.
38. A ground fault occurs when the current in the hot wire is greater than the current in the neutral wire.
39. Anti-static materials, ionizers, and radioactive neutralizers reduce electrical static buildup.
40. Fuses and circuit breakers open the circuit with excess amperage.
41. Double insulation increases electrical safety.
42. Lightning hazard control includes using lightning rods, avoiding tall objects and flammable materials, not touching conductive materials, not using the telephone, not touching the walls of metal buildings, and not standing near open doors and windows during electrical storms.
43. Interlocks automatically open the circuit when an unsafe condition is detected.
44. It is better to design in safety for the electrical system than to deal effectively with accidents.

KEY TERMS AND CONCEPTS

AC	DC
Alternating current	Direct current
Amperes	Double insulation
Amps	Effective current
Anti-static materials	Electrical current
Arc	Electrical hazards
Bonding	Electrical system grounding
Circuit tester	Electricity
Conductors	Electrons
Continuity tester	Electrostatic hazards

Electrostatic neutralizers
Equipment failure
Freeze
Frequency
Fuses
GFCI
GFI
Ground fault
Ground fault circuit interrupter
Ground fault interrupter
Ground wire
Grounded conductor
Grounding conductor
Hazardous locations
Hertz
Hot wire
Humidification
Improper wiring
Insulation failure
Insulators
Interlocks
Ionizers
Isolating the hazard
Jump the ground wire
Let-go current
Lightning
Load
Magnetic circuit breaker
National Board of Fire Underwriters

National Electrical Code
NEC
Negative charge
Neutral wire
Neutrons
Ohms
Ohm's law
Open ground
Positive charge
Potential difference
Power
Protons
Radioactive neutralizers
Receptacle wiring tester
Resistance
Reversed polarity
Semiconductor
Separate equipment grounding
Shock
Short circuit
Spark
Static electricity
Thermal circuit breaker
Underwriters Laboratories
Voltage
Warning devices
Watts
Zero potential

REVIEW QUESTIONS

1. Define *zero potential*. Explain the relationship between zero potential and grounding.
2. Explain what each of the following terms measure: volt, amp, ohm, hertz, and watt.
3. Briefly discuss the difference between DC and AC using the concept of effective current.

4. What is the relationship between the NEC and UL?
5. Briefly state the relationship among potential difference, lightning, and grounding.
6. Define *open ground*.
7. List at least five lightning hazard control measures.
8. Explain the relationship between circuit load and short circuits.
9. How do ionizers, radioactive neutralizers, and anti-static materials work? Why does humidification work?
10. Discuss how bonding and grounding work together to increase electrical safety.
11. How do continuity testers, circuit testers, and receptacle wiring testers operate?
12. Explain freeze and let-go current.
13. Describe the structure of an atom.
14. Discuss the proper wiring of a three-wire circuit.
15. Why is jumping the ground wire a hazard?
16. Explain reversed polarity.
17. Why are warning devices less effective than designed-in safety precautions?

ENDNOTES

1. Hammer, W. *Occupational Safety Management and Engineering* (Upper Saddle River, NJ: Prentice Hall, 1989), p. 367.
2. Ibid., p. 368.
3. Kavianian, H. R., and C. A. Wentz, Jr. *Industrial Safety* (New York: Van Nostrand Reinhold, 1990), p. 231.
4. Asfahl, C. Ray. *Industrial Safety and Health Management* (Upper Saddle River, NJ: Prentice Hall, 1990), p. 347.
5. OSHA, *An Illustrated Guide to Electrical Safety* (Washington, D.C.: U.S. Department of Labor, 1983).
6. Kavianian and Wentz, *Industrial Safety*, p. 214.
7. Ibid., p. 218.
8. Hammer, *Occupational Safety Management and Engineering*, p. 372.
9. Kavianian and Wentz, *Industrial Safety*, p. 232.

CHAPTER THIRTEEN:

Fire Hazards and Life Safety

The assistant manager of a fuel storage plant was killed while trying to repair a broken-down piece of equipment. According to the *NFPA Journal,*

> The victim had been called to investigate a strong odor of gasoline that had been detected by a gasoline tank driver. . . . It was suspected that the facility's vapor-recovery system, which captures gasoline vapors displaced from tank trucks being filled with product, had malfunctioned. Soon after the manager entered the area, a violent explosion occurred. The victim's severely burned body and repair tools were found near the damaged equipment.[1]

In another recent example, facility damage was held to a minimum and human injury was avoided when a sprinkler system suppressed a fire that broke out on the second floor of a polyurethane foam manufacturing plant in North Carolina. According to the *NFPA Journal,*

> The fire broke out just after 4:00 P.M., an hour after all employees but one had left for the day. The ballast of a ceiling-mounted fluorescent light fixture short-circuited and cracked open, allowing burning ballast material to drop to the floor and the conveyor below. The burning material ignited scrap urethane.[2]

Fortunately, in this case, the company had a properly functioning automatic fire suppression system, and tragedy was avoided.

Two employees were killed and another received second- and third-degree burns over 45 percent of his body when an explosion and fire occurred at a chemical plant in Ohio. According to the *NFPA Journal,* "The employees inserted a long metal rod into the hopper and attempted to dislodge the blockage. As the men moved the rod around, an explosion occurred inside the hopper."[3]

What all of these tragedies have in common is that they did not have to happen. These incidents could have been prevented. The damage and injuries that can result from fire can be both physically and psychologically damaging. The resulting trauma can affect even those employees who are not physically injured. Therefore, modern safety and health professionals should be familiar with fire hazards and their prevention.

FIRE HAZARDS DEFINED

Fire hazards are conditions that favor fire development or growth. Three elements are required to start and sustain fire: (1) oxygen, (2) fuel, and (3) heat. Since **oxygen** is naturally present in most earth environments, fire hazards usually involve the mishandling of **fuel** or **heat.**

Fire, or **combustion**, is a chemical reaction between oxygen and a combustible fuel. Combustion is the process by which fire converts fuel and oxygen into energy, usually in the form of heat. By-products of combustion include light and smoke. For the reaction to start, a **source of ignition,** such as a spark or open flame, or a sufficiently high temperature is needed. Given a sufficiently high temperature, almost every substance will burn. The **ignition temperature** or **combustion point** is the temperature at which a given fuel can burst into flame.

Fire is a chain reaction. For combustion to continue, there must be a constant source of fuel, oxygen, and heat (see Figure 13–1). **Exothermic** chemical reactions create heat. Combustion and fire are exothermic reactions and can often generate large quantities of heat. **Endothermic** reactions consume more heat than they generate. An ongoing fire usually provides its own sources of heat. It is important to remember that cooling is one of the principal ways to control a fire or put it out.

All chemical reactions involve forming and breaking chemical bonds between atoms. In the process of combustion, materials are broken down into basic elements.

Figure 13–1
The fire triangle.

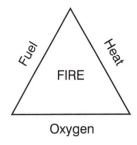

Loose atoms form bonds with each other to create molecules of substances that were not originally present.

Carbon is found in almost every flammable substance. When a substance burns, the carbon is released and then combines with the oxygen that *must* be present to form either **carbon dioxide** or **carbon monoxide.**

Carbon dioxide is produced when there is more oxygen than the fire needs. It is not toxic, but it can be produced in such volumes as to reduce seriously the concentration of oxygen in the air surrounding the fire site. Carbon monoxide—a colorless, odorless, deadly gas—is the result of incomplete combustion of a fuel. It is produced when there is insufficient oxygen to burn the fuel present efficiently. In general, most fires have insufficient oxygen and therefore produce large quantities of carbon monoxide. It is important in any intentional industrial fire that the fuel be consumed as completely as possible. This will reduce ash and minimize smoke and gases, including carbon monoxide.

Hydrogen, found in most fuels, combines with oxygen to form water. Synthetic **polymers,** found in plastics and vinyls, often form deadly fumes when they are consumed by fire, or when they melt or disintegrate from being near fire or high heat. Burning, melting, or disintegrating plastic at a fire site should be presumed to be releasing toxic fumes.

Liquids and solids, such as oil and wood, do not burn directly but must first be converted into a flammable **vapor** by heat. Hold a match to a sheet of paper, and the paper will burst into flames. Look closely at the paper, and you will see that the paper is not burning. The flames reside in a vapor area just above the surface of the sheet.

Vapors will only burn at a specific range of mixtures of oxygen and fuel, determined by the composition of the fuel. At the optimum mixture, a fire would burn, generate heat and some light, and produce no other by-products. In an unintentional fire, the mixture is constantly changing as more or less oxygen is brought into the flames and more or less heat is generated, producing more or less vapors and flammable gases.

Remove the fire's access to fuel or remove the oxygen, and the fire dies. Although a spark, flame, or heat may start a fire, the heat that a fire produces is necessary to sustain it. Therefore, a fire may be put out by removing the fuel source, starving it of oxygen, or cooling it below the combustion point. Even in an oxygen-rich, combustible environment, such as a hospital oxygen tent, fire can be avoided by controlling heat and eliminating sparks and open flames (see Figure 13–2).

An **explosion** is a very rapid, contained fire. When the gases produced exceed the pressure capacity of the vessel, a rupture or explosion must result. The simplest example

Figure 13–2
The broken triangle.

is a firecracker. The fuse, which usually contains its own source of oxygen, burns into the center of a firecracker. The surrounding powder ignites, and the heat produced vaporizes the balance of the explosive material and ignites it. The tightly wrapped paper of the firecracker cannot contain the expanding gases. The firecracker explodes, in much less time than was required to read about it. Explosives and toxic materials are discussed further in Chapter 14.

Heat always flows from a higher temperature to a lower temperature, never from a lower temperature to a higher temperature without an outside force being applied. Fires generate heat, which is necessary to sustain the fire. Excess heat is then transferred to surrounding objects, which may ignite, explode, or decompose. **Heat transfer** is accomplished by three means, usually simultaneously: (1) conduction, (2) radiation, and (3) convection.

Conduction is direct **thermal energy** transfer. On a molecular level, materials near a source of heat absorb the heat, raising their kinetic energy. **Kinetic energy** is the energy resulting from a moving object. Energy in the form of heat is transferred from one molecule to the next. Materials conduct heat at varying rates. Metals are very good conductors of heat. Concrete and plastics are poor conductors, hence good insulators. Nevertheless, a heat buildup on one side of a wall will transfer to the other side of the wall by conduction.

Radiation is electromagnetic wave transfer of heat to a solid. Waves travel in all directions from the fire and may be reflected off a surface, as well as be absorbed by it. Absorbed heat may raise the temperature beyond a material's combustion point, and then a fire erupts. Heat may also be conducted through a vessel to its contents, which will expand and may explode. An example is the spread of fire through an oil tank field. A fire in one tank can spread to nearby tanks through radiated heat, raising the temperature and pressure of the other tank contents.

Convection is heat transfer through the movement of hot gases. The gases may be the direct products of fire, the results of chemical reaction, or additional gases brought to the fire by the movement of air and heated at the fire surfaces by conduction. Convection determines the general direction of the spread of a fire. Convection causes fires to rise as heat rises and to move in the direction of the prevailing air currents.

All three forms of heat transfer are present at a campfire. A metal poker left in a fire gets red hot at the flame end. Heat is conducted up the handle, which gets progressively hotter until the opposite end of the poker is too hot to touch. People around the fire are warmed principally by radiation, but only on the side facing the fire. People farther away from the fire will be warmer on the side facing the fire than the backs of people closer to the fire. Marshmallows toasted above the flames are heated by convection (see Figure 13–3).

Spontaneous combustion is rare, but it must not be considered impossible. Organic compounds decompose through natural chemical processes. As they degrade, they release methane gas (natural gas), an excellent fuel. The degradation process, a chemical reaction, produces heat. In a forest, the concentrations of decomposing matter are relatively minimal, and both the gas and the heat vent naturally.

An example of spontaneous combustion is the classic pile of oil-soaked rags. A container of oil will seldom ignite spontaneously. A collection of clean fabrics will seldom

Figure 13–3
Campfire.

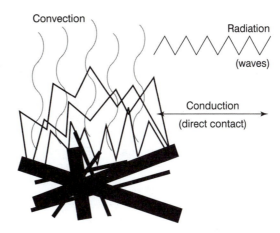

Convection

Radiation

(waves)

Conduction
(direct contact)

burst into flames. Rags soaking completely within oil are usually safe. One oil-soaked rag is unlikely to cause a problem. But in a pile of oil-soaked rags, especially in a closed container, the chemistry is quite different.

The fibers of the rags expose a large surface area of oil to oxidation. The porous nature of rags allow oxygen in to replace that which is consumed. When the temperature rises sufficiently, the surfaces of the oil on the rags will vaporize. Fuel, oxygen, heat—the fire triangle.

Hypergolic reactions occur when mixing fuels. Oxidizers produce just such a rapid heat buildup, causing immediate combustion at room temperature with no apparent source of ignition. Although the term *hypergolic* originated with rocket propellants, the phenomenon has been around for a long time. **Pyrophor hypergolic fuels** are those that self-ignite in the presence of oxygen found at normal atmospheric concentrations. One example is white phosphorus, which is kept underwater. If it starts to dry out, the phosphorus will erupt in flames.

SOURCES OF FIRE HAZARDS

Almost everything in an industrial environment will burn. Metal furniture, machines, plaster, and concrete block walls are usually painted. Most paints and lacquers will easily catch fire. Oxygen is almost always present. Therefore, the principal method of fire suppression is passive—the absence of sufficient heat. Within our environment, various conditions elevate the risk of fire and so are termed *fire hazards*.

For identification, fires are classified according to their properties, which relate to the nature of the fuel. The properties of the fuel directly correspond to the best means of combating a fire (see Figure 13–4).

Without a source of fuel, there is no fire hazard. However, almost everything in our environment could be a fuel. Fuels occur as solids, liquids, vapors, and gases.

Solid fuels include wood, building decorations and furnishings such as fabric curtains and wall coverings, and synthetics used in furniture. What would an office be with-

Class A fires	Solid materials such as wood, plastic, textiles, and their products: paper, housing, clothing.
Class B fires	Flammable liquids and gases.
Class C fires	Electrical (referring to live electricity situations, not including fires in other materials started by electricity).
Class D fires	Combustible, easily oxidized metals such as aluminum, magnesium, titanium, and zirconium.
Special categories	Extremely active oxidizers or mixtures, flammables containing oxygen, nitric acid, hydrogen peroxide, solid missile propellants.

Figure 13–4
Classes of fire.

out paper? What would most factories be without cardboard and packing materials such as Styrofoam molds and panels, shredded or crumpled papers, bubble wrap, and shrink wrap? All of these materials easily burn.

Few solid fuels are or can be made fireproof. Even fire walls do not stop fires, although they are defined by their ability to slow the spread of fire. Wood and textiles can be treated with fire- or flame-retardant chemicals to reduce their flammability.

Solid fuels are involved in most industrial fires, but mishandling **flammable liquids** and **flammable gases** is a major cause of industrial fires. Two often-confused terms applied to flammable liquids are *flash point* and *fire point*. The **flash point** is the lowest temperature for a given fuel at which vapors are produced in sufficient concentrations to flash in the presence of a source of ignition. The **fire point** is the minimum temperature at which the vapors will continue to burn, given a source of ignition. The **auto-ignition temperature** is the lowest point at which the vapors of a liquid or solid will self-ignite *without* a source of ignition.

Flammable liquids have a flash point below 100°F. **Combustible liquids** have a flash point at or higher than 100°F. Both flammable and combustible liquids are further divided into the three classifications shown in Figure 13–5.

As the temperature of any flammable liquid increases, the amount of vapor generated on the surface also increases. Safe handling, therefore, requires both a knowledge of the properties of the liquid and an awareness of ambient temperatures in the work or storage place. The **flammable range,** or **explosive range,** defines the concentrations of a vapor or gas in air that can ignite from a source. The auto-ignition temperature is the lowest temperature at which liquids spontaneously ignite.

Most flammable liquids are lighter than water. If the flammable liquid is lighter than water, water cannot be used to put the fire out.[4] The application of water floats the fuel and spreads a gasoline fire. Crude oil fires will burn even while floating on fresh or sea water.

Unlike solids, which have a definite shape and location, and unlike liquids, which have a definite volume and are heavier than air, gases have no shape. Gases expand to fill

Flammable Liquids

Class I–A	Flash point below 73°F, boiling point below 100°F.
Class I–B	Flash point below 73°F, boiling point at or above 100°F.
Class I–C	Flash point at or above 73°F, but below 100°F.

Combustible Liquids

Class II	Flash point at or above 100°F, but below 140°F.
Class III–A	Flash point at or above 140°F, but below 200°F.
Class III–B	Flash point at or above 200°F.

Figure 13–5
Classes of flammable and combustible liquids.
Source: NFPA, *Fire Protection Handbook,* 15th ed. (Quincy, MA: National Fire Protection Association, 1981).

the volume of the container in which they are enclosed, and they are frequently lighter than air. Released into air, gas concentrations are difficult to monitor due to the changing factors of air, current direction, and temperature. Gases may stratify in layers of differing concentrations but will often collect near the top of whatever container in which they are enclosed. Concentrations may have been sampled as being safe at workbench level but, at the same time, be close to or exceed flammability limits just above head height.

The **products of combustion** are gases, flame (light), heat, and smoke. Smoke is a combination of gases, air, and suspended particles, which are the products of incomplete combustion. Many of the gases present in smoke and at a fire site are toxic to humans. Other, usually nontoxic, gases may replace the oxygen normally present in air. Most fatalities associated with fire are from breathing toxic gases and smoke and from being suffocated because of oxygen deprivation. Gases that may be produced by a fire include acrolein, ammonia, carbon monoxide, carbon dioxide, hydrogen bromide, hydrogen cyanide, hydrogen chloride, hydrogen sulfide, sulfur dioxide, and nitrogen dioxide. Released gases are capable of traveling across a room and randomly finding a spark, flame, or adequate heat source, **flashing back** to the source of the gas.

The National Fire Protection Association (NFPA) has devised a system, NFPA 704, for the quick identification of hazards presented when substances burn. The NFPA's red, blue, yellow, and white diamond is used on product labels, shipping cartons, and buildings. Ratings within each category are 0 to 4, where zero represents no hazard; 4, the most severe hazard level. The colors refer to a specific category of hazard:

Red = flammability
Blue = health
Yellow = reactivity
White = special information

Figure 13–6 discusses this hazard identification system.

Flammability has a red background and is the top quarter of the diamond.

0   No hazard. Materials are stable during a fire and do not react with water.

1   Slight hazard. Flash point well above normal ambient temperature.

2   Moderate hazard. Flash point is slightly above normal ambient temperature.

3   Extreme fire hazard. Gases or liquids that can ignite at normal temperature.

4   Extremely flammable gases or liquids with very low flash points.

Health has a blue background and is the left quarter of the diamond.

0   No threat to health.

1   Slight health hazards. Respirator is recommended.

2   Moderate health hazard. Respirator and eye protection required.

3   Extremely dangerous to health. Protective clothing and equipment is required.

4   Imminent danger to health. Breathing or skin absorption may cause death. A fully encapsu-lating suit is required.

Reactive has a yellow background and is the right quarter of the diamond.

0   No hazard. Material is stable in a fire and does not react with water.

1   Slight hazard. Materials can become unstable at higher temperatures, or react with water to produce a slight amount of heat.

2   Moderate or greater hazard. Materials may undergo violent chemical reaction, but will not explode; or materials that react violently with water directly or form explosive mixtures with water.

3   Extreme hazard. Materials may explode given an ignition source, or have violent reactions with water.

4   Constant extreme hazard. Materials may polymerize, decompose, explode, or undergo other hazardous reactions on their own. Area should be evacuated in event of a fire.

Special information has a white background and is the bottom quarter of the diamond.

This area is used to note any special hazards presented by the material.

Figure 13–6
Identification of fire hazards.

Although we do not think of electricity as burning, natural and generated electricity play a large role in causing fires. Lightning strikes cause many fires every year. In the presence of a flammable gas or liquid mixture, one spark could produce a fire.

Electrical lines and equipment can cause fires either by a short circuit that provides an ignition spark, by arcs, or by resistances generating a heat buildup. Electrical switches and relays commonly arc as contact is made or broken.

Another source of ignition is heat in the form of hot surfaces. It is easy to see the flame hazard present when cooking oil is poured on a very hot grill. The wooden broom handle leaning up against the side of a hot oven may not be so obvious a hazard. Irons used in textile manufacturing and dry-cleaning plants also pose a heat hazard.

Space heaters frequently have hot sides, tops, backs, and bottoms, in addition to the heat-generating face. Hotplates, coffee pots, and coffee makers often create heated surfaces. Many types of electric lighting generate heat, which is transferred to the lamp housing.

Engines produce heat, especially in their exhaust systems. Compressors produce heat through friction, which is transferred to their housings. Boilers produce hot surfaces, as do steam lines and equipment using steam as power. Radiators, pipes, flues, and chimneys all have hot surfaces. Metal stock that has been cut by a blade heats up as the blade does. Surfaces exposed to direct sunlight become hot surfaces and transmit their heat by conduction to their other side. Heated surfaces are a potential source of fire.

FIRE DANGERS TO HUMANS

Direct contact with flame is obviously dangerous to humans. Flesh burns, as do muscles and internal organs. The fact that we are 80 percent water, by some estimations, does not mitigate the fact that virtually all of the other 20 percent burns. Nevertheless, burns are not the major cause of death in a fire.

National Fire Protection Association statistics show that most people die in fires from suffocating or breathing smoke and toxic fumes. Carbon dioxide can lead to suffocation because it can be produced in large volumes, depleting oxygen from the air. Many fire extinguishers use carbon dioxide because of its ability to starve the fire of oxygen while simultaneously cooling the fire. The number one killer in fires is carbon monoxide, which is produced in virtually all fires involving organic compounds. Carbon monoxide is produced in large volumes and can quickly reach lethal dosage concentrations.

Figure 13–7 shows the major chemical products of combustion. Other gases may be produced under some conditions. Not all of these gases are present at any particular fire site. Many of these compounds will further react with other substances often present at a fire. For example, sulfur dioxide will combine with water to produce sulfuric acid. Oxides

---

DISCUSSION CASE

What Is Your Opinion?

"It must be terrible to die in a fire," said the man to his wife as they checked into their hotel room. "That's why I always ask for a room near the elevator. I know if I'm near the elevator, I'll always have time to escape. In fact, as long as I don't see flames, I'll stay put and wait for rescue personnel. I'd rather breathe a little smoke than get burned by hot flames." Is this man's assessment of his situation accurate? What is your opinion?

Product	Fuels	Pathology
Acrolein	Cellulose, fatty substances, woods and paints	Highly toxic irritant to eyes and respiratory system.
Ammonia (NH_3)	Wool, silk, nylon, melamine, refrigerants, hydrogen-nitrogen compounds	Somewhat toxic irritant to eyes and respiratory system.
Carbon dioxide (CO_2)	All carbon and organic compounds	Not toxic, but depletes available oxygen.
Carbon monoxide (CO)	All carbon and organic compounds	Can be deadly.
Hydrogen chloride (HCN)	Wool, silk, nylon, paper, polyurethane, rubber, leather, plastic, wood	Quickly lethal asphyxiant.
Hydrogen sulfide (H_2S)	Sulfur-containing compounds, rubber, crude oil	Highly toxic gas. Strong odor of rotten eggs, but quickly destroys sense of smell.
Nitrogen dioxide (NO_2)	Cellulose nitrate, celluloid, textiles, other nitrogen oxides	Lung irritant, causing death or damage.
Sulfur dioxide (SO_2)	Sulfur and sulfur-containing compounds	Toxic irritant.

Figure 13–7
Major chemical products of combustion.

of nitrogen may combine with water to produce nitric acid. Sulfuric acid and nitric acid can cause serious acid burns.

DETECTION OF FIRE HAZARDS

Many automatic fire detection systems are used in industry today. Many systems can warn of the presence of smoke, radiation, elevated temperature, or increased light intensity. **Thermal expansion detectors** use a heat-sensitive metal link that melts at a predetermined temperature to make contact and ultimately sound an alarm. Heat-sensitive insulation can be used, which will melt at a predetermined temperature, thereby initiating a short circuit and activating the alarm.

Photoelectric fire sensors detect changes in infrared energy that is radiated by smoke, often by the smoke particles obscuring the photoelectric beam. A relay is open under acceptable conditions and closed to complete the alarm circuit when smoke interferes.

Ionization or **radiation sensors** use the tendency of a radioactive substance to ionize when exposed to smoke. The substance becomes electrically conductive with the smoke exposure and permits the alarm circuit to be completed.

Ultraviolet or **infrared detectors** sound an alarm when the radiation from fire flames is detected. When rapid changes in radiation intensities are detected, a fire alarm signal is given.

OSHA has mandated the monthly and annual inspection and recording of the condition of fire extinguishers in industrial settings. A hydrostatic test to determine the integrity of the fire extinguisher metal shell is recommended according to the type of fire extinguisher. The hydrostatic test measures the capability of the shell to contain internal pressures and the pressure shifts expected to be encountered during a fire.

REDUCING FIRE HAZARDS

The best way to reduce fires is to prevent their occurrence. A major cause of industrial fires is hot, poorly insulated machinery and processes. One means of reducing a fire hazard is the **isolation** of the three triangle elements: fuel, oxygen, and heat. In the case of fluids, closing a valve may stop the fuel element.

Fires may also be prevented by the proper storage of flammable liquids. Liquids should be stored as follows:

- In flame-resistant buildings that are isolated from places where people work. Proper drainage and venting should be provided for such buildings.
- In tanks below ground level.
- On the first floor of multi-story buildings.

Substituting less flammable materials is another effective technique for fire reduction. A catalyst or fire inhibitor can be employed to create an endothermic energy state that eventually will smother the fire. Several ignition sources can be eliminated or isolated from fuels:

- Smoking should be prohibited near any possible fuels.
- Electrical sparks from equipment, wiring, or lightning should not be close to fuels.
- Open flames should be kept separate from fuels. These may include welding torches, heating elements, or furnaces.
- Tools or equipment that may produce mechanical or static sparks must also be isolated from fuels.

Other strategies for reducing the risk of fires are as follows:

- Clean up spills of flammable liquids as soon as they occur, and properly dispose of the materials used in the clean-up.
- Keep work areas free from extra supplies of flammable materials (e.g., paper, rags, boxes, and so on). Have only what is needed on hand with the remaining inventory properly stored.
- Run electrical cords along walls rather than across aisles or in other trafficked areas. Cords that are walked on can become frayed and dangerous.
- Turn off the power and completely de-energize equipment before conducting maintenance procedures.

- Don't use spark- or friction-prone tools near combustible materials.
- Routinely test fire extinguishers.

Fire Extinguishing Systems

In larger or isolated industrial facilities, an employee fire brigade may be created. **Standpipe and hose systems** provide the hose and pressurized water for fire fighting. Hoses for these systems usually vary from 1 inch to 2.5 inches in diameter.[5]

Automatic sprinkler systems are an example of a **fixed extinguishing system** since the sprinklers are fixed in position. Water is the most common fluid released from the sprinklers. Sprinkler supply pipes may be kept filled with water in heated buildings; in warmer climates, valves are used to fill the pipes with water when the sprinklers are activated. When a predetermined heat threshold is breached, water flows to the heads and is released from the sprinklers.

Portable fire extinguishers are classified by the types of fire that they can most effectively reduce. Figure 13–8 describes the four major fire extinguisher classifications. Blocking or shielding the spread of fire can include covering the fire with an inert foam, inert powder, nonflammable gas, or water with a thickening agent added. The fire may suffocate under such a covering. Flooding a liquid fuel with nonflammable liquid can dilute this fire element. Figures 13–9, 13–10, and 13–11 are photographs of effective fire prevention equipment.

Fire Class	Extinguisher Contents	Mechanism	Disadvantages
A	Foam, water, dry chemical	Cooling, smothering, dilution, breaks the fire, reaction chain	Freezing if not kept heated
B	Dry chemical, bromotrifluoromethane, and other halogenated compounds, foam, CO_2, dry chemical	Chain-breaking smothering, cooling, shielding	Halogenated compounds are toxic
C	Bromotrifluoromethane, CO_2, dry chemical	Chain-breaking smothering, cooling, shielding	Halogenated compounds are toxic; fires may ignite after CO_2 dissipates
D	Specialized powders such as graphite, sand	Cooling, smothering	Expensive cover of powder may be broken with resultant reignition

Figure 13–8
Fire extinguisher characteristics.

Figure 13–9
Fireproof storage cabinet.
Courtesy of JUSTRITE®.

Figure 13–10
Fire-protective drums on a
poly spill pallet.
Courtesy of JUSTRITE®.

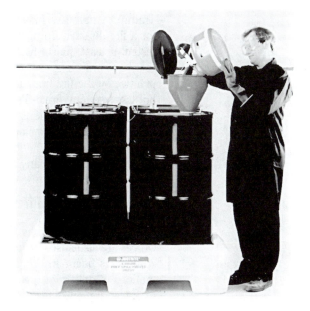

Figure 13–11
Fire-protective drums stored in an outside storage cabinet.
Courtesy of JUSTRITE®.

Disaster Preparations

Training employees may be the most successful life-saving preparation for a fire disaster. Company fire brigade members should be trained and tested at least quarterly. Disaster preparation initially requires management commitment and planning and continued response and recovery practice by the fire brigade on a regular basis. Also necessary are regular, but less frequent, fire drills for all personnel.

Disaster preparations also include the integration of company planning with community plans. Community disaster relief agencies such as the police, fire department, Red Cross, and hospitals should be consulted and informed of company disaster preparation plans.

Preventing Office Fires

The shop floor is not the only part of the plant where fire hazards exist. Offices are also susceptible to fires. According to Vogel, "Every year about 7,000 fires occur in office buildings, which cause injuries, deaths, and millions of dollars in fire damages."[6] The following strategies are helpful in preventing office fires:

■ Confine smoking to designated areas that are equipped with nontip ashtrays and fire-resistant furnishings.

- Periodically check electrical circuits and connections. Replace frayed or worn cords immediately.

- Make sure that extension cords and other accessories are UL approved and used only as recommended.

- Make sure there is plenty of air space left around copying machines and other office machines that might overheat.

- Locate heat-producing appliances away from the wall or anything else that could ignite.

- Frequently inspect personal appliances such as hot plates, coffee pots, and cup warmers. Assign responsibility for turning off such appliances every day to a specific person.

- Keep aisles, stairwells, and exits clear of paper, boxes, and other combustible materials.[7]

DEVELOPMENT OF FIRE SAFETY STANDARDS

The purpose of modern fire safety standards is the protection of life and the prevention of property damage. However, the impetus for developing standards has always been and continues to be the occurrence of major disasters. Typically, standards are developed after a major tragedy occurs in which property is damaged on a large scale and lives are lost. Public shock turns into an outcry for action. A flurry of political activity follows, and agencies/organizations that develop standards are called on to develop new standards.

According to Arthur E. Cote,

> There are approximately 89,000 standards in the United States. Of these, 50,000 of them actually are procurement specifications developed and used by the government. The remaining 39,000 are voluntary standards developed in the U.S. Most codes and standards in the fire protection field are developed by three organizations: the National Fire Protection Association (NFPA), the American Society for Testing Materials (ASTM), and Underwriter's Laboratories (UL). In addition, two model code organizations—the International Conference of Building Officials and Code Administration (BOCA), and the Southern Building Code Congress International (SBCC)—develop model building and fire prevention codes.[8]

The trend in fire safety standards is toward performance-based standards and away from the traditional specification-based approach. An example of each type of standard will help to illustrate the difference. A specification-based standard might require that brick, concrete, or steel material be used in a given type of building. A performance-based standard would specify that materials used have a one-, two-, or four-hour fire resistance rating.[9] Advances in the testing of engineering materials will help overcome most of the barriers to full development and implementation of performance-based standards.

Cote summarizes his views on the future of fire safety standards as follows:

> Codes and standards will survive in the 21st century. They may, however, be considerably different from the codes and standards we now have. They certainly will be based more on standardized fire tests, models, data, and related science and engineering than on consensus judgment. How much more will depend on the extent to which there is widespread acceptance of the anticipated breakthroughs in fire science and in related modeling and calculation methods.[10]

OSHA FIRE STANDARDS

OSHA standards for fire protection appear in 29 C.F.R. (Subpart L). This subpart contains the standards for fire brigades, fixed fire suppression equipment, and other fire protection systems. Employers are not required to form fire brigades, but those that choose to do so must meet a number of specific requirements. There are other fire-related requirements in other subparts. For example, fire exits, emergency action plans, and means of egress are covered in Subpart E. The standards in Subpart L are as follows:

Fire Protection

1910.155	Scope, application, and definitions
1910.156	Fire brigades

Portable Fire Suppression Equipment

1910.157	Portable fire extinguishers
1910.158	Standpipe and hose systems

Fixed Fire Suppression Equipment

1910.159	Automatic sprinkler systems
1910.160	Fixed extinguishing systems, general
1910.161	Fixed extinguishing systems, dry chemical
1910.162	Fixed extinguishing systems, gaseous agent
1910.163	Fixed extinguishing systems, water spray and foam

Other Fire Protection Systems

1910.164	Fire detection systems
1910.165	Employee alarm systems

FIRE SAFETY MYTHS

A number of myths relating to fire safety have grown up over the years. Employees who believe these myths can endanger their lives and those of their fellow workers should a fire occur. Therefore, it is important for safety and health managers to understand and dispel the myths. Kirk Birginal lists the following as being among the most common fire safety myths:

- Fire will light the exit route.
- There is time to escape.
- It's the flames that kill.
- Wait inside for rescue.
- You cannot prepare for a fire.[11]

Rather than lighting the exit route, a fire is more likely to cloud it with dark smoke and soot. There is never enough time when it comes to escaping a fire. Time is critical because, in less than a minute, a small flame can turn into an inescapable conflagration. Flames can kill, of course. However, the smoke and lethal gases produced by the fire more often are the cause of death. Waiting inside for rescue is a mistake. Employees in a burning building should act immediately to save themselves. Finally, not only can preparations be made that will minimize the harmful effects of a fire, preparation can be the difference between a tragedy and a tragedy averted.

Figure 13–12 summarizes major fire prevention and suppression suggestions.

LIFE SAFETY

Life safety involves protecting the vehicles, vessels, and lives of people in buildings and structures from fire. The primary reference source for life safety is the **Life Safety Code,** published by the National Fire Protection Association. The code applies to new and existing buildings. It addresses the construction, protection, and occupancy features necessary to minimize the hazards of fire, smoke fumes, and/or panic. A major part of the code is devoted to the minimum requirements for design of egress necessary to ensure that occupants can quickly evacuate a building or structure.

Basic Requirements[12]

In this section, the term *structure* refers to a structure or building.

- Every structure, new and existing, that is to be occupied by people must have a **means of egress** and other fire protection safeguards that, together, meet the following criteria: (1) ensure that occupants can promptly evacuate or be adequately protected without evacuating; and (2) provide sufficient backup safeguards to ensure that human life is not endangered if one system fails.

- Every structure must be constructed or renovated, maintained, and operated in such a way that occupants are

  (1) protected from fire, smoke, or fumes; (2) protected from fire-related panic; (3) protected long enough to allow a reasonable amount of time for evacuation; and (4) protected long enough to defend themselves without evacuating.

- In providing structures with means of egress and other fire protection safeguards, the following factors must be considered: (1) character of the occupancy; (2) capabilities of occupants; (3) number of occupants; (4) available fire protection; (5) height of the structure; (6) type of construction; and (7) any other applicable concerns.

- No lock or other device may be allowed to obstruct egress in any part of a structure at any time that it is occupied. The only exceptions to this requirement are mental health detention and correctional facilities. In these, the following criteria are required: (1) responsible personnel must be available to act in the case of fire or a similar emergency; and (2) procedures must be in place to ensure that occupants are evacuated in the event of an emergency.

- Use the least flammable materials whenever possible.
- Analyze the company to determine types of potential fires and provide appropriate sprinklers and/or extinguishers.
- Develop a database of the flammability of materials available in the company.
- Store containers or flammable materials away from sources of heat or sparks and away from humans.
- Do not permit smoking near flammable materials.
- Include a venting mechanism in storage containers and locate them near a drain.
- Minimize fuel storage container size to reduce the size of the fire that may involve those fuels.
- Isolate fuels from sources of heat.
- Include a smoke detection system and portable fire extinguishers in the facility. Extinguishers should be easily available to every workstation.
- Make sure sources of heat have controlling mechanisms and are near fire detection equipment.
- Check fire extinguishing equipment regularly.
- Perform periodic inspections for fire hazards and reappraisal of fire hazards.
- Train plant personnel in basic fire prevention, which should include periodic fire drills.
- Make sure fire brigade personnel are well trained, tested, and regularly practice fire control.
- Stress cleanliness and an organized method of disposal of flammable materials.

Figure 13–12
Fire prevention and suppression summary.

- All exits in structures must satisfy the following criteria: (1) be clearly visible or marked in such a way that an unimpaired individual can readily discern the route of escape; (2) all routes to a place of safety must be arranged or clearly marked; (3) any doorway and passageway that might be mistaken as a route to safety must be arranged or clearly marked in such a way as to prevent confusion in an emergency; and (4) all appropriate steps must be taken to ensure that occupants do not mistakenly enter a dead-end passageway.
- Egress routes and facilities must be included in the lighting design wherever artificial illumination is required in a structure.
- Fire alarm systems must be provided in any facility that is large enough or so arranged that a fire itself might not adequately warn occupants of the danger. Fire alarms should alert occupants to initiate appropriate emergency procedures.
- In any structure or portion of a structure in which a single means of egress might be blocked or overcrowded in an emergency situation, at least two means of egress must

be provided. The two means of egress must be arranged in such a way as to minimize the possibility of both becoming impossible in the same emergency situation.

- All stairs, ramps, and other means of moving from floor to floor must be enclosed (or otherwise protected) to afford occupants protection when used as a means of egress in an emergency situation. These means of vertical movement should also serve to inhibit the spread of fire, fumes, and smoke from floor to floor.

- Compliance with the requirements summarized herein does not eliminate or reduce the need to take other precautions to protect occupants from fire hazards, nor does it permit the acceptance of any condition that could be hazardous under normal occupancy conditions.

The information in this section is a summary of the broad fundamental requirements of the *Life Safety Code* of the National Fire Protection Association. More specific requirements relating to means of egress and features of fire protection are explained in the sections that follow.

Means of Egress[13]

This section explains some of the more important issues in the *Life Safety Code* relating to means of egress. Students and practitioners who need more detailed information are encouraged to refer to the *Life Safety Code*.

1. *Doors.* Doors that serve as exits must be designed, constructed, and maintained in such a way that the means of egress is direct and obvious. Windows that could be mistaken for doors in an emergency situation must be made inaccessible to occupants.

2. *Capacity of means of egress.* The means of egress must have a capacity sufficient to accommodate the occupant load of the structure calculated in accordance with the requirements of the *Life Safety Code*.

3. *Number of means of egress.* Any component of a structure must have a minimum of two means of egress (with exceptions as set forth in the code). The minimum

---

SAFETY MYTH

Fire Ratings for Doors, Walls, and Floors

A three-hour fire wall will provide protection for three hours. Right? Not necessarily. In fact, any relationship between a fire rating and the reality of fire resistance may be little more than coincidental. The problem is that the factors used to determine fire ratings are outdated. They are based on materials that were in use many years ago and no longer have any relevance. The materials used for constructing and furnishing buildings today are radically different from those upon which fire ratings are based.

number of means of egress from any story or any part of a story is three for occupancy loads of 500 to 1,000 and four for occupancy loads of more than 1,000.

4. *Arrangement of means of egress.* All exits must be easily accessible at all times in terms of both location and arrangement.

5. *Measurement of travel distance to exits.* The travel distance to at least one exit must be measured on the walking surface along a natural path of travel beginning at the most remote occupied space and ending at the center of the exit. Distances must comply with the code.

6. *Discharge from exits.* All exits from a structure must terminate at a public way or at yards, courts, or open spaces that lead to the exterior of the structure.

7. *Illumination of means of egress.* All means of egress shall be illuminated continuously during times when the structure is occupied. Artificial lighting must be used as required to maintain the necessary level of illumination. Illumination must be arranged in such a way that no area is left in darkness by a single lighting failure.

8. *Emergency lighting.* Emergency lighting for all means of egress must be provided in accordance with the code. In cases where maintaining the required illumination depends on changing from one source of power to another, there shall be no appreciable interruption of lighting.

9. *Marking of means of egress.* Exits must be marked by readily visible, approved signs in all cases where the means of egress is not obviously apparent to occupants. No point in the exit access corridor shall be more than 100 feet from the nearest sign.

10. *Special provisions for high hazard areas.* If an area contains contents that are classified as highly hazardous, occupants must be able to exit by traveling no more than 75 feet. At least two means of egress must be provided, and there shall be no dead-end corridors.

The requirements summarized in this section relate to the fundamental specifications of the *Life Safety Code* relating to means of egress. For more detailed information concerning general requirements, means of egress, and other factors such as fire protection and fire protection equipment, refer to the actual code.

FLAME-RESISTANT CLOTHING[14]

For employees who work in jobs in which flames or electric arcs may occur, wearing **flame-resistant clothing** can be a life saver. Electric arcs are the result of electricity passing through ionized air. Although electric arcs last for only a few seconds, during that time they can produce extremely high levels of heat and flash flame.

OSHA's standards relating to flame-resistant clothing are found in C.F.R. 1910.269, paragraph 1. Key elements of paragraph 1 explain the employer's responsibilities regarding personal protective equipment and flame-resistant clothing.

Apparel[15]

C.F.R. 1910.269, paragraph 1(6) reads as follows:

(i)   When work is performed within reaching distance of exposed energized parts of equipment, the employer shall ensure that each employee removes or renders nonconductive all exposed conductive articles, such as key or watch chains, rings, or wrist watches or bands, unless such articles do not increase the hazards associated with contact with the energized parts.

(ii) The employer shall train each employee who is exposed to the hazards or flames or electric arcs in the hazards involved.

(iii) The employer shall ensure that each employee who is exposed to the hazards of flames or electric arcs does not wear clothing that, when exposed to flames or electric arcs, could increase the extent of injury that would be sustained by the employee.

   Note: Clothing made from the following types of fabrics, either alone or in blends, is prohibited by this paragraph, unless the employer can demonstrate that the fabric has been treated to withstand the conditions that may be encountered or that the clothing is worn in such a manner as to eliminate the hazard involved: acetate, nylon, polyester, rayon.

(iv) *Fuse Handling.* When fuses must be installed or removed with one or both terminals energized at more than 300 volts or with exposed parts energized at more than 50 volts, the employer shall ensure that tools or gloves rated for the voltage are used. When expulsion-type fuses are installed with one or both terminals energized at more than 300 volts, the employer shall ensure that each employee wears eye protection meeting the requirements of Subpart I of this Part, uses a tool rated for the voltage, and is clear of the exhaust path of the fuse barrel.

SUMMARY

1.   Fire hazards are conditions that favor fire development or growth.
2.   The three elements required to start and sustain a fire are oxygen, fuel, and heat.
3.   Fire, or combustion, is a chemical chain reaction between oxygen and a fuel.
4.   The product of combustion is energy in the form of heat.
5.   By-products of combustion include light and smoke.
6.   For a fire to start, there must be either a source of ignition or a sufficiently high temperature for the fuel.
7.   Fire is an exothermic chemical reaction. Exothermic reactions generate heat. Endothermic reactions consume more heat than they generate.
8.   Chemical reactions in a fire break materials down into basic elements.
9.   Loose atoms bond with each other to create substances that were not originally present.
10. Cooling is one of the principal ways to control a fire or put it out.
11. Carbon is found in almost every flammable substance.
12. In a fire, released carbon atoms combine with oxygen to form either carbon dioxide or carbon monoxide.

13. Carbon dioxide can deplete oxygen concentrations in the air near the fire.
14. Carbon monoxide is a colorless, odorless, deadly gas.
15. Hydrogen, found in most fuels, combines with oxygen to form water.
16. Synthetic polymers in plastics and vinyls often form deadly toxic fumes when they are consumed, melted, or disintegrated in the presence of fire or high heat.
17. Liquid and solid fuels are first converted to a vapor before they burn.
18. The trend with regard to safety standards is away from the traditional specifications-based approach to a performance-based approach.
19. Removing the fuel, starving the fire of oxygen, or cooling it below the combustion point may put out a fire.
20. An explosion is a very rapid, contained fire.
21. Heat always travels from a higher temperature to a lower one.
22. Excess heat is transferred to other objects by conduction, radiation, or convection.
23. Conduction is direct thermal energy heat transfer, molecule to molecule, through solids and liquids.
24. Radiation is electromagnetic wave heat transfer through air in a straight line to surrounding solids.
25. Convection is heat transfer through the movement of hot gases.
26. Spontaneous combustion is rare, but not impossible.
27. Almost everything in the industrial environment will burn.
28. Fires are classified according to their properties, which relate to the fuels.
29. Class A fires involve solid fuels.
30. Class B fires involve flammable liquids and gases.
31. Class C fires involve live electricity.
32. Class D fires involve combustible metals.
33. Special categories include extremely active oxidizers and flammables containing oxygen.
34. All common packing materials burn easily.
35. Fire walls are defined by their ability to slow the spread of fire.
36. Wood and textiles can be treated to reduce their flammability.
37. The flash point is the lowest temperature at which vapors are produced in sufficient concentration to burn, given a source of ignition.
38. The fire point is the lowest temperature at which vapors will continue to burn, given a source of ignition.
39. The auto-ignition temperature is the lowest point at which the vapors of a liquid or solid will self-ignite without a source of ignition.
40. Flammable liquids have a flash point below 100°F.
41. Combustible liquids have a flash point at or above 100°F.
42. Flammable and combustible liquids are each divided into three classifications.
43. Most flammable liquids are lighter than water; therefore, water cannot be used to put these fires out.
44. Most gases are lighter than air.
45. Many of the gases present in smoke and at a fire site are toxic to humans.

46. Most fatalities from fire result from breathing toxic gases and smoke, or from suffocating because of a lack of oxygen.
47. A red, blue, yellow, and white diamond label is used to identify hazards present when a substance burns.
48. Natural and generated electricity play a large role in causing fires.
49. Heat, in the form of hot surfaces, can be a source of ignition.
50. Automatic fire detection systems employ different means of detecting a fire.
51. The best way to reduce fires is to prevent their occurrence.

KEY TERMS AND CONCEPTS

Auto-ignition temperature
Automatic sprinkler systems
Carbon
Carbon dioxide
Carbon monoxide
Combustible liquids
Combustion
Combustion point
Conduction
Convection
Endothermic
Exothermic
Explosion
Explosive range
Fire
Fire point
Fixed extinguishing system
Flame-resistant clothing
Flammable gases
Flammable liquids
Flammable range
Flashing back
Flash point
Fuel
Heat

Heat transfer
Hydrogen
Hypergolic
Ignition temperature
Infrared detectors
Ionization sensors
Isolation
Kinetic energy
Life Safety Code
Means of egress
Oxygen
Photoelectric fire sensors
Polymers
Products of combustion
Pyrophor hypergolic fuels
Radiation
Radiation sensors
Source of ignition
Spontaneous combustion
Standpipe and hose systems
Thermal energy
Thermal expansion detectors
Ultraviolet detectors
Vapor

REVIEW QUESTIONS

1. What are the three elements of the fire triangle?
2. Fire is a chemical reaction. What is going on?
3. Where is carbon found?
4. Compare and contrast carbon monoxide and carbon dioxide.
5. How is combustion of liquids and solids different from gases?
6. What are the three methods of heat transfer? Describe each.
7. What can happen to a pile of oil-soaked rags in a closed container? Describe the process.
8. Which directions does a fire normally travel?
9. Name something in this room that will not burn.
10. What are the classes of fires?
11. What property do almost all packing materials share?
12. What are the differences among flash point, fire point, and auto-ignition temperature?
13. Which are more stable: combustible liquids or flammable liquids?
14. Which way do gases usually travel?
15. Describe the NFPA hazards identification system.
16. In what four ways can electricity cause a fire?
17. What are the leading causes of fire-related deaths?
18. What are some of the toxic chemicals often produced by fires?
19. What are some of the systems utilized by smoke detectors?
20. What is the most successful life-saving preparation for a fire disaster?
21. Explain the best ways to prevent an office fire.
22. What is the trend with regard to future fire safety standards?
23. Define the term *life safety.*
24. What types of fabrics are prohibited in environments that are flame-or arc-prone?

ENDNOTES

1. "Worker Killed in Vapor Cloud Ignition at Tank Storage Facility," *NFPA Journal,* May/June 1991, Vol. 85, No. 3, p. 29.
2. Ibid., pp. 29–30.
3. Ibid., p. 31.
4. Kavianian, H. R., and Wentz, C. A., Jr. *Fire Safety Manual* (New York: Van Nostrand Reinhold, 1990), p. 175.
5. Asfahl, C. R. *Industrial Safety and Management* (Upper Saddle River, NJ: Prentice Hall, 1990), p. 220.
6. Vogel C. "Fires Can Raze Office Buildings," *Safety & Health,* September 1991, Vol. 144, No. 3, p. 27.
7. Ibid., pp. 26–27.
8. Cote, A. E. "Will Fire Safety Standards Survive in the 21st Century?" *NFPA Journal,* July/August 1991, p. 37.
9. Ibid., p. 42.

10. Ibid.
11. Birginal, K. "Extinguish Fire-Safety Myths," *Safety & Health,* September 1991, Vol. 144, No. 3, pp. 76–77.
12. National Fire Protection Association. *Life Safety Code* (NFPA 101), 1997, pp. 101–19
13. Ibid., pp. 101-26 through 101–50.
14. Neal, Thomas E. "The Need for Flame-Resistant Protective Clothing," *Occupational Safety,* May 1997, p. 96.
15. C.F.R. 1910.269, paragraph 1(6).

CHAPTER FOURTEEN

Industrial Hygiene: Toxic Substances, Explosive Materials, and Confined Spaces

Industrial hygiene is an area of specialization within the broader field of industrial safety and health. This chapter provides prospective and practicing safety and health professionals with the information they need to know about this area of specialization.

OVERVIEW OF INDUSTRIAL HYGIENE

Industrial hygiene is an area of specialization in the field of industrial safety and health that is concerned with predicting, recognizing, assessing, controlling, and preventing environmental stressors in the workplace that can cause sickness or serious discomfort to workers. An environmental stressor is any factor in the workplace that can cause enough discomfort to result in lost time or illness. Common stressors include gases, fumes, vapors, dusts, mists, noise, and radiation.

The Code of Ethics of the American Academy of Industrial Hygiene describes the responsibilities of industrial hygienists:

■ To ensure the health of employees

■ To maintain an objective approach in recognizing, assessing, controlling, and preventing health hazards regardless of outside pressure and influence

■ To help employees understand the precautions that they should take to avoid health problems

■ To respect employers' honesty in matters relating to industrial hygiene

■ To make the health of employees a higher priority than obligations to the employer[1]

Role of the Safety and Health Professional

The role of modern safety and health professionals vis-à-vis industrial hygiene often depends on the size of the company employing them. Large companies often employ professionals who specialize in industrial hygiene. These specialists have titles such as occupational physician, industrial hygienist, industrial toxicologist, and health physicist. In smaller companies, safety and health professionals often have responsibility for all safety and health matters, including industrial hygiene.

In companies that employ specialists, their recommendations are used by safety and health professionals to develop, implement, monitor, and evaluate the overall safety and health program. If specialists are not employed, safety and health professionals are responsible for seeking the advice and assistance necessary to predict, recognize, assess, control, and overcome environmental stressors that might cause sickness or serious discomfort to employees.

OSHAct AND INDUSTRIAL HYGIENE

The principal piece of federal legislation relating to industrial hygiene is the Occupational Safety and Health Act of 1970 (OSHAct) as amended in 1994. The OSHAct sets forth the following requirements relating to industrial hygiene:

- Use of warning labels and other means to make employees aware of potential hazards, symptoms of exposure, precautions, and emergency treatment
- Prescription of appropriate personal protective equipment and other technological preventive measures (29 C.F.R. Subpart I. 1910.133 and 1910.134)
- Provision of medical tests to determine the effect on employees of exposure to environmental stressors
- Maintenance of accurate records of employee exposures to environmental stressors that are required to be measured or monitored
- Accessibility of monitoring tests and measurement activities to employees
- Availability of monitoring tests and measurement activities records to employees on request
- Notification of employees who have been exposed to environmental stressors at a level beyond the recommended threshold and corrective action being taken

OSHA Process Safety Standard

The **OSHA Process Safety Standard** is found in 29 C.F.R. 1910.119. Its purpose is to prevent *catastrophic* accidents caused by major releases of highly hazardous chemicals. To comply with this standard, companies must have written operating procedures, mechanical integrity programs, and formal incident investigation procedures. Other key elements are as follows:

1. *Coverage.* Although the process safety standard is typically associated with large chemical and petrochemical processing plants, its coverage is actually much broader than this. Any company is covered that uses the threshold amount of a chemical listed in the standard—or 10,000 pounds or more of a flammable material at one site in one location.

2. *Employee participation.* Section (c) of the standard requires that employees be involved in all aspects of the process safety management program. In addition, employees must be given access to information developed as part of the program.

3. *Process safety information (PSI).* Section (d) of the standard requires organizations to establish and maintain process safety information files. Information included in the files includes chemical, process, and equipment data.

4. *Process hazard analyses (PHAs).* Section (e) of the standard requires that companies conduct process hazard analyses for all processes covered by the standard. Like any other hazard analysis, the PHAs are supposed to identify potential problems so that prompt corrective action or preventive measures can be taken.

5. *Standard operating procedures (SOPs).* Section (f) of the standard requires employees to establish and maintain written standard operating procedures for using chemicals safely. The requirement applies to handling, processing, transporting, and storing chemicals.

6. *Requirements for contractors.* Section (h) of the standard describes the special requirements imposed on companies that contract portions of their work to other com-

panies. Complying with the standard is a matter of making sure that contractors comply. The following requirements are imposed by Section (h):

■ Screen contractors before issuing a contract to ensure that they have a comprehensive safety and health program in place.

■ Orient contractors concerning the chemicals with which they might be required to work or be around, the emergency action plan, and other pertinent information.

■ Evaluate contractors periodically to ensure that their safety performance is acceptable.

■ Maintain an OSHA injury and illness log for the contractor that is separate from, and in addition to, that of the host company.

OSHA Regulation for Chemical Spills

OSHA issues a special regulation dealing with chemical spills. The standard (29 C.F.R. 1910.120) is called the Hazardous Waste Operations and Emergency Response, or **HAZWOPER,** Standard. HAZWOPER gives organizations two options for responding to a chemical spill. The first is to evacuate all employees in the event of a spill and to call in professional emergency response personnel. Employers who use this option must have an emergency action plan (EAP) in place in accordance with 29 C.F.R. 1010.38(a). The second option is to respond internally. Employers using this must have an emergency response plan that is in accordance with 29 C.F.R. 1010.120.

1. *Emergency action plans (EAPs)*. An **emergency action plan** should have at least the following elements: alarm systems, evacuation plan, a mechanism or procedure for emergency shutdown of the equipment, and a procedure for notifying emergency response personnel.

2. *Emergency response plan*. Companies that opt to respond internally to chemical spills must have an **emergency response plan** that includes the provision of comprehensive training for employees. OSHA Standard 29 C.F.R. 1910.120 specifies the type and amount of training required, ranging from awareness to in-depth technical training for employees who will actually deal with the spill. It is important to note that OSHA forbids the involvement of untrained employees in responding to a spill. The following topics are those covered in the HAZWOPER seminar provided by Environmental Safety Awareness, a safety and health company in Fort Walton Beach, Florida. These topics are typical of those covered in up-to-date HAZWOPER courses.

■ Summary of key federal laws
■ Overview of impacting regulations
■ Classification and categorization of hazardous waste
    Definition of hazardous waste
    Characteristics
    TCLP
    Lists of hazardous wastes

■ Hazardous waste operations

Definitions
Levels of response

■ Penalties for noncompliance
Civil penalty policy

■ Responses to spills
Groundwater contamination
Sudden releases
Clean-up levels
Risk assessment
Remedial action

■ Emergency response
Workplan
Site evaluation and control
Site specific safety and health plan
Information and training program
Personal protective equipment
Monitoring
Medical surveillance
Decontamination procedures
Emergency response
Other provisions

■ Contingency plans
Alarm systems
Action plan

■ Personal protective equipment
Developing a PPE program
Respiratory equipment
Protective clothing
Donning PPE
Doffing PPE

■ Material safety data sheets
Introduction
Preparing MSDSs
MSDS information
Hazardous ingredients
Physical/chemical characteristics

Fire and explosion hazard data
Reactivity data
Health hazard data
Precautions for safe handling and use
Control measures

■ Site control
Site maps
Site preparation
Work zones
Buddy system
Site security
Communications
Safe work practices

■ Hazardous waste containers
Emergency control
Equipment
Tools
Safety

■ Decontamination
Types
Decontamination plan
Prevention of contamination
Planning
Emergencies
Physical injury
Heat stress
Chemical exposure
Medical treatment area
Decontamination of equipment
Decontamination procedures
Sanitation of PPE
Disposal of contaminated materials

In a given year, industrial facilities in the United States release almost 6 billion pounds of toxic substances into the environment. The Environmental Protection Agency (EPA) began measuring such emissions in 1987. Since that time, the trend in toxic emissions has been downward. Even with the downward trend, however, in a typical year, toxic substances will be released into the environment in the following amounts:

- 2.4 billion pounds released into the air
- 1.2 billion pounds injected into underground wells
- 916 million pounds placed in treatment and disposal facilities
- 551 million pounds placed in municipal wastewater treatment plants
- 445 million pounds placed in landfills
- 101 million pounds released into the water[2]

People are exposed to a variety of substances every day in the home and at work—paints, paint remover, detergent, cleaning solvents, antifreeze, and motor oil, to name just a few. Most substances with which we interact are not dangerous in small amounts or limited exposure. However, high levels of exposure to certain substances in high concentrations can be dangerous. Levels of exposure and concentration, as well as how we interact with substances, help to determine how hazardous substances are. In addition, some substances used frequently in certain industrial settings can explode. Following is some information that modern safety and health professionals need to know about toxic and explosive substances.

HAZARDS IN THE WORKPLACE

The environmental stressors on which industrial hygiene focuses can be divided into the following broad categories: chemical, physical, biological, and ergonomic hazards. Typical **chemical hazards** include mists, vapors, gases, dusts, and fumes. Chemical hazards are either inhaled or absorbed through the skin or both. **Physical hazards** include noise, vibration, extremes of temperature, and excessive radiation (electromagnetic or ionizing). **Biological hazards** come from molds, fungi, bacteria, and insects. Bacteria may be introduced into the workplace through sewage, food waste, water, or insect droppings. **Ergonomic hazards** are related to the design and condition of the workplace. Poorly designed workstations and/or tools are ergonomic hazards. Conditions that put workers in awkward positions and impair their visibility are also hazards.

Material Safety Data Sheets

Employees should be warned of chemical hazards by labels on containers or material safety data sheets **(MSDSs)**. MSDSs are special sheets that summarize all pertinent information about a specific chemical. The hazard communication standard of the OSHAct requires that chemical suppliers provide users with an MSDS for each chemical covered by the standard.

An MSDS should contain the following information as appropriate: manufacturer's name, address, and telephone number; a list of hazardous ingredients; physical/chemical characteristics; fire and explosion hazard information; reactivity information; health hazard information; safety precautions for handling; and recommended control procedures.

An MSDS must contain specified information in eight categories:

Section I: General information. This section contains directory information about the manufacturer of the substance, including the following: manufacturer's name and address, telephone number of an emergency contact person, a nonemergency telephone number for information, and a dated signature of the person who developed or revised the MSDS.

Section II: Hazardous ingredients. This section should contain the common name, chemical name, and Chemical Abstracts Service (CAS) number for the substance. Chemical names are the scientific designations given in accordance with the nomenclature system of the International Union of Pure and Applied Chemistry. The CAS number is the unique number for a given chemical that is assigned by the Chemical Abstracts Service.

Section III: Physical and chemical characteristics. Data relating to the vaporization characteristics of the substance are contained in this section.

Section IV: Fire and explosive hazard data. Data relating to the fire and explosion hazards of the substance are contained in this section. Special fire-fighting procedures are also included in this section.

Section V: Reactivity data. Information concerning the stability of the substance as well as the potential for hazardous decomposition or polymerization of the substance is contained in this section.

Section VI: Health hazards. This section contains a list of the symptoms that might be suffered as a result of overexposure to the substance. Emergency first aid procedures are also explained in this section.

Section VII: Safe handling and use. This section explains special handling, storage, spill, and disposal methods and precautions relating to the substance.

Section VIII: Control measures. The types of ventilation, personal protective equipment, and special hygienic practices recommended for the substance are explained in this section.

Environmental Stressors

Noise is sound that is unwanted or that exceeds safe limits. It can cause problems ranging from annoyance to hearing loss. Acceptable levels of noise have been recommended by OSHA, NIOSH, and the Environmental Protection Agency (EPA). OSHA recommends that an employee's exposure level be limited to 90 dBA calculated as an eight-hour, time-weighted average. NIOSH and the EPA recommend limiting exposure to 85 dBA. Applying the OSHA recommendation, 14 percent of the working population are employed in situations that expose them to excess noise levels.[4]

Temperature control is the most basic way to eliminate environmental hazards:

People function efficiently only in a very narrow body temperature range, a "core" temperature measured deep inside the body, not on the skin or at body extremities. Fluctuations in core temperatures exceeding about 2°F below, or 3°F above, the normal core temperature of 37.6°C (99.6°F), which is 37°C mouth temperature (98.6°F mouth temperature), impair performance markedly. If this five-degree range is exceeded, a health hazard exists.[5]

Radiation hazards are increasingly prevalent in the age of high technology. In the category of **ionizing radiation,** safety and health professionals are concerned with five kinds of **radiation** (alpha, beta, X-ray, gamma, and neutron). Of these, alpha radiation is the least penetrating, making shielding simple, whereas the others are more difficult to shield against. Meters and other instruments are available to measure radiation levels in the workplace. The greatest risk for nonionizing radiation in the modern workplace comes from lasers. Shielding requirements for lasers are described in the construction regulations of the OSHAct.[6]

Extremes of pressure also represent a potential hazard in the workplace. According to Olishifski,

> One of the most common troubles encountered by workers under compressed air is pain and congestion in the ears from inability to ventilate the middle ear properly during compression and decompression. As a result, many workers subjected to increased air pressures suffer from temporary hearing loss; some have permanent hearing loss. This damage is believed to be caused by obstruction of the Eustachian tubes, which prevents proper equalization of pressure from the throat to the middle ear. The effects of reduced pressure on the worker are much the same as the effects of decompression from a high pressure. If pressure is reduced too rapidly, decompression sickness and ear disturbances similar to the diver's conditions can result.[7]

Biological hazards from various biological organisms can lead to disease in workers. The now-famous outbreak of what has come to be known as Legionnaire's disease is an example of what can result from biological hazards. This disease first surfaced at a convention where numerous participants became sick and soon died. The cause was eventually traced back to bacteria that grew in the cooling/air-moving systems serving the convention center. That bacteria has since been named *Legionnella*.

Ergonomic hazards are conditions that require unnatural postures and unnatural movement. The human body can endure limited amounts of unnatural postures or motions. However, repeated exposure to such conditions can lead to physical stress and injury. Design of tools, workstations, and jobs can lead to or prevent ergonomic hazards.

TOXIC SUBSTANCES DEFINED

A **toxic substance** is one that has a negative effect on the health of a person or animal. Toxic effects are a function of several factors, including the following: (1) properties of the substance, (2) amount of the dose, (3) level of exposure, (4) route of entry, and (5) resistance of the individual to the substance. Smith describes the issue of toxic substances as follows:

> When a toxic chemical acts on the human body, the nature and extent of the injurious response depends upon the dose received—that is, the amount of the chemical that actually enters the body or system and the time interval during which this dose was administered. Response can vary widely and might be as little as a cough or mild respiratory irritation or as serious as unconsciousness and death.[8]

ENTRY POINTS FOR TOXIC AGENTS

The development of preventive measures to protect against the hazards associated with industrial hygiene requires first knowing how toxic agents enter the body. A toxic substance must first enter the bloodstream to cause health problems. The most common routes of entry for toxic agents are inhalation, absorption, injection, and ingestion (see Figure 14–1). These routes are explained in the following paragraphs.

Inhalation

The route of entry about which safety and health professionals should be most concerned is **inhalation.** Airborne toxic substances such as gases, vapors, dust, smoke, fumes, aerosols, and mists can be inhaled and pass through the nose, throat, bronchial tubes, and lungs to enter the bloodstream. The amount of a toxic substance that can be inhaled depends on the following factors: (1) concentration of the substance, (2) duration of exposure, and (3) breathing volume.
According to Olishifski,

> Inhalation, as a route of entry, is particularly important because of the rapidity with which a toxic material can be absorbed in the lungs, pass into the bloodstream, and reach the brain. Inhalation is the major route of entry for hazardous chemicals in the work environment.[9]

Absorption[10]

The second most common route of entry in an industrial setting is **absorption,** or passage through the skin and into the bloodstream. The human skin is a protective barrier against many hazards. However, certain toxic agents can penetrate the barrier through absorption. Of course, unprotected cuts, sores, and abrasions facilitate the process, but even healthy skin will absorb certain chemicals. Humans are especially susceptible to absorbing such chemicals as organic lead compounds, nitro compounds, organic phosphate pesticides, TNT, cyanides, aromatic amines, amides, and phenols.

Figure 14–1
Common routes of entry of
toxic substances.

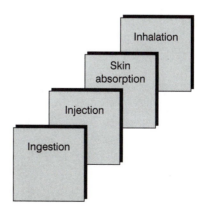

With many substances, the rate of absorption and, in turn, the hazard levels increase in a warm environment. The extent to which a substance can be absorbed through the skin depends on the factors shown in Figure 14–2. Another factor is body site. Different parts of the body have different absorption capabilities. For example, the forearms have a lower absorption potential than do the scalp and forehead.

Ingestion[11]

Ingestion, not a major concern in an industrial setting, is entry through the mouth. An ingested substance is swallowed. It moves through the stomach into the intestines and from there into the bloodstream. Toxic agents sometimes enter the body by ingestion when they are accidentally consumed by workers eating lunch or a snack. Airborne contaminants can also rest on food or the hands and, as a result, be ingested during a meal or snack. The possibility of ingesting toxic agents makes it critical to confine eating and drinking to sanitary areas away from the worksite and to make sure that workers practice good **personal hygiene** such as washing their hands thoroughly before eating or drinking.

As it moves through the gastrointestinal tract, the toxic substance's strength may be diluted. In addition, depending on the amount and toxicity of the substance, the liver may be able to convert it to a nontoxic substance. The liver can, at least, decrease the level of toxicity and pass some of the substance along to the kidneys, where some of the substance is eliminated in the urine.

Figure 14–2
Substances can be absorbed through the skin.

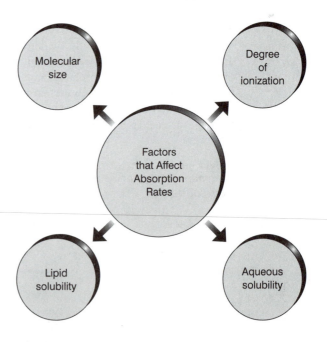

Injection

Injection involves introducing a substance into the body using a needle and syringe. Consequently, this is not often a route of entry for a toxic substance in the workplace. Injection is sometimes used for introducing toxic substances in experiments involving animals. However, this approach can produce misleading research results because the needle bypasses some of the body's natural protective mechanisms.

EFFECTS OF TOXIC SUBSTANCES

The effects of toxic substances vary widely, as do the substances themselves. However, all of the various effects and exposure times can be categorized as being either acute or chronic.

Acute effects/exposures involve a sudden dose of a highly concentrated substance. They are usually the result of an accident (a spill or damage to a pipe) that results in an immediate health problem ranging from irritation to death. Acute effects/exposures are (1) sudden, (2) severe, (3) typically involve just one incident, and (4) cause immediate health problems. Acute effects/exposures are not the result of an accumulation over time.

Chronic effects/exposures involve limited continual exposure over time. Consequently, the associated health problems develop slowly. The characteristics of chronic effects/exposures are (1) continual exposure over time, (2) limited concentrations of toxic substances, (3) progressive accumulation of toxic substances in the body and/or progressive worsening of associated health problems, and (4) little or no awareness of exposures on the part of affected workers.

When a toxic substance enters the body, it eventually affects one or more body organs. Part of the liver's function is to collect such substances, convert them to non-toxics, and send them to the kidneys for elimination in the urine. However, when the dose is more than the liver can handle, toxics move on to other organs, producing a variety of different effects. The organs that are affected by toxic substances are the blood, kidneys, heart, brain, central nervous system, skin, liver, lungs, and eyes. Figure 14–3 lists some of the more widely used toxic substances and the organs that they endanger most.

RELATIONSHIP OF DOSES AND RESPONSES

Safety and health professionals are interested in predictability when it comes to toxic substances. How much of a given substance is too much? What effect will a given dose of a given substance produce? These types of questions concern dose-response relationships. A **dose** of a toxic substance can be expressed in a number of different ways depending on the characteristics of the substance, for example, amount per unit of body

Blood	Kidneys	Heart	Brain
Benzene	Mercury	Aniline	Lead
Carbon monoxide	Chloroform		Mercury
Arsenic			Benzene
Aniline			Manganese
Toluene			Acetaldehyde

Eyes	Skin	Lungs	Liver
Cresol	Nickel	Asbestos	Chloroform
Acrolein	Phenol	Chromium	Carbon tetrachloride
Benzyl chloride	Trichloroethylene	Hydrogen sulfide	Toluene
Butyl alcohol		Mica	
		Nitrogen dioxide	

Figure 14–3
Selected toxic substances and the organs that they endanger.

weight, amount per body-surface area, or amount per unit of volume of air breathed. Smith expresses the dose-response relationship mathematically as follows:[12]

$$(C) \times (T) = K$$

where:

C = concentration
T = duration (time) of exposure
K = constant

Note that in this relationship, C times T is *approximately* equal to K. The relationship is not exact.

Three important concepts to understand relating to doses are dose threshold, lethal dose, and lethal concentration. These concepts are explained in the following paragraphs.

Dose Threshold

The **dose threshold** is the minimum dose required to produce a measurable effect. Of course, the threshold is different for different substances. In animal tests, thresholds are established using such methods as (1) observing pathological changes in body tissues, (2) observing growth rates (are they normal or retarded?), (3) measuring the level of food intake (has there been a loss of appetite?), and (4) weighing organs to establish body weight to organ weight ratios.

Lethal Dose

A **lethal dose** of a given substance is the dose that is highly likely to cause death. Such doses are established through experiments on animals. When lethal doses of a given substance are established, they are typically accompanied by information that is of value to medical professionals and industrial hygienists. Such information includes the type of animal used in establishing the lethal dose, how the dose was administered to the animal, and the duration of the administered dose. Lethal doses do not apply to inhaled substances. With these substances, the concept of lethal concentration is applied.

Lethal Concentration

A **lethal concentration** of an inhaled substance is the concentration that is highly likely to result in death. With inhaled substances, the duration of exposure is critical because the amount inhaled increases with every unprotected breath.

AIRBORNE CONTAMINANTS[13]

It is important to understand the different types of **airborne contaminants** that may be present in the workplace. Each type of contaminant has a specific definition that must be understood in order to develop effective safety and health measures to protect against it. The most common types of airborne contaminants are dusts, fumes, smoke, aerosols, mists, gases, and vapors (Figure 14–4).

Dusts

Dusts are various types of solid particles that are produced when a given type of organic or inorganic material is scraped, sawed, ground, drilled, handled, heated, crushed, or

Figure 14–4
Common airborne contaminants.

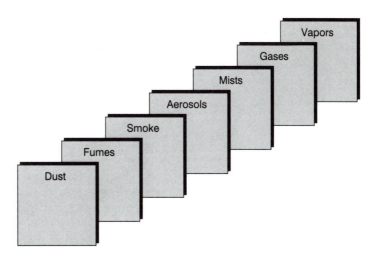

otherwise deformed. The degree of hazard represented by dust depends on the toxicity of the parent material and the size and level of concentration of the particles.

Fumes

The most common causes of **fumes** in the workplace are such manufacturing processes as welding, heat treating, and metalizing, all of which involve the interaction of intense heat with a parent material. The heat volatilizes portions of the parent material, which then condenses as it comes in contact with cool air. The result of this reaction is the formation of tiny particles that can be inhaled.

Smoke

Smoke is the result of the incomplete combustion of carbonaceous materials. Because combustion is incomplete, tiny soot and/or carbon particles remain and can be inhaled.

Aerosols

Aerosols are liquid or solid particles that are so small they can remain suspended in air long enough to be transported over a distance. They can be inhaled.

Mists

Mists are tiny liquid droplets suspended in air. Mists are formed in two ways: when vapors return to a liquid state through condensation, and when the application of sudden force or pressure turns a liquid into particles.

Gases

Unlike other airborne contaminants that take the form of either tiny particles or droplets, **gases** are formless. Gases are actually formless fluids. Gases become particularly hazardous when they fill a confined, unventilated space. The most common sources of gases in an industrial setting are from welding and the exhaust from internal combustion engines.

Vapors

Certain materials that are solid or liquid at room temperature and at normal levels of pressure turn to **vapors** when heated or exposed to abnormal pressure. Evaporation is the most common process by which a liquid is transformed into a vapor.

In protecting workers from the hazards of airborne contaminants, it is important to know the permissible levels of exposure for a given contaminant and to monitor continually the level of contaminants using accepted measurement practices and technologies. The topic of exposure thresholds is covered later in this chapter.

EFFECTS OF AIRBORNE TOXICS

Airborne toxic substances are also classified according to the type of effect they have on the body. The primary classifications are shown in Figure 14–5 and explained in the paragraphs that follow. With all airborne contaminants, concentration and duration of exposure are critical concerns.

Irritants

Irritants are substances that cause irritation to the skin, eyes, and the inner lining of the nose, mouth, throat, and upper respiratory tract. However, they produce no irreversible damage. According to Smith,

> Irritants can be subdivided into primary and secondary irritants. A primary irritant is a material that exerts little systemic toxic action, either because the products formed on the tissues of the respiratory tract are nontoxic or because the irritant action is far in excess of any systemic toxic action. A secondary irritant produces irritant action on mucous membranes, but this effect is overshadowed by systemic effects resulting from absorption. Normally, irritation is a completely reversible phenomenon.[14]

Asphyxiants

Asphyxiants are substances that can disrupt breathing so severely that suffocation results. Asphyxiants may be simple or chemical in nature. A simple asphyxiant is an inert gas that dilutes oxygen in the air to the point that the body cannot take in enough air to satisfy its needs for oxygen. Common simple asphyxiants include carbon dioxide, ethane, helium, hydrogen, methane, and nitrogen. Chemical asphyxiants, by chemical action, interfere with the passage of oxygen into the blood or the movement of oxygen from the lungs to body tissues. Either way, the end result is suffocation due to insufficient or no oxygenation. Common chemical asphyxiants include carbon monoxide, hydrogen cyanide, and hydrogen sulfide.

Figure 14–5
Airborne toxic substances.

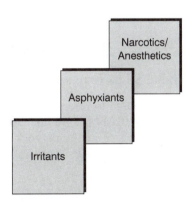

Narcotics/Anesthetics

Narcotics and anesthetics are similar in that carefully controlled dosages can inhibit the normal operation of the central nervous system without causing serious or irreversible effects. This makes them particularly valuable in a medical setting. Dentists and physicians use narcotics and anesthetics to control pain before and after surgery. However, if the concentration of the dose is too high, narcotics and anesthetics can cause unconsciousness and even death. When this happens, death is the result of asphyxiation. Widely used narcotics and anesthetics include acetone, methyl-ethyl-ketone, acetylene hydrocarbons, ether, and chloroform.

EFFECTS OF CARCINOGENS

A **carcinogen** is any substance that can cause a malignant tumor or a **neoplastic growth.** A *neoplasm* is cancerous tissue or tissue that might become cancerous. Other terms used synonymously for carcinogen are *tumorigen, oncogen,* and *blastomogen.*

According to Smith,

> It is well established that exposure to some chemicals can produce cancer in laboratory animals and man. There are a number of factors that have been related to the incidence of cancer—the genetic pattern of the host, viruses, radiation including sunshine, and hormone imbalance, along with exposure to certain chemicals. Other factors such as cocarcinogens and tumor accelerators are involved. It is also possible that some combination of factors must be present to induce cancers. There is pretty good clinical evidence that some cancers are virus-related. It may be that a given chemical in some way inactivates a virus, activates one, or acts as a cofactor.[15]

Medical researchers are not sure exactly how certain chemicals cause cancer. However, there are a number of toxic substances that are either known, or are strongly suspected, to be carcinogens. These include coal tar, pitch, creosote oil, anthracene oil, soot, lamp black, lignite, asphalt, bitumen waxes, paraffin oils, arsenic, chromium, nickel compounds, beryllium, cobalt, benzene, and various paints, dyes, tints, pesticides, and enamels.[16]

SAFETY FACT

Respirable Silica Exposure and Lung Disease

Employees exposed to respirable silica are at risk of developing the lung disease, silicosis. Particularly hazardous activities are hammering, drilling, chipping, crushing, loading, and dumping rock. Employees involved in abrasive blasting are also at risk. Symptoms of silicosis include coughing, difficulty in breathing, fever, and night sweats.

ASBESTOS HAZARDS

The Environmental Protection Agency (EPA) estimates that approximately 75 percent of the commercial buildings in use today contain asbestos in some form.[17] Asbestos was once thought to be a miracle material because of its many useful characteristics including fire resistance, heat resistance, mechanical strength, and flexibility. As a result, asbestos was widely used in commercial and industrial construction between 1900 and the mid-1970s[18] (see Figure 14–6).

In the mid-1970s, medical research clearly tied asbestos to respiratory cancer, scarring of the lungs (now known as *asbestosis)*, and cancer of the chest or abdominal lining *(mesothelioma)*.[19] It was finally banned by the EPA in 1989.

The following quote from *Occupational Hazards* on **friable asbestos** shows why asbestos is still a concern even though its further use has been banned:

> When asbestos becomes friable (crumbly), it can release fibers into the air that are dangerous when inhaled. As asbestos-containing material (ACM) ages, it becomes less viable and more friable. Asbestos can be released into the air if it is disturbed during renovation or as a result of vandalism.[20]

OSHA has established an exposure threshold known as the permissible exposure limit (PEL) for asbestos. The PEL for asbestos is 0.2 fibers per cubic centimeter of air for an eight-hour time-weighted average.[21] The addresses for OSHA and other sources of further information about asbestos are given in Figure 14-7.

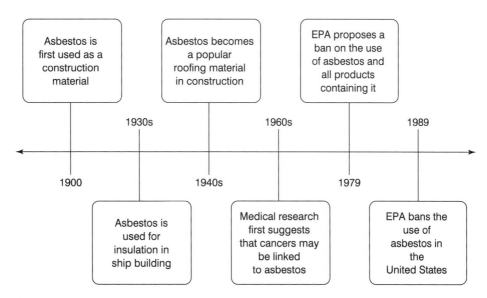

Figure 14–6
Asbestos use, 1900 to present.

Asbestos Action Program
EPA
Mail Code TS-799
401 M Street, S.W.
Washington, D.C. 20460
202-382-3949

Cancer Information Service
National Cancer Institute
Bldg. 31, Room 10A24
9000 Rockville Pike
Bethesda, MD 20892
800-4CANCER

NIOSH Publications Office
4676 Columbia Pkwy.
Cincinnati, OH 45226
513-533-8287

OSHA Publications Office
Room N–3101
200 Constitution Avenue, N.W.
Washington, D.C. 20210
202-523-9649

Asbestos Abatement Council of AWCI
1600 Cameron Street
Alexandria, VA 22314
703-684-2924

Asbestos Information Association of
  North America
1745 Jefferson Davis Highway, Suite 509
Arlington, VA 22202
703-979-1150

The National Asbestos Council
1777 Northeast Expressway, Suite 150
Atlanta, GA 30329
404-633-2622

Figure 14–7
Sources of information about asbestos in the workplace.

Removing and Containing Asbestos

When an industrial facility is found to contain asbestos, safety and health professionals are faced with the question of whether to remove it or contain it. According to Hughes, before making this decision, the following factors should be considered:

- Is there evidence that the ACM is deteriorating? What is the potential for future deterioration?
- Is there evidence of physical damage to the ACM? What is the potential for future damage?
- Is there evidence of water damage to the ACM or spoilage? What is the potential for future damage or spoilage?[22]

Several approaches can be used for dealing with asbestos in the workplace. The most widely used are removal, enclosure, and encapsulation. These methods are explained in the following paragraphs.

Removal[23]

Asbestos removal is also known as *asbestos abatement.* The following procedures are recommended for removal of asbestos: (1) the area in question must be completely enclosed in walls of tough plastic; (2) the enclosed area must be ventilated by high-efficiency particle absolute **(HEPA)** filtered negative air machines (these machines work somewhat like a vacuum cleaner in eliminating asbestos particles from the enclosed area); (3) the ACM must be covered with a special liquid solution to cut down on the release of asbestos fibers; and (4) the ACM must be placed in leakproof containers for disposal.

Enclosure[24]

Enclosure of an area containing ACMs involves completely encapsulating the area in air-tight walls. The following procedures are recommended for enclosing asbestos: (1) use HEPA-filtered negative air machines in conjunction with drills or any other tools that may penetrate or otherwise disturb ACMs; (2) construct the enclosing walls of impact-resistant and airtight materials; (3) post signs indicating the presence of ACMs within the enclosed area; and (4) note the enclosed area on the plans of the building.

Encapsulation[25]

Encapsulation of asbestos involves spraying the ACMs with a special sealant that binds them together, thereby preventing the release of fibers. The sealant should harden into a tough, impact-resistant skin. This approach is generally used only on acoustical plaster and similar materials.

Personal Protective Equipment for Asbestos Removal

It is important to use the proper types of personal protective clothing and respiratory devices. Clothing should be disposable and should cover all parts of the body.[26] According to Hughes, respirators used when handling asbestos should be "high-efficiency cartridge filter type (half-and-full-face types); any powered-air purifying respirator; any type C continuous-flow supplied-air, pressure-demand respirator, equipped with an auxiliary positive pressure self-contained breathing apparatus."[27]

Medical Records and Examinations[28]

It is important that employees who handle ACMs undergo periodic medical monitoring. Medical records on such employees should be kept current and maintained for at least 20 years. They should contain a complete medical history on the employee. These records must be made available on request to employees, past employees, health-care professionals, employee representatives, and OSHA personnel.

Medical examinations, conducted at least annually, should also be required for employees who handle ACMs. These examinations should include front and back chest

X-rays, that are at least 7 inches by 14 inches. The examination should also test pulmonary function, including forced vital capacity and forced expiratory volume at one second.

VENTILATION AND THE "SICK-BUILDING" SYNDROME[29]

The Bartow (Florida) County Courthouse cost $37 million by the time the ribbon was cut at its grand opening ceremony. Just five years later, the courthouse had to be evacuated, and employees had to be set up in temporary quarters. The problem? What has come to be known as the **"sick-building" syndrome.** In reality, a "sick" building is one that makes people sick because it has become infested with mold, mildew, spores, and other airborne micro-organisms. Although much is still unknown about "sick-building" syndrome, the Environmental Protection Agency (EPA) estimates that as many as 30 percent of the buildings in the United States have air quality problems.

Poor indoor air quality can cause a variety of health problems ranging from the temporary to the long-term. Health problems commonly associated with poor indoor air quality include allergic reactions, respiratory problems, eye irritation, sinusitis, bronchitis, and pneumonia. Often, the cause of poor indoor air quality can be slipshod maintenance such as failure to keep fans, ductwork, and filters clean. Other contributors are the particles and gases that can be released by office equipment, carpets, paints, cleaning solvents, and office supplies.

One of the keys to preventing "sick-building" syndrome is air exchange. Important factors in a building's ability to eliminate contaminated air and bring in fresh air are as follows:

- Ventilation
- Air infiltration rates
- Airflow rates in ducts
- Airflow patterns
- Fume exhaust

The most accurate methods available for measuring these factors fall under the broad heading of **tracer gas techniques.** A tracer gas is any gas or vapor not normally found in a building. The best have the following characteristics:

- Nontoxic
- Nonallergic at the levels used
- Chemically inert
- Odorless and tasteless
- Nonflammable and nonexplosive
- Easily transported
- Easily dispersed as an atmospheric gas
- Easily and economically measured with a high degree of reliability

The most widely used tracer gases are sulfur hexafluoride, halogenated refrigerants, and perfluorocarbons. To perform a tracer gas test, the following materials and equipment are needed:

- A suitable tracer gas
- A device for measuring tracer gas concentrations
- An air-sampling system
- A tracer gas injection system
- A data acquisition/control system

There are several different types of tracer gas tests including the following: tracer decay, constant concentration, buildup/decay, CO_2 measurement, and re-entrainment/recirculation. Regardless of the type of test used, the testing process involves the following steps:

1. Inject the tracer gas into the building.
2. Measure the concentration of the tracer gas in different parts of the building at different times over a period of time.

The data collected during a tracer gas test can give safety and health professionals the following types of information:

- Total air exchange rate for the building
- Air change rate due to the operation of the building's HVAC system
- Air change rate due to air infiltration and leakage
- Percent of outside air supplied by the building's HVAC system
- Effectiveness of the ventilation system in removing contaminants
- The distribution of the ventilation air throughout the building

With this type of information available, safety and health personnel can determine if there are pockets where contaminated air is trapped; if the ventilation and air infiltration rates are sufficient; if airflow rates through ducts are sufficient; if airflow patterns are what they should be; and if fume hoods are performing as they should. This type of information is needed to detect and prevent indoor air quality problems.

SAFETY FACT

Indoor Air Quality Information

Information about indoor air quality is available from the Environmental Protection Agency (EPA). The EPA's information hotline is open from 9:00 A.M. to 5:00 P.M., Eastern Standard Time, Monday–Friday. The hotline number is 1-800-438-4318.

EXPLOSIVE HAZARDS

In June 1991, 39 people were injured, six fatally, when an explosion occurred at a chemical plant in Charleston, South Carolina. The explosion occurred as employees were mixing chemicals used in making a fire retardant. In June 1992, an explosion and fire in a refinery in Cheyenne, Wyoming, injured six employees. In May 1991, a fire and explosion in a Sterlington, Louisiana, fertilizer plant killed eight workers and injured more than 100. What all of these incidents have in common is that toxic substances unexpectedly reached the flash point, causing an explosion and fire. As a result, people were killed and injured.

Many chemical and toxic substances used in modern industry are flammable or combustible. Consequently, under certain conditions, they will explode. Working in these conditions involves hazards that require special precautions for handling, storing, transporting, and using such substances.

Explosives-Related Concepts[30]

Safety relating to explosive materials is a highly specialized field. This section discusses terms and concepts used in this field of safety and health with which modern safety and health professionals should be familiar.

- A **flammable substance** is any substance with a flash point below 100°F (37.8°C) and a vapor pressure of less than 40 pounds per square inch at 100°F. Such liquids are also known as Class I liquids. They tend to be compositions of hydrogen and carbon such as crude oil and its numerous by-products.
- A **combustible substance** is any substance with a flash point of 100°F (37.8°C) or higher. Such liquids are known as Class II liquids. They also tend to be compositions of hydrogen and carbon such as crude oil and its numerous by-products.
- The **flash point** is the lowest temperature at which a substance gives off sufficient vapors to combine with air to form an ignitable mixture. Ignition can be precipitated by a spark.
- The **auto-ignition temperature** is the lowest temperature at which a vapor-producing substance or a flammable gas will ignite even without the presence of a spark or a flame. This is sometimes known as *spontaneous ignition*.
- In most cases, a certain amount of oxygen must be present in a vapor-air mixture for an explosion to occur. The amount that must be present for a given substance is the **oxygen limit** for that substance.
- **Volatility** is the evaporation (vaporization) capability of a given substance. The greater the tendency of a substance to vaporize, the more volatile it is.

Common Uses of Flammable/Combustible Substances[31]

Flammable and combustible substances are widely used in modern industry. Therefore, the hazards associated with them are not limited just to the industries producing such materials and substances. The National Safety Council lists the following as common uses of flammable and combustible substances in modern industry and specific related precautions that should be taken with each.

Dip Tanks

Dipping operations involving flammable or combustible substances should take place in a stand-alone one-story building constructed of noncombustible materials. The building should be (1) well ventilated, (2) clearly marked as a hazardous area, (3) free of ignition sources, (4) and large. The dip tank itself should be covered and should contain an automatic fire extinguishing system.

Japanning and Drying Ovens

Ovens used to evaporate varnish, japan enamel, and any other combustible substance should be (1) well ventilated, (2) equipped with an automatic fire protection system, and (3) have a shutdown system that activates automatically in the event of a fire or explosion.

Oil Burners

Selecting the proper type of fuel for use in an oil burner is the best precaution to prevent the accumulation and potential ignition of soot. The safest fuel to use in an oil burner is one that meets the following criteria: (1) flash point higher than 100°F (37.8°C), (2) hydrocarbon-based, and (3) acid- and grit-free. In addition, the supply tank should be located outside of the building housing the oil burner and should be underground. The top of the storage tank should be lower than all pipes entering it. Finally, the oil burner should have an automatic system for preventing the discharge of unburned oil into a hot firebox.

Cleaning Metal Parts

Many of the solvents used to clean metal parts are combustible. The primary precaution when cleaning metal parts is selecting substances that are not easily ignited. Additional precautions include ventilation and the selection of a cleaning area that is free of ignition sources.

Internal Combustion Engines

Internal combustion engines are widely used in modern industry for powering such equipment as forklifts and lift trucks. Because they are typically fueled with gasoline or diesel fuel, there are fire and explosive hazards associated with this operation. Precautions include the following: (1) proper maintenance, (2) good housekeeping, (3) shutdown of engines and cooling of exhaust pipes before filling fuel tanks, and (4) a well-ventilated area for filling fuel tanks.

Spray Painting Booths

The hazard associated with spray painting booths is that an explosive mixture of paint vapor and air can occur. To prevent such occurrences, proper ventilation is critical. Regular cleaning of the booth to remove accumulated spray deposits is also important. Paint booths should be equipped with automatic fire protection systems.

The preceding lists only some of the ways that potentially explosive materials are used in modern industry. There are many others. The precautions set forth in the following section apply to the processes that they accompany. General precautions that apply more broadly are explained in the next section.

Other Health Hazards of Explosive Materials

The health hazards associated with explosions and fires are well known. The potential for serious injury or death from the force of a blast or from burns is very high. However, there are other hazards associated with explosive and combustible materials. These include skin irritation, intoxication, and suffocation.

Irritation can occur when the skin comes in contact with hazardous substances. The degree of irritation can range from minor to severe, depending on the type of substance, its concentration, and the duration of contact. Intoxication can occur when an employee breathes the vapors of combustible substances. This can cause impaired judgment, performance, and reaction time, which can, in turn, result in an accident. Finally, the vapors from combustible materials can accumulate in confined spaces. When this happens, the air becomes contaminated and is both toxic and explosive. In such cases, the hazard of suffocation must be added to those associated with explosives.[32]

CONFINED SPACES HAZARDS

A **confined space** is any area with limited means of entry and exit that is large enough for a person to fit into but is not designed for occupancy. Examples of confined spaces include vaults, vats, silos, ship compartments, train compartments, sewers, and tunnels. What makes confined spaces hazardous, beyond those factors that define the concept, is their potential to trap toxic and/or explosive vapors and gases.

In addition to the toxic and explosive hazards associated with confined spaces, there are often physical hazards. For example, tunnels often contain pipes that can trip an employee or that can leak and cause a fall. Empty liquid or gas storage vessels may contain mechanical equipment or pipes that must be carefully maneuvered around, often in the dark.

THRESHOLD LIMIT VALUES

How much exposure to a toxic substance is too much? How much is acceptable? Guidelines that answer these questions for safety and health professionals are developed and issued annually by the American Conference of Governmental Industrial Hygienists (ACGIH). The guidelines are known as **threshold limit values,** or **TLVs.** The ACGIH describes threshold limit values as follows:

> Threshold limit values refer to airborne concentrations of substances and represent conditions under which it is believed that nearly all workers may be repeatedly exposed day after

day without adverse effect. Because of wide variation in individual susceptibility, however, a small percentage of workers may experience discomfort from some substances at concentrations at or below the threshold limit; a smaller percentage may be affected more seriously by aggravation of preexisting condition or by development of an occupational illness.

Threshold limits are based on the best available information from industrial experience, from experimental human and animal studies, and, when possible, from a combination of the three. The basis on which the values are established may differ from substance to substance; protection against impairment of health may be a guiding factor for some, whereas reasonable freedom from irritation, narcosis, nuisance, or other forms of stress may form the basis for others.[33]

EXPOSURE THRESHOLDS

An **exposure threshold** is a specified limit on the concentration of selected chemicals. Exposure to these chemicals that exceeds the threshold may be hazardous to a worker's health. Threshold recommendations are revised and updated frequently as new chemicals are introduced and as more is learned about existing chemicals. Since exposure threshold data are prone to frequent change, actual values for given chemicals are not presented in this section. Further, it is not the purpose of this section to relate any specific threshold values. Rather, this section will help prospective and practicing safety and health professionals understand the language used to express threshold values so that they can interpret threshold data, regardless of the format in which they are published.

The three most important concepts for understanding exposure thresholds are (1) time-weighted average (TWA), (2) short-term exposure limit, and (3) exposure ceiling.

- The **time-weighted average (TWA)** is the average concentration of a given substance to which employees may be safely exposed over an 8-hour workday or a 40-hour work week.[34]

- A **short-term exposure limit** is the maximum concentration of a given substance to which employees may be safely exposed for up to 15 minutes without suffering irritation, chronic or irreversible tissue change, or narcosis to a degree sufficient to increase the potential for accidental injury, impair the likelihood of self-rescue, or reduce work efficiency.[35]

- The exposure ceiling refers to the concentration level of a given substance that should not be exceeded at any point during an exposure period.[36]

Calculating a TWA

Olishifski gives the following formula for calculating the TWA for an eight-hour day:

$$TWA = \frac{CaTa + CbTb + \cdots CnTn}{8}$$

where:

Ta = time of the first exposure period during the eight-hour shift

Ca = concentration of substances in period a
Tb = another time period during the same shift
Cb = concentration during period b
Tn = nth or final time period in the eight-hour shift
Cn = concentration during period n[37]

HAZARD RECOGNITION AND EVALUATION

Understanding the degree and nature of the hazard is needed before effective hazard control procedures can be developed. This involves recognizing that a hazard exists and then making judgments about its magnitude with regard to chemical, physical, biological, and/or ergonomic stresses.

Questions that can be used for recognizing hazards in the workplace are as follows:

1. What is produced?
2. What raw materials are used in the process?
3. What additional materials are used in the process?
4. What equipment is used?
5. What operational procedures are involved?
6. What dust control procedures are involved?
7. How are accidental spills cleaned up?
8. How are waste by-products disposed?
9. Is there adequate ventilation?
10. Are processes equipped with exhaust devices?
11. How does the facility layout contribute to employee exposure?
12. Are properly working personal protective devices available?
13. Are safe operating procedures recorded, made available, monitored, and enforced?[38]

Olishifski recommends that all industrial processes be subjected to the following **hazard recognition** procedures:

■ Determine the exposure threshold for each hazardous substance identified when applying the questions just listed, including airborne contaminants.
■ Determine the level of exposure to each hazardous substance.
■ Determine which employees are exposed to each hazardous material, how frequently, and for how long.
■ Calculate the TWAs to the exposure thresholds identified earlier.[39]

For **hazard evaluation,** the following considerations are important: the nature of the material/substance involved, the intensity of the exposure, and the duration of the exposure. Key factors to consider are how much exposure is required to produce injury or illness; the likelihood that enough exposure to produce injury or illness will take place; the rate of generation of airborne contaminants; the total duration of exposure; and the prevention/control measures used.[40]

The textile industry has a long history of dealing with industrial hygiene problems. In 1980, the textile industry was ranked fifth on a list of 43 industries in terms of preventing workplace injuries and fatalities. By 1987, this ranking had improved to first. Much of the improvement is attributed to hazard evaluation and recognition efforts within the industry to comply with the OSHA cotton-dust standard (1910.1043).[41]

The textile industry has become a leader in the area of industrial safety and health. Even so, there are still problems. According to *Safety & Health,*

> Although the textile industry has stopped the use of cancer-associated dyes, as many as 50,000 current and former industry workers have had significant major exposure during their work lives and are at high risk for developing bladder cancer.[42]

One of the ways in which textile companies are dealing with their industrial hygiene problems is by adding more safety and health professionals to their staffs. According to Bone, there will be "a growth of safety positions at the plant level. Each plant is going to need one safety-and-health professional to keep up with regulations."[43]

Allied-Signal, Inc., includes industrial hygiene as a major component in its overall safety and health program. A corporate-wide environmental auditing program for recognizing and evaluating hazards is the mainstay of Allied-Signal's industrial hygiene effort. According to Moretz,

> The company's board of directors became involved in an environmental auditing program, in part, . . . after its involvement with Kepone, made for the company when it was called Allied Chemical Co. [known as the Kepone Tragedy]. The story broke in the mid-1970s, when it was found that workers at Life Sciences Products Co., Hopewell, Va., suffered from Kepone poisoning. Possible long-term effects included liver damages, sterility, and cancer. Environmental concerns were raised.[44]

Each year Allied-Signal conducts about 50 environmental audits, of which eight or nine focus on industrial hygiene. Functional areas audited include occupational health, medical programs, air pollution control, water pollution control and spill prevention, solid waste disposal, hazardous waste disposal, safety/loss prevention, and product safety.[45] The hazard evaluation and recognition that result from an audit give Allied-Signal a basis on which to develop prevention and control strategies.

PREVENTION AND CONTROL

Most prevention and control strategies can be placed in one of the following four categories: (1) engineering controls, (2) ventilation, (3) personal protective equipment, and (4) administrative controls.[46] Examples of strategies in each category are given in the following paragraphs.

Engineering Controls

The category of **engineering controls** includes such strategies as replacing a toxic material with one that is less hazardous or redesigning a process to make it less stressful or to reduce exposure to hazardous materials or conditions. Other engineering controls are isolating a hazardous process to reduce the number of people exposed to it and introducing moisture to reduce dust.[47]

Ventilation

Exhaust ventilation involves trapping and removing contaminated air. This type of ventilation is typically used with such processes as abrasive blasting, grinding, polishing, buffing, and spray painting/finishing. It is also used in conjunction with open-surface tanks. Dilution ventilation involves simultaneously removing and adding air to dilute a contaminant to acceptable levels.[48]

Personal Protection from Hazards

When the work environment cannot be made safe by any other method, **personal protective equipment (PPE)** is used as a last resort. PPE imposes a barrier between the worker and the hazard but does nothing to reduce or eliminate the hazard. Typical equipment includes safety goggles, face shields, gloves, boots, earmuffs, earplugs, full-body clothing, barrier creams, and respirators.[49]

Occasionally, in spite of an employee's best efforts in wearing PPE, his or her eyes or skin will be accidentally exposed to a contaminant. When this happens, it is critical to wash away or dilute the contaminant as quickly as possible. Specially designed eye wash and emergency wash stations such as those shown in Figures 14-8, 14-9, and 14-10 should be readily available and accessible in any work setting where contaminants may be present.

Figure 14-8
Haws Model 700 BT Pedestal
OMNI-FLO eye-face station.
Courtesy of Haws Drinking
Faucet Company.

Figure 14-9
Haws Model 700BT Wall-
Mounted Eyewash Station.
Courtesy of Haws Drinking
Faucet Company.

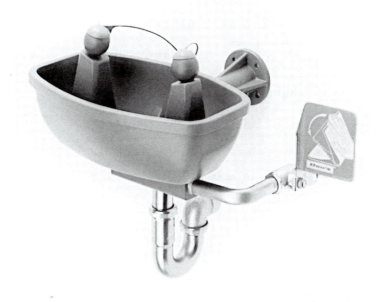

Figure 14-10
Haws Model 8730 Emer-
gency Polar Shower Booth.
Courtesy of Haws Drinking
Faucet Company.

Administrative Controls

Administrative controls involve limiting the exposure of employees to hazardous conditions using such strategies as the following: rotating schedules, required breaks, work shifts, and other schedule-oriented strategies.[50]

Additional Strategies

The type of prevention and control strategies used will depend on the evaluation of the specific hazards present in the workplace. The Society of Manufacturing Engineers recommends the following list of generic strategies that apply regardless of the setting:

- Practicing **good housekeeping,** including workplace cleanliness, waste disposal, adequate washing and eating facilities, healthful drinking water, and control of insects and rodents
- Using special control methods for specific hazards, such as reduction of exposure time, film badges and similar monitoring devices, and continuous sampling with preset alarms
- Setting up medical programs to detect intake of toxic materials
- Providing training and education to supplement engineering controls[51]

Self-Protection Strategies

One of the best ways to protect employees from workplace hazards is to teach them to protect themselves. Modern safety and health professionals should ensure that all employees are familiar with the following rules of **self-protection:**[52]

1. *Know the hazards in your workplace.* Take the time to identify all hazardous materials/conditions in your workplace and know the safe exposure levels for each.

2. *Know the possible effects of hazards in your workplace.* Typical effects of workplace hazards include respiratory damage, skin disease/irritation, injury to the reproductive system, and damage to the blood, lungs, central nervous system, eyesight, and hearing.

3. *Use personal protective equipment properly.* Proper use of personal protective equipment means choosing the right equipment, getting a proper fit, correctly cleaning and storing equipment, and inspecting equipment regularly for wear and damage.

4. *Understand and obey safety rules.* Read warning labels before using any contained substance, handle materials properly, read and obey signs, and do only authorized work.

5. *Practice good personal hygiene.* Wash thoroughly after exposure to a hazardous substance, shower after work, wash before eating, and separate potentially contaminated work clothes from others before washing them.

NIOSH AND INDUSTRIAL HYGIENE

The National Institute for Occupational Safety and Health (NIOSH) is part of the Department of Health and Human Services (HHS). This agency is important to industrial hygiene professionals. The main focus of the agency's research is on toxicity levels and human tolerance levels of hazardous substances. NIOSH prepares recommendations for OSHA standards dealing with hazardous substances, and NIOSH studies are made available to employers.

The areas of research of NIOSH's four major divisions—Biomedical and Behavioral Science; Respiratory Disease Studies; Surveillance, Hazard Evaluations, and Field Studies; and Training and Manpower Development—were discussed in Chapter 4. The results of these divisions' studies and their continually updated lists of toxic materials and recommended tolerance levels are extremely helpful to industrial hygienists concerned with keeping the workplace safe.

NIOSH GUIDELINES FOR RESPIRATORS

The **respirator** is one of the most important types of personal protective equipment available to individuals who work in hazardous environments. Because the performance of a respirator can mean the difference between life and death, the National Institute for Occupational Safety and Health (NIOSH) publishes strict guidelines regulating the manufacture of respirators. The standard with which manufacturers must comply is 42 C.F.R. Part 84. In addition, safety and health professionals must ensure that employees are provided respirators that meet all of the specifications set forth in 42 C.F.R. Part 84.

There are two types of respirators: air filtering and air supplying. **Air-filtering respirators** filter toxic particulates out of the air. To comply with 42 C.F.R. Part 84, an air-filtering respirator must protect its wearer from the most penetrating aerosol size of particle, which is 0.3 microns aerodynamic mass in median diameter. The particulate filters

---

SAFETY MYTH

Contact Lenses and Respirators

Fire fighters and other personnel who wear respirators should not wear contact lenses. Right? Not necessarily. OSHA Standard 29 C.F.R. 1910.134(e)(5)(ii) forbids the practice of wearing contact lenses with a respirator, but the standard is based on old information relating to hard contact lenses. The problem with hard contacts is their tendency to pop out. However, most contact lens wearers now use the soft variety which stay in place better. Employees who wear soft contact lenses can use respirators.

used in respirators are divided into three classes, each class having three levels of efficiency as follows:

	Class N (Not Oil Resistant)	Class R (Oil Resistant)	Class P (Oil Proof)
Efficiency	95%	95%	95%
Efficiency	99%	99%	99%
Efficiency	100%	100%	100%

Class N respirators may be used only in environments that contain no oil-based particulates. They may be used in atmospheres that contain solid or non-oil contaminants. Class R respirators may be used in atmospheres containing any contaminant. However, the filters in Class R respirators must be changed after each shift if oil-based contaminants are present. Class P respirators may be used in any atmosphere containing any particulate contaminant.

If there is any question about the viability of an air-filtering respirator in a given setting, employees should use **air-supplying respirators.** This type of respirator works in much the same way as an air tank for a scuba diver. Air from the atmosphere is completely blocked out, and fresh air is provided via a self-contained breathing apparatus.

Air Safety Program Elements

Companies with facilities in which fumes, dust, gases, vapors, or other potentially harmful particulates are present should have an **air safety program** as part of their overall safety and health program. The program should have at least the following elements:

- Accurate hazard identification and analysis procedures to determine what types of particulates are present and in what concentration
- Standard operating procedures (in writing) for all elements of the air safety program
- Respirators that are appropriate in terms of the types of hazards present and that are 42 C.F.R. Part 84 approved
- Training including fit testing, limitations, use, and maintenance of respirators
- Standard procedures for routine procedures and storage of respirators

STANDARDS AND REGULATIONS

Standards and regulations relating to toxic substances come from a number of different sources, prominent among these are OSHA and the EPA. Important standards and regulations in this area include the OSHA Chemical Process Standard, the EPA Clean Air Act,

Title III of the Superfund Amendments and Reauthorization Act of 1986 (SARA), the Hazardous Materials Transportation Uniform Safety Act of 1990, and the Toxic Substances Control Act. These and other standards/regulations are explained in this section.

OSHA Chemical Process Standard

Accidental chemical releases and the explosions and fires that can subsequently result were the driving force behind the development of the **OSHA Chemical Process Standard.** An 1989 explosion and fire that occurred at Phillips Petroleum's Pasadena, Texas, plant killing 23 and injuring 100 people has been used as an example of the types of tragedies that OSHA is trying to prevent with this standard.[53]

The standard requires chemical producers to analyze their processes to identify potentially hazardous situations and to assess the extent of the hazard. Having done so, they must accommodate this knowledge in their emergency response plans (see Chapter 17) and take action to minimize the hazards. Specific additional requirements include the following:

- Compiling process safety information
- Maintaining safe operating procedures
- Training and educating employees
- Maintaining equipment
- Conducting incident investigations
- Developing emergency response plans
- Conducting safety compliance audits[54]

EPA Clean Air Act

President Bush signed the **Clean Air Act** amendments into law in November 1990, thereby renewing and extending the original Clean Air Act of 1970. LaBar describes the updated act as follows:

> Relying on technology-driven market-based strategies, the law is designed to reduce air pollution—in the form of hazardous air pollutants, acid rain, and smog—by 56 billion lb per year. This includes a 75 percent reduction in air toxics, a 50 percent cut in acid rain, and a 40 percent decrease in smog over the next 20 years or so.[55]

A key element of the Clean Air Act is its requirement that companies use the **maximum achievable control technology (MACT).** Such technologies represent the current state of the art in pollution control on an ever-changing, ever-improving basis. This means that as technological improvements occur, companies will have to upgrade their pollution control systems to stay in compliance.

The standard also focuses on preventing accidental releases of toxic substances. In addition to identifying potential hazards and taking steps to control them, companies are required to develop plans for minimizing the damage that results when, in spite of controls, releases occur.

The Clean Air Act should enhance a downward trend in air pollution that was already occurring when it was renewed in 1990. That trend saw the amount of toxics released into the air being reduced at a rate of approximately 650 million pounds per year. Of course, this is far short of the 56 billion pounds per year reduction required by the Clean Air Act Amendments of 1990, but it does represent movement in the right direction.

Superfund Amendments and Reauthorization Act

Title III of the Superfund Amendments and Reauthorization Act of 1986 **(SARA)** is also known as the Emergency Planning and Community Right-to-Know Act. This law is designed to allow individuals to obtain information about hazardous chemicals in their communities so that they can protect themselves in case of an emergency. It applies to all companies that use, make, transport, or store chemicals.

Safety and health professionals involved in developing emergency response plans for their companies should be familiar with SARA and its requirements relating to emergency planning. The major components of the Emergency Planning and Community Right-to-Know Act are discussed in the following paragraphs.

Emergency Planning (Sections 301–303)

Communities are required to form local **emergency planning** committees (LEPCs), and states are required to form state emergency response commissions (SERCs). LEPCs must develop emergency response plans for their local communities, host public forums, select a planning coordinator for the community, and work with the coordinator in developing local plans. SERCs must oversee LEPCs and review their emergency response plans. Plans for individual companies in a given community should be part of that community's larger plan. Local emergency response professionals should use their community's plan as the basis for simulating emergencies and practicing their response.

Emergency Notification (Section 304)

Chemical spills or releases of toxic substances that exceed established allowable limits must be reported to appropriate LEPCs and SERCs. Immediate **emergency notification** may be verbal, provided a written notification is filed promptly thereafter. The report must contain at least the following information: (1) names of the substances released, (2) where the release occurred, (3) when the release occurred, (4) the estimated amount of the release, (5) known hazards to people and property, (6) recommended precautions, and (7) name of a contact person in the company.

Information Requirements (Section 311)

Local companies are required to keep their LEPCs and SERCs—and, through them, the public—informed about the hazardous substances that they store, handle, transport, and/or use. These information requirements include keeping comprehensive, up-to-date records of the substances on file and readily available; providing copies of material safety data sheets for all hazardous substances; recording general storage locations for all hazardous substances; estimating the amount of each hazardous substance on hand on a given day; and estimating the average annual amount of hazardous substances kept on hand.

Toxic Chemical Release Reporting (Section 313)

Local companies must report the total amount of toxic substances released into the environment as either emissions or hazardous waste. Toxic chemical release reports go to the Environmental Protection Agency and the state-level environmental agency. Section 313 applies to companies that meet the following criteria:

- Employ ten or more full-time personnel
- Produce or process more than 25,000 pounds of a given toxic substance or use more than 10,000 pounds of a given toxic substance in any capacity
- Conduct 50 percent or more of their business in areas defined by Standard Industrial Classification (SIC) codes 20–39

Section 313 of SARA requires companies that produce, store, use, and/or transport chemicals to estimate and report their releases of toxic substances. According to LaBar, this requirement has led some companies to "adopt waste reduction, pollution prevention, and environmental stewardship as guiding principles."[56] He cites Monsanto, BP Chemicals, and Alcoa as examples.[57]

Hazardous Materials Transportation and Uniform Safety Act

Transportation of toxic substances always involves a certain amount of hazard. According to the National Safety Council, in a given year,

> ... hazardous materials incidents caused 165 injuries, 17 deaths and more than $21 million in damages. Most of these happened on the nation's highways where more than 60 percent of the materials are transported.... Human error caused most of the incidents. Package failure caused another 30 percent of incidents and vehicle crashes caused 8 percent.[58]

The Hazardous Materials Transportation and Uniform Safety Act of 1990 was passed in response to statistics such as these. The basic provisions of the act are shown in Figure 14–11. In addition to these provisions, the act makes companies that transport hazardous materials partially liable for damages when an accident occurs and the carrier does not have a satisfactory rating from the Department of Transportation.

Figure 14–11
Basic provisions of the Hazardous Materials Transportation and Uniform Safety Act.

- Training relating directly to specific functions for private manufacturers, shippers, and carriers.
- Training for employees of public agencies.
- National registration with registration fees used to help pay for the costs of emergency response training.
- A national permit system for motor carriers.

Schueth recommends the following steps for minimizing the risks associated with transporting hazardous materials:

- Hire personnel who are certified and experienced in dealing with hazardous materials. Then make sure to brief them properly on the hazardous materials that they will be handling and keep them up to date on the latest federal regulations.

- Establish a training program that covers at least the following topics: safety equipment, identification of hazardous shipments, routine handling procedures, and emergency procedures.

- At each location where hazardous materials will be loaded and/or unloaded, name one person who is responsible for making periodic inspections, instructing new employees, meeting with representatives of companies that ship major hazardous materials, maintaining comprehensive records, filing all necessary reports, and contacting the shipper and cosigner when there is a problem.

- Identify all hazardous materials and post a diamond-shaped placard on the outside of all vehicles used to transport hazardous materials.[59]

GENERAL SAFETY PRECAUTIONS[60]

Following are a number of general safety precautions that apply in any settings where explosive and combustible materials are present.

1. *Prohibit smoking.* Smoking should be prohibited in any areas of a plant where explosive and combustible materials are present. Eliminating potential sources of ignition is a standard safety precaution in settings where explosions and fire are possible. In the past, such areas were marked off as restricted, and No Smoking signs were posted. It is becoming common practice to prohibit smoking on the premises altogether or to restrict smoking to designated areas that are well removed from hazard areas.

2. *Eliminate static electricity.* Static electricity occurs when dissimilar materials come into contact and then separate. If these materials are combustible or are near other materials that are combustible, an explosion can occur. Therefore, it is important to eliminate static electricity. The potential for the occurrence of static electricity can be reduced substantially by the processes of grounding and bonding. **Bonding** involves eliminating the difference in static charge potential between materials. **Grounding** involves eliminating the difference in static charge potential between a material and the ground. Figures 14–12, 14–13, and 14–14 illustrate these concepts.

Figure 14–12
How static electricity occurs.

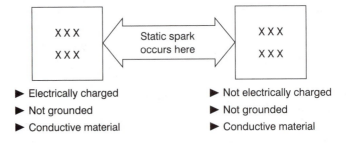

Figure 14–13
Bonding prevents static spark.

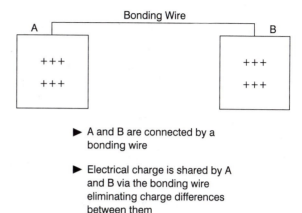

▶ A and B are connected by a bonding wire

▶ Electrical charge is shared by A and B via the bonding wire eliminating charge differences between them

3. *Use spark-resistant tools.* Sparks from tools used in settings where explosive and combustible materials are present represent a threat that must be addressed. In such settings, spark-resistant tools should be used to the maximum extent possible. Wooden, leather-faced, and rubber-covered tools can help prevent sparks that might ignite volatile materials.

OSHA CONFINED SPACE STANDARD

The OSHA standard relating to confined spaces is found in 29 C.F.R. 1910.146. This standard mandates that entry permits be required before employees are allowed to enter a potentially hazardous confined space. This means that an employee must have a written permit to enter a confined space. Before the permit is issued, a supervisor, safety/health professional, or some other designated individual should do the following:

Figure 14–14
Grounding prevents static spark.

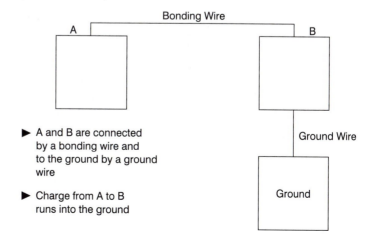

▶ A and B are connected by a bonding wire and to the ground by a ground wire

▶ Charge from A to B runs into the ground

1. *Shutdown equipment/power.* Any equipment, steam, gas, power, or water in the confined space should be shut off and locked or tagged to prevent its accidental activation.

2. *Test the atmosphere.* Test for the presence of airborne contaminants and to determine the oxygen level in the confined space. Fresh, normal air contains 20.8 percent oxygen. OSHA specifies the minimum and maximum safe levels of oxygen as 19.5 percent and 23.5 percent, respectively. Atmospheric tests indicate whether a respirator is required and, if so, what type, classification, and level. Figures 14-15, 14-16, 14-17, and 14-18 are examples of devices used for checking the atmosphere.

3. *Ventilate the space.* Spaces containing airborne contaminants should be purged to remove them. Such areas should also be ventilated to keep contaminants from building up again while an employee is working in the space.

4. *Have rescue personnel stand by.* Never allow an employee to enter a confined space without having rescue personnel standing by in the immediate vicinity. These personnel should be fully trained and properly equipped. It is not uncommon for an untrained, improperly equipped employee to be injured or killed trying to rescue a colleague who gets into trouble in a confined space.

5. *Maintain communication.* An employee outside of the confined space should stay in constant communication with the employee inside. Communication can be visual, verbal, or electronic (radio, telephone) depending on the distance between the employee inside and the entry point.

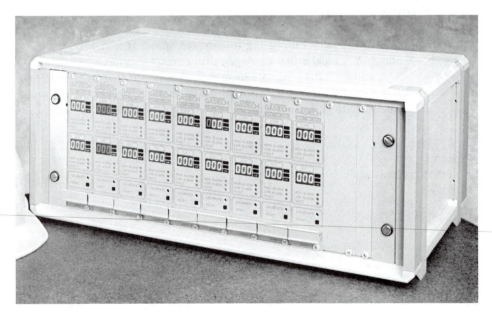

Figure 14-15
SAFE T NET 2000 detection device.
Courtesy of Gas Tech, Inc.

Figure 14-16
GT Land Surveyor detection
device.
Courtesy of Gas Tech, Inc.

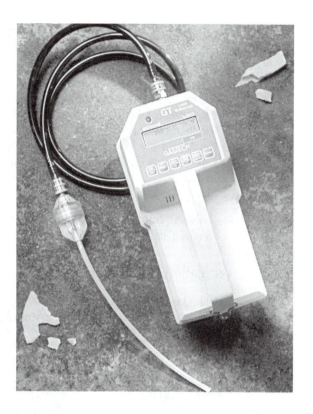

6. *Use a lifeline.* A lifeline attached to a full-body harness and a block and tackle will ensure that the employee who is inside can be pulled out should he or she lose consciousness. The apparatus should be rigged so that one employee working alone can pull an unconscious employee out of the confined space.

Ventilation of Confined Spaces[61]

Before allowing employees to enter a confined space, it is important to make the space as safe as possible. One of the most effective strategies for doing so is ventilation. Because confined spaces vary in size, shape, function, and hazard potential, there must be a number of different methods for ventilating them.

Before ventilating a confined space, it should be *purged*. Purging is the process of initially clearing the space of contaminants. Once the area has been purged, ventilation can begin. Ventilation is the process of continually moving fresh air through a space. Ventilation, when properly done, will accomplish the following:

■ Dilute and replace airborne contaminants that might still be present in the confined space.

■ Ensure an adequate supply of oxygen (between 19.5 and 23.5 percent).

DISCUSSION CASE

What Is Your Opinion?

"You'll be fine as long as you wear your respirator," said the supervisor to the employee as he entered the empty vessel. "There might be some residual toxic gas left over, but there won't be much. If you don't breathe it, the gas can't hurt you. Get in, do the inspection, and get out. It shouldn't take five minutes." Is this supervisor giving the employee accurate advice? What is your opinion?

■ Exhaust contaminants produced by work performed in the confined space (e.g., welding, painting, and so on).

Ventilation and Local Exhaust

Providing ventilation in a confined space can maintain a comfortable temperature, it can remove odors, and it can dilute contaminants. However, never depend solely on general

Figure 14-17
STM 2100 detection device.
Courtesy of Gas Tech, Inc.

Figure 14-18
SAFE T NET 150 detection device.
Courtesy of Gas Tech, Inc.

ventilation to remove toxic contaminants from the air. To eliminate the hazards posed by toxic contaminants such as solvent vapors and welding fumes, it is necessary to exhaust the confined space aggressively. The combination of initial purging, local exhaust, and ventilation is the ideal approach. If contaminant concentrations remain too high even with this approach, employees should wear an appropriate respirator.

Rescue Preparation[62]

The time to think about getting injured employees out of a confined space is well before they enter the space in the first place. Every year, employees are killed trying to save an injured colleague inside a confined space. In an attempt to save injured colleagues, well-meaning employees, who are neither properly trained nor adequately equipped, often fall victim to the toxic atmosphere themselves and die. This is a tragic circumstance made even more so because it is unnecessary and avoidable.

With the right amount of planning and training, employees can be quickly and effectively rescued from confined spaces. Planning should answer the following questions:

- What types of injuries/incidents might occur in a given space?
- What types of hazards might be present in the space?
- What precautions should be taken by rescue personnel entering the space (e.g., lifelines, hoist, respirator, and so on)?
- How much room is there to maneuver in the confined space?
- What if the victim needs first aid before he or she can be moved?

All of these questions should be answered in the organization's emergency action plan. In addition, all members of the rescue team should have received the training necessary to respond quickly, safely, and effectively. An effective response is one that is appropriate to the magnitude of the incident and is carried out safely.

OSHA STANDARDS FOR TOXIC AND HAZARDOUS MATERIALS

The OSHA standards for hazardous materials are contained in 29 C.F.R. (Subpart H). Nine of the standards apply to specific materials. Four of the standards have broader applications. The standards applying to specific materials are as follows:

Hazardous Materials (Specific Standards)

1910.101	Compressed gases
1910.102	Acetylene
1910.103	Hydrogen
1910.104	Oxygen
1910.105	Nitrous oxide
1910.108	Dip tanks
1910.109	Explosives and blasting agents. (An amendment to the OSHAct passed in 1992 now requires that manufacturers of explosives and pyrotechnics must observe the requirements of the Process Safety Management Standards in 1910.119 in addition to this standard.)
1910.110	Liquefied petroleum gases
1910.111	Anhydrous ammonia

In addition to these specific standards, Subpart H contains four standards that have broad applications. Standard 1910.106 parallels the National Fire Protection Association's NFPA 30: *Flammable and Combustible Liquids Code.* Standard 1910.107 regulates processes in which paint is applied by compressed air, electrostatic steam, or other continuous or intermittent processes. Standard 1910.119 regulates process safety management relating to 125 specific chemicals. Standard 1910.120 regulates both hazardous waste operations and spills or accidental releases.

Toxic and hazardous substances are covered in 29 C.F.R. (Subpart Z). The standards in this subpart establish permissible exposure limits (PELS) for over 450 toxic and hazardous substances. Each standard deals with a specific substance or substances. The standards contained in Subpart Z begin with 1910.1000 and run through 1910.1500.

=========== SUMMARY ============

1. The responsibilities of industrial hygienists include the following: ensuring the health of employees; objectively recognizing, assessing, controlling, and preventing health hazards; helping employees understand precautions; and making the health of employees a high priority.

2. The OSHAct established the following requirements relating to industrial hygiene: use of warning labels, use of personal protective equipment, medical testing, records maintenance, accessibility of information about monitoring activities open to employees, availability of such records to employees, and notification of exposure to environmental stressors.

3. The most prominent hazards in the workplace are chemical, physical, biological, and ergonomic.

4. The main routes of entry for toxic agents are inhalation, absorption, and ingestion.

5. The most common types of airborne contaminants are dusts, fumes, smoke, aerosols, mists, gases, and vapors.

6. Asbestos, once thought to be a miracle material, is now known to be an extremely hazardous substance. It has been tied to respiratory cancer, scarring of the lungs, and cancer of the chest or abdominal lining. When identified in the workplace, asbestos should be handled by removal, enclosure, or encapsulation.

7. The three most important concepts to understand concerning exposure thresholds are time-weighted average (TWA), short-term exposure limit, and exposure ceiling.

8. Hazard recognition procedures include the following: Determine the exposure threshold for each hazardous substance in the workplace, determine the level of exposure to each, determine which employees are exposed and for how long, and calculate the TWAs.

9. General prevention and control strategies include the following: substitution, process changes, isolation, moisture to reduce dust, exhaust methods, ventilation, personal protective equipment, good housekeeping, special control methods, medical programs, and education/training.

10. The National Institute for Occupational Safety and Health (NIOSH) is part of the Department of Health and Human Services (HHS). Its two broad functions are research and education in the areas of toxic materials and human tolerance levels.

11. A toxic substance is one that has a negative effect on the health of a person or animal. The effect produced by a toxic substance depends on its properties, the amount of the dose, the level of exposure, and the individual's resistance.

12. The route of entry of a toxic substance is an important consideration. Common routes of entry include ingestion, injection, absorption, and inhalation.

13. The dose threshold is the minimum dose of a toxic substance required to produce a measurable effect. A lethal dose is one that is highly likely to cause death. A lethal concentration of an inhaled substance is the concentration that is likely to cause death.

14. Exposures to toxic substances are either acute or chronic. Acute exposure involves sudden exposure to high concentrations of the substance in question. Chronic exposure involves limited but continual exposure to the substance in question.

15. Airborne contaminants are classified according to the type of effect that they have on the body. There are irritants, asphyxiants, and narcotics/anesthetics.

16. A carcinogen is any substance that can cause a malignant tumor or a neoplastic growth. Other terms used synonymously for carcinogen are *tumorigen, oncogen,* and *blastomogen.*

17. Pertinent standards relating to toxic substances include the OSHA Chemical Process Standard, the EPA Clean Air Act, the Superfund Amendments and Reauthorization Act, and the Hazardous Materials Transportation and Uniform Safety Act.

18. Threshold limit value (TLV) refers to airborne concentrations of substances and represents conditions under which it is believed that nearly all workers may be repeatedly exposed day after day, without adverse effect. TLVs are expressed as time-weighted average, short-term exposure limit, and ceiling.

19. Material safety data sheets are an excellent source of help for safety and health professionals concerned about the potential hazards of a given toxic substance. Information in an MSDS is presented in eight sections: general information, hazardous ingredients, physical and chemical characteristics, fire and explosive hazard data, reactivity data, health hazards, safe handling and use, and control measures.

20. Important concepts relating to explosive materials include flammable substance, combustible substance, flash point, auto-ignition temperature, oxygen limits, and volatility.

21. Common uses of flammable/combustible substances include dip tanks, japanning and drying ovens, oil burners, metal parts-cleaning processes, internal combustion engines, and spray painting booths.

22. General safety precautions that can be applied in any setting where explosives and combustibles are present include prohibiting smoking, eliminating static electricity, and using spark-resistant tools.

KEY TERMS AND CONCEPTS

Absorption	Asbestos removal
ACM	Asphyxiants
Acute effects/exposure	Auto-ignition temperature
Aerosols	Biological hazards
Airborne contaminants	Bonding
Air-filtering respirators	Carcinogen
Air safety program	Ceiling
Air-supplying respirators	Chemical hazards
Anesthetics	Chemical Process Standard

Chronic effects/exposure

Clean Air Act

Combustible substance

Confined space

Dose

Dose threshold

Dusts

Emergency action plan

Emergency notification

Emergency planning

Emergency response plan

Encapsulation

Enclosure

Engineering controls

Ergonomic hazards

Exposure ceiling

Exposure thresholds

Flammable substance

Flash point

Friable asbestos

Fumes

Gases

Good housekeeping

Grounding

Hazard recognition

HAZWOPER

HEPA

Industrial hygiene

Ingestion

Inhalation

Injection

Ionizing radiation

Irritants

Isolating or enclosing

Lethal concentration

Lethal dose

Local exhaust

Maximum achievable control technology (MACT)

Mists

MSDS

Narcotics

Neoplastic growth

Noise

Nonionizing radiation

OSHA Process Safety Standard

Oxygen limit

Personal hygiene

Personal protective equipment (PPE)

Physical hazards

Radiation

Respirator

Route of entry

SARA

Self-protection strategies

Short-term exposure limit

"Sick-building" syndrome

Smoke

Static electricity

Substituting

Temperature control

Threshold limit value (TLV)

Time-weighted average (TWA)

Toxic substance

Tracer gas techniques

Vapors

Ventilation

Volatility

Wet methods

REVIEW QUESTIONS

1. Define the term *industrial hygiene.*
2. Briefly explain the responsibilities of the modern industrial hygienist.
3. What is the role of the safety and health professional regarding industrial hygiene?
4. List five OSHA requirements relating to industrial hygiene.
5. Briefly explain the typical categories of hazards in the workplace.
6. What are the most common routes of entry for toxic agents?
7. Describe the following types of airborne contaminants: dusts, fumes, smoke, mists, and gases.
8. What factors should be considered in deciding whether to remove or contain asbestos?
9. Explain the following ways of dealing with asbestos in the workplace: removal, enclosure, and encapsulation.
10. What types of medical examinations should be required of employees who handle ACMs?
11. Briefly explain the following concepts relating to exposure thresholds: time-weighted average, short-term exposure limit, and exposure ceiling.
12. List the most important considerations when evaluating hazards in the workplace.
13. List five generic prevention and control strategies that can be used in any workplace.
14. Give an example of a prevention/control strategy in each of the following categories: engineering controls, ventilation, and personal protective equipment.
15. Explain five self-protection strategies that employees can use in the workplace.
16. How does NIOSH relate to industrial hygiene?
17. What is a toxic substance?
18. List the factors that determine the effect that a toxic substance will have.
19. Describe the most common routes of entry for toxic substances.
20. Explain the mathematical expression of the dose-response relationship.
21. Define the following terms: dose threshold, lethal dose, and lethal concentration.
22. Differentiate between acute and chronic effects/exposures.
23. List and describe the various classifications of airborne toxics.
24. What is a carcinogen?
25. Describe the basic provisions of the following standards: OSHA Chemical Process Standard, EPA Clean Air Act, and SARA.
26. What is a threshold limit value?
27. Define the following terms: *time-weighted average* and *ceiling.*
28. Differentiate between flammable and combustible substances.
29. Define the following terms: *flash point, auto-ignition temperature,* and *volatility.*
30. List and explain the general precautions that should be applied in any setting where explosive and combustible materials are present.
31. Explain the three NIOSH categories of respirators.
32. What is the "sick-building" syndrome?
33. Explain the major tenets of the OSHA Confined Space Standard.

===== ENDNOTES =====

1. Olishifski, J. B. "Overview of Industrial Hygiene," in *Fundamentals of Industrial Hygiene,* 3rd ed. (Chicago: National Safety Council, 1988), p. 3.
2. "EPA Releases 1989 TRIDATA," *Occupational Hazards,* July 1991, p. 28.
3. Smith, R. G., and Olishifski, J. B. (Revised by C. Zuez.) *Industrial Hygiene* (Chicago: National Safety Council, 1988), p. 382.
4. Olishifski, J. B. "Overview of Industrial Hygiene," pp. 164–65.
5. Ibid., p. 10.
6. Ibid., pp. 12–14.
7. Ibid., p. 14.
8. Smith, R. G., and Olishifski, J. B. *Industrial Hygiene*, p. 383.
9. Ibid., p. 17.
10. Ibid.
11. Ibid., p. 14.
12. Smith, R. G., and Olishifski, J. B. *Industrial Hygiene,* pp. 366–67.
13. Olishifski, J. B. "Overview of Industrial Hygiene," pp. 18–19.
14. Smith, R. G., and Olishifski, J. B. *Industrial Hygiene,* p. 367.
15. Ibid., p. 368.
16. Ibid., p. 369.
17. "Asbestos: Facts and Sources," *Occupational Hazards*, September 1989, p. 58.
18. Ibid.
19. Ibid.
20. Ibid.
21. Ibid.
22. Hughes, A. "Asbestos: Remove or Contain, That Is the Question," *Safety & Health,* May 1988, Vol. 137, No. 5, pp. 47–48.
23. Ibid., p. 48.
24. Ibid.
25. Ibid.
26. Ibid., p. 49.
27. Ibid.
28. Ibid.
29. Grot, Richard A., and Lagus, Peter L., "Tracing Building Ventilation," *Occupational Hazards,* February 1996, pp. 45–48.
30. National Safety Council. *Accident Prevention Manual for Industrial Operations,* 9th ed. (Chicago: National Safety Council, 1988), p. 423.
31. Ibid., p. 406.
32. Ibid., p. 424.
33. American Conference of Governmental Industrial Hygienists. Preface to List of Threshold Limit Values.
34. Olishifski, "Overview of Industrial Hygiene," p. 21.
35. Ibid.

36. Ibid.
37. Ibid.
38. Ibid.
39. Ibid.
40. Ibid., p. 22.
41. Bone, J. "Textile Industry Weaves a Safety Future," *Safety & Health,* September 1991, Vol. 144, No. 3, p. 17.
42. Ibid., p. 52.
43. Ibid., p. 53.
44. Moretz, S. "Industrial Hygiene Auditing: Allied-Signal Takes the Extra Step," *Occupational Hazards,* May 1989, p. 73.
45. Ibid., p. 74.
46. Olishifski, J. B. "Overview of Industrial Hygiene," p. 25.
47. Ibid.
48. Ibid., p. 26.
49. Ibid.
50. Ibid., p. 27.
51. Olishifski, J. B. "Overview of Industrial Hygiene," p. 24.
52. Florida Department of Labor and Employment Security, Division of Safety, Toxic Substances Information Center. "What You Should Know About On-the-Job Health," (Tallahassee: 1991), pp. 3–15.
53. Sheridan, P. J. "OSHA's Chemical Process Standard Sparks Controversy," *Occupational Hazards,* March 1991, p. 21.
54. Ibid., p. 22.
55. LaBar, G. "Seeing Past the Clean Air Act," *Occupational Hazards,* March 1991, p. 29.
56. Ibid., p. 27.
57. Ibid., pp. 27–29.
58. Schueth, T. "Stay on the Move with HAZMAT Transportation," *Today's Supervisor,* January 1992, pp. 12–13.
59. Ibid., p. 35.
60. National Safety Council. *Accident Prevention Manual for Industrial Operations,* p. 425.
61. Rekus, John F. "Confined Space Ventilation," *Occupational Hazards,* March 1996, pp. 35–38.
62. Rekus, John F. "Confined Space Rescue Planning," *Occupational Hazards,* March 1996, pp. 45–48.

Radiation Hazards

The widow of a construction worker who helped build the British Nuclear Fuels (BNF) Sellafield plant was awarded $286,500 when it was determined that his death from chronic myeloid leukemia was the result of overexposure to radiation. Sellafield was constructed for the purpose of separating uranium from used fuel rods. Working at the plant for approximately nine months, the victim received a total cumulative dose of almost 52 millisieverts of radiation, which exceeded the established limit for an entire 12-month period. BNF compensated the victim's wife and the families of 20 additional workers who died from causes related to radiation.

Radiation hazards in the workplace fall into one of two categories: ionizing or non-ionizing. This chapter provides prospective and practicing safety and health professionals with the information they need concerning radiation hazards in both categories.

IONIZING RADIATION: TERMS AND CONCEPTS

An ion is an electrically charged atom (or group of atoms) that became charged when a neutral atom (or group of atoms) lost or gained one or more electrons as a result of a chemical reaction. If an electron is lost during this process, a positively charged ion is produced; if an electron is gained, a negatively charged ion is produced. To ionize is to become electrically charged or to change into ions. Therefore, **ionizing radiation** is radiation that becomes electrically charged or changed into ions. Types of ionizing radiation, as shown in Figure 15–1, include alpha particles, beta particles, neutrons, X-radiation, gamma radiation, high-speed electrons, and high-speed protons.

To understand the hazards associated with radiation, safety and health professionals need to understand the basic terms and concepts summarized in the following paragraphs, adapted from C.F.R. (Code of Federal Regulations) 1910.96.

- **Radiation** consists of energetic nuclear particles and includes alpha rays, beta rays, gamma rays, X-rays, neutrons, high-speed electrons, and high-speed protons.
- **Radioactive material** is material that emits corpuscular or electromagnetic emanations as the result of spontaneous nuclear disintegration.
- A **restricted area** is any area to which access is restricted in an attempt to protect employees from exposure to radiation or radioactive materials.
- An **unrestricted area** is any area to which access is not controlled because there is no radioactivity hazard present.
- A **dose** is the amount of ionizing radiation absorbed per unit of mass by part of the body or the whole body.

Figure 15–1
Types of ionizing radiation.

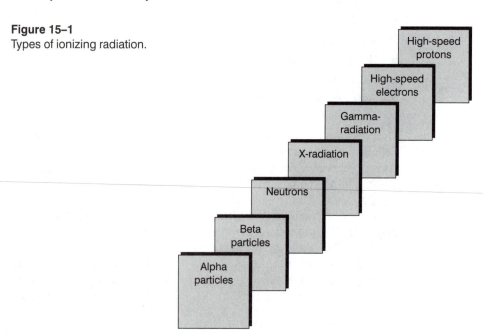

High-speed protons

High-speed electrons

Gamma-radiation

X-radiation

Neutrons

Beta particles

Alpha particles

- **Rad** is a measure of the dose of ionizing radiation absorbed by body tissues stated in terms of the amount of energy absorbed per unit of mass of tissue. One rad equals the absorption of 100 ergs per gram of tissue.

- **Rem** is a measure of the dose of ionizing radiation to body tissue stated in terms of its estimated biological effect relative to a dose of one roentgen (r) of X-rays.

- **Air dose** means that an instrument measures the air at or near the surface of the body where the highest dosage occurs to determine the level of the dose.

- **Personal monitoring devices** are devices worn or carried by an individual to measure radiation doses received. Widely used devices include film badges, pocket chambers, pocket dosimeters, and film rings.

- A **radiation area** is any accessible area in which radiation hazards exist that could deliver doses as follows: (1) within one hour, a major portion of the body could receive more than 5 millirems; or (2) within five consecutive days, a major portion of the body could receive more than 100 millirems.

- A **high-radiation area** is any accessible area in which radiation hazards exist that could deliver a dose in excess of 100 millirems within one hour.

EXPOSURE OF EMPLOYEES TO RADIATION

The **exposure** of employees to radiation must be carefully controlled and accurately monitored. Figure 15–2 shows the maximum doses for individuals in one **calendar quarter.** Employers are responsible for ensuring that these dosages are not exceeded.

There are exceptions to the amounts shown in Figure 15–2. According to OSHA, an employer may permit an individual in a restricted area to receive doses to the whole body greater than those shown in Figure 15–2 as long as the following conditions are met:

- During any calendar quarter, the dose to the whole body does not exceed three rems.

Figure 15–2
Ionizing radiation exposure
limits of humans.

Body/Body Region	Rems Per Calendar Quarter
Whole body	1.25
Head and trunk	1.25
Blood-forming organs	1.25
Lens of eyes	1.25
Gonads	1.25
Hands and forearms	18.75
Feet and ankles	18.75
Skin of whole body	7.50

■ The dose to the whole body, when added to the accumulated occupational dose to the whole body, shall not exceed 5(*N – 18*) rems, where N *is the employee's age in years at the last birthday.*

■ The employer maintains up-to-date past and current exposure records, which show that the addition of such a dose does not cause the employee to exceed the specified doses.[1]

Employers must ensure even more careful controls with individuals under 18 years of age. Such individuals may receive only doses that do not exceed 10 percent of those specified in Figure 15–2 in any calendar quarter.

OSHA is not the only agency that regulates radiation exposure. The **Nuclear Regulatory Commission (NRC)** is also a leading agency in this area. The NRC's regulations specify that the total internal and external dose for employees may not exceed five rems per year. This same revision established a total exposure limit of 0.6 rems over the entire course of a pregnancy for female employees. According to the NRC, the average radiation exposure of nuclear plant workers is less than 400 millirems annually.[2]

PRECAUTIONS AND PERSONAL MONITORING

Personal monitoring precautions are important for employees of companies that produce, use, release, dispose of, or store radioactive materials or any other source of ionizing radiation. Accordingly, OSHA requires the following precautions:

■ Employers must conduct comprehensive surveys to identify and evaluate radiation hazards present in the workplace from any and all sources.

■ Employers must provide appropriate personal monitoring devices such as film badges, pocket chambers, pocket dosimeters, and film rings.

■ Employers must require the use of appropriate personal monitoring devices by the following: (1) any employee who enters a restricted area where he or she is likely to receive a dose greater than 25 percent of the total limit of exposure specified for a calendar quarter; (2) any employee 18 years of age or less who enters a restricted area where he or she is likely to receive a dose greater than 5 percent of the total limit of exposure specified for a calendar quarter; and (3) any employee who enters a high-radiation area.[3]

CAUTION SIGNS AND LABELS

Caution signs and **labels** have always been an important part of safety and health programs. This is particularly true in companies where radiation hazards exist. The universal color scheme for caution signs and labels warning of radiation hazards is purple or magenta superimposed on a yellow background.

Both OSHA and the NRC require caution signs in radiation areas, high-radiation areas, airborne radiation areas, areas containing radioactive materials, and containers in which radioactive materials are stored and/or transported.[4]

Figure 15–3 shows the universal symbol for radiation. Along with the appropriate warning words, this symbol should be used on signs and labels. Figures 15–4, 15–5, 15–6, and 15–7 show the types of warning signs and labels that might be used in various radioactive settings. On containers, labels should also include the following information: (1) quantity of radioactive material, (2) kinds of radioactive materials, and (3) date on which the contents were measured.[5]

EVACUATION WARNING SIGNAL

Companies that produce, use, store, and/or transport radioactive materials are required to have a signal-generating system that can warn of the need for evacuation.[6] OSHA describes the **evacuation warning signal** system as follows:

> The signal shall be a mid-frequency complex sound wave amplitude modulated at a subsonic frequency. The complex sound wave in free space shall have a fundamental frequency (f_1) between 450 and 500 hertz (Hz) modulated at a subsonic rate between 4 and 5 hertz. The signal generator shall not be less than 75 decibels at every location where an individual may be present whose immediate, rapid, and complete evacuation is essential.[7]

In addition to this basic requirement, OSHA also stipulates the following:

- A sufficient number of signal generators must be installed to cover all personnel who may need to be evacuated.
- The signal shall be unique, unduplicated, and instantly recognizable in the plant where it is located.
- The signal must be long enough in duration to ensure that all potentially affected employees are able to hear it.
- The signal generator must respond automatically without the need for human activation, and it must be fitted with backup power.[8]

Figure 15–3
Universal radiation symbol.

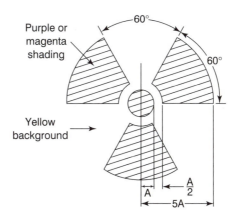

Figure 15–4
Sample warning sign.

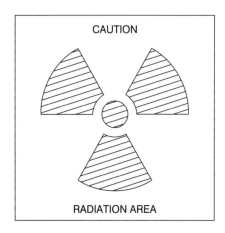

INSTRUCTING/INFORMING PERSONNEL

It is critical that companies involved in producing, using, storing, handling, and/or transporting radioactive materials keep employees informed concerning radiation hazards and the appropriate precautions for minimizing them. Consequently, OSHA has established specific requirements along these lines. They are summarized as follows:

■ All employees must be informed of existing radiation hazards and where they exist; the extent of the hazards; and how to protect themselves from the hazards (precautions and personal protective equipment).

■ All employees must be advised of any reports of radiation exposure requested by other employees.

■ All employees must have ready access to C.F.R. 1910.96(i) and any related company operating procedures.[9]

Figure 15–5
Sample warning sign.

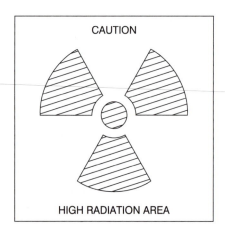

Figure 15–6
Sample warning sign.

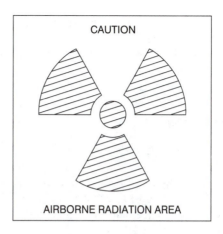

These requirements apply to all companies that do not have superseding require-ments (i.e., companies regulated by the Atomic Energy Commission and companies in states with their own approved state-level OSHA plans).

Instruction and information are important in all safety and health programs. They are especially important in settings in which radiation hazards exist. Employees in these settings must be knowledgeable about radiation hazards and how to minimize them. Periodic updating instruction for experienced workers is as important as initial instruc-tion for new employees and should not be overlooked. Often, it is the overly comfortable, experienced worker who overlooks a precaution and thereby causes an accident.

STORAGE AND DISPOSAL OF RADIOACTIVE MATERIAL

Radioactive materials that are stored in restricted areas must be appropriately labeled, as described earlier in this chapter. Radioactive materials that are stored in unrestricted

Figure 15–7
Sample warning sign.

areas "shall be secured against unauthorized removal from the place of storage."[10] This requirement precludes the handling and transport, intentional or inadvertent, of radioactive materials by persons who are not qualified to move them safely.

A danger inherent in storing radioactive materials in unrestricted areas is that an employee, such as a maintenance worker, might unwittingly attempt to move the container and damage it in the process. This could release doses that exceed prescribed acceptable limits.

The disposal of radioactive material is also a regulated activity. There are only three acceptable ways to dispose of radioactive waste: (1) transfer to an authorized recipient; (2) transfer in a manner approved by the Atomic Energy Commission; or (3) transfer in a manner approved by any state that has an agreement with the Atomic Energy Commission pursuant to Section 27(b)42 U.S.C. 2021(b) of the Atomic Energy Act.[11] States having such agreements are listed in Figure 15–8.

NOTIFICATION OF INCIDENTS

A radiation-related incident must be reported if employees meet a specific set of requirements. An **incident** is defined by OSHA as follows:

> Exposure of the whole body of any individual to 25 rems or more of radiation; exposure of the skin of the whole body of any individual to 150 rems or more of radiation; or exposure of the feet, ankles, hands, or forearms of any individual to 375 rems or more of radiation.[12]

Figure 15–8
States having agreements with the Atomic Energy Commission.

The following states have agreements with the Atomic Energy Commission to dispose of radioactive waste pursuant to 27(b) 42 U.S.C. 2021(b) of the Atomic Energy Act:

Alabama	Mississippi
Arizona	New Hampshire
Arkansas	New York
Colorado	North Carolina
Florida	North Dakota
Georgia	Oregon
Idaho	South Dakota
Kansas	Tennessee
Kentucky	Texas
Louisiana	Washington
Maryland	

The release of radioactive material in concentrations which, if averaged over a period of 24 hours, would exceed 5,000 times the limit specified. . . . [13]

If an incident meeting one of these criteria occurs, the employer must notify the proper authorities immediately. Companies regulated by the Atomic Energy Commission are to notify the commission. Companies in states that have agreements with the Atomic Energy Commission (Figure 15–8) are to notify the state designee. All other companies are to notify the U.S. assistant secretary of labor.[14] Telephone or telegraph notifications are sufficient to satisfy the immediacy requirement.

The notification requirements are eased to 24 hours in cases where whole body exposure is between 5 and 24 rems; exposure of the skin of the whole body is between 30 and 149 rems; or exposure of the feet, ankles, hands, or forearms is between 75 and 374 rems.[15]

REPORTS AND RECORDS OF OVEREXPOSURE

In addition to the immediate and 24-hour notification requirements explained in the previous section, employers are required to follow up with a written report within 30 days. Written reports are required when an employee is exposed as set forth in the previous section, or when radioactive materials are on hand in concentrations greater than specified limits. Each report should contain the following material, as applicable: extent of exposure of employees to radiation or radioactive materials; levels of radiation and concentration of radiation involved; cause of the exposure; levels of concentrations; and corrective action taken.[16]

Whenever a report is filed concerning the overexposure of an employee, the report should also be given to that employee. The following note should be placed prominently on the report or in a cover letter: "You should preserve this report for future reference."[17]

Records of the doses of radiation received by all monitored employees must be maintained and kept up to date. Records should contain cumulative doses for each monitored employee. Figure 15–9 is an example of a cumulative exposure report form of the kind that can be used to satisfy reporting requirements. Notice that the maximum quarterly dose per body or body region is indicated for that region. Such records must be shared with monitored employees at least annually.[18] A better approach is to advise monitored employees of their cumulative doses as soon as the new cumulative amount is recorded for that period.

Cumulative radiation records must be made available to former employees upon request. Upon receiving a request from a former employee, employers must provide the information requested within 30 days.[19] According to OSHA,

> Such report shall be furnished within 30 days from the time the request is made, and shall cover each calendar quarter of the individual's employment involving exposure to radiation or such lesser period as may be requested by the employee. The report shall also include the results of any calculations and analysis of radioactive material deposited in the body of the employee. The report shall be in writing and contain the following statement: "You should preserve this report for future reference."[20]

Employee _____

Dates Covered _____

	Rems in 1st Quarter	Rems in 2nd Quarter	Rems in 3rd Quarter	Rems in 4th Quarter	Total
Whole body: head and trunk; blood-forming organs; lens of eyes; or gonads (1¼ rems max/quarter)					
Hands and forearms; feet and ankles (18 rems max/quarter)					
Skin of whole body (7½ rems max/quarter)					

Figure 15–9
Cumulative radiation exposure record.

NONIONIZING RADIATION

Nonionizing radiation is that radiation on the electromagnetic spectrum that has a frequency (hertz, cycles per second) of 10^{15} or less and a wavelength in meters of 3×10^{-7} or less. This encompasses visible, ultraviolet, infrared, microwave, radio, and AC power frequencies. Radiation at these frequency levels does not have sufficient energy to shatter atoms and ionize them.[21] However, such radiation can cause blisters and blindness.

DISCUSSION CASE

What Is Your Opinion?

"We are moving out of this house today!" said the wife to her husband. "What are you talking about? You love this house." "Yes, I do," said the wife. "But I love our kids more, and I don't intend to stay here and see them die of cancer." "Cancer? What do you mean?," asked the husband, feeling increasingly lost. "It's the power lines that run along our back lot line," said the wife. "I read an article today that says power lines cause cancer." "I've heard that," said the husband. "But I don't believe it. That's just a myth." Who is correct in this discussion? What is your opinion?

In addition, there is mounting evidence of a link between nonionizing radiation and cancer. The warning symbol for radio frequency radiation is shown in Figure 15–10.

The greatest concerns about nonionizing radiation relate to the following sources: visible radiation, ultraviolet radiation, infrared radiation, lasers, and video display terminals. The main concerns in each of these areas are explained in the following paragraphs. The following section deals specifically with electromagnetic radiation (EMR) from power lines and other sources.

- **Visible radiation** comes from light sources that create distortion. This can be a hazard to employees whose jobs require color perception. For example, 8 percent of the male population is red color-blind and cannot properly perceive red warning signs.[22]

- The most common source of **ultraviolet radiation** is the sun. Potential problems from ultraviolet radiation include sunburn, skin cancer, and cataracts. Precautionary measures include special sunglasses treated to block out ultraviolet rays and protective clothing. Other sources of ultraviolet radiation include lasers, welding arcs, and ultraviolet lamps.[23]

- **Infrared radiation** creates heat. Consequently, the problems associated with this kind of nonionizing radiation involve heat stress and dry skin and eyes. Primary sources of infrared radiation are high-temperature processes such as the production of glass and steel.[24]

- **Lasers** are being used increasingly in modern industry. The hazards of lasers consist of a thermal threat to the eyes and the threat of electrocution from power sources. In addition, the smoke created by lasers in some processes can be toxic.[25]

- **Video display terminals** (VDTs) are widely used in the modern workplace. They emit various kinds of nonionizing radiation. Typically, the levels are well below

Figure 15–10
Warning symbol for radio frequency radiation.

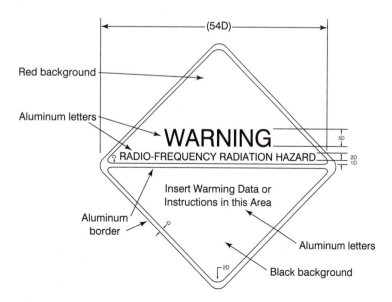

established standards. However, concerns persist about the long-term effects of pro-longed and continual exposure to VDT-based radiation.[26] The National Institute for Occupational Safety and Health (NIOSH) undertook a study of this issue in 1984. NIOSH specifically looked into the potential impact of VDT use on pregnant users. The study found "that women who work with VDTs have no greater risk of miscarriage than those who do not."[27]

ELECTROMAGNETIC FIELDS IN THE WORKPLACE

The first studies of the potential effects on worker health of occupational exposure to electric and magnetic fields were published in the literature of the former Soviet Union in the mid-1960s. In these studies, electric switchyard workers reported a variety of subjective complaints, including problems with their cardiovascular, digestive, and central nervous systems.[28] Since then, numerous studies of the effects of **electromagnetic fields (EMF)** on humans have been done. Although much of the research has been inconclusive in establishing a clear cause-and-effect relationship between EMF and health problems, the case for a clear link between EMF and a variety of health problems is strong. Consequently, safety and health professionals must be prepared to take appropriate precautions in the workplace.

According to Savita, " . . . associations reported between electric occupations and leukemia and brain cancer seem too consistent to be attributable to chance."[29] Occupations with a higher-than-normal incidence of leukemia and brain cancer are as follows:

- Telephone operators
- Electrical manufacturing workers
- Power plant workers
- Telecommunication workers
- Electrical engineers
- Electrical line workers
- Power station operators
- Electricians
- Cable splicers

The health problems most frequently associated with EMF exposure are brain cancer, acute myeloid leukemia, leukemia, and lymphatic leukemia.

An Appropriate Response for Safety Professionals

The research into the possible cause-and-effect relationship between EMF and health problems is inconclusive. On the other hand, the anecdotal and circumstantial evidence strongly suggest a link. How then, is the safety and health professional to respond? According to William E. Feero,

The emergence of new electrical technologies has produced an extremely complex electric and magnetic field environment in which we must live and work. Although there presently exists no universally accepted human exposure guidelines, current biological research may someday produce such limits. It is in anticipation of these exposure limits that electric and magnetic field management techniques are being investigated. At low frequencies, two categories of field management exist: cancellation and shielding. The particular class of control used will depend on several functions of the field and its source, and will be determined finally on a case-by-case basis.[30]

Cancellation Approach

Cancellation is an attenuation technique in which the magnetic fields produced by sources of electricity are, in effect, canceled out. It works as follows: Phase currents flowing through a given conductor are canceled out or drastically reduced by phase currents flowing in the opposite conductors. The cancellation approach can be used for both single-phase and balanced multi-phase systems.

According to Feero,

Cancellation fields can be set up in some cases with very little cost penalty. In many cases, a principal source of magnetic fields is found to be the conductor systems leading to tools or power apparatus. In such relatively simple cases, these fields could be canceled via compaction of the conductor systems. For example, a low voltage device, either a 120 or 240V service, is typically supplied by a power cord and the fields any distance from the cord are rapidly attenuated. In the situation where the cords or leads have to be very close to the worker, the added precaution of twisting or interleaving of these leads with each other will further reduce the field. The net effect of twisting a pair of conductors is that each individual conductor appears to occupy the same space. Therefore, the fields are much more effectively canceled because the spacing between the conductors is reduced to nearly zero.

Electrical apparatus that consume considerable energy would generally be supplied by three-phase cables rather than single-phase cables. If single-phase cables are used, care in bundling of the cables within cable trays and the routing of the cable trays can be a very effec-

---

SAFETY FACT

Measuring Magnetic Flux Density and Electric Fields

The units used to measure the density of the magnetic field to which an employee might be exposed have changed over time. Consequently the density may be measured in either of the following units: milligauss (mG) or tesla (T). These units can be converted as follows:

$$1 \text{ milligauss (mG)} = 0.001 \text{ gauss(G)} = 0.0000001 \text{ Tesla (T)}$$
$$1 \text{ milligauss (mG)} = 0.1 \text{ micro Tesla} = 100 \text{ gamma}$$

The intensity of an electric field is measured in volts per meter, or V/m. In situations involving extremely high voltage, the prefix (K) for kilo or 1,000 volts is used. For example: KV/m. In situations involving extremely low voltage, the measurement is volts per centimeter, or V/cm.

tive control technology. In many instances where strong fields have been found near transformer vaults in buildings, the source of the magnetic field is the cable system leading to and from the transformer vault.

The magnetic field produced directly from most apparatus exhibits the characteristics of loop current source fields. Therefore, a simple control technique may be to move the device, i.e., a compressor motor in a refrigeration unit, to the back of the unit's housing. If the device's function does not permit it to be moved (hand held tools, for instance) then more sophisticated and possibly expensive techniques must be employed.[31]

Shielding Approach[32]

Shielding is another approach available for decreasing exposure to EMF. According to Feero,

> Shielding of magnetic fields requires either that the magnetic fields be diverted around the volume considered to be sensitive to the magnetic fields, or the magnetic fields be contained within the device that produces the fields. Effectively accomplishing shielding at either the source or the subject requires extreme care in choosing the shielding material. The electrical properties of ferromagnetic materials are very complex functions of magnetic field frequencies and magnitudes. For strong magnetic fields, the highly non-linear saturation characteristics of ferromagnetic materials have been widely recognized and reasonably adjusted to achieve source shielding. However, only a few engineers and physicists are aware of the effects of coerciveness at very low magnetizing forces. Subject shielding invariably involves weak magnetic fields. Ferromagnetic materials that are normally considered to have very high permeability may exhibit quite low permeability if being used to attempt to shield milligauss field levels. Thus, the problem of dynamic range encountered when trying to apply cancellation techniques reappears in a different form when attempting to utilize shielding techniques.[32]

Both cancellation and shielding are highly technical approaches requiring specialized knowledge. Safety and health professionals who are not specialists in electromag-

---

SAFETY FACT

Reducing EMF Exposure in the Workplace

Employees are often exposed to electromagnetic fields in the workplace without even knowing it. It is important to inform employees about EMF and the associated hazards and to take steps to reduce exposure levels. Strategies for reducing exposure include the following:

- Magnetic fields drop off significantly approximately three feet from the source. Increase the distance between employees and sources of EMF.
- Substitute low EMF-producing equipment (power supplies) wherever possible.
- Reduce the amount of time that employees are exposed to EMF sources.

netic fields may find it necessary to consult with EMF experts before attempting to implement either approach.

OSHA STANDARDS FOR HEALTH AND ENVIRONMENTAL CONTROLS

OSHA's standards relating to radiation hazards are contained in 29 C.F.R. 1910 (Subpart G). These standards are as follows:

1910.94	Ventilation
1910.95	Occupational noise exposure
1910.96	Ionizing radiation
1910.97	Nonionizing radiation
1910.98	Effective dates
1910.99	Sources of standards
1910.100	Standards organizations

═════ SUMMARY ═════

1. Widely used terms relating to ionizing radiation are radiation, radioactive material, restricted area, unrestricted area, dose, rad, rem, air dose, personal monitoring devices, radiation area, and high-radiation area.
2. Exposure of individuals to radiation must be carefully controlled and accurately monitored. Doses are typically measured in rems.
3. Employers must require the use of personal monitoring devices such as film badges, pocket chambers, pocket dosimeters, and film rings. Employers are also required to conduct comprehensive surveys to identify and evaluate radiation hazards present in the workplace.
4. Caution signs are required by both OSHA and the NRC in the following areas and situations: radiation areas, high-radiation areas, airborne radiation areas, areas containing radioactive materials, and containers in which radioactive materials are stored and/or transported.
5. Companies that produce, use, store, and/or transport radioactive materials are required to have a signal-generating system to warn workers to evacuate immediately. There must be a sufficient number of signal generators to cover all personnel who need to be evacuated.
6. Employees must be informed and instructed regarding potential radiation hazards, precautions that they should take, and records of exposure.
7. Radioactive materials that are stored in restricted areas must be appropriately labeled. Materials stored in unrestricted areas must be secured against unauthorized removal.
8. Radiation incidents that involve exposure beyond prescribed limits must be reported immediately or within 24 hours, depending on the dose. Such incidents must be reported in writing within 30 days.

9. Reports of overexposure should contain the following information: extent of exposure, levels of radiation, concentration of radiation, cause of exposure, and corrective action taken.
10. Nonionizing radiation is radiation on the electromagnetic spectrum that lacks sufficient energy to ionize atoms. This encompasses visible, ultraviolet, infrared, microwave, radio, and AC power frequencies.
11. Electromagnetic fields, or EMF, encompasses radiation from power lines and a long list of electrical appliances. Concerns about, and evidence of, a link between EMF and cancer exist.

KEY TERMS AND CONCEPTS

Air dose	Nonionizing radiation
Calendar quarter	Nuclear Regulatory Commission (NRC)
Caution signs	Personal monitoring devices
Dose	Rad
Electromagnetic fields (EMF)	Radiation
Evacuation warning signal	Radiation area
Exposure	Radioactive material
High-radiation area	Rem
Incident	Restricted area
Infrared radiation	Ultraviolet radiation
Ionizing radiation	Unrestricted area
Labels	Video display terminals (VDTs)
Lasers	Visible radiation

REVIEW QUESTIONS

1. Define the following terms relating to ionizing radiation: *radiation, restricted area, dose, rem, radiation area*.
2. What is the maximum dose of radiation allowed to the whole body during a calendar quarter for a seventeen-year-old individual?
3. Explain the requirements of employers regarding the use of personal monitoring devices.
4. List the situations in which caution signs are required of employers.
5. Describe, in OSHA's language, the required evacuation warning signal.
6. How should radioactive material be treated when stored in a nonrestricted area?
7. Describe the incidents that require immediate notification concerning radiation overexposure.
8. List the required contents in a written report of radiation overexposure.
9. Define the term nonionizing radiation.

10. What are the primary sources of concern regarding nonionizing radiation?
11. Describe the anecdotal and scientific evidence linking EMF to cancer.

ENDNOTES

1. 29 C.F.R. [Code of Federal Regulations] 1910.96(b)(2).
2. "Workers' Radiation Protection Improved by NRC Chairman Carr," *Occupational Health & Safety Letter,* December 26, 1990, Vol. 20, No. 26, p. 212.
3. 29 C.F.R. 1910.96(d).
4. 29 C.F.R. 1910.96(e).
5. 29 C.F.R. 1910.96(e)(6)(iv).
6. 29 C.F.R. 1910.96(g)(1)(2)(3).
7. 29 C.F.R. 1910.96(f)(i)(ii).
8. 29 C.F.R. 1910.96(f)(iii-vi)(2)(ii).
9. 29 C.F.R. 1910.96(i).
10. 29 C.F.R. 1910.96(j).
11. 29 C.F.R. 1910.96(k) and (i).
12. 29 C.F.R. 1910.96(l)(i).
13. 29 C.F.R. 1910.96(l)(ii).
14. 29 C.F.R. 1910.96(l)(1).
15. 29 C.F.R. 1910.96(l)(2).
16. 29 C.F.R. 1910.96(m)(1).
17. 29 C.F.R. 1910.96(m)(2).
18. 29 C.F.R. 1910.96(n)(1).
19. 29 C.F.R. 1910.96(o)(1).
20. Ibid.
21. "New Concerns About Nonionizing Radiation," *Occupational Hazards,* August 1989, p. 41.
22. Ibid., p. 42.
23. Ibid., p. 43.
24. Ibid.
25. Ibid.
26. Ibid.
27. Castelli, J. "NIOSH Releases Results of VDT Study," *Safety & Health,* June 1991, p. 69.
28. Murray, William E., Jr., and Patterson, Robert M. "Electric and Magnetic Fields: What Do We Know?" *American Industrial Hygiene Association Journal,* April 1997, Vol. 54, No. 4, p. 164.
29. Savitz, David A. "Overview of Epidemiologic Research on Electric and Magnetic Fields and Cancer," *American Industrial Hygiene Journal,* April 1993, Vol. 54, No. 4, p. 202.
30. Feero, William E. "Electric and Magnetic Field Management," *American Industrial Hygiene Journal,* April 1993, Vol. 54, No. 4, p. 205.
31. Ibid., pp. 207–208.
32. Ibid., pp. 208–209.

CHAPTER SIXTEEN:

Noise and Vibration Hazards

The modern industrial worksite can be a noisy place. This poses two safety- and-health related problems. First, there is the problem of distraction. Noise can distract workers and disrupt their concentration, which can lead to accidents. Second, there is the problem of hearing loss. Exposure to noise that exceeds prescribed levels can result in permanent hearing loss.

Modern safety and health professionals need to understand the hazards associated with noise and vibration, how to identify and assess these hazards, and how to prevent injuries related to them. This chapter provides the information that prospective and practicing safety and health professionals need in order to do so.

CHARACTERISTICS OF SOUND

Sound is any change in pressure that can be detected by the ear. Typically, sound is a change in **air pressure**. However, it can also be a change in water pressure or any other pressure-sensitive medium. **Noise** is unwanted sound. Consequently, the difference between noise and sound is in the perception of the person hearing it (e.g., loud rock music may be considered sound by a rock fan but noise by a shift worker trying to sleep). Olishifski describes what occurs physiologically with sound as follows:

The generation and propagation of sound are easily visualized by means of a simple model. Consider a plate suspended in midair. When struck, the plate vibrates rapidly back and forth. As the plate travels in either direction, it compresses the air, causing a slight increase in pressure. When the plate reverses direction, it leaves a partial vacuum, or rarefaction, of the air. These alternate compressions and rarefactions cause small but repeated fluctuations in the atmospheric pressure that extend outward from the plate. When these pressure variations strike an eardrum, they cause it to vibrate in response to the slight changes in atmospheric pressure. The disturbance of the eardrum is translated into a neural sensation in the inner ear and is carried to the brain where it is interpreted as sound.[1]

Sound and vibration are very similar. Sound typically relates to a sensation that is perceived by the inner ear as hearing. **Vibration**, on the other hand, is inaudible and is perceived through the sense of touch. Sound can occur in any medium that has both mass and elasticity (air, water, and so on). It occurs as elastic waves that cross over (above and below) a line representing normal atmospheric pressure (Figure 16–1).

Normal atmospheric pressure is represented in Figure 16-1 by a straight horizontal line. Sound is represented by the wavy line that crosses above and below the line. The more frequently the sound waves cross the normal atmospheric pressure line (the shorter the cycle), the higher the pitch of the sound. The greater the vertical distance above and below the atmospheric pressure line (distance X), the louder or more intense the sound.

The unit of measurement used for discussing the level of sound and, correspondingly, what noise levels are hazardous is the **decibel**, or one-tenth of a *bel*. One decibel represents the smallest difference in the level of sound that can be perceived by the human ear. Figure 16–2 shows the decibel levels for various common sounds. The weakest sound that can be heard by a healthy human ear in a quiet setting is known as the **threshold of hearing** (10 dBA). The maximum level of sound that can be perceived without experiencing pain is known as the **threshold of pain** (140 dBA).

The three broad types of industrial noise are described by McDonald as follows: **Wide band noise** is noise that is distributed over a wide range of **frequencies**. Most noise from manufacturing machines is wide band noise. **Narrow band noise** is noise

Figure 16–1
Sound waves.

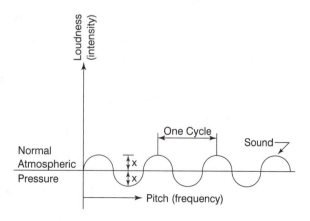

Figure 16–2
Selected sound levels
* Decibels measured on the "A" weighted network (an international standardized characteristic used in sound pressure weighting).

Source	Decibels (dBA)*
Whisper	20
Quiet office	50
Normal conversation	60
Noisy office	80
Power saw	90
Chain saw	90
Grinding operations	100
Passing truck	100
Jet aircraft	150

that is confined to a narrow range of frequencies. The noise produced by power tools is narrow band noise. Finally, **impulse noise** consists of transient pulses that can occur repetitively or nonrepetitively. The noise produced by a jackhammer is nonrepetitive impulse noise.[2]

HAZARD LEVELS AND RISKS

The fundamental hazard associated with excessive noise is hearing loss. Exposure to excessive noise levels for an extended period of time can damage the inner ear so that the ability to hear high-frequency sound is diminished or lost altogether. Additional exposure can increase the damage until even lower frequency sounds cannot be heard.[3]

In addition to hearing loss, there is evidence that excessive noise can cause other physiological problems. According to McDonald,

> Although research on the effects of noise on health is not yet complete, it appears excessive noise can cause quickened pulse, increased blood pressure, and constriction of blood vessels. These additional stresses place more burden on the heart and may lead to heart disease.[4]

A number of different factors affect the risk of hearing loss associated with exposure to excessive noise. The most important of these are

- Intensity of the noise **(sound pressure level)**
- Type of noise (wide band, narrow band, or impulse)
- **Duration** of daily exposure
- Total duration of exposure (number of years)
- Age of the individual
- Coexisting hearing disease
- Nature of environment in which exposure occurs

- Distance of the individual from the source of the noise
- Position of the ears relative to the sound waves[5]

Of these various factors, the most critical are the sound level, frequency, duration, and distribution of noise (Figure 16–3). The unprotected human ear is at risk when exposed to sound levels exceeding 115 dBA. Exposure to sound levels below 80 dBA is generally considered safe. Prolonged exposure to noise levels higher than 80 dBA should be protected against through the use of appropriate personal protective devices.

To decrease the risk of hearing loss, exposure to noise should be limited to a maximum eight-hour time-weighted average of 90 dBA. McDonald provides the following general rules for dealing with noise in the workplace:

- Exposures of less than 80 dBA may be considered safe for the purpose of risk assessment.
- A level of 90 dBA should be considered the maximum limit of continuous exposure over eight-hour days without protection.
- Continuous exposure to levels of 115 dBA and higher should not be allowed.
- Impulse noise should be limited to 140 dBA per eight-hour day for continuous exposure.[6]

STANDARDS AND REGULATIONS

The primary sources of standards and regulations relating to noise hazards are OSHA and the American National Standards Institute (ANSI). OSHA regulations require the implementation of hearing conservation programs under certain conditions. OSHA's regulations should be considered as minimum standards. ANSI's standard provides a way to determine the effectiveness of hearing conservation programs such as those required by OSHA. The ANSI standard and OSHA regulations are discussed in the following sections.

Figure 16–3
Critical noise risk factors.

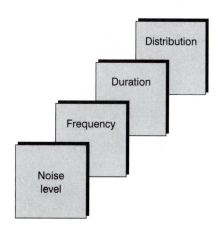

DISCUSSION CASE

What Is Your Opinion?

The father and son have had this argument before. The son wants to go to a rock concert in which four different bands will play. The *Megablast* concert will last eight hours. It is being held in the civic auditorium in a nearby city. The father is happy that his son is attending the concert but wants him to wear earplugs to guard against excess noise. According to the father, " . . . eight hours of continuous loud music can damage your hearing." However, the son is adamant in refusing to wear earplugs. "I'm going to hear the music." Is the father in this case overreacting? What is your opinion?

ANSI Standard

In 1991, the American National Standards Institute (ANSI) published ANSI standard **S12.13—1991**. Entitled "Evaluation of Hearing Conservation Programs," this standard is designed to help safety and health professionals determine if hearing conservation programs work the way they are intended.[7] According to Ruck,

> The draft standard may provide a means by which firms can evaluate the effectiveness of worker audiometric testing programs, pinpoint problems among particular groups of workers or in specific parts of a plant, promote the use of protective hearing devices and provide a defense in workers' compensation claims.[8]

Federal regulations require that employees be protected from excessive noise in the workplace. However, they provide no methodology for determining the effectiveness of hearing conservation programs. The primary reason for the development of ANSI S12.13—1991 was because hearing conservation programs were not really protecting employees but were actually only recording their steadily declining hearing ability.

The working group that developed the standard used **audiometric database analysis (ADBA)** to identify procedures for measuring variability in hearing threshold levels. According to Ruck, the two procedures identified were as follows:

1. *Percent worse sequential.* This procedure identifies the percentage of subjects who show a deterioration of 15 dBA or more in their ability to hear at least one test frequency (500 through 6,000 Hz) in either ear between two sequential audiograms.

2. *Percent better or worse sequential.* This procedure identifies the percentage of subjects who show either a deterioration or an improvement of 15 dBA or more in thresholds for at least one test frequency (500 through 6,000 Hz) in either ear between two sequential audiograms.[9]

There are important differences between the traditional approaches to measuring the effectiveness of hearing conservation programs and those required by ANSI S12.13—1991. Ruck summarizes the ANSI approach as follows:

■ Results of tests are compared in sequence. For example, the results of year 4 are compared with those of year 3. The results of year 3 are compared with those of year 2 and so on. In this way, a current audiogram is compared against an earlier audiogram. The results of the earlier test are used as a baseline for comparison.

■ Test results from several employees in a given work unit are examined individually and compared with past results sequentially. If enough employees show hearing loss, it might be concluded that the work unit's hearing conservation program is ineffective.[10]

OSHA Regulation

In 1983, OSHA adopted a Hearing Conservation Amendment to OSHA 29 C.F.R. 1910.95 that requires employers to implement **hearing conservation** programs in any work setting where employees are exposed to an eight-hour time-weighted average of 85 dBA and above.[11] Employers are required to implement hearing conservation procedures in settings where the noise level exceeds a time-weighted average of 90 dBA. They are also required to provide personal protective devices for any employee who shows evidence of hearing loss, regardless of the noise level at his or her worksite.

In addition to concerns over noise levels, the OSHA regulation also addresses the issue of duration of exposure. LaBar explains the duration aspects of the regulation as follows:

> Duration is another key factor in determining the safety of workplace noise. The regulation has a 50 percent 5 dBA logarithmic tradeoff. That is, for every 5 decibel increase in the noise level, the length of exposure must be reduced by 50 percent. For example, at 90 decibels (the sound level of a lawn-mower or shop tools), the limit of "safe" exposure is 8 hours. At 95 dBA, the limit on exposure is 4 hours, and so on. For any sound that is 106 dBA and above—this would include such things as a sandblaster, rock concert, or jet engine—exposure without protection should be less than 1 hour, according to OSHA's rule.[12]

Figure 16–4 shows the basic requirement of OSHA's Hearing Conservation Standard. These requirements are explained here:

1. *Monitoring noise levels.* Noise levels should be monitored on a regular basis. Whenever a new process is added, an existing process is altered, or new equipment is purchased, special monitoring should be undertaken immediately.

2. *Medical surveillance.* The **medical surveillance** component of the regulation specifies that employees who will be exposed to high noise levels be tested upon being hired and again at least annually.

3. *Noise controls.* The regulation requires that steps be taken to control noise at the source. **Noise controls** are required in situations where the noise level exceeds 90 dBA. Administrative controls are sufficient until noise levels exceed 100 dBA. Beyond 100 dBA, engineering controls must be used.

4. *Personal protection.* **Personal protection** devices are specified as the next level of protection when administrative and engineering controls do not reduce noise hazards to acceptable levels. They are to be used in addition to, rather than instead of, administrative and engineering controls.

Figure 16–4
Requirements of OSHA's
Hearing Conservation Stan-
dard.

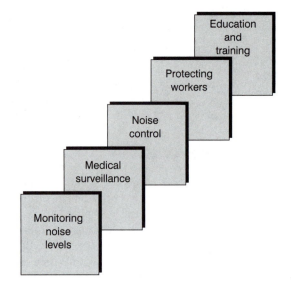

5. *Education and training.* The regulation requires the provision of education and training to do the following: Ensure that employees understand (1) how the ear works, (2) how to interpret the results of audiometric tests, (3) how to select personal protective devices that will protect them against the types of noise hazards to which they will be exposed, and (4) how to use personal protective devices properly.[13]

WORKERS' COMPENSATION AND NOISE HAZARDS

Hearing loss claims are being covered by state workers' compensation laws. Some states have actually written hearing loss into their workers' compensation law. Others are covering claims whether hearing loss is in the law or not. The National Safety Council summarizes how hearing loss claims are being handled:

> While there is no standardized procedure for negotiating levels of compensation, there has been a steady, upward trend in cost-per-case during the past decade. For instance, the average maximum award for total loss in both ears in 1979 was $21,700, and the average had increased to almost $36,000 by the mid-1980's. The same type of hearing loss among federally administered programs was $135,000 in 1977 and more than $181,000 in 1983.[14]

Medical professionals have established a procedure for determining if there is a causal relationship between workplace noise and hearing loss. In making determinations of such relationships, physicians consider the following factors:

1. Onset and progress of the employee's history of hearing loss
2. The employee's complete work history
3. Results of the employee's otological examination
4. Results of hearing studies that have been performed

5. Determination of whether causes of hearing loss originated outside the workplace[15]

Olishifski describes the situation with workers' compensation relating to hearing loss:

> It has been estimated that 1.7 million workers in the U.S. between 50 and 59 years of age have compensable noise-induced hearing loss. Assuming that only 10 percent of these workers file for compensation and that the average claim amounts to $3,000 the potential cost to industry could exceed $500 million.[16]

Since approximately 15 percent of all working people are exposed to noise levels exceeding 90 dBA, hearing loss may be as significant in workers' compensation costs in the future as back injuries, carpal tunnel syndrome, and stress are now.

IDENTIFYING AND ASSESSING HAZARDOUS NOISE CONDITIONS

Identifying and assessing hazardous noise conditions in the workplace involve the following: (1) conducting periodic noise surveys, (2) conducting periodic audiometric tests, (3) record keeping, and (4) follow-up action. Each of these components is covered in the following sections.

Noise Surveys

Conducting **noise surveys** involves measuring noise levels at different locations in the workplace. The devices that are most widely used to measure noise levels are sound level meters and dosimeters. A **sound level meter** produces an immediate reading that represents the noise level at a specific instant in time. A **dosimeter** provides a time-weighted average over a period of time such as one complete work shift.[17] The dosimeter is the most widely used device because it measures total exposure, which is what OSHA and ANSI standards specify. Using a dosimeter in various work areas and attaching a personal dosimeter to one or more employees is the recommended approach to ensure dependable, accurate readings.

Audiometric Testing

Audiometric testing measures the hearing threshold of employees. Tests conducted according to ANSI S12.13—1991 can detect changes in the hearing threshold of the employee. A negative change represents hearing loss within a given frequency range.

According to Breisch,

> The audiogram measures the noise threshold at which a subject responds to different test frequencies. These frequencies range from low to high tones. Taking less than half an hour once a year, an audiometric examination can identify changes in hearing thresholds. It is important to note that significant threshold shifts or loss of hearing might be attributed to overexposure to noise, either on or off the job. The purpose of conducting the noise survey is to identify employees who are exposed to high levels of noise and those who have had past hazardous

exposure to noise should be tested as well. An employee's work history, health and job classification, along with current noise survey records for the specific area where the worker is employed also is important data to be obtained at the time of testing.[18]

The initial **audiogram** establishes a baseline hearing threshold. After that, audiometric testing should occur at least annually. Testing should not be done on an employee who has a cold, an ear infection, or who has been exposed to noise levels exceeding 80 dBA within 14 to 16 hours prior to a test. Such conditions can produce invalid results.[19]

When even small changes in an employee's hearing threshold are identified, more frequent tests should be scheduled and conducted as specified in ANSI S12.13—1991. "For those employees found to have standard threshold shift—a loss of 10 dBA or more averaged at 2,000, 3,000, and 4,000 hertz (Hz) in either ear—the employer is required to fill out an OSHA 200 form in which the loss is recorded as a worktime illness."[20]

Record Keeping

Figure 16–5 is an example of an audiometric form that can be used to record test results for individual employees. Such forms should be completed and kept on file to allow for

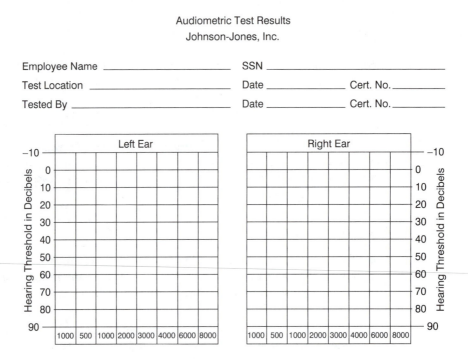

Figure 16–5
Sample audiometric test form.

sequential comparisons. It is also important to retain records containing a worker's employment history, including all past positions and the working conditions in those positions.

Follow-Up

According to Breisch, "One of the most significant deficiencies associated with audiometric monitoring in occupational settings is failure to take appropriate action when the earliest stage of noise-induced hearing loss is observable."[21] Hearing loss can occur without producing any evidence of physiological damage. Therefore, it is important to follow up on even the slightest evidence of a change in an employee's hearing threshold.

Follow-up can take a number of different forms. The following would all be appropriate follow-up responses:

- Administering a retest to verify the hearing loss.
- Changing or improving the type of personal protection used.
- Conducting a new noise survey in the employee's work area to determine if engineering controls are sufficient.
- Testing other employees to determine if the hearing loss is isolated to the one employee in question or if other employees have been affected.

NOISE CONTROL STRATEGIES

Figure 16–6 illustrates the three components of a noise hazard. Noise can be reduced by engineering and/or administrative controls applied to one or more of these components. The most desirable noise controls are those that reduce noise at the source. The second priority is to reduce noise along its path. The last resort is noise reduction at the receiver using personal protective devices. The latter approach should never be substituted for the two former approaches.

The following paragraphs explain widely used strategies for reducing workplace noise at the source, along its path, and at the receiver:

Figure 16-6
Three parts of a noise hazard.

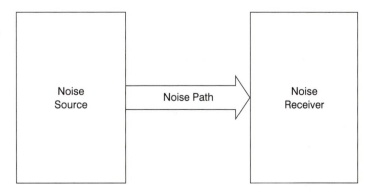

Noise Source

Noise Path

Noise Receiver

- Noise can be reduced at its **source** by enclosing the source, altering the acoustical design at the source, substituting equipment that produces less noise, making alterations to existing equipment, or changing the process so that less noisy equipment can be used.
- Noise can be reduced along its **path** by moving the source farther away from receivers and improving the acoustical design of the path so that more sound is absorbed as it travels toward receivers.
- Noise can be reduced at the **receiver** by enclosing the worker, using personal protective devices, and changing job schedules so that exposure time is reduced.

Some of the noise reduction strategies explained in the preceding paragraphs are engineering controls; others are administrative controls. For example, enclosing a noise source and substituting less noisy equipment are both examples of engineering controls. Changing job schedules is an example of an administrative control. Safety and health professionals should be familiar with both types of controls.

Engineering Controls

The National Safety Council describes **engineering controls** as "procedures other than administrative or personal protection procedures that reduce the sound level at the source or within the hearing zone of the workers."[22] What follows are commonly used engineering controls. All of these controls are designed to reduce noise at the source, along its path, or at the receiver. They focus primarily on the noise rather than the employees who are exposed to it.

Maintenance
- Replacement or adjustment of worn, loose, or unbalanced parts of machines
- Lubrication of machine parts and use of cutting oils
- Use of properly shaped and sharpened cutting tools

Substitution of Machines
- Larger, slower machines for smaller, faster ones
- Step dies for single-operation dies
- Presses for hammers
- Rotating shears for square shears
- Hydraulic presses for mechanical presses
- Belt drives for gears

Substitution of Processes
- Compression riveting for impact riveting
- Welding for riveting
- Hot working for cold working
- Pressing for rolling or forging

Reduce the Driving Force of Vibrating Surfaces by

■ Reducing the forces
■ Minimizing rotational speed
■ Isolating

Reduce the Response of Vibrating Surfaces by

■ Damping
■ Additional support
■ Increasing the stiffness of the material
■ Increasing the mass of vibrating members
■ Changing the size to change resonance frequency

Reduce the Sound Radiation from the Vibrating Surfaces by

■ Reducing the radiating area
■ Reducing overall size
■ Perforating surfaces

Reduce the Sound Transmission Through Solids by Using

■ Flexible mounting
■ Flexible sections in pipe runs
■ Flexible-shaft couplings
■ Fabric sections in ducts
■ Resilient flooring

Reduce the Sound Produced by Gas Flow by

■ Using intake and exhaust mufflers
■ Using fan blades designed to reduce turbulence
■ Using large, low-speed fans instead of smaller, high-speed fans
■ Reducing the velocity of fluid flow (air)
■ Increasing the cross-section of streams
■ Reducing the pressure
■ Reducing the air turbulence

Reduce Noise by Reducing Its Transmission Through Air by

■ Using sound-absorptive material on walls and ceiling in work areas
■ Using sound barriers and sound absorption along the transmission path
■ Completely enclosing individual machines
■ Using baffles
■ Confining high-noise machines to insulated rooms[23]

Administrative Controls

Administrative controls are controls that reduce the exposure of employees to noise rather than reducing the noise. The National Safety Council describes administrative controls as follows:

> There are many operations in which the exposure of employees to noise can be controlled administratively, that is, production schedules can simply be changed or jobs can be rotated so that exposure times are reduced. This includes such measures as transferring employees from a job location with a high noise level to a job location with a lower one if this procedure would make the employee's daily noise exposure acceptable.[24]

Administrative controls should be considered a second-level approach, with engineering controls given top priority. Smaller companies that cannot afford to reduce noise through engineering measures may use administrative controls instead. However, this approach should be avoided if at all possible.

Hearing Protection Devices

In addition to engineering and administrative controls, employees should be required to use appropriate hearing protection devices (HPDs). The following four classifications of HPDs are widely used: enclosures, earplugs, superaural caps, and earmuffs.

Enclosures are devices that completely encompass the employee's head, much like the helmets worn by jet pilots. **Earplugs** (also known as *aurals)* are devices that fit into the ear canal. Custom-molded earplugs are designed and molded for the individual employee. Premolded earplugs are generic in nature, are usually made of a soft rubber or plastic substance, and can be reused. **Formable earplugs** can be used by anyone. They are designed to be formed individually to a person's ears, used once, and then disposed.

Superaural caps fit over the external edge of the ear canal and are held in place by a headband. Earmuffs, also known as *circumaurals,* cover the entire ear with a cushioned cup that is attached to a headband. Earplugs and earmuffs are able to reduce noise by 20 to 30 decibels. By combining earplugs and earmuffs, an additional three to five decibels of blockage can be gained.

Figures 16-7 through 16-10 are examples of various types of HPDs. Figure 16-7 illustrates ear, face, and head protection combined into one comprehensive device. Figure 16-8 contains two types of superaural caps. Figure 16-9 shows an earmuff-style HPD equipped with an FM radio capability. Figure 16-10 displays soft, moldable earplugs.

The effectiveness of HPDs can be enhanced through the use of technologies that reduce noise levels. These **Active Noise Reduction (ANR)** technologies reduce noise by manipulating sound and signal waves. Such waves are manipulated by creating an electronic mirror image of sound waves that tends to cancel out the unwanted noise in the same way that negative numbers cancel out positive numbers in a mathematical equation. Using ANR in conjunction with enclosure devices or earmuffs can be an especially effective strategy.

Traditional, or passive, HPDs can distort or muffle sounds at certain frequencies, particularly high-pitched sounds. **Flat-attenuation HPDs** solve this problem by using

Figure 16-7
Earmuff style HPD.
Courtesy of ELVEX Corporation.

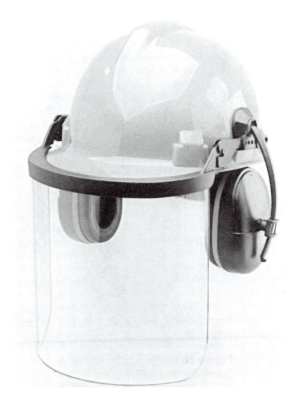

electronic devices to block all sound frequencies equally. This eliminates, or at least reduces, the distortion and muffling problems. Flat-attenuation HPDs are especially helpful for employees in settings where high-pitched sound is present and should be heard and for employees who have already begun to lose their ability to hear such sounds. The ability to hear high-pitched sounds is significant because warning signals and human voices can be high pitched.

A benefit of ANR technologies is **optimization.** The amount of noise protection can be adjusted so that employees can hear as much as they should, but not too much. Too much noise can cause employees to suffer hearing loss. Too little noise can mean that they may not hear warning signals.

VIBRATION HAZARDS

Vibration hazards are closely associated with noise hazards because tools that produce vibration typically also produce excessive levels of noise. The strategies for protecting employees against the noise associated with vibrating tools are the same as those presented so far in this chapter. This section focuses on the other safety and health hazards associated with vibration.

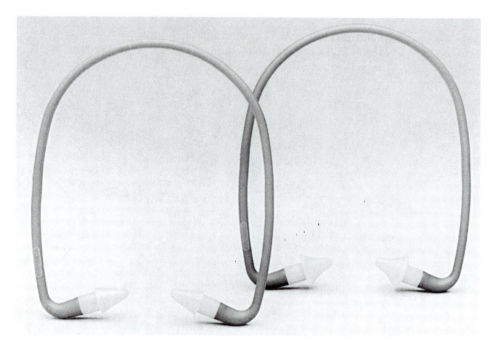

Figure 16-8
Canal-cap style HPD.
Courtesy of ELVEX Corporation.

Eastman explains the problems associated with vibration as follows:

> Vibration related problems are not only serious, they are widespread. Donald Wasserman, author of *Human Aspects of Occupational Vibration*, says that up to 8 million workers are exposed to some type of vibration hazard. Of these, it has been estimated that more than half will show some signs of injury.[25]

SAFETY FACT

Do You Have a Noise Problem?

Taking measurements is an effective way to determine if a noise problem exists in the workplace, but there are also other ways. Safety and health professionals can learn a great deal by simple observation. Walk through the workplace at different times during the day. Do employees have to raise their voices in order to hear each other? If so, noise control and hearing protection strategies are in order. At the end of an eight-hour shift, do employees have trouble distinguishing among similar-sounding words? If so, workers should be using HPDs, and both engineering and administrative controls should be adopted, as appropriate.

Figure 16-9
Earmuff style HPDs with FM
radio capability.
Courtesy of ELVEX Corporation.

Figure 16-10
Earplug style HPDs.
Courtesy of ELVEX Corporation.

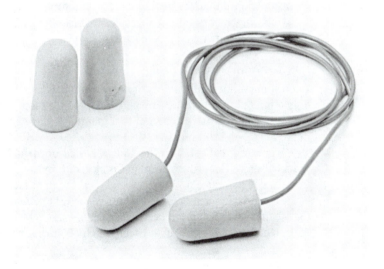

The types of injuries associated with vibration depend on its source. For example, workers who operate heavy equipment often experience vibration over the whole body. This can lead to problems ranging from motion sickness to spinal injury. However, the most common vibration-related problem is known as **hand-arm vibration syndrome**, or **HAV**. Eastman describes HAV as follows:

> The condition, a form of Reynaud's Syndrome, strikes an alarming number of workers who use vibrating power tools day in and day out as part of their jobs. For HAV sufferers, . . . the sensations in their hands are more than just minor, temporary discomforts. They are symptoms of the potentially irreversible damage their nerves and blood vessels have suffered. As the condition progresses, it takes less and less exposure to vibration or cold to trigger the symptoms, and the symptoms themselves become more severe and crippling.[26]

Environmental conditions and worker habits can exacerbate the problems associated with vibration. For example, working with vibrating tools in a cold environment is more dangerous than working with the same tools in a warm environment. Gripping a vibrating tool tightly will lead to problems sooner than using a loose grip. Smoking and excessive noise also increase the potential for HAV and other vibration-related injuries. What all of these conditions/habits have in common is that they constrict blood vessels which in turn restricts blood flow to the affected part of the body.[27]

Injury Prevention Strategies

Modern safety and health professionals should know how to prevent vibration-related injuries. Prevention is especially important with HAV because the disease is thought to be irreversible. This does not mean that HAV cannot be treated. It can, but the treatments developed to date only reduce the symptoms. They do not cure the disease.

Following are prevention strategies that can be used by safety and health professionals in any company regardless of its size.

Purchase Low-Vibration Tools

Interest in producing **low-vibration tools** is relatively new but growing. Since the 1960s, only a limited number of manufacturers produced low-vibration tools. However, a lawsuit filed against three prominent tool manufacturers generated a higher level of interest in producing low-vibration tools.

According to Eastman, the suit was filed on behalf of 300 employees of General Dynamic's Electric Boat Shipyard in Connecticut, most of whom now suffer from HAV. The case claimed that three predominant tool manufacturers failed to (1) warn users of their tools of the potential for vibration-related injury; and (2) produce low-vibration tools even though the technology to do so has been available for many years.[28] As a result of this lawsuit and the potential for others like it, low-vibration tools are becoming more commonplace.

Limit Employee Exposure

Although a correlation between cumulative exposure to vibration and the onset of HAV has not been scientifically quantified, there is strong suspicion in the safety and health community that such a link exists. For example, NIOSH recommends that companies limit the exposure of their employees to no more than four hours per day, two days per week.[29] Until the correlation between cumulative exposure and HAV has been quantified, safety and health professionals are well advised to apply the NIOSH recommendation.

Change Employee Work Habits

Employees can play a key role in protecting themselves if they know how. Safety and health professionals should teach employees who use vibration-producing tools the work habits that will protect them from HAV and other injuries. These work habits include the following: (1) wearing properly fitting thick gloves that can partially absorb vibration; (2) taking periodic breaks (at least ten minutes every hour); (3) using a loose grip on the tool and holding it away from the body; (4) keeping tools properly maintained (i.e., replacing vibration-absorbing pads regularly); (5) keeping warm; and (6) using vibration-absorbing floor mats and seat covers as appropriate.[30]

Modern safety and health professionals should also encourage higher management to require careful screening of applicants for jobs involving the use of vibration-producing tools and equipment. Applicants who smoke or have other conditions that constrict blood vessels should be guided away from jobs that involve excessive vibration.

OTHER EFFECTS OF NOISE HAZARDS

Hearing loss is the principal concern of safety and health professionals relating to noise hazards. However, hearing loss is not the only detrimental effect of excess noise. Noise can also cause communication problems, isolation, and productivity problems. According to S. L. Smith,

> Workers in noisy environments, even those where the noise is not loud enough to require hearing protection, often have trouble communicating with co-workers and supervisors. It is difficult for them to hear warning bells and signals, and they generally can't hold conversations with co-workers.[31]

Noise can also be detrimental to productivity by interfering with an employee's ability to think, reason, and solve problems. Not all employees respond to noise in this way, but many do. Can you concentrate on your studies in a noisy room? Some students can, whereas others cannot. If excessive noise makes it difficult for you to study, you will probably have the same problem on the job.

Isolation is another problem for some employees in a noisy environment. According to S. L. Smith, "While some workers benefit from the isolation provided by noise, it can have a significantly negative impact on employees who must act as a group or who must communicate frequently with supervisors or co-workers."[32] They can begin to feel left out and uninformed, the antithesis of the goal of a modern teamwork-oriented organization.

===== SUMMARY =====

1.   Sound is any change in pressure that can be detected by the ear. Typically, sound is a change in air pressure. However, it can also be a change in water pressure or any other pressure-sensitive medium. The unit of measurement for sound is the decibel, or one-tenth of a bel.

2.   The threshold of hearing is the weakest sound that can be detected by the human ear. The maximum sound that can be perceived without experiencing pain is known as the threshold of pain.

3.   Noise is excessive sound. Industrial noise is classified as wide band noise, narrow band noise, and impulse noise.

4.   Several different factors affect the risk of hearing loss from excessive noise: (a) intensity of the noise, (b) type of noise, (c) duration of daily exposure, (d) total duration of exposure, (e) age of the individual, (f) coexisting disease, (g) nature of the environment, (h) distance of the receiver from the noise source, and (i) position of the ears relative to the sound waves.

5.   Noise levels of less than 80 dBA are considered safe. A level of 90 dBA should be considered the maximum limit of continuous exposure over eight hours without protection.

6.   Important standards and regulations related to noise hazards are (a) ANSI S12.13—1991, Evaluation of Hearing Conservation Programs, and (b) the Hearing Conservation Amendment to OSHA 29 C.F.R. 1910.95.

7.   With workers' compensation claims for hearing loss on the rise, the medical profession has established a procedure for identifying causal relationships between industrial noise and hearing loss. The procedure requires that physicians consider the following factors: (a) onset and progress of hearing loss, (b) the employee's complete work history, (c) results of otological examination, (d) results of hearing studies, and (e) ruling out hearing loss from sources outside of the workplace.

8.   Methods for identifying and assessing hazardous noise conditions include the following: (a) noise surveys, (b) audiometric tests, (c) record keeping, and (d) follow-up.

9.   Noise reduction strategies are of three types: those that reduce noise at the source, those that reduce noise along its path, and those that reduce noise at the receiver.

10. Noise can be reduced by applying either engineering or administrative controls. Engineering controls attempt to reduce noise. Administrative controls limit human exposure to noise.

11. Vibration can cause physical problems ranging from motion sickness to spinal injury to hand-arm vibration syndrome (HAV). HAV is the most widespread of these problems. Because the physical damage associated with HAV may be irreversible, prevention is especially important.

12. HAV-prevention strategies include purchasing low-vibration tools, limiting employee exposure, and changing employee work habits.

=========== KEY TERMS AND CONCEPTS ===========

Active Noise Reduction (ANR)

Administrative controls

Air pressure

ANSI S12.13—1991

Audiogram

Audiometric database analysis (ADBA)

Audiometric testing

Decibel

Dosimeter

Duration

Earplugs

Enclosures

Engineering controls

Follow-up

Formable earplugs

Frequency

Hand-arm vibration syndrome (HAV)

Hearing conservation

Hearing protection devices (HPDs)

Impulse noise

Low-vibration tools

Medical surveillance

Narrow band noise

Noise

Noise controls

Noise surveys

Optimization

Path

Percent better or worse sequential

Percent worse sequential

Personal protection

Receiver

Sound

Sound level meter

Sound pressure level

Source

Superaural caps

Threshold of hearing

Threshold of pain

Vibration

Wide band noise

=========== REVIEW QUESTIONS ===========

1. Define the term *sound*.
2. What is the difference between sound and noise?
3. Differentiate between sound and vibration.
4. Describe the relationship between the pitch of sound and the cycle of sound waves.
5. List and briefly explain the three broad types of industrial noise.
6. Describe the various physiological problems associated with excessive noise.
7. List four factors that affect the risk of hearing loss from exposure to excessive noise.
8. At what sound level is it necessary to begin using some type of personal protection?
9. Give a brief description of ANSI S12.13—1991.

10. Give a brief description of OSHA 29 C.F.R. 1910.95, Hearing Conservation Amendment.
11. What factors do medical professionals consider in determining causal relationships of hearing loss?
12. Define the following terms: follow-up, noise survey, and audiometric testing.
13. List three appropriate follow-up activities when an audiometric test reveals hearing loss in an employee.
14. Differentiate between engineering and administrative controls.
15. What is HAV? How can it be prevented?
16. Explain the four classifications of HPDs that are widely used.

ENDNOTES

1. Olishifski, J. B. (Revised by J. J. Standard.) *Fundamentals of Industrial Hygiene* (Chicago: National Safety Council, 1988), pp. 165–66.
2. McDonald, O. F. "Noise: How Much Is Too Much," *Safety & Health,* November 1987, Vol. 136, No. 5, p. 37.
3. Ibid.
4. Ibid.
5. Olishifski, *Fundamentals of Industrial Hygiene,* p. 171.
6. McDonald, "Noise: How Much Is Too Much," p. 38.
7. Ruck, D. "ANSI Dams the Hearing Loss Tide," *Safety & Health,* July 1991, Vol. 144, No. 1, pp. 40–43.
8. Ibid., p. 40.
9. Ibid., p. 42.
10. Ibid.
11. LaBar, G. "Sound Policies for Protecting Workers' Hearing," *Occupational Hazards,* July 1989, p. 46.
12. Ibid.
13. Ibid., p. 48.
14. National Safety Council. "Claims on the Rise," *Safety & Health,* April 1989, Vol. 139, No. 4, p. 56.
15. Olishifski, *Fundamentals of Industrial Hygiene,* pp. 163–64.
16. Ibid., p. 164.
17. Breisch, S. L. "Hear Today and Hear Tomorrow," *Safety & Health,* June 1989, Vol. 139, No. 6, p. 40.
18. Ibid., p. 41.
19. Ibid., p. 42.
20. Ibid.
21. Ibid., p. 43.
22. Olishifski, *Fundamentals of Industrial Hygiene,* pp. 178–79.
23. Ibid., p. 179.
24. Ibid.

25. Eastman, M. "Vibration Shakes Workers," *Safety & Health*, May 1991, Vol. 143, No. 5, p. 32.
26. Ibid.
27. Ibid., pp. 32–33.
28. Ibid., p. 34.
29. Ibid., p. 33.
30. Ibid., p. 35.
31. Smith, S. L. "The 'Other' Effects of Noise," *Occupational Hazards,* January 1997, p. 79.
32. Ibid.

Preparing for Emergencies

Despite the best efforts of all involved, emergencies do sometimes occur. It is very important to respond to such emergencies in a way that minimizes harm to people and damage to property. To do so requires plans that can be implemented without delay. This chapter provides prospective and practicing safety and health professionals with the information they need to prepare for emergencies in the workplace.

RATIONALE FOR EMERGENCY PREPARATION

An *emergency* is a potentially life-threatening situation, usually occurring suddenly and unexpectedly. Emergencies may be the result of natural and/or human causes. Have you ever witnessed the timely, organized, and precise response of a professional emergency medical crew at an automobile accident? While passers-by and spectators may wring their hands and wonder what to do, the emergency response professionals quickly organize, stabilize, and administer. Their ability to respond in this manner is the result of preparation. As shown in Figure 17–1, preparation involves a combination of **planning, practicing, evaluating,** and **adjusting** to specific circumstances.

Figure 17–1
Elements of emergency
preparation.

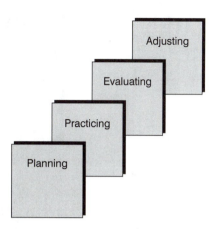

When an emergency occurs, immediate reaction is essential. Speed in responding can mean the difference between life and death or between minimal damage and major damage. Ideally, all those involved should be able to respond properly with a minimum of hesitation. This can happen only if all exigencies have been planned for and planned procedures have been practiced, evaluated, and improved.

A quick and proper response—which results because of proper preparation—can prevent panic, decrease the likelihood of injury and damage, and bring the situation under control in a timely manner. Since no workplace is immune to emergencies, preparing for them is critical. An important component of preparation is planning.

EMERGENCY PLANNING AND COMMUNITY RIGHT-TO-KNOW ACT

Title III of the Superfund Amendments and Reauthorization Act of 1986 (SARA) is also known as the Emergency Planning and Community Right-to-Know Act. This law is designed to make information about hazardous chemicals available to a community where they are being used so that residents can protect themselves in the case of an emergency. It applies to all companies that use, make, transport, or store chemicals.

Safety and health professionals involved in developing emergency response plans for their companies should be familiar with the act's requirements for emergency planning. As shown in Figure 17–2, the **Emergency Planning and Community Right-to-Know Act** includes the four major components discussed in the following paragraphs.

Emergency Planning

The emergency planning component requires that communities form **local emergency planning committees (LEPCs)** and that states form **state emergency response commissions (SERCs)**. LEPCs are required to develop emergency response plans for the local communities, host public forums, select a planning coordinator for the community, and work with the coordinator in developing local plans. SERCs are required to

Figure 17–2
Parts of an emergency
response plan.

oversee LEPCs and review their emergency response plans. Plans for individual companies in a given community should be part of that community's larger plan. Local emergency response professionals should use their community's plan as the basis for simulating emergencies and practicing their responses.

Emergency Notification

The **emergency notification** component requires that chemical spills or releases of toxic substances that exceed established allowable limits be reported to appropriate LEPCs and SERCs. Immediate notification may be verbal as long as a written notification is filed promptly. Such reports must contain at least the following information: (1) the names of the substances released, (2) where the release occurred, (3) when the release occurred, (4) the estimated amount of the release, (5) known hazards to people and property, (6) recommended precautions, and (7) the name of a contact person in the company.

Information Requirements

Information requirements mean that local companies must keep their LEPCs and SERCs and, through them, the public informed about the hazardous substances that the companies store, handle, transport, and/or use. This includes keeping comprehensive records of such substances on file, up to date, and readily available; providing copies of material safety data sheets for all hazardous substances; giving general storage locations for all hazardous substances; providing estimates of the amount of each hazardous substance on hand on a given day; and estimating the average annual amount of hazardous substances kept on hand.

Toxic Chemical Release Reporting

The **toxic chemical release reporting** component requires that local companies report the total amount of toxic substances released into the environment as either emissions or hazardous waste. Reports go to the Environmental Protection Agency and the state-level environmental agency.

ORGANIZATION AND COORDINATION

Responses to emergencies are typically from several people or groups of people including medical, fire-fighting, security, and safety personnel as well as specialists from a variety of different fields. People in each of these areas have different but interrelated and often interdependent roles to play in responding to the emergency. Because of their disparate backgrounds and roles, organization and coordination are critical.

A company's **emergency response plan** should clearly identify the different personnel/groups that respond to various types of emergencies and, in each case, who is in charge. One person should be clearly identified and accepted by all emergency responders as the **emergency coordinator**. This person should be knowledgeable, at least in a general sense, of the responsibilities of each individual emergency responder and how each relates to those of all other responders. This knowledge must include the **order of response** for each type of emergency set forth in the plan.

A company's safety and health professional is the obvious person to organize and coordinate emergency responses. However, regardless of who is designated, it is important that (1) one person is in charge, (2) everyone involved knows who is in charge, and (3) everyone who has a role in responding to an emergency is given ample opportunities to practice in simulated conditions that come as close as possible to real conditions.

Union Carbide's Texas City plant employs 1,500 workers and several hundred additional contract personnel. Safety is a high priority. Consequently, emergency planning is a fundamental part of the company's safety and health program. Emergency responders include emergency medical technicians, nurses, physicians, and production workers who are assigned specific emergency response duties. They are coordinated by a designated emergency director.[1]

Union Carbide's administrator of health services stresses the need for organization and coordination of emergency response teams. According to Marley, "Often staff physicians and nurses don't know the quickest, safest route to an emergency site. That's why emergency medical teams need to be organized and trained."[2]

Workers in the plant are obviously also knowledgeable about traffic flow, access routes, exit points, emergency approaches, and shortcuts. Because of this, Union Carbide's approach to emergency response "provides a method for dispatching EMTs and ambulance drivers to the scene while nursing and medical staff remain in radio communication and available for consultation or to receive patients if necessary."[3]

Another unique aspect of emergency planning and response at Union Carbide is how responses are coordinated: "The planning and coordination of all emergency response belongs to the **emergency response management team (ERMT).** Composed of shift emergency directors and a full-time fire chief, this group unifies all emergency groups and equipment into a single, coordinated effort."[4]

OSHA STANDARDS

There are no OSHA standards dedicated specifically to the issue of planning for emergencies. However, all OSHA standards are written for the purpose of promoting a safe, healthy, accident-free, and hence emergency-free workplace. Therefore, **OSHA standards** do play a role in emergency prevention and should be considered when developing emergency plans. For example, exits are important considerations when planning for emergencies. Getting medical personnel in and employees and injured workers out quickly is critical when responding to emergencies. The following sections of OSHA's standards deal exclusively with exit requirements:

Exit arrangements	29 C.F.R. 1910.37(e)
Exit capacity	29 C.F.R. 1910.37(c),(d)
Exit components	29 C.F.R. 1910.37(a)
Exit workings	29 C.F.R. 1910.37(q)
Exit width	29 C.F.R. 1910.37(c)
Exterior exit access	29 C.F.R. 1910.37(g)

A first step for companies developing emergency plans is to review these OSHA standards. This can help safety and health personnel identify and correct conditions that may exacerbate emergency situations before they occur.

FIRST AID IN EMERGENCIES

Workplace emergencies often require a medical response. The immediate response is usually first aid. First aid consists of life-saving measures taken to assist an injured person until medical help arrives. According to Kelly,

> This gap in time between when the injury happens and when medical care can be administered is when first aid proves itself. A trained co-worker or supervisor can begin cardiopulmonary resuscitation (CPR), stop bleeding at a pressure point or use the Heimlich maneuver on a choking employee during lunch. In each of these cases, if first aid were not given within four to six minutes, the victim could die.[5]

Since there is no way to predict when first aid may be needed, providing **first aid training** to employees should be part of preparing for emergencies. In fact, in certain cases, OSHA requires that companies have at least one employee on-site who has been trained in first aid (C.F.R. 1910.151). Figure 17–3 contains a list of the topics that may be covered in a first aid class for industrial workers.

Figure 17–3
Sample course outline for first aid class.

> **Basic First Aid**
>
> - Cardiopulmonary resuscitation
> - Severe bleeding
> - Broken bones and fractures
> - Burns
> - Choking on an obstruction
> - Head injuries/concussion
> - Cuts and abrasions
> - Electric shock
> - Heart attack
> - Stroke recognition
> - Moving an injured person
> - Drug overdose
> - Unconscious victim
> - Eye injuries
> - Chemical burns
> - Rescue

First Aid Training Program

First aid programs are usually available in most communities. The continuing education departments of community colleges and universities typically offer first aid training. Classes can often be provided on-site and customized to meet the specific needs of individual companies.

The American Red Cross provides training programs in first aid specifically geared toward the workplace. For more information about these programs, safety and health professionals may contact the national office of the American Red Cross at (202) 639-3200.

The National Safety Council also provides first aid training materials. Its First Aid and Emergency Care Teaching Package contains a slide presentation, overhead transparencies, a test bank, and an instructor's guide. The council also produces a book entitled *First Aid Essentials*. For more information about these materials, safety and health professionals may contact the National Safety Council at (800) 832-0034.

Beyond Training

Training employees in first aid techniques is an important part of preparing for emergencies. However, there is more to being prepared to administer first aid than just training. In addition, it is important to do the following:

1. *Have well-stocked first aid kits available.* First aid kits should be placed throughout the workplace in clearly visible, easily accessible locations. They should be properly and fully stocked and periodically checked to ensure that they stay fully stocked. Figure 17–4 lists the minimum recommended contents for a workplace first aid kit.

2. *Have appropriate personal protective devices available.* With the concerns about AIDS and hepatitis, administering first aid has become more complicated than in the past. The main concerns are with bleeding and other body fluids. Consequently, a properly stocked first aid kit should contain rubber surgical gloves and face masks or mouthpieces for CPR.

3. *Post emergency telephone numbers.* The advent of 911 service has simplified the process of calling for medical care, police, or fire-fighting assistance. If 911 services are not available, emergency numbers for ambulance, hospital, police, fire department, LEPC, and appropriate internal personnel should be posted at clearly visible locations near all telephones in the workplace.

4. *Keep all employees informed.* Some companies require all employees to undergo first aid training; others choose to train one or more employees in each department. Regardless of the approach used, it is important that all employees be informed and keep up to date concerning basic first aid information. Figures 17–5 and 17–6 are first aid fact sheets of the type used to keep employees informed.

■ Sterile gauze dressings (individually wrapped)	■ Scissors
■ Triangular bandages	■ Tweezers
■ Roll of gauze bandages	■ Needles
■ Assorted adhesive bandages	■ Sharp knife or stiff-backed razor blades
■ Adhesive tape	■ Medicine dropper (eye dropper)
■ Absorbent cotton	■ Measuring cup
■ Sterile saline solution	■ Oral thermometer
■ Mild antiseptic for minor wounds	■ Rectal thermometer
■ Ipecac syrup to induce vomiting	■ Hot water bag
■ Powdered activated charcoal to absorb swallowed poisons	■ Wooden safety matches
■ Petroleum jelly	■ Flashlight
■ Baking soda (bicarbonate of soda)	■ Rubber surgical gloves
■ Aromatic spirits of ammonia	■ Face masks or mouthpieces

Figure 17–4
Minimum recommended contents of workplace first aid kits.

Global Technologies

500 Anchors Street • Industrial Park • Fort Walton Beach, Florida 32748

First Aid Fact Sheet No. 16
Moving an Injured Person

If a victim has a neck or back injury, do not move him unless it must be done to prevent additional injuries. If it is absolutely essential to move the victim, remember the following rules of thumb:

1. Call for professional medical help.

2. Always pull the body lengthwise, never sideways.

3. If there is time, slip a blanket under the victim and use it to pull him to safety.

4. If the victim must be lifted, support all parts of the body so that it does not bend or jackknife.

Figure 17–5
First aid fact sheet.

Global Technologies

500 Anchors Street • Industrial Park • Fort Walton Beach, Florida 32748

First Aid Fact Sheet No. 12
ABC's of First Aid

If a fellow employee is injured and you are the first person to respond, remember the ABC's of first aid.

A = Airway
Is the airway blocked? If so, clear it quickly.

B = Breathing
Is the victim breathing? If not, begin administering artificial respiration.

C = Circulation
Is the victim bleeding severely? If so, stop the bleeding.
Is there a pulse? If not, begin administering CPR.

Figure 17–6
First aid fact sheet.

DISCUSSION CASE

What Is Your Opinion?

"Forget it! We are not going to train our employees in first aid. Emergencies not withstanding, I don't want a bunch of amateur doctors running around the company doing more harm than good." Mary VoDinh, safety director for Gulf Coast Manufacturing, was getting nowhere trying to convince her boss that the company should have employees trained in first aid in the event of an emergency. "But John, we have had three hurricanes in just two years. Tornadoes are not uncommon here on the coast." "I will repeat myself just one more time," said her boss. "No first aid training." Who is right in this case? What is your opinion?

HOW TO PLAN FOR EMERGENCIES

Developing an **emergency action plan (EAP)** is a major step in preparing for emergencies. A preliminary step is to conduct a thorough analysis to determine the various types of emergencies that may occur. For example, depending on geography and the types of products and processes involved, a company may anticipate such emergencies as the following: fires, chemical spills, explosions, toxic emissions, train derailments, hurricanes, tornadoes, lightning, floods, earthquakes, or volcanic eruptions.

A company's EAP should be a collection of small plans for each anticipated or potential emergency. These plans should have the following components:

1. *Procedures.* Specific, step-by-step emergency response procedures should be developed for each potential emergency.

2. *Coordination.* All cooperating agencies and organizations and emergency responders should be listed along with their telephone number and primary contact person.

3. *Assignments/responsibilities.* Every person who will be involved in responding to a given emergency should know his or her assignment. Each person's responsibilities should be clearly spelled out and understood. One person may be responsible for conducting an evacuation of the affected area, another for the immediate shutdown of all equipment, and another for telephoning for medical, fire, or other types of emergency assistance. When developing this part of the EAP, it is important to assign a back-up person for each area of responsibility. Doing so will ensure that the plan will not break down if a person assigned a certain responsibility is one of the victims.

4. *Accident prevention strategies.* The day-to-day strategies to be used for preventing a particular type of emergency should be summarized in this section of the EAP. In this way, the strategies can be reviewed, thereby promoting prevention.

5. *Schedules.* This section should contain the dates and times of regularly scheduled practice drills. It is best to vary the times and dates so that practice drills don't become predictable and boring. Figure 17–7 is a checklist that can be used for developing an EAP.

Type of Emergency

_____ Fire	_____ Explosion
_____ Chemical spill	_____ Toxic emission
_____ Train derailment	_____ Hurricane
_____ Tornado	_____ Lightning
_____ Flood	_____ Earthquake
_____ Volcanic eruption	

Procedures for Emergency Response

1. Controlling and isolating?
2. Communication?
3 Emergency assistance?
4. First aid?
5. Shut-down/evacuation/protection of workers?
6. Protection of equipment/property?
7. Egress, ingress, exits?
8. Emergency equipment (e.g., fire extinguishers)?
9. Alarms?
10. Restoration of normal operations?

Coordination

1. Medical care providers?
2. Fire service providers?
3. LEPC personnel?
4. Environmental protection personnel?
5. Civil defense personnel (in the case of public evacuations)?
6. Police protection providers?
7. Communication personnel?

Figure 17–7
Emergency planning checklist.

CUSTOMIZING PLANS TO MEET LOCAL NEEDS

Emergency plans must be **location-specific**. General plans developed centrally and used at all plant locations will have limited effectiveness. The following rules of thumb can be used to ensure that EAPs are location-specific:

1. *A map in the plan.* A map of the specific plant will help localize an EAP. The map should include the locations of exits, access points, evacuation routes, alarms,

Assignments/Responsibilities

1. Who cares for the injured?
2. Who calls for emergency assistance?
3. Who shuts down power/operations?
4. Who coordinates communication?
5. Who conducts the evacuation?
6. Who meets emergency responders and guides them to the emergency site?
7. Who contacts coordinating agencies and organizations?
8. Who is responsible for ensuring the availability and upkeep of fire extinguishers?
9. Who is responsible for ensuring that alarms are in proper working order?
10. Who is responsible for organizing cleanup activities?

Accident Prevention Strategies

1. Periodic safety inspections?
2. Industrial hygiene strategies?
3. Personal protective equipment?
4. Ergonomic strategies?
5. Machine safeguards?
6. Hand/portable power tool safeguards?
7. Material handling and storage strategies?
8. Electrical safety strategies?
9. Fire safety strategies?
10. Chemical safety strategies?

Schedules

1. Dates of practice drills: _____
2. Times of practice drills: _____
3. Duration of practice drills: _____

emergency equipment, a central control or command center, first aid kits, emergency shut-downs buttons, and any other important elements of the EAP.

2. *Chain of command.* An organizational chart illustrating the **chain of command**—who is responsible for what and who reports to whom—will also help to localize an EAP. The chart should contain the names and telephone numbers (internal and external) of everyone involved in responding to an emergency. It is critical to keep the organizational chart up to date as personnel changes occur. It is also important to have a designated back-up person shown for every position on the chart.

3. *Coordination information.* All telephone numbers and contact names of people in agencies with which the company coordinates emergency activities should be listed. Periodic contact should be maintained with all of these people so that the EAP can be updated as personnel changes occur.

4. *Local training.* All training should be geared toward the types of emergencies that may occur in the plant. In addition, practice drills should take place on-site and in the specific locations where emergencies are most likely to happen.

EMERGENCY RESPONSE

An **emergency response team (ERT)** is a special team that responds "to general and localized emergencies to facilitate personnel evacuation and safety, shut down building services and utilities, as needed, work with responding civil authorities, protect and salvage company property, and evaluate areas for safety prior to reentry."[6] The ERT is typically composed of representatives from several different departments such as the following: maintenance, security, safety and health, production/processing, and medical. The actual composition of the team will depend on the size and type of company in question. The ERT should be contained in the assignments/responsibilities section of the EAP.

Not all ERTs are company-based. Communities also have ERTs for responding to emergencies that occur outside of a company environment. Such teams should be included in a company's EAP in the coordinating organizations section. This is especially important for companies that use hazardous materials.

According to Geyer, "Although the most dangerous spills occur on the road, the greatest number of hazmat [hazardous materials] incidents occur on the manufacturer's loading dock. A national Tank Truck Carriers, Inc. study attributes more than 80 percent of chemical releases to errors in loading and unloading procedures."[7] Community-based teams typically include members of the police and fire departments who have had special training in handling hazmat emergencies.

Another approach to ERTs is the **emergency response network (ERN)**. An ERN is a network of ERTs that covers a designated geographical area and is typically responsible for a specific type of emergency. For example, "In 1972, the Chlorine Institute (CI), an association of more than 130 firms that produce chlorine and manufacture related products, launched its Chlorine Emergency Plan (CHLOREP), a network for responding to chlorine emergencies. Sixty-two teams from 28 companies are assigned geographic areas for which they could be called upon to respond to chlorine emergencies."[8]

The following example shows how a CHLOREP team responds to a chlorine-related emergency. In June 1986, a chlorine leak was discovered in a cement pit near Arlington National Cemetery. The pit had once been used to chlorinate water for irrigating the cemetery. Initial reports were that as many as five of ten 150-pound chlorine cylinders were leaking. Local emergency response teams asked CHLOREP for help.[9] The CHLOREP team covering Washington, D.C., and vicinity responded and found that only one of the ten cylinders was leaking. They took the following action:

> . . . the team used a recently designed recovery vessel, a coffin-like steel container for enclosing leaking cylinders. After the nine unaffected cylinders were moved aside, the 64-inch-long

recovery vessel was lowered into the pit. The team slid the leaking cylinder into the coffin and secured it.[10]

Regardless of whether the ERT is a local company team or a network of teams covering a geographical region, it should be included in the EAP. In-house teams are included in the assignments/responsibilities section. Community-based teams and networks are included in the coordinating organizations section.

COMPUTERS AND EMERGENCY RESPONSE

Advances in chemical technology have made responding to certain types of emergencies particularly complicated. The following scenario makes the point:

> An employee accidentally crashes a lift into several barrels of various chemicals. While you have been generally trained to handle certain spills, you cannot find specific information on this combined spill, what protective equipment should be used, and who should be contacted. Your company has an emergency plan, but it is 1½ inches thick and you need to take action now. The spilled material may be highly toxic and its plume should reach the fence line shortly.[11]

Fortunately, the complications brought by technology can be simplified by technology. Expert computer systems especially programmed for use in emergency situations can help to meet the challenge of responding to a mixed-chemical emergency or any other type of emergency involving multiple hazards interacting.

An **expert system** is a computer programmed to solve problems. Such systems "rely on a database of knowledge about a very particular subject area, an understanding of the problems addressed in that area, and a skill at solving problems."[12] Talking to an expert system is like sitting at a terminal and keying in questions to an expert who is sitting at

---

SAFETY FACT

Responding to Chemical Spills

To respond quickly and effectively to a chemical spill, the emergency response team must have the right equipment. Standard equipment should include a portable **spill cart** that can be quickly rolled to the site of a chemical spill. This cart should contain at least the following items:

- Spill suppression/absorption materials such as pads, blankets, and/or pillows
- Mops and brooms
- Acid neutralizers
- Barricade devices or materials such as mesh or tape
- Appropriate personal protective gear (e.g., gloves, eye protection, aprons, coveralls, and so on)

another terminal responding to the questions. The expert in this case is a computer program that pulls information from a database and uses it to make decisions based on heuristics or suppositional rules stated in an if-then format.[13]

Human thought processes work in a similar manner. For example, if our senses provide the brain with input suggesting the stovetop is hot, the decision is made not to touch it. In an if-then format, this might read as follows:

IF hot, THEN do not touch.

This similarity to human thought processes is why the science on which expert systems are based is known as *artificial intelligence.*

A modern expert system used for responding to chemical emergencies provides such information as the following:

- Personal protective equipment needed for controlling and cleaning up
- Methods to be used in cleaning up the spill or toxic release
- Decontamination procedures
- Estimation of the likelihood that employees or the community will be exposed to hazard
- Reactions that might result from interaction of chemicals
- Combustibility of chemicals and other materials on hand
- Evacuation information
- Impact of different weather conditions on the situation
- Recommended first aid procedures[14]

Expert systems can be user friendly so that computer novices have no difficulty interacting with them. Kelly summarizes the advantages of expert systems for emergency response as follows:

> Expert systems . . . do not jump to conclusions nor try to defend bad decisions in the face of contrary evidence. Their decisions are not biased, they are detail-oriented and consider all possible solutions; they do not have bad days, and they do not get tired.[15]

DEALING WITH THE PSYCHOLOGICAL TRAUMA OF EMERGENCIES

In addition to the physical injuries and property damage that can occur in emergencies, modern safety and health professionals must also be prepared to deal with potential psychological damage. Psychological trauma among employees involved in workplace disasters is as common as it is among combat veterans. According to Johnson, "Traumatic incidents do not affect only immediate survivors and witnesses. Most incidents result in layers of victims that stretch far beyond those who were injured or killed."[16]

Trauma is psychological stress. It occurs as the result of an event, typically a disaster or some kind of emergency, that is so shocking it impairs a person's sense of security or well-being. Johnson calls trauma response "the normal reactions of normal people to an abnormal event."[17] **Traumatic events** are typically unexpected and shocking, and they involve the reality and/or threat of death.

Dealing with Emergency-Related Trauma

The typical approach to an emergency can be described as follows: control it, take care of the injured, clean up the mess, and get back to work. Often, the psychological aspect is ignored. This leaves witnesses and other coworkers to deal on their own with the trauma they've experienced. "Left to their own inadequate resources, workers can become ill or unable to function. They may develop resentment toward the organization which can lead to conflicts with bosses and co-workers, high employee turnover—even subconscious sabotage."[18]

It is important to respond to trauma quickly, within 24 hours if possible and within 72 hours in all cases. The purpose of the response is to help employees get back to normal by enabling them to handle what they have experienced. This is best accomplished by a team of people who have had special training. Such a team is typically called the **trauma response team (TRT)**.

Trauma Response Team

A company's trauma response team may consist of safety and health personnel who have undergone special training or fully credentialed counseling personnel, depending on the size of the company. In any case, the TRT should be included in the assignments/responsibilities section of the EAP.

The job of the TRT is to intervene as early as possible, help employees acknowledge what they have experienced, and give them opportunities to express how they feel about it to people who are qualified to help. The *qualified to help* aspect is very important. TRT members who are not counselors or mental health professionals should never attempt to provide care that they are not qualified to offer. Part of the trauma training that safety and health professionals receive involves recognizing the symptoms of employees who need professional care and referring them to qualified care providers.

In working with employees who need to deal with what they have experienced, but are not so traumatized as to require referral for outside professional care, a group approach is best. According to Johnson, the group approach offers several advantages, including the following:

- It facilitates public acknowledgment of what the employees have experienced.
- It keeps employees informed, thereby cutting down on the number of unfounded rumors and horror stories that will inevitably make the rounds.
- It encourages employees to express their feelings about the incident. This alone is often enough to get people back to normal and functioning properly.
- It allows employees to see that they are not alone in experiencing traumatic reactions (e.g., nightmares, flashbacks, shocking memories, and so on) and that these reactions are normal.[19]

Convincing Companies to Respond

Modern safety and health professionals may find themselves having to convince higher management of the need to have a TRT. Some corporate officials may not believe that trauma even exists. Others may acknowledge its presence but view trauma as a personal problem that employees should handle on their own.

In reality, psychological trauma that is left untreated can manifest itself as **post-traumatic stress disorder**, the same syndrome experienced by some veterans of Vietnam and other wars. This disorder is characterized by "intrusive thoughts and flashbacks of the stressful event, the tendency to avoid stimulation, paranoia, concentration difficulties, and physiological symptoms such as rapid heartbeat and irritability."[20]

The American Psychiatric Association included post-traumatic stress disorder in its *Diagnostic and Statistical Manual* as far back as 1980.[21] Jeffrey T. Mitchell, president of the American Critical Incident Stress Foundation, likens preventing post-traumatic stress disorder to working with cement. Wet cement can be molded, shaped, manipulated, and even washed away. However, once it hardens, there is not much one can do with it.[22] This is the rationale for early intervention.

In today's competitive marketplace, companies need all of their employees operating at peak performance levels. Employees experiencing trauma-related disorder will not be at their best. This is the rationale safety and health professionals should use when it is necessary to convince higher management of the need to provide a company-sponsored trauma response team.

=== **SUMMARY** ===

1. An emergency is a potentially life-threatening situation, usually occurring suddenly and unexpectedly. Emergencies may be the result of natural and/or human causes.
2. Preparing for emergencies involves planning, practicing, evaluating, and adjusting. An immediate response is critical in emergencies.
3. The Emergency Planning and Community Right-to-Know Act has the following four main components: emergency planning, emergency notification, information requirements, and toxic chemical release reporting.
4. For proper coordination of the internal emergency response, it is important that one person be in charge and that everyone involved knows who that person is.
5. Since there is no way to predict when first aid might be needed, part of preparing for emergencies should include training employees to administer first aid. In certain cases, OSHA requires that companies have at least one employee on-site who has been trained in first aid.
6. In addition to providing first aid training, it is important to have well-stocked first aid kits readily available, have personal protective devices available, post emergency telephone numbers, and keep all employees informed.

7. A company's emergency action plan should be a collection of small plans for each anticipated emergency. These plans should have the following components: procedures, coordination, assignments/responsibilities, accident prevention strategies, and schedules.

8. EAPs should be customized so that they are location-specific by including a map, an organization chart, local coordination information, and local training schedules.

9. An emergency response team is a special team to handle general and localized emergencies to facilitate evacuation and shutdown, protect and salvage company property, and work with civil authorities.

10. An emergency response network is a network of emergency response teams that covers a designated geographical area.

11. Computers can help simplify some of the complications brought by advances in technology. Expert systems mimic human thought processes in making decisions on an if-then basis regarding emergency responses.

12. Trauma is psychological stress. It typically results from exposure to a disaster or emergency so shocking that it impairs a person's sense of security or well-being. Trauma left untreated can manifest itself as post-traumatic stress disorder. This disorder is characterized by intrusive thoughts, flashbacks, paranoia, concentration difficulties, rapid heartbeat, and irritability.

KEY TERMS AND CONCEPTS

Adjusting

Chain of command

Coordinator

Emergency action plan (EAP)

Emergency notification

Emergency planning

Emergency Planning and Community Right-to-Know Act

Emergency response management team (ERMT)

Emergency response network (ERN)

Emergency response plan

Emergency response team (ERT)

Evaluating

Expert system

First aid training

Heuristics

Information requirements

Local emergency planning committee (LEPC)

Location-specific

Order of response

OSHA standards

Planning

Post-traumatic stress disorder

Practicing

Spill cart

State emergency response commission (SERC)

Toxic chemical release reporting

Trauma

Trauma response team (TRT)

Traumatic event

REVIEW QUESTIONS

1. Define the term *emergency*.
2. Explain the rationale for emergency preparation.
3. List and explain the four main components of the Emergency Planning and Community Right-to-Know Act.
4. Describe how a company's emergency response effort should be coordinated.
5. How do OSHA standards relate to emergency preparation?
6. Explain how you would provide first aid training if you were responsible for setting up a program at your company.
7. What are the steps relating to first aid that a company should take beyond providing training?
8. Describe the essential components of an EAP.
9. How can a company localize its EAP?
10. Define the following emergency-response concepts: ERT, ERN, and TRT.
11. What is an expert system? How can one be used in responding to an emergency?
12. What is trauma?
13. Why should a company include trauma response in its EAP?
14. Describe how a company might respond to the trauma resulting from a workplace emergency.

ENDNOTES

1. Marley, L. "Emergency Medical Teams Need Organization and Administration," *Safety & Health*, November 1990, Vol. 142, No. 5, p. 28.
2. Ibid.
3. Ibid.
4. Ibid., p. 30.
5. Kelly, S. "First Aid Is Emergency Care," *Safety & Health,* May 1989, Vol. 139, No. 5, p. 47.
6. National Safety Council. *Introduction to Occupational Health and Safety* (Chicago: National Safety Council, 1986), p. 341.
7. Geyer, S. "Emergency-Response Teams Halt Hazmat Destruction," *Safety & Health*, June 1991, Vol. 143, No. 6, p. 45.
8. "Emergency Response Network in Action," *Occupational Hazards,* November 1988, p. 41.
9. Ibid.
10. Ibid.
11. Kelly, R. B. "In Case of an Emergency Call a Computer," *Safety and Health*, August 1988, Vol. 138, No. 2, p. 37.
12. Ibid.
13. Ibid., p. 38.
14. Ibid., p. 39.
15. Ibid.

16. Johnson, E. "Where Disaster Strikes," *Safety & Health*, February 1992, Vol. 145, No. 2, p. 29.
17. Ibid., p. 28.
18. Ibid., p. 29.
19. Ibid., p. 30.
20. National Safety Council, "Trained for Trauma," *Safety & Health*, February 1992, Vol. 145, No. 2, p. 32.
21. Ibid.
22. Ibid.

CHAPTER EIGHTEEN:

Safety Analysis and Prevention

There is a saying that an ounce of prevention is worth a pound of cure. This is certainly the case with workplace safety and health. Every accident that can be prevented should be. Every hazard that can be identified should be corrected or at least minimized through the introduction of appropriate safeguards. Careful analysis of potential hazards in the workplace has led to many of today's widely used safety measures and practices.

The key to preventing accidents is identifying and eliminating hazards. A hazard may be defined as follows:

> *A hazard is a condition or combination of conditions that, if left uncorrected, may lead to an accident, illness, or property damage.*

This chapter provides prospective and practicing safety and health professionals with the information they need to analyze the workplace, identify hazards that exist there, and take the preventative measures necessary to neutralize the hazards.

OVERVIEW OF HAZARD ANALYSIS

If a hazard is a condition that could lead to an injury or illness, **hazard analysis** is a systematic process for identifying hazards and recommending corrective action. There are two approaches to hazard analysis: preliminary and detailed (Figure 18–1). A **prelimi-**

nary hazard analysis (PHA) is conducted to identify potential hazards and prioritize them according to the (1) likelihood of an accident or injury being caused by the hazard; and (2) severity of injury, illness, and/or property damage that could result if the hazard caused an accident.

The Society of Manufacturing Engineers has this to say about preliminary hazard analysis:

> Each hazard should be ranked according to its probability to cause an accident and the severity of that accident. The ranking here is relative. Some hazards identified might be placed into a category for further analysis. A useful practice for early sorting is to place catastrophic severities together, followed by critical, marginal, and nuisance hazards, respectively. Then, with each item, indicate the probability of occurrence—considerable, probable, or unlikely. Rate the correction of these hazards next by cost.[1]

Key terms in this quote are those that describe the likelihood of an accident occurring (considerable, probable, or unlikely) and those that rate the probable level of injuries that could occur (catastrophic, critical, marginal, and nuisance). A key step is rating the cost of correcting hazards. All of these concepts are covered later in this chapter.

Whereas a preliminary analysis may involve just observation or pilot testing of new equipment and systems, the **detailed hazard analysis** involves the application of analytical, inductive, and deductive methods. Figure 18–2 lists some of the more widely used methods for conducting a detailed hazard analysis. Each of these methods is covered at length later in this chapter.

PRELIMINARY HAZARD ANALYSIS

It is not always feasible to wait until all the data are compiled from a detailed analysis before taking steps to identify and eliminate hazards. For example, when a new system or piece of equipment is installed, management is likely to want to bring it on line as soon as possible. In such cases, a preliminary hazard analysis is in order. The PHA can serve two purposes: (1) it can expedite bringing the new system on line, but at a substantially reduced risk of injuring workers; and (2) it can serve as a guide for a future detailed analysis.

Preliminary hazard analysis amounts to forming an ad hoc team of experienced personnel who are familiar with the equipment, material, substance, and/or process being

Figure 18–1
Two approaches to hazard analysis.

Figure 18–2
Detailed hazard analysis
methods.

- Failure mode and effects analysis (FMEA)
- Fault tree analysis (FTA)
- Hazard and operability review (HAZOP)
- Human error analysis (HEA)
- Risk analysis
- Technic of operation review (TOR)

analyzed. **Experience** and **related expertise** are important factors in conducting a preliminary review. For example, say, a new piece of equipment has been installed such as a computer numerically controlled (CNC) machining center. The safety and health professional may form a team that would include an experienced machinist, an electrician, a materials expert, and a computer control specialist.

All members of the team would be asked to look over the machining center for obvious hazards relating to their respective areas of expertise. Then, they would work together as a group to play devil's advocate. Each team member would ask the others a series of "what if" questions: What if a cutting bit breaks? What if the wrong command is entered? What if the material stock is too long? Depending on the nature of the process being analyzed, personnel from adjacent and/or related processes should be added to the team.

Figure 18–3 is an example of a job hazard analysis survey adapted from one developed by the National Institute for Occupational Safety and Health (NIOSH). A preliminary analysis team would use this form to identify potential hazards associated with a spray painting process. Key elements include the substances to which workers will be exposed, the form that those substances will take, the probable route of entry, and recommended hazard control strategies. A similar form could be developed for any process or operation that might be the focus of a preliminary hazard analysis.

Cost/Benefit in Hazard Analysis

Every hazard typically has several different remedies. Every remedy has a corresponding **cost** and corresponding **benefit**. Management is not likely to want to apply $10 solutions to $1 problems. Therefore, it is important to factor in cost when recommending corrective action regarding hazards. This amounts to listing all of the potential remedies along with their respective costs and then estimating the extent to which each will reduce the hazard (its benefit).

Going back to the earlier example of the CNC machining center, assume that the analysis team identified the following potential hazards:

- Lubricants sprayed on the machine operator or floor
- Flying metal chips hitting the operator or other workers
- Jammed metal stock kicking back into the operator

Operation:									Date:							
Number of Employees	Job Title	Exposure Substance	Form (Type of hazard)						Route of Entry		Control Methods						
			Dust	Liquid	Vapor	Gas	Fume	Mist	Skin	Inhaled	Local Ventilation	General Ventilation	Respirator	Gloves	Face Protection	Other Protection

Figure 18–3

Sample job hazard analysis survey.

Adapted from survey instrument in *Safety Guide for Textile Machinery Manufacturers*. NIOSH, U. S. Department of Health, Education, and Welfare, January 1978, p. 11.

Figure 18–4 is a matrix that may be developed by the analysis team to illustrate the cost of each hazard versus the benefit of each remedy. After examining this matrix, the remedy that makes the most sense from the perspective of both cost and impact on the hazards is the Plexiglass door. It eliminates two of the hazards and reduces the third. The flexible curtain costs less but does not have a sufficient impact on the hazards. The third and fourth options cost more and have less impact on the hazards.

DISCUSSION CASE

What Is Your Opinion?

Mike Chinchar is the new safety director at MicroTel Corporation. He completed his college degree just six weeks ago. At the moment, he is wishing he could transport himself back in time and be a college student again. Mike Chinchar is on the *hot seat*. "Chinchar, you seem to think that this company should simply stop functioning every time you identify a hazard. This situation you are proposing will be expensive! Isn't there some other way to solve the problem? Did you do a cost/benefit analysis before arriving at this recommendation?" What types of factors should Chinchar have considered before recommending a solution to a hazardous situation? What is your opinion?

	Cost/Benefit Analysis Matrix			
		Impact on Hazard		
Possible Remedy	**Estimated Cost**	**Spraying Lubricants**	**Metal Chips**	**Jammed Stock**
Plexiglass door	$250	E	E	R
Flexible curtain	75	R	R	N
Acme chip/jam guard	260	N	E	E
Acme spray guard	260	N	E	E

R = Reduces the hazard
E = Eliminates the hazard
N = No effect on the hazard
 I = Increases the hazard
C = Creates new hazard

Figure 18–4
Sample cost/benefit analysis matrix.

DETAILED HAZARD ANALYSIS

Typically, a preliminary hazard analysis is sufficient. However, in cases where the potential exists for serious injury, multiple injuries, or catastrophic illness, a detailed hazard analysis is conducted. A number of different methods can be used for conducting detailed analyses. The most widely used of these are as follows:

- Failure mode and effects of analysis (FMEA)
- Hazard and operability review (HAZOP)
- Technic of operations review (TOR)
- Human error analysis (HEA)
- Fault tree analysis (FTA)
- Risk analysis

Failure Mode and Effects Analysis (FMEA)

FMEA is a formal step-by-step analytical method that is a spin-off of reliability analysis, a method used to analyze complex engineering systems. FMEA proceeds as follows:

1. Critically examine the system in question.
2. Divide the system into its various components.

3. Examine each individual component and record all of the various ways in which the components may fail. Rate each potential failure according to the degree of hazard posed (0 = No hazard, 1 = Slight, 2 = Moderate, 3 = Extreme, 4 = Severe).

4. Examine all potential failures for each individual component of the system and decide what effect the failures could have.

Figure 18–5 is an example of an FMEA conducted on a direct extrusion process. The process/system is broken down into seven components: die backer, die, billet, dummy block, pressing stem, container liner, and container fillet. The types of failures that may occur are identified as corrosion, cracking, shattering, bending, and surface wear. Of the various components, only the dummy block poses an extreme hazard and a corresponding hazard to workers.

An FMEA produces an extensive analysis of a specific process or system, as illustrated in Figure 18–5. However, FMEAs have their limitations. First, the element of **human error** is missing. This is a major weakness because human error is more frequently at the heart of a workplace accident than is system/process failure. This weakness can be overcome by coupling human error analysis, which is covered later in this chapter, with an FMEA. Second, FMEAs focus on the components of a given system as if the components operate in a vacuum. They do not take into account the interface mechanisms between components or between systems. It is at these interface points that problems often occur.

Hazard and Operability Review (HAZOP)

Hazard and operability review (HAZOP) is an analysis method that was developed for use with new processes in the chemical industry. Its strength is that it allows problems to be identified even before a body of experience has been developed for a given process/system. Although originally intended for use with new processes, it need not be limited to new operations. HAZOP works equally well with old processes/systems.

HAZOP consists of forming a team of experienced, knowledgeable people from a variety of backgrounds relating to the process/system and having team members brainstorm about potential hazards. The safety and health professional should chair the team and serve as a facilitator. The chair's role is to elicit and record the ideas of team members, make sure that one member does not dominate or intimidate other members, encourage maximum participation from all members, and assist members in combining ideas where appropriate to form better ideas.

A variety of approaches can be used with HAZOP. The one recommended by the American Institute of Chemical Engineers (AICHE) is probably the most widely used. AICHE recommends the following guidewords: *no, less, more, part of, as well as, reverse,* and *other than.*[2]

These guidewords relate to the operation of a specific component in the system or a specific part of an overall operation. They describe ways in which the component may deviate from its design or its intended mode of operation. For example, if a component that should rotate 38L in a cycle fails to rotate at all, the *no* guideword applies. If it

Plastics Extrusions, Inc.
17 Industrial Boulevard
Fort Walton Beach, Florida 32548

Department _____ Manufacturing _____ Process/System _____ Direct Extrusion _____ Date _____ November 12, 1992 _____

| Component | Type of Potential Failure | Potential Effect On | | | | 0 | 1 | 2 | 3 | 4 | H | M | L | U | Examination Method | Recommendation |
		Component	Related Components	Process/System	Workers											
Diebacker	Corrosion	Shutdown to replace	None	Shutdown to replace	None	✓								✓	Visual	Periodic checks for corrosion
Die	Cracking	Shutdown to replace	Damage to die backer	Shutdown to replace	None			✓				✓			Visual	Periodic checks for cracks
Billet	- - -	- - -	- - -	- - -	- - -	- - -									- - -	- - - - -
Dummy block	Shattering	Shutdown to replace	Could damage others	Shutdown all	Injuries from flying metal				✓			✓			Visual	Inspect and replace periodically
Pressing stem	Bending	Shutdown to replace	None	Shutdown to replace	None	✓							✓		Visual	Inspect and replace periodically
Container liner	Surface wear	Shutdown to replace	None	None	None	✓								✓	Visual	Periodic checks for wear
Container fillet	- - -	- - -	- - -	- - -	- - -	- - -									- - -	- - - - -

Analysis Conducted by: _____

Figure 18–5
Sample FMEA.

rotates fewer than 38L, the *less* guideword applies. *More* would apply if the component's rotation exceeded 38L. *Reverse* would be used if the component rotated 38L in a direction opposite of the one intended. *As well as* is similar to *more* in that it indicates an increase in an intended amount. *Other than* is used when what actually occurs is something completely different from what was intended. For example, if the component fell off rather than rotating 38L, *other than* would be used.

A HAZOP proceeds in a step-by-step manner. These steps are summarized as follows:

1. Select the process/system to be analyzed.
2. Form the team of experts.
3. Explain the HAZOP process to all team members.
4. Establish goals and timeframes.
5. Conduct brainstorming sessions.
6. Summarize all input.

Figure 18–6 is an example of a form that can be used to help organize and focus brainstorming sessions. It can also be used for summarizing the results of the brainstorming sessions. This particular example involves a plastic mixing process. Only one component in the process (flow-gate number 1) has been analyzed. If the flow-gate does not work as intended, there will be no flow, too little flow, or too much flow. Each condition will result in a specific problem. Action necessary to correct each situation has been recommended. Every critical point, sometimes referred to as a *node*, in the process would be analyzed in a similar manner.

HAZOPs have the same weaknesses as FMEAs—they do not factor human error into the equation. HAZOPs predict problems associated with system/process failures. However, these are technological failures. Since human error is so often a factor in accidents, this weakness must be addressed. The next section sets forth guidelines for analyzing human error.

Human Error Analysis (HEA)

Human error analysis (HEA), as described in the content of this chapter, is used to predict human error and not as an after-the-fact process. Although the records of past accidents can be studied to identify trends that can, in turn, be used to predict accidents, this should be done as part of an accident investigation (see Chapter 19). HEA should be used to identify hazards before they cause accidents.

Two approaches to HEA can be effective: (1) observing employees at work and noting hazards (the task analysis approach), and (2) actually performing job tasks to get a firsthand feel for hazards. Regardless of how the HEA is conducted, it is a good idea to perform it in conjunction with FMEAs and HAZOPs. This will enhance the effectiveness of all three processes.

Technic of Operation Review (TOR)

Technic of operation review (TOR) is an analysis method that allows supervisors and employees to work together to analyze workplace accidents, failures, and incidents. It

Anderson Chemical Company
1512 Airport Road
Crestview, Florida 32536
HAZOP SUMMARY FORM

Department _____ Composites _____ System/Process _____ Plastic Mix _____ Date _____ January 7, 1993 _____

System/Process Component	Factor Analyzed	Guide-word	Resulting Difference	Potential Problem	Recommended Remedy
Flow-gate number one	Amount of flow	No	No flow	No mix	Make sure flow-gate is open
		Less	Insufficient flow	Weak mix	Troubleshoot and repair the flow-gate
		More	Excess flow	Too strong mix	Troubleshoot and repair flow-gate

Figure 18–6
Sample HAZOP.

answers the question, "Why did the system allow this incident to occur?" Like FMEA and HAZOP, this approach seeks to identify systemic causes. It does not seek to assign blame.

TOR is not new. It was originally developed in the early 1970s by D. A. Weaver of the American Society of Safety Engineers. However, for 20 years, user documentation on TOR was not readily available. Consequently, wide-scale use did not occur until the early 1990s when documentation began to be circulated. Richard G. Hallock describes TOR as follows:

> TOR is a hands-on analytical methodology designed to determine the root system causes of an operation failure. Because it uses a work sheet written in simple-to-understand terms and follows an uncomplicated yes/no decision-making sequence, it can be used even at the lowest levels of the firm. TOR is triggered by an incident occurring at a specific time and place and involving specific people. It is not a hypothetical process. It demands careful and systematic evaluation of the real circumstances surrounding the incident, and results in isolating the specific ways in which the organization failed to prevent the occurrence.[3]

A weakness of TOR is that it is designed as an after-the-fact process. It is triggered by an accident or incident. A strength of TOR is its involvement of line personnel in the analysis. The process proceeds as follows:

1. Establish the TOR team. It should consist of workers who were present when the accident/incident occurred, the supervisor, and the safety and health professional. The safety and health professional should chair the team and serve as a facilitator.
2. Conduct a roundtable discussion to establish a common knowledge base among team members. At the beginning of the discussion, five team members may have five different versions of the accident/incident. At the end, there should be a consensus.
3. Identify one major systematic factor that led to, or played a significant role in, causing the accident/incident. This one TOR statement, about which there must be consensus, serves as the starting point for further analysis.
4. Use the group consensus to respond to a sequence of yes/no options. Through this process, the team identifies a number of factors that contributed to the accident/incident.
5. Evaluate identified factors carefully to make sure that there is a team consensus about each. Then, prioritize the contributing factors beginning with the most serious.
6. Develop **corrective/preventive strategies** for each factor. Include them in a final report that is forwarded through normal channels for appropriate action.

Fault Tree Analysis (FTA)

Fault tree analysis (FTA) can be used to predict and prevent accidents or as an investigative tool after the fact. **FTA** is an analytical methodology that uses a graphic model to display the analysis process visually. A fault tree is built of special symbols, some derived from Boolean algebra. Consequently, the resultant model resembles a logic diagram or a flow chart. Figure 18–7 shows and describes the symbols used in constructing fault trees. Figure 18–8 shows how these symbols may be used to construct a fault tree. The

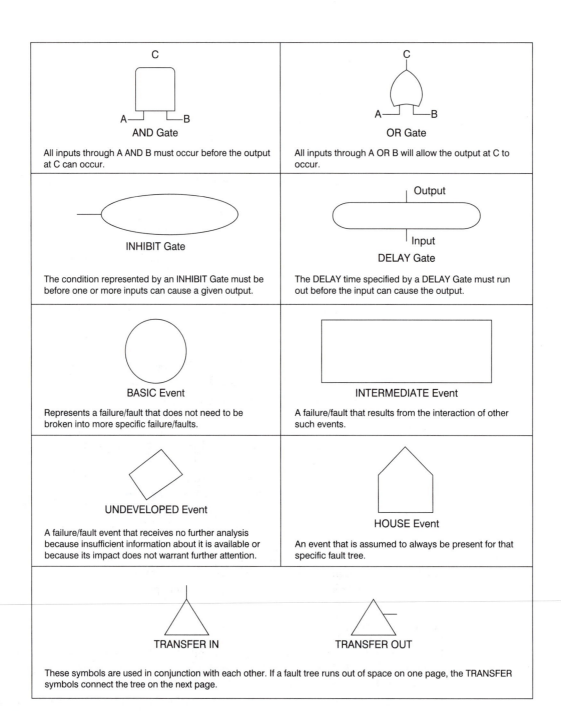

Figure 18-7
Symbols used in fault tree analysis.

top box in a fault tree represents the accident/incident that either could occur or has occurred.

All symbols below the top box represent events that contribute in some way to the ultimate accident/incident. The sample fault tree shown in Figure 18–8 is qualitative in nature. Fault trees can be made quantitative by assigning probability figures to the various events below the top box. However, this is rarely done since reliable probability figures are seldom available. A fault tree is developed using the following steps:

1. Decide on the accident/incident to be placed at the top of the tree.
2. Identify the broadest level of failure/fault event that could contribute to the top event. Assign the appropriate symbols.
3. Move downward through successively more specific levels until basic events are identified.

Experience, deliberate care, and systematic analysis are very important in constructing fault trees. Once a fault tree has been constructed, it is examined to determine the various combinations of failure/fault events that could lead to the top event. With simple fault trees, this can be accomplished manually; with more complex trees, this step is difficult. However, there are computer programs available to assist in accomplishing this step. The final step involves making recommendations for preventive measures.

Risk Analysis

Where are we at risk? Where are we at greatest risk? These are important questions for safety and health professionals involved in analyzing the workplace for the purpose of identifying and overcoming hazards. **Risk analysis** is an analytical methodology nor-

Figure 18-8
Sample fault tree.

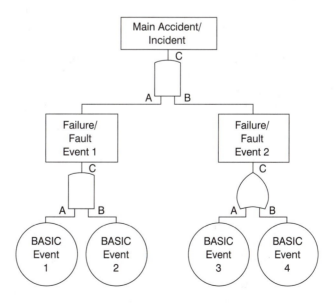

mally associated with insurance and investments. However, risk analysis can be used to analyze the workplace, identify hazards, and develop strategies for overcoming these hazards. The risk analysis process focuses on two key questions:

1. How *frequently* does a given event occur?
2. How *severe* are the consequences of a given event?

The fundamental rule-of-thumb of risk analysis is that risk is decreased by decreasing the frequency and severity of hazard-related events.

Safety and health professionals should understand the relationship that exists between the **frequency** and **severity** factors relating to accidents. Historical data on accidents, injuries, and illness show that the less severe an injury/illness, the more frequently it is likely to occur. Correspondingly, the more severe an injury/illness, the less frequently it is likely to occur. For example, there are many more minor scrapes, bumps, and abrasions experienced in the workplace than major debilitating injuries such as amputations or broken bones.

A number of different approaches can be used in conducting a risk analysis. One of the most effective is that developed by Chapanis.[4] Chapanis's approach to risk analysis considers both **probability** and **impact**.

Probability levels and corresponding frequency of occurrence ratings are as follows: 1 = Impossible (frequency of occurrence 10^{-8}/day); 2 = Extremely unlikely (frequency of occurrence 10^{-6}/day); 3 = Remote (frequency of occurrence 10^{-5}/day); 4 = Occasional (frequency of occurrence 10^{-4}/day); 5 = Reasonably probable (frequency of occurrence 10^{-3}/day); 6 = Frequent (frequency of occurrence 10^{-2}/day).[5]

The lowest rating (1) means it is impossible that a given error will be committed or a given failure will occur. The highest rating (6) means it is very likely that a given error will be committed frequently or a given failure will occur frequently. Notice the quantification of frequency levels for each level of probability. For example, the expected frequency of occurrence for a probability level of remote is ten to the negative fifth power per day.

Severity levels can also be rated, with the likely consequence of an accident/failure event of that severity. The least severe incidents (1) are not likely to cause an injury or damage property. The most severe incidents (4) are almost certain to cause death or serious property damage. Critical accidents (3) may cause severe injury or major loss. Marginal accidents (2) may cause minor injury, minor occupational illness, or minor damage.[6]

HAZARD PREVENTION/DETERRENCE

All of the methods and procedures discussed in this chapter have been concerned with identifying potential hazards. This section deals with using the information learned during analysis to prevent accidents and illnesses. The Society of Manufacturing Engineers recommends the following hazard control methods:

- Eliminate the source of the hazards.

SAFETY FACT

What to Include in a Hazards Inventory

An excellent tool for getting the work of safety and health personnel organized, prioritized, and properly focused is the **hazards inventory.** Such an inventory is a comprehensive list of all hazards associated with all processes and work tasks in a company. A hazards inventory should include at least the following information:

- Process descriptions
- Associated hazards
- Controls relating to the hazards
- Department location of each process
- Names of supervisors of all personnel who work each process (including telephone numbers)
- Number of employees who work each process
- Medical information relating to the hazards
- Historical information about the process and related hazards

---

- Substitute a less hazardous equivalent.
- Reduce the hazards at the source.
- Remove the employee from the hazard (e.g., substitute a robot or other automated system).
- Isolate the hazards (e.g., enclose them in barriers).
- Dilute the hazard (e.g., ventilate the hazardous substance).
- Apply appropriate management strategies.
- Use appropriate personal protective equipment.
- Provide employee training.
- Practice good housekeeping.[7]

For every hazard identified during the analysis process, one or more of these hazard control methods will apply. Figure 18–9 shows the steps involved in implementing hazard control methods. The first step involves selecting the method or methods that are most likely to produce the desired results. Once selected, the method is applied and monitored to determine if the expected results are being achieved.

Monitoring and observing are informal procedures. They should be followed by a more formal, more structured assessment of the effectiveness of the method. If the method selected is not producing the desired results, adjustments should be made. This may mean changing the way in which the method is applied or dropping it and trying another.

Figure 18-9
Steps for implementing hazard control measures.

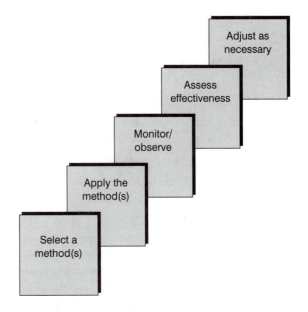

The example of Crestview Container Corporation's (CCC) problems with toxic paint illustrates how the process works. CCC produces airtight aluminum containers for transporting electromechanical devices. The containers must be painted as the last step in the production process. Although the specified paint was supposed to be only slightly toxic—a problem that should have been resolved by using personal protective equipment—paint station operators complained frequently of various negative side effects.

The CCC safety and health professional, working with management, solved the problem by applying the following steps:

1. *Select a method.* Of the various methods available, the one selected involved eliminating the source of the hazard (the toxic paint). CCC personnel were tasked with testing various nontoxic paints until one was found that could match the problem paint in all categories (e.g., ease of application, drying time, quality of surface finish, etc.) After 40 different paints were tested, a nontoxic substitute was found.

2. *Apply the method.* The new paint was ordered and used on a partial shipment of containers.

3. *Monitor/observe.* The safety and health professional, along with CCC's painting supervisors, monitored both employee performance and employee complaints concerning the paint.

4. *Assess effectiveness.* To assess effectiveness, employee complaints were tabulated. The number of complaints was down to a negligible amount and not serious in nature. Productivity was also assessed. It was found that the new paint had had no noticeable effect on productivity, negative or positive.

5. *Adjust as necessary.* CCC found that no adjustments were necessary.

OSHA PROCESS SAFETY STANDARD

The OSHA Process Safety Standard has relevance from the perspective of hazard prevention, relating specifically to chemical hazards. OSHA's standard for process safety is found in 29 C.F.R. 1910.119. Its purpose is to prevent *catastrophic* accidents caused by major releases of highly hazardous chemicals. To comply with this standard, companies must have written operating procedures, mechanical integrity programs, and formal incident investigation procedures. Other key elements are as follows:

1. *Coverage.* Although the Process Safety Standard is typically associated with large chemical and petrochemical processing plants, its coverage is actually much broader than this. Any company is covered that uses the threshold amount of a chemical listed in the standard, or 10,000 pounds or more of a flammable material on-site in one location.

2. *Employee participation.* Section (c) of the standard requires that employees be involved in all aspects of the process safety management program. In addition, employees must be given access to information developed as part of the program.

3. *Process Safety Information (PSI).* Section (d) of the standard requires organizations to establish and maintain process safety information files. Information included in the files includes chemical, process, and equipment data.

4. *Process Hazard Analyses (PHAs).* Section (e) of the standard requires that companies conduct process hazard analyses for all processes covered by the standard. Like any other hazard analysis, the PHAs are supposed to identify potential problems so that prompt corrective action or preventive measures can be taken.

5. *Standard Operating Procedures (SOPs).* Section (f) of the standard requires employees to establish and maintain written standard operating procedures for using chemicals safely. The requirement applies to handling, processing, transporting, and storing chemicals.

6. *Requirements for contractors.* Section (h) of the standard describes the special requirements imposed on companies that contract portions of their work to other companies. Complying with the standard is a matter of making sure that contractors comply. The following requirements are imposed by Section (h):
- Screen contractors before issuing a contract to ensure that they have a comprehensive safety and health program in place.
- Orient contractors concerning the chemicals with which they may be required to work or be around, the emergency action plan, and other pertinent information.
- Evaluate contractors periodically to ensure that their safety performance is acceptable.
- Maintain an OSHA injury and illness log for the contractor that is separate from, and in addition to, that of the host company.

OSHA's Regulation for Chemical Spills

OSHA issues a special regulation dealing with chemical spills. The standard (29 C.F.R. 1910.120) is called the Hazardous Waste Operations and Emergency Response, or HAZWOPER, standard. HAZWOPER gives organizations two options for responding to a chemical spill. The first is to evacuate all employees in the event of a spill and call in professional emergency response personnel. Employers who use this option must have an emergency action plan (EAP) in place in accordance with 29 C.F.R. 1010.38(a). The second option is to respond internally. Employers who use this option must have an emergency response plan in place that is in accordance with 29 C.F.R. 1010.120.

1. *Emergency action plans (EAPs).* An emergency action plan should have at least the following elements: alarm systems, evacuation plan, a mechanism or procedure for emergency shutdown of the equipment, and a procedure for notifying emergency response personnel.

2. *Emergency response plan.* Companies that opt to respond internally to chemical spills must have an emergency response plan that includes the provision of comprehensive training for employees. OSHA Standard 29 C.F.R. 1910.120 specifies the type and amount of training required, ranging from awareness to in-depth technical training for employees who will actually deal with the spill. It is important to note that OSHA forbids the involvement of untrained employees in responding to a spill. The following topics are those covered in the HAZWOPER seminar provided by Environmental Safety Awareness, a safety and health company in Fort Walton Beach, Florida. These topics are typical of those covered in up-to-date HAZWOPER courses.

Summary of Key Federal Laws

Overview of Impacting Regulations

Classification and Categorization of Hazardous Waste
- Definition of Hazardous Waste
- Characteristics
- TCLP
- Lists of Hazardous Wastes

Hazardous Waste Operations
- Definitions
- Levels of Response

Penalties for Noncompliance
- Civil Penalty Policy

Responding to Spills
- Groundwater Contamination

- Sudden Releases
- Clean-Up Levels
- Risk Assessment
- Remedial Action

Emergency Response

- Workplan
- Site Evaluation and Control
- Site Specific Safety and Health Plan
- Information and Training Program
- Personal Protective Equipment
- Monitoring
- Medical Surveillance
- Decontamination Procedures
- Emergency Response
- Other Provisions

Contingency Plans

- Alarm Systems
- Action Plan

Personal Protective Equipment

- Developing a PPE Program
- Respiratory Equipment
- Protective Clothing
- Donning PPE
- Doffing PPE

Material Safety Data Sheets

- Introduction
- Preparing MSDSs
- MSDS Information
- Hazardous Ingredients
- Physical/Chemical Characteristics
- Fire and Explosion Hazard Data
- Reactivity Data
- Health Hazard Data
- Precautions for Safe Handling and Use
- Control Measures

Site Control

- Site Maps
- Site Preparation
- Work Zones
- Buddy System
- Site Security
- Communications
- Safe Work Practices

Hazardous Waste Containers

- Emergency Control
- Equipment
- Tools
- Safety

Decontamination

- Types
- Decontamination Plan
- Prevention of Contamination
- Planning
- Emergencies
- Physical Injury
- Heat Stress
- Chemical Exposure
- Medical Treatment Area
- Decontamination of Equipment
- Decontamination Procedures
- Sanitation of PPE
- Disposal of Contaminated Materials

SUMMARY

1. A hazard is a condition or combination of conditions that, if left uncorrected, may lead to an accident, illness, or property damage.
2. Hazard analysis is a systematic process for identifying hazards and recommending corrective action. There are two approaches to hazard analysis: preliminary and detailed.
3. Hazards can be ranked as potentially catastrophic, critical, marginal, and nuisance.
4. A preliminary hazard analysis involves forming an ad hoc team of experienced personnel who are familiar with the equipment, material substance, and/or process

being analyzed. Experience and related expertise are critical in conducting a preliminary hazard analysis.

5. Failure mode and effects analysis (FMEA) is a detailed hazard analysis methodology that involves dividing a system into its various components, examining each component to determine how it may fail, rating the probability of failure, and deciding what effect these failures would have.

6. Hazard and operability review (HAZOP) is a detailed hazard analysis methodology that was developed for use in the chemical industry. It involves forming a team of experts and brainstorming.

7. Human error analysis (HEA) is used to predict human error and its potential effects. It can be used in conjunction with FMEA and HAZOP to strengthen those approaches.

8. Technic of operation review (TOR) is a hazard analysis methodology that allows workers and supervisors to conduct the analysis. It uses a simple worksheet that allows team members to respond to a sequence of yes/no options.

9. Fault tree analysis (FTA) is a hazard analysis methodology that uses a graphic model to display the analysis process visually. The model resembles a logic diagram.

10. Risk analysis, although more commonly associated with the insurance industry, can be used for hazard/safety analysis. The process revolves around answering two questions: How frequently does a given event occur? How severe are the consequences of a given event? The fundamental rule-of-thumb of risk analysis is that risk is decreased by decreasing the frequency and severity of hazard-related events.

11. The fundamentals of hazard prevention and deterrence include the following strategies: eliminate the source of the hazard, substitute a less hazardous substance, reduce the hazard at the source, remove the employee from the hazard, isolate the hazard, dilute the hazard, apply appropriate management strategies, use personal protective equipment, provide employee training, and practice good housekeeping.

KEY TERMS AND CONCEPTS

Corrective/preventive strategies	Hazard prevention/deterrence
Cost/benefit	HAZWOPER
Detailed hazard analysis	Human error
Experience	Human error analysis (HEA)
Failure mode and effects analysis (FMEA)	Impact
Fault tree analysis (FTA)	Preliminary hazard analysis
Frequency	Probability
Hazard	Related expertise
Hazard analysis	Risk analysis
Hazard and operability review (HAZOP)	Severity
Hazard inventory	Technic of operation review (TOR)

REVIEW QUESTIONS

1. Define the term *hazard*.
2. What is the purpose of preliminary hazard analysis?
3. Explain why experience and related expertise are so important when conducting a preliminary hazard analysis.
4. Why is cost/benefit analysis so critical a part of hazard analysis and prevention?
5. Briefly describe the following detailed hazard analysis methodologies: FMEA, HAZOP, HEA, FTA, and TOR.
6. What is the most fundamental weakness of both FMEA and HAZOP? How can it be overcome?
7. Name and briefly explain two approaches to HEA.
8. Why did it take so long for TOR to be adopted?
9. What is the most important strength of TOR?
10. Name five widely applicable hazard prevention strategies.
11. Explain the two options given to organizations by HAZWOPER for responding to a chemical spill.

ENDNOTES

1. Society of Manufacturing Engineers. *Tool and Manufacturing Engineers Handbook*, Vol. 5. (Dearborn, MI: Society of Manufacturing Engineers, 1988), pp. 12–17.
2. American Institute of Chemical Engineers. *Guidelines for Hazard Evaluation Procedures* (Chicago: American Institute of Chemical Engineers, 1985), p. 13..
3. Hallock, R. G. "Technic of Operations Review Analysis: Determine Cause of Accident/Incident," *Safety & Health*, August 1991, Vol. 60, No. 8, pp. 38–39, 46.
4. Chapanis, A. "To Err Is Human, to Forgive, Design." Proceedings of the ASSE Annual Professional Development Conference, New Orleans, 1986, p. 6.
5. Ibid.
6. Ibid.
7. Society of Manufacturing Engineers, *Tool and Manufacturing Engineers Handbook,* pp. 12–20.

CHAPTER NINETEEN

Accident Investigation and Reporting

- When to Investigate
- What to Investigate
- Who Should Investigate
- Conducting the Investigation
- Interviewing Witnesses
- Reporting Accidents

When an accident occurs, it is important that it be investigated thoroughly. The results of a comprehensive accident report can help safety and health professionals pinpoint the cause of the accident. This information can then be used to prevent future accidents, which is the primary purpose of accident investigation.

The Society of Manufacturing Engineers describes the importance of thoroughly investigating accidents:

> The primary reason for investigating an accident is not to identify a scapegoat, but to determine the cause of the accident. The investigation concentrates on gathering factual information about the details that led to the accident. If investigations are conducted properly, there is the added benefit of uncovering problems that did not directly lead to the accident. This information benefits the ongoing effort of reducing the likelihood of accidents. As problems are revealed during the investigation, action items and improvements that can prevent similar accidents from happening in the future will be easier to identify than at any time.[1]

This chapter gives prospective and practicing safety and health professionals the information they need to conduct thorough, effective accident investigations and prepare comprehensive accident reports.

WHEN TO INVESTIGATE

Of course, the first thing to do when an accident takes place is to implement **emergency procedures.** This involves bringing the situation under control and caring for the injured worker. As soon as all emergency procedures have been accomplished, the accident investigation should begin. Waiting too long to complete an investigation can harm the results. This is an important rule-of-thumb to remember. Another is that *all* accidents, no matter how small, should be investigated. Evidence suggests that the same factors that cause minor accidents cause major accidents.[2] Further, a near miss should be treated like an accident and investigated thoroughly.

There are several reasons why it is important to conduct investigations immediately. First, immediate investigations are more likely to produce accurate information. Conversely, the longer the time span between an accident and an investigation, the greater the likelihood of important facts becoming blurred as memories fade. Second, it is important to collect information before the accident scene is changed and before witnesses begin comparing notes. Human nature encourages people to change their stories to agree with those of other witnesses.[3] Finally, an immediate investigation is evidence of management's commitment to preventing future accidents. An immediate response shows that management cares.[4]

WHAT TO INVESTIGATE

The purpose of an **accident investigation** is to collect facts. It is not to find fault. It is important that safety and health professionals make this distinction known to all involved. **Fault finding** can cause reticence among witnesses who have valuable information to share. **Causes** of the accident should be the primary focus. The investigation should be guided by the following words: *who, what, when, where, why,* and *how.*

This does not mean that mistakes and breaches of precautionary procedures by workers are not noted. Rather, when these things are noted, they are recorded as facts instead of faults. If fault must be assigned, that should come later, after all the facts are in. The distinction is a matter of emphasis. The National Safety Council summarizes this approach as follows:

> As you investigate, don't put the emphasis on identifying who could be blamed for the accident. This approach can damage your credibility and generally reduce the amount and accuracy of information you receive from workers. This does not mean you ignore oversights or mistakes on the part of employees nor does it mean that personal responsibility should not be determined when appropriate. It means that the investigation should be concerned with only the facts. In order to do a quality job of investigating accidents you must be objective and analytical.[5]

In attempting to find the facts and identify causes, certain questions should be asked, regardless of the nature of the accident. The Society of Manufacturing Engineers recommends using the following questions when conducting accident investigations:

1.   What type of work was the injured person doing?

2. Exactly what was the injured person doing or trying to do at the time of the accident?
3. Was the injured person proficient in the task being performed at the time of the accident? Had the worker received proper training?
4. Was the injured person authorized to use the equipment or perform the process involved in the accident?
5. Were there other workers present at the time of the accident? If so, who are they and what were they doing?
6. Was the task in question being performed according to properly approved procedures?
7. Was the proper equipment being used including personal protective equipment?
8. Was the injured employee new to the job?
9. Was the process/equipment/system involved new?
10. Was the injured person being supervised at the time of the accident?
11. Are there any established safety rules or procedures that were clearly not being followed?
12. Where did the accident take place?
13. What was the condition of the accident site at the time of the accident?
14. Has a similar accident occurred before? If so, were corrective measures recommended? Were they implemented?
15. Are there obvious solutions that would have prevented the accident?[6]

The answers to these questions should be carefully and copiously recorded. You may find it helpful to dictate your findings into a microcassette recorder. This approach allows you to focus more time and energy on investigating and less on taking written notes.

Regardless of how the findings are recorded, it is important to be thorough. What may seem like a minor unrelated fact at the moment could turn out to be a valuable fact later when all of the evidence has been collected and is being analyzed.

WHO SHOULD INVESTIGATE

Who should conduct the accident investigation? Should it be the responsible supervisor? The safety and health professional? A higher-level manager? An outside specialist? There is no simple answer to this question, and there is disagreement among professional people of good will.

In some companies, the supervisor of the injured worker conducts the investigation. In others, a safety and health professional performs the job. Some companies form an investigative team; others bring in outside specialists. There are several reasons for the various approaches used. Factors considered in deciding how to approach accident investigations include the following:

- Size of the company
- Structure of the company's safety and health program
- Type of accident
- Seriousness of the accident

- Number of times that similar accidents have occurred
- Company's management philosophy
- Company's commitment to safety and health

After considering all of the variables just listed, it is difficult to envision a scenario in which the safety and health professional would not be involved in conducting an accident investigation. If the accident in question is very minor, the injured employee's supervisor may conduct the investigation, but the safety and health professional should at least study the accident report and be consulted regarding recommendations for corrective action.

If the accident is so serious that it has widespread negative implications in the community and beyond, responsibility for the investigation may be given to a high-level manager or corporate executive. In such cases, the safety and health professional should assist in conducting the investigation. If a company prefers the team approach, the safety and health professional should be a member of the team and, in most cases, should chair it. Regardless of the approach preferred by a given company, the safety and health professional should play a leadership role in collecting and analyzing the facts and developing recommendations.

CONDUCTING THE INVESTIGATION

The questions in the previous section summarize what to look for when conducting accident investigations. Figure 19–1 lists five steps to follow in conducting an accident investigation.[7] These steps are explained in the following paragraphs.

Figure 19–1
Steps in conducting an accident investigation.

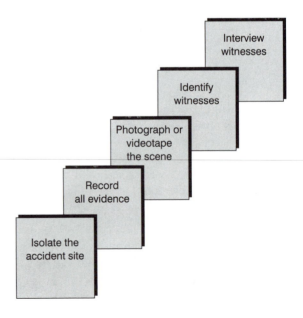

Isolate the Accident Scene

You may have seen a crime scene that had been sealed off by the police. The entire area surrounding such a scene is typically blocked off by barriers or heavy yellow tape. This is done to keep curious onlookers from removing, disturbing, or unknowingly destroying vital evidence. This same approach should be used when conducting an accident investigation. As soon as emergency procedures have been completed and the injured worker has been removed, the accident scene should be **isolated** until all pertinent evidence has been collected or observed and recorded. Further, nothing but the injured worker should be removed from the scene. If necessary, a security guard should be posted to maintain the integrity of the **accident scene.** The purpose of isolating the scene is to maintain as closely as possible the conditions that existed at the time of the accident.

Record All Evidence

It is important to make a permanent record of all pertinent **evidence** as quickly as possible. There are three reasons for this: (1) certain types of evidence may be perishable; (2) the longer an accident scene must be isolated, the more likely it is that evidence will be disturbed, knowingly or unknowingly; and (3) if the isolated scene contains a critical piece of equipment or a critical component in a larger process, pressure will quickly mount to get it back in operation. Evidence can be recorded in a variety of ways including written notes, sketches, photography, videotape, dictated observations, and diagrams. In deciding what to record, a good rule-of-thumb is *if in doubt, record it.* It is better to record too much than to skip evidence that may be needed later after the accident scene has been disturbed.

Photograph and/or Videotape the Scene

This step is actually an extension of the previous step. Modern photographic and videotaping technology has simplified the task of observing and recording evidence. Safety and health professionals should be proficient in the operation of a camera, even if it is just an instant camera, and a videotaping camera.

The advent of the digital camera has introduced a new meaning for the concept of "instant photographs." Using a digital camera in conjunction with a computer, photographs of accident scenes can be viewed immediately and transmitted instantly to numerous different locations. Digital camera equipment is especially useful when photographs of accident scenes in remote locations are needed.

Both still and video cameras should be on hand, loaded, and ready to use immediately should an accident occur. As with the previous step, a good rule-of-thumb in photographing and videotaping is *if in doubt, shoot it.* When recording evidence, it is better to have more shots than necessary than it is to risk missing a vital piece of evidence.

A problem with photographs is that, by themselves, they don't always reveal objects in their proper perspective. To overcome this shortcoming, the National Safety Council recommends the following technique:

When photographing objects involved in the accident, be sure to identify and measure them to show the proper perspective. Place a ruler or coin next to the object when making a close-up photograph. This technique will help to demonstrate the object's size or perspective.[8]

Identify Witnesses

In **identifying witnesses,** it is important to compile a witness list. Names on the list should be recorded in three categories: (1) **primary witnesses,** (2) **secondary witnesses,** and (3) **tertiary witnesses** (Figure 19–2). When compiling the witness list, ask employees to provide names of all three types of witnesses.

Interview Witnesses

Every witness on the list should be interviewed, preferably in the following order: primary witnesses first, secondary next, and tertiary last. Once all witnesses have been interviewed, it may be necessary to reinterview witnesses for clarification and/or corroboration. Interviewing witnesses is so specialized a process that the next major section is devoted to it.

INTERVIEWING WITNESSES

The techniques used for interviewing accident witnesses are designed to ensure that the information is objective, accurate, as untainted by the personal opinions and feelings of witnesses as possible, and able to be corroborated. For this reason, it is important to understand the *when, where,* and *how* of interviewing the accident witnesses.

When to Interview

Immediacy is important. Interviews should begin as soon as the witness list has been compiled and, once begun, should proceed expeditiously. There are two main reasons for

---

SAFETY MYTH

Where to Conduct Accident Interviews

To ensure that employees are willing to give accurate information, safety and health professionals should conduct accident interviews in the privacy of their office. Right? Not necessarily. Experience has shown that the best way to promote accuracy is to interview witnesses at the site of the accident. This puts the accident interview *in context* in a setting that will help stimulate the memory. To ensure privacy and confidentiality, interview witnesses one at a time, at the accident site.

Figure 19-2
Categories of accident witnesses.

- Primary witnesses are eyewitnesses to the accident.
- Secondary witnesses are witnesses who did not actually see the accident happen, but were in the vicinity and arrived on the scene immediately or very shortly after the accident.
- Tertiary witnesses are witnesses who were not present at the time of the accident nor afterward but may still have relevant evidence to present (e.g., an employee who had complained earlier about a problem with the machine involved in the accident).

this. The first is that a witness's best recollections will be right after the accident. The more time that elapses between the accident and the interview, the more blurred the witness's memory will become. The second reason for immediacy is the possibility of witnesses comparing notes and, as a result, changing their stories. This is just human nature, but it is a tendency that can undermine the value of testimony given and, in turn, the facts collected. Recommendations based on questionable facts are not likely to be valid.

Where to Interview

The best place to interview is at the **accident scene.** If this is not possible, interviews should take place in a private setting elsewhere. It is important to ensure that all distractions are removed, interruptions are guarded against, and the witness is not accompanied by other witnesses. All persons interviewed should be allowed to have their say without fear of contradiction or influence by other witnesses or employees. It is also important to select a neutral location in which witnesses will feel comfortable. Avoid the **principal's office syndrome** by selecting a location that is not likely to be intimidating to witnesses.

How to Interview

The key to getting at the facts is to put the witness at ease and to listen. Listen to what is said, how it is said, and what is not said. Ask questions that will get at the information listed earlier in this chapter, but phrase them in an **open-ended** format. For example, instead of asking, "Did you see the victim pull the red lever?" phrase your question as follows: "Tell me what you saw." Don't lead witnesses with your questions or influence them with gestures, facial expressions, tone of voice, or any other form of nonverbal communication. Interrupt only if absolutely necessary to seek clarification on a critical point. Remain nonjudgmental and objective.

An accident investigation is similar to a police investigation of a crime in that the information being sought can be summarized as **who, what, when, where, why,** and

how (Figure 19–3). As information is given, it may be necessary to take notes. If you can keep your note-taking to a minimum during the interview, your chances of getting uninhibited information are increased. Note-taking can distract and even frighten a witness.

An effective technique is to listen during the interview and make mental notes of critical information. At the end of the interview, summarize what you have heard and have the witness verify your summary. After the witness leaves, develop your notes immediately.

A question that sometimes arises is, "Why not tape the interview?" Safety and health professionals disagree on the effectiveness and advisability of taping. Those who favor taping claim it allows the interviewer to concentrate on listening without having to worry about forgetting a key point or having to interrupt the witnesses to jot down critical information. It also preserves everything that is said for the record as well as the tone of voice in which it is said. A complete transcript of the interview also ensures that information is not taken out of context.

Those opposed to taping say that taping devices tend to inhibit witnesses so that they are not as forthcoming as they would be without taping. Taping also slows down the investigation while the taped interview is transcribed and the interviewer wades through voluminous testimony trying to separate critical information from irrelevant information.

Figure 19-3
Questions to ask when interviewing witnesses.

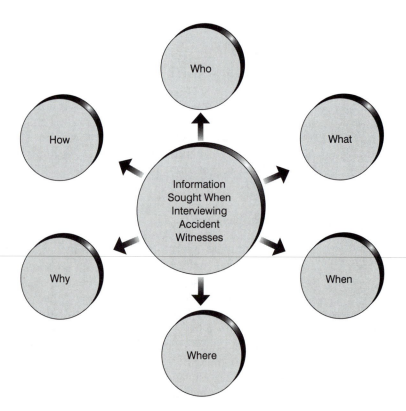

In any case, if the interview is to be taped, the following rules-of-thumb should be applied:

- Use the smallest, most unobtrusive taping device available, such as a microcassette recorder.
- Inform the witness that the interview will be taped.
- Make sure the taping device is working properly and that the tape it contains can run long enough so that you don't have to interrupt the witness to change it.
- Take time at the beginning of the interview to discuss unrelated matters long enough to put the witness at ease and overcome the presence of the taping device.
- Make sure the personnel are available to transcribe the tapes immediately.
- Read the transcripts as soon as they are available and highlight critical information.

An effective technique to use with **eyewitnesses** is to ask them to reenact the accident for you. Of course, the effectiveness of this technique is enhanced if the reenactment can take place at the accident site. However, even when this is not possible, an eyewitness reenactment can yield valuable information.

In using the **reenactment** technique, a word of caution is in order. If an eyewitness does exactly what the victim did, there may be another accident. Have the eyewitnesses explain what they are going to do before letting them do it. Then, have them *simulate* rather than actually perform the steps that led up to the accident.

REPORTING ACCIDENTS

An accident investigation should culminate in a comprehensive **accident report.** The purpose of the report is to record the findings of the accident investigation, the cause or causes of the accident, and recommendations for corrective action.

DISCUSSION CASE

What Is Your Opinion?

"Find out who is at fault and get rid of him," demanded the CEO. "I'm not going to have a careless employee running up our health care costs. It's tough enough trying to make a profit without some careless employee causing accidents. There will be a lawsuit, just you wait and see. We are going to be sued!" Gordon Jasperton, Clark Processing Company's safety director, bit his tongue and just listened. He had learned to let his boss vent before making a counter-proposal. When the time seemed right, Jasperton said, "Sir, if we focus on finding an employee to blame, it's just going to make matters worse. I'll never get to the root of it that way." Whose approach is best in this case? What is your opinion?

OSHA has established requirements for reporting and record keeping. According to OSHA document 2056,

> Employers of 11 or more employees must maintain records of occupational injuries and illnesses as they occur. Employers with 10 or fewer employees are exempt from keeping such records unless they are selected by the Bureau of Labor Statistics (BLS) to participate in the Annual Survey of Occupational Injuries and Illnesses.[9]

All injuries and illnesses are supposed to be recorded, regardless of severity, if they result in any of the outcomes shown in Figure 19–4. If an accident results in the death of an employee or hospitalization of five or more employees, a report must be submitted to the nearest OSHA office within 48 hours. This rule applies regardless of the size of the company.

Accident report forms vary from company to company. However, the information contained in them is fairly standard. Regardless of the type of form used, an accident report should contain at least the information needed to meet the record-keeping requirements set forth by OSHA. This information includes at least the following, according to the National Safety Council:

- Case number of the accident
- Victim's department or unit
- Location and date of the accident or date that an illness was first diagnosed
- Victim's name, social security number, sex, age, home address, and telephone number
- Victim's normal job assignment and length of employment with the company
- Victim's employment status at the time of the accident (i.e., temporary, permanent, full-time, part-time)
- Case numbers and names of others injured in the accident
- Type of injury and body part(s) injured (e.g., burn to right hand; broken bone, lower right leg) and severity of injury (i.e., fatal, first aid only required, hospitalization required)
- Name, address, and telephone number of the physician called
- Name, address, and telephone number of the hospital to which the victim was taken
- Phase of the victim's workday when the accident occurred (e.g., beginning of shift, during break, end of shift, and so on)

Figure 19–4
OSHA record-keeping requirements.

Injuries/illnesses must be recorded if they result in any of the following:

- Death
- One or more lost workdays
- Restriction of motion or work
- Loss of consciousness
- Transfer to another job
- Medical treatment (more than first aid)

■ Description of the accident and how it took place, including a step-by-step sequence of events leading up to the accident

■ Specific tasks and activities with which the victim was involved at the time of the accident (e.g., task: mixing cleaning solvent; activity: adding detergent to the mixture)

■ Employee's posture/proximity related to his or her surroundings at the time of the accident (e.g., standing on a ladder; bent over at the waist inside the robot's work envelope)

■ Supervision status at the time of the accident (i.e., unsupervised, directly supervised, indirectly supervised)

■ Causes of the accident

■ Corrective actions that have been taken so far

■ Recommendations for additional corrective action[10]

In addition to these items, you may want to record such additional information as the list of witnesses; dates, times, and places of interviews; historical data relating to similar accidents; information about related corrective actions that were made previously but had not yet been followed up on; and any other information that might be relevant. Figure 19–5 is an example of an accident report form that meets the OSHA record-keeping specifications.

Why Some Accidents Are Not Reported

In spite of OSHA's reporting specifications, some accidents still go unreported. According to Cunningham and Kane,

> The majority of accidents are not being reported! Articles in the Wall Street Journal testify to this fact. Many firms failed to report OSHA recordable incidents, presumably either to avoid OSHA inspections that result from poor incident rates, or to achieve statistical goals. The saddest part of non-reporting of accidents is that they are not investigated to determine and eliminate the causes.[11]

There are several reasons why accidents go unreported. Be familiar with these reasons so that you can do your part to overcome them. Cunningham and Kane list the main reasons as follows:

1. *Red tape.* Some people see the paperwork involved in accident reporting as red tape and, therefore, don't report accidents as a way to avoid paperwork.

2. *Ignorance.* Not all managers and supervisors are as knowledgeable as they should be about the reasons for accident reporting. Many are not familiar with OSHA's reporting specifications.

3. *Embarrassment.* Occasionally, people will let an accident go unreported because they are embarrassed by their part in it. A supervisor who did not properly supervise or a manager who has not provided the proper training for employees may be embarrassed to file a report.

4. *Record-spoiling.* Some accidents go unreported just to preserve a safety record, such as the record for days worked without an accident.

ACCIDENT REPORT FORM*

Fairmont Manufacturing Company
1501 Industrial Park Road
Fort Walton Beach, Florida 32548
904-725-4041

Victim-Related Information

Person completing report _____ Case no. _____

Gender _____ Age _____

Date of accident/illness _____

Victim's home address/telephone _____

_____

Victim's assignment at the time of the accident and length of time in that assignment:

_____

_____

_____

Victim's normal job and length of time in that job: _____

_____

_____

Time of injury/illness and phase of victim's workday: _____

_____

_____

Severity of the injury (e.g., hospitalization required, first aid required, etc.): _____

_____

_____

Type of injury and body part(s) injured: _____

_____

Exact location of the accident (which facility, department, place within the department):

_____

_____

Physician and hospital: _____

_____

Figure 19–5
Sample accident report form.
*One form for each injured worker.

Accident-Related Information

Accident description with step-by-step sequence of events: _____

_____

_____

Task and specific activity at the time of the accident: _____

_____

_____

Posture/proximity of employee at the time of the accident: _____

_____

_____

Supervision status at the time of the accident: _____

_____

_____

Apparent causes including conditions, actions, events, and activities and other contributing factors: _____

_____

_____

Recommendations for corrective action: _____

_____

_____

Case numbers and names of other persons injured in the accident: _____

_____

_____

Witnesses to the accident, and dates/places of their interviews:

_____   _____

_____   _____

_____   _____

_____   _____

_____   _____
Name   Date

5. *Repercussions/fear.* Some accidents go unreported because the people involved are afraid of being found at fault, being labeled accident prone, and being subjected to other negative repercussions.

6. *No feedback.* Some accidents go unreported because those involved feel filing a report would be a waste of time. This typically happens when management does not respond to recommendations made in earlier accident reports.[12]

Clearly these reasons for not reporting accidents present safety and health professionals with a challenge. To overcome these inhibitors, it will be necessary to develop a simple reporting system that will not be viewed as too much bureaucratic paperwork with which to bother. Safety and health professionals will have to educate personnel at all levels concerning the purpose of accident reporting and why it is important. An important step will be to communicate the fact that fault finding is not the purpose. Another important step is to follow up to ensure that recommendations are acted on or that employees are made aware of why they aren't. This will help ensure the integrity of the process.

Discipline and Accident Reporting

Fault finding is not the purpose of an accident investigation. However, an investigation will sometimes reveal that an employee has violated or simply overlooked safety regulations. Should such violations be condoned? According to Kane and Cunningham,

> Many companies condone nonconformance to safety rules as long as no injury results. However, if the nonconformance results in an accident involving an injury, the disciplinary boom is promptly lowered. This inconsistency inevitably leads to resentment and failure to report accidents and a hiding of accident problems.[13]

There is a built-in dilemma here that modern safety and health professionals must be prepared to handle. On the one hand, it is important that fault finding not be seen as the purpose of an accident investigation. Such a perception will limit the amount of information that can be collected. On the other hand, if those workers whose behavior leads to accidents are not disciplined, the credibility of the safety program is undetermined. Kane and Cunningham recommend the following procedures for handling this dilemma: *Never* discipline an employee because he or she had an accident. *Always* discipline employees for noncompliance with safety regulations.[14]

Such an approach applied with consistency will help maintain the integrity of both the accident investigation process and the overall safety program.

=== SUMMARY ===

1. Accidents are investigated for the purpose of identifying causal factors that could lead to other accidents if not corrected. The purpose is not to assign blame.

2. It is important to begin an accident investigation as soon as possible after an accident occurs so that evidence and the memories of witnesses are still fresh.

3. Facts to be uncovered in an accident investigation can be summarized as who, what, when, where, why, and how.

4. Who conducts the accident investigation can vary according to circumstances. However, regardless of how it is done, the safety and health professional should play an active role in the process.

5. Steps for conducting an accident investigation are as follows: (a) isolate the accident scene, (b) record all evidence, (c) photograph and/or videotape the accident scene, (d) identify witnesses, and (e) interview witnesses.

6. Witnesses to accidents fall into one of three categories: primary (eye-witnesses); secondary (were present at the scene, but did not see the accident); and tertiary (were not present but have information that might be relevant).

7. Interviews should take place at the accident site whenever possible. When this isn't practical, interviews should take place at a neutral location that is private and with which the witness is comfortable.

8. The keys to getting at the facts in an interview are as follows: (a) put the witness at ease, (b) ask open-ended questions, and (c) listen. Interrupt only if absolutely necessary.

9. When possible, let eyewitnesses reenact the accident through simulation at the job site. Do not let them actually perform the tasks that led up to the accident.

10. The purpose of an accident report is to record the findings of the accident investigation, the cause or causes of the accident, and recommendations for corrective action. Report forms should meet the record-keeping specifications of OSHA.

KEY TERMS AND CONCEPTS

Accident investigation	Open-ended question
Accident report	Primary witness
Accident scene	Principal's office syndrome
Causes	Reenactment
Emergency procedures	Secondary witness
Evidence	Tertiary witness
Eyewitness	What
Fault finding	When
How	Where
Identify witnesses	Who
Immediacy	Why
Isolate the scene	

REVIEW QUESTIONS

1. Explain the rationale for investigating accidents.
2. When should an investigation be reported? Why?
3. What are the terms that should guide the conduct of an accident investigation?
4. What role should the safety and health professional play in the conduct of an accident investigation?
5. List and explain the steps for conducting an accident investigation.
6. Why is it important to record all pertinent evidence relating to an accident immediately after an accident has occurred?
7. How can you help put close-up photographs in the proper perspective?
8. List and differentiate among the three categories of witnesses to an accident.
9. Briefly explain the *when* and *where* of interviewing witnesses.
10. Briefly explain the *how* of interviewing witnesses.
11. What is the purpose of an accident report?

ENDNOTES

1. Society of Manufacturing Engineers. *The Manufacturing Engineer's Handbook*, Vol. 5, (Dearborn, MI: Society of Manufacturing Engineers, 1988), pp. 12–21.
2. Ibid.
3. Ibid.
4. Ibid.
5. National Safety Council. *Supervisor's Safety Manual* (Chicago: National Safety Council, 1991), pp. 69–70.
6. Society of Manufacturing Engineers. *The Manufacturing Engineer's Handbook*, pp. 12–21.
7. National Safety Council. *Supervisor's Safety Manual*, p. 71.
8. Ibid.
9. OSHA 2056, 1991 (Revised), U.S. Department of Labor, p. 11.
10. National Safety Council. *Supervisor's Safety Manual*, pp. 76–77.
11. Cunningham, J., and Kane, A. "Accident Reporting—Part I: Key to Prevention," *Safety & Health*, April 1989, Vol. 139, No. 4, p. 70.
12. Ibid., pp. 70–71.
13. Kane, A., and Cunningham, J. "Accident Reporting—Part II: Consistent Discipline Is Vital," *Safety & Health*, May 1989, Vol. 139, No. 5, p. 78.
14. Ibid.

Promoting Safety

One of the best ways to promote safety is to design it into the tools, machines, and technologies with which people interact in the workplace. Safety analysis can also be effective by eliminating hazards before they cause accidents/illnesses. However, even the best design/analysis cannot completely eliminate the potential for accidents. For this reason, it is important to have accident prevention procedures and make sure that employees follow them.

The purpose of safety promotion is to keep employees focused on doing their work the safe way, every day. This chapter provides prospective and practicing safety and health professionals with information that will enable them to promote safety effectively.

COMPANY SAFETY POLICY

Promoting safety begins with having a published company **safety policy.** The policy should make it clear that safe work practices are expected of all employees at all levels at all times. The safety policy serves as the foundation upon which all other promotional efforts are built.

Figure 20–1 is an example of a company safety policy. This policy briefly and succinctly expresses the company's **commitment** to safety. It also indicates clearly that employees are expected to perform their duties with safety foremost in their minds. With such a policy in place and clearly communicated to all employees, other efforts to promote safety will have solid backing.

A company's safety policy need not be long. In fact, a short and simple policy is better. Regardless of its length or format, a safety policy should convey at least the following messages:

1. The company is committed to safety and health.
2. Employees are expected to perform their duties in a safe and healthy manner.
3. The company's commitment extends beyond the walls of its plant to include customers and the community.

Promoting Safety by Example

Once a safety policy has been implemented, its credibility with employees will be determined by the example set by management, from supervisors through executives. It is critical that managers follow the company safety policy in both letter and spirit. Managers who set a poor example undermine all of the company's efforts to promote safety. The "do as I say, not as I do" approach will not work with modern employees.

Where positive examples tend to break down most frequently is under the pressure of deadlines. To meet a deadline, supervisors may encourage their team members to take shortcuts or, at least, look the other way when they do. This type of behavior conveys the message that safety is not really important—something we talk about, but not some-

Figure 20-1
Sample company safety policy.

Okaloosa Poultry Processing, Inc.

414 Baker Highway
Crestview, Florida 36710

Safety Policy

It is the policy of this company to ensure a safe and healthy workplace for employees, a safe and healthy product for customers, and a safe and healthy environment for the community. OPP, Inc. is committed to safety on the job and off. Employees are expected to perform their duties with this commitment in mind.

thing we believe in. The issue of setting a positive example will be discussed further in the next section.

SAFETY RULES AND REGULATIONS

A company's safety policy is translated into everyday action and behavior by rules and regulations. Rules and regulations define behavior that is acceptable and unacceptable from a safety and health perspective. From a legal point of view, an employer's obligations regarding **safety rules** can be summarized as follows:

1. Employers must have rules that ensure a safe and healthy workplace.
2. Employers must ensure that all employees are knowledgeable about the rules.
3. Employers must ensure that safety rules are enforced objectively and consistently.

The law tends to view employers who do not meet these three criteria as being *negligent*. Having the rules is not enough. Having rules and making employees aware of them is not enough. Employers must develop appropriate rules, familiarize all employees with them, and enforce the rules. It is this final step—enforcement—from which most negligence charges arise.

Although it is acceptable to prioritize rules and assign different levels of punishment for failing to observe them, it is unacceptable to ignore rules. If the punishment for failure to observe Rule X is a letter of reprimand, then every person who fails to observe Rule X should receive such a letter every time. Of course, repeat violations should lead to more than a letter of reprimand.

Objectivity and consistency are critical when enforcing rules. **Objectivity** means that rules are enforced equally regardless of who commits an infraction—from the newest employee to the chief executive officer. **Consistency** means that the rules are enforced in the same manner every time with no regard to any outside factors. This means that the same punishment is assigned regardless of who commits the infraction. Failure to be objective and consistent can undermine the credibility and effectiveness of a company's efforts to promote safety.

Setting a positive example of always following safety rules is the best way to ensure that they are followed. Just how effective positive role modeling can be is seen in the example of the eye protection program at Westinghouse Electric Corporation's Electro-Mechanical Division. Although eye hazards are not especially high in this division, they are taken seriously. The management team has a rule that makes eye protection mandatory. This means that all workers, regardless of their job titles, must wear eye protection when on the plant floor. To promote the use of eye protection, the company provides prescription glasses in approximately 40 men's and women's styles. The frames are attractive enough that some employees wear the safety glasses as their regular prescription glasses. Consequently, of the 40 OSHA recordable accidents in the division during 1988, only two were eye-related.[1]

Figure 20–2 contains guidelines to follow when developing safety rules. These guidelines will help ensure a safe and healthy workplace without unduly inhibiting workers in the performance of their jobs. This is an important point for prospective and prac-

Figure 20-2
Guidelines for developing
safety rules and regulations.

- Minimize the number of rules to the extent possible. Too many rules can result in rule *overload*.

- Write rules in clear and simple language. Be brief and to the point, avoiding ambiguous or overly technical language.

- Write only the rules that are necessary to ensure a safe and healthy workplace. Do not *nitpick*.

- Involve employees in the development of rules that apply to their specific areas of operation.

- Develop only rules that can and will be enforced.

- Use common sense in developing rules.

ticing safety and health professionals to understand. Fear of negligence charges can influence an employer in such a way that the book of safety rules becomes a multivolume nightmare that is beyond the comprehension of most employees.

Such attempts to avoid costly litigation, penalties, or fines by regulating every move that employees make and every breath that they take are likely to backfire. Remember, employers must do more than just write rules. They must also familiarize all employees with them and enforce them. This will not be possible if the rulebook is as thick as an unabridged dictionary. Apply common sense when writing safety rules.

EMPLOYEE PARTICIPATION IN PROMOTING SAFETY

One of the keys to successfully promoting safety is to involve employees. They usually know better than anyone where hazards exist. In addition, they are the ones who must follow safety rules. A fundamental rule of management is *if you want employees to make a commitment, involve them from the start*. One of the most effective strategies for getting employees to commit to the safety program is to involve them in the development of it. This way, your program becomes their program.

According to Payne, **employee participation and involvement** is the key to the success of the safety program at Dover Products, a leading manufacturer of tungsten and molybdenum wire.[2] Writing for *Safety & Health,* Payne said,

> In the Dover Program, employees are given not only the responsibility to identify safety and health concerns, but the authority to take action on them. Employees focus on hazards and problems in their specific areas of work and write up work orders for maintenance on what needs to be done to resolve their concerns in a timeframe.[3]

Employee involvement at Dover Products can serve as an instructive example. According to Payne,

> It has produced one of the most accident-free workplaces in the state of Ohio. It has been eight and one-half years and more than two million work hours since the plant experienced a lost-time injury. Only four recordable occupational injuries or illnesses have occurred there in the last five years.[4]

What the success story at Dover Products shows is that employees should be involved in all aspects of a safety program including development, implementation, monitoring, and follow-up. In all phases, employees should be empowered to take action to improve safety. The most effective safety program will be the one that employees view as *their* program.

SAFETY TRAINING

One of the best ways to promote safety in the workplace is to provide all employees with ongoing **safety training.** Initial safety training should be part of the orientation process for new employees. Subsequent safety training should be aimed at developing new, more specific, and more in-depth knowledge and at renewing and updating existing knowledge.

Safety training is covered at length in Chapter 21. This chapter emphasizes the importance of promoting safety by providing training on a continual basis. Training serves a dual purpose in the promotion of safety. First, it ensures that employees know how to work safely and why doing so is important. Second, it shows that management is committed to safety. Refer to Chapter 21 to see how safety training programs are organized and which topics should be covered at the different levels of training.

SUGGESTION PROGRAMS

Suggestion programs, if properly handled, promote safety and health. Well-run suggestion programs offer two advantages: (1) they solicit input from the people most likely to know where hazards exist; and (2) they involve and empower employees which, in turn, gives them **ownership** of the safety program.

Suggestion programs must meet certain criteria to be effective:

- All suggestions must receive a formal response.
- All suggestions must be answered immediately.
- Management must monitor the performance of each department in generating and responding to suggestions.
- System costs and savings must be reported.
- Recognition and awards must be handled promptly.
- Good ideas must be implemented.
- Personality conflicts must be minimized.[5]

Suggestion programs that meet these criteria are more likely to be successful than those that don't. Figure 20–3 is an example of a suggestion form that may be used as part of a company's safety program. Note that all of the following must be recorded: the date that the suggestion was submitted, the date that the suggestion was logged in, and the date that the employee received a response.

This form satisfies the **formal response** and **immediate response** criteria. It also makes it easier to monitor response. Jones Petroleum Products (in Figure 20–3) publishes system costs and savings in its monthly newsletter for employees, implements

---

Jones Petroleum Products, Inc.

Highway 90 East
DeFuniak Springs, Florida 32614

Suggestion Form

Name of employee: _____ Date of suggestion: _____

Department: _____

Suggested improvement: _____

_____

_____

_____

Date logged in: _____ Time: _____

Logged in by: _____

Action taken: _____

_____

Current status: _____

_____

Date of response to employee: _____

Person responding: _____

(Signature)

Figure 20–3
Sample safety suggestion form.

good ideas, and recognizes employees with a variety of awards ranging from certificates to cash at a monthly recognition ceremony. This company's suggestion program is an example of one that promotes not just safety, but continual improvements in quality, productivity, and competitiveness.

VISUAL AWARENESS

We tend to be a visual society. This is why television and billboards are so effective in marketing promotions. Making a safety and health message **visual** can be an effective way to get the message across. Figure 20–4 is a sign that gives machine operators a visual reminder to use the appropriate machine guards. Such a sign is placed on or near the machine in question. If operators cannot activate their machines without first reading this sign, they will be reminded to use the safe way every time they operate the machine.

Figure 20–5 may be placed on the door leading into the hard hat area or on a stand placed prominently at the main point of entry if there is no door. Such a sign will help

Figure 20–4
Sample safety reminder sign.

prevent inadvertent slip-ups when employees are in a hurry or are thinking about something else. Figures 20–6, 20–7, and 20–8 are additional examples of signs/posters that make a safety message visual.

Several rules-of-thumb can help ensure the effectiveness of efforts to make safety visual:

- Change signs, posters, and other visual aids periodically. Visual aids left up too long begin to blend into the background and are no longer noticed.
- Involve employees in developing the messages that will be displayed on signs and posters. Employees are more likely to notice and heed their own messages than those of others.
- Keep visual aids simple and the message brief.
- Make visual aids large enough to be seen easily from a reasonable distance.
- Locate visual aids for maximum effect. For example, the sign in Figure 20–4 should be located on the machine in question, preferably near the on-off switch so that the operator cannot activate the machine without seeing it.
- Use color whenever possible to attract attention to the visual aid.

SAFETY COMMITTEES

Another way to promote safety through employee involvement is the safety committee. **Safety committees** provide a formal structure through which employees and

Figure 20–5
Sample safety reminder sign.

Figure 20–6
Sample safety reminder sign.

management can funnel concerns and suggestions about safety and health issues. The composition of the safety committee can be a major factor in the committee's success or failure.

The most effective committees are those that are composed of a broad cross-section of workers representing all departments. This offers two advantages: (1) it gives each member of the committee a constituent group for which he or she is responsible; and (2) it gives all employees a representative voice on the committee.

There is disagreement over whether an executive-level manager should serve on the safety committee. On the one hand, an executive-level participant can give the committee credibility, visibility, and access. On the other hand, the presence of an executive manager can inhibit the free flow of ideas and concerns. The key to whether an executive manager's participation will be positive or negative lies in the personality and management skills of the executive in question.

An executive who knows how to put employees at ease, interact in a nonthreatening manner, and draw people out will add to the effectiveness of the committee. An executive with a threatening attitude will render the committee useless. Consequently, the author recommends the involvement of a very carefully selected executive manager on the safety committee.

The safety and health professional should be a member of the committee serving as an advisor, facilitator, and catalyst. Committee members should select a chairperson from the membership and a recording secretary for taking minutes and maintaining committee records. Neither the executive manager nor the safety and health professional should serve as chairperson, but either can serve as recording secretary. Excluding executive managers and safety and health professionals from the chair will give employees more ownership in the committee.

Figure 20–7
Sample safety reminder sign.

Figure 20–8
Sample safety reminder sign.

NOTICE
SAFETY SHOES ARE
REQUIRED
IN THIS AREA

Safety committees will work only if members are truly empowered to identify hazards and take steps to eliminate them. Consequently, Peters recommends that members of such committees be trained in cause-and-effect analysis and group problem-solving techniques.[6] Such training can help ensure that safety committee members identify actual problems rather than just deal with symptoms.

GAINING A PERSONAL COMMITMENT

If every employee is committed to working safely every day, workplace safety will take care of itself. But how does a company gain this type of **personal commitment** from its employees? One way is to have employees commit themselves to safety by signing on the bottom line. According to Fettig,

> Most safety superstars have formal safety programs—teams, committees, whatever the personal term. Many of these organizations also have signed commitments from their employees. They have their people's signatures on the dotted line indicating that they bought into the safety program. A signature on the dotted line is serious business. When our founding fathers signed the Declaration of Independence, they pledged their lives, their fortunes, and their sacred honor. How do you sell safety? One way is to get that signature on the dotted line.[7]

Fettig's approach to gaining a personal commitment from employees has merit. Ours is a society that revolves around the written signature. We sign countless docu-

---

DISCUSSION CASE

What Is Your Opinion?

"I want our management team to develop a program to promote better safety practices in this company," said the CEO. "If we develop the program, it will show employees that we are committed to safety." "I like the idea of showing executive-level commitment," said the vice president for engineering. "But the program might be more readily accepted if we involve employees in developing it." Executive commitment and employee involvement. Which is the better approach? What is your opinion?

ments in our lives from credit statements to bank loans to home mortgages to college registration forms. In each of these cases, our signature is a written pledge of our commitment to meet certain responsibilities.

According to Fettig, companies gain the following three advantages from making signing on the dotted line part of their program to promote safety:

1. By their signature, employees make a personal commitment.
2. By their signature, employees promise to interact positively with fellow workers when they see them ignoring safety precautions.
3. By their signature, employees give fellow workers permission to correct them when they ignore safety precautions.[8]

UNION/MANAGEMENT PARTICIPATION

An excellent way to promote safety is to secure the cooperation of management and labor. For a company's safety program to succeed, the **union/management participation** and support is critical. Fortunately, union/management agreement on workplace safety is commonplace. In fact, workplace safety is one of the few issues on which labor and management agree.

When disagreement over a safety procedure does surface, the issue at the heart is usually money. The union is likely to favor most procedures that enhance workplace safety regardless of cost. Management, on the other hand, is likely to want to weigh the cost vs. the benefits of safety improvement strategies. However, it is not always management that questions such strategies; occasionally, unions will balk.

An example of a union questioning a safety enhancement strategy is the Sign Up for Safety campaign conducted by safety consultant Art Fettig at the Northwestern Region of Consumer Power Company in Muskegon, Michigan.[9] In an attempt to gain a personal commitment to safety, Fettig asks employees to sign a declaration that they will work in a safe manner. This is a technique that has met with a great deal of success.

However, in attempting to sell employees of Consumer Power Company on the strategy, Fettig ran into resistance from a union steward. The steward refused to sign his name, claiming the company's management team would use it against him. Here is Fettig's account of how he handled the situation and what eventually resulted:

> I pointed out that the document was not for the company, but for the workers. Very seldom do managers fall from their chairs, but linemen fall from ladders and poles. And I have yet to hear of a manager who was electrocuted by a computer, but I often read of such accidents that involve linemen and power lines. He (the union steward) still wasn't sold. He said the declaration should hang at the union hall and not at the company. He's right since the declaration is for the employees. But management is also part of the team and should have a copy for the company as well. He finally signed the declaration. In fact, all of the employees signed the declaration, which now hangs in the union hall.[10]

It was the union steward's eventual willingness to sign the safety declaration in this case that made the program work for Consumer Power Company. With management and labor on the same team, the safety program is much more likely to succeed.

INCENTIVES

If properly used, **incentives** can help promote safety. However, the proper use of incentives is a widely misunderstood concept. Tim Puffer has this to say about the use of incentives in the modern workplace:

> Most companies have no problem developing incentive programs for their sales forces. . . . But with the shift toward a service economy, a growing number of companies are becoming just as concerned with non-sales performance issues such as productivity and customer service.[11]

Safety is another issue that is being confronted using incentives. To promote safety effectively, incentives must be properly structured. Puffer recommends the following strategies for enhancing the effectiveness of incentive programs:

1. *Define objectives.* Begin by deciding what is supposed to be accomplished by the incentive program.

2. *Develop specific criteria.* On what basis will the incentives be awarded? This question should be answered during the development of the program. Specific criteria define the type of behavior and level of performance that is to be rewarded as well as guidelines for measuring success.

3. *Make rewards meaningful.* For an incentive program to be effective, the rewards must be meaningful to the recipients. Giving an employee a reward that he or she does not value will not produce the desired results. To determine what types of rewards will be meaningful, it is necessary to involve employees.

4. *Recognize that only employees who will participate in an incentive program know what incentives will motivate them.* In addition, employees must feel it is *their* program. This means that employees should be involved in the planning, implementation, and evaluation of the incentive program.

5. *Keep communications clear.* It is important for employees to understand fully the incentive program and all of its aspects. Communicate with employees about the program, ask for continual feedback, listen to the feedback, and act on it.

6. *Reward teams.* Rewarding teams can be more effective than rewarding individuals. This is because work in the modern industrial setting is more likely to be accomplished by a team than an individual. When this is the case, other team members may resent the recognition given to an individual member. Such a situation can cause the incentive program to backfire.[12]

COMPETITION

Competition is another strategy that can be used to promote safety. However, if this approach is not used wisely, it can backfire and do more harm than good. To a degree, most people are competitive. A child's competitive instinct is nurtured through play and reinforced by sports and school activities. Safety and health professionals can use the adult's competitive instinct when trying to motivate employees, but competition on the job should be carefully organized, closely monitored, and strictly controlled. Competition that is allowed to get out of hand can lead to cheating and hard feelings among fellow workers.

Competition can be organized between teams, shifts, divisions, or even plants. Here are some tips that will help safety and health professionals use competition in a positive way while ensuring that it does not get out of hand:

- Involve the employees who will compete in planning programs of competition.
- Where possible, encourage competition among groups rather than individuals, while simultaneously promoting individual initiative within groups.
- Make sure that the competition is fair by ensuring that the resources available to competing teams are equitably distributed and that human talent is as appropriately spread among the teams as possible.

The main problem with using competition to promote safety is that it can induce competing teams to cover up or fail to report accidents just to win. Safety and health professionals should be particularly attentive to this situation and watch carefully for evidence that accidents are going unreported. If this occurs, the best approach is to confront the situation openly and frankly. Employees should be reminded that improved safety is the first priority and winning the competition is second. Failing to report an accident should be grounds for eliminating a team from competition.

COMPANY-SPONSORED WELLNESS PROGRAMS

Faced with mounting health-care and workers' compensation costs, some employers are looking for innovative ways to keep their workers safe and healthy. The annual cost of health insurance for business and industry in this country is approximately $150 billion. One innovation that is gaining in popularity as a way to promote safety and health is the company-sponsored wellness program. A **wellness program** is any program designed to help and encourage employees to adopt a healthier lifestyle.

A typical wellness program includes diet and exercise under the supervision of an appropriately qualified professional; stress management activities; and special activities designed to help high-risk employees overcome such lifestyle-related behaviors as smoking and overeating.

Industry in the United States began to experiment with company-sponsored wellness programs in the early 1980s. However, hard data on tangible financial benefits were not available in the early years, and wellness programs failed to gain broad-based acceptance, particularly among small to mid-sized companies. However, as health-care costs continued to rise and industry continued to lose valuable employees to heart attacks and strokes, corporate leaders began searching for ways to measure objectively the costs vs. the benefits of company-sponsored wellness programs.

A major breakthrough in this area resulted when the Institute for Preventive Medicine (IPM), located in Houston, Texas, developed a computer-based method for measuring the success of preventive health-care programs. The method was pilot-tested on a large Houston-based corporation.[13]

This company agreed to spend $46,000 to fund an in-house wellness program that included weight control, stress management, and smoking cessation, exercise, and cardiac rehabilitation activities. After six months, IPM assessed the costs vs. benefits of the program and arrived at the following conclusions:

> Approximately 200 employees participated in the six-month health management program, and the results were dramatic. Comparing insurance claims and absenteeism for the six months before the program and those for the six months after its inception, the IPM staff recorded an 18 percent decline in sick pay hours and a 15 percent fall in sick costs. Overall, the company saw health care costs drop 39 percent.[14]

The IPM study showed that company-sponsored wellness programs can be an effective way to promote employee safety and health. According to Myerson,

> Corporate wellness works. Participants improve their health and quality of life, and employers cut costs—while at the same time generating goodwill. It's a combination that a growing number of organizations are finding hard to resist.[15]

This raises a question of how many companies are sponsoring wellness programs for their employees and how many employees are taking advantage of the programs. To answer this and related questions, the National Safety Council conducted a survey in April 1991.[16] This study revealed the following: (1) 66 percent of respondents work for companies that sponsor a wellness program for employees; (2) only 36 percent of respondents participate in their company's wellness program; and (3) of the companies that sponsor wellness programs, 43 percent are provided on-site and 57 percent off-site.[17]

TEAMWORK APPROACH TO PROMOTING SAFETY

Increasingly, teamwork is stressed as the best way to get work done in the contemporary workplace. Consequently, it follows that the teamwork approach is an excellent way to promote safety. Teamwork is a fundamental component of the *Total Safety Management,* or TSM, approach explained in Chapter 28. Consequently, this section will be limited to covering teamwork as it relates specifically to the promotion of safety.

Characteristics of Effective Teams

Effective teams share several common characteristics: supportive environment, team player skills, role clarity, clear direction, team-oriented rewards, and accountability.

Supportive Environment

The characteristics of a team-supportive environment are well known. These characteristics are

- Open communication
- Constructive, nonhostile interaction
- Mutually supportive approach to work

■ Positive, respectful climate

Team Player Skills

Team player skills are personal characteristics of individuals that make them good team players. They include the following:

■ Honesty
■ Selflessness
■ Initiative
■ Patience
■ Resourcefulness
■ Punctuality
■ Tolerance
■ Perseverance

Role Clarity

On any team, different members play different roles. Consider the example of a football team. When the offensive team is on the field, each of the 11 team members has a specific role to play. The quarterback plays one role; the running backs, another; the receivers, another; the center, another; and the linemen, another. Each of these roles is different but important to the team. When each of these players executes his role effectively, the team performs well.

But what would happen if the center suddenly decided he wanted to pass the ball? What would happen if one of the linemen suddenly decided that he wanted to run the ball? Of course, chaos would ensue. A team cannot function if team members try to play roles that are assigned to other team members. **Role clarity** means that all members understand their respective roles on the team and play those roles.

Clear Direction

What is the team's purpose? What is the team supposed to do? What are the team's responsibilities? These are the types of questions that people ask when they are assigned to teams. The team's charter should answer such questions. The various components of a team's charter are as follows:

1. *Mission.* The team's mission statement defines its purpose and how the team fits into the larger organization. In the case of a safety promotion team, it explains the team's role in the organization's overall safety program.

2. *Objectives.* The team's objectives spell out exactly what the team is supposed to accomplish in terms of the safety program.

3. *Accountability measures.* The team's accountability measures spell out how the team's performance will be evaluated.

Figure 20-9 is an example of a team charter for the safety promotion team in a manufacturing company. This charter clearly defines the committee's purpose, where it fits

Figure 20-9
Sample team charter.

> *Team Charter*
>
> **Safety Promotion Team**
> **MTC Corporation**
>
> **Mission**
> The mission of the *Safety Promotion Team* at MTC Corporation is to make all employees at all levels of the company aware of the importance of safety and health, and, having made them aware, to keep them aware.
>
> **Objectives**
> 1. Identify innovative, interesting ways to communicate the company's safety rules and regulations to employees.
> 2. Develop a company-wide suggestion system to solicit safety-related input from employees.
> 3. Identify eye-catching approaches for making safety a *visible* issue.
> 4. Develop appropriate safety competition activities.
>
> **Accountability Measures**
> The quality of participation in all of the activities of this team will be assessed by the team leader and included in the annual performance appraisal of each team member. Team members are expected to be consistent in their attendance, punctual, cooperative, and mutually supportive.

into the overall organization, what it is supposed to accomplish, and how the committee's success will be measured.

Team-Oriented Rewards

One of the most commonly made mistakes in organizations is attempting to establish a teamwork culture while maintaining an individual-based reward system. If teams are to function fully, the organization must adopt team-oriented rewards, incentives, and recognition strategies. For example, teams function best when the financial rewards of its members are tied at least partially to team performance. Performance appraisals that contain criteria relating to team performance, in addition to individual performance, promote teamwork. The same concept applies to recognition activities.

Accountability

There is a rule-of-thumb in management that says, "If you want to improve performance, measure it." **Accountability** is about being held responsible for accomplishing specific objectives or undertaking specific actions. The most effective teams know what their responsibilities are and how their success will be measured.

Potential Benefits of Teamwork in Promoting Safety

Teamwork can have both direct and indirect benefits for an organization. Through teamwork, counterproductive internal competition and internal politics are replaced by collaboration. When this happens, the following types of benefits typically accrue:

- Better understanding of safety rules/regulations
- Visibility for safety
- Greater employee awareness
- Positive, productive competition
- Continual improvement
- Broader employee input and acceptance

Potential Problems with Teams

Teamwork can yield important benefits, but as with any concept, there are potential problems. The most pronounced potential problems with teams are as follows:

- It can take a concerted effort over an extended period of time to mold a group into an effective team, but a team can fall apart quickly.
- Personnel changes are common in organizations, but personnel changes can disrupt a team and break down team cohesiveness.
- Participative decision making is inherent in teamwork. However, this approach to decision making takes time, and time is often in short supply.
- Poorly motivated and lazy employees can use a team to blend into the crowd, to avoid participation. If one team member sees another slacking, he or she may respond in kind.

These potential problems can be prevented, of course. The first step in doing so is recognizing them. The next step is ensuring that all team members fulfill their responsibilities to the team and to one another.

Responsibilities of Team Members

Accountability in teamwork amounts to team members fulfilling their individual responsibilities to the team and to each other. These responsibilities are as follows:

- Active participation in all team activities
- Punctuality in attendance of meetings
- Honesty and openness toward fellow team members
- Making a concerted effort to work well with team members
- Being a good listener for other team members
- Being open to the ideas of others

SAFETY FACT

Team Building through Training

The United States is the land of the *rugged individualist*. Even people who grow up playing team sports learn early on about standing out as an individual or as the Most Valuable Player. Consequently, employees may not be natural team players. That's the bad news. The good news is that with the proper training, employees can learn to be good team players. An initial teamwork training program should cover the following topics:

- Rationale for team training
- Direction and understanding (team mission and objectives)
- Characteristics of team players
- Accountability measures
- Team building

If individual team members fulfill these responsibilities to each other and the team, the potential problems with teams can be overcome, and the benefits of teamwork can be fully realized. It is important for members of the safety team to understand these responsibilities, accept them, and set an example of fulfilling them. If this happens, the benefits to the organization will go well beyond just safety and health.

SUMMARY

1. A company's safety policy should convey the following messages: (a) a company-wide commitment; (b) expectation that employees will perform their duties in a safe manner; and (c) the company's commitment includes customers and the community.
2. From a legal perspective, an employer's obligations regarding safety rules can be summarized as follows: (a) employers must have rules that ensure a safe and healthy workplace; (b) employers must ensure that all employees are knowledgeable about the rules; and (c) employers must ensure that safety rules are enforced objectively and consistently.
3. A fundamental rule of management is: If you want employees to make a commitment, involve them from the start. This is especially important when formulating safety rules.
4. Safety training should be a fundamental part of any effort to promote safety. Safety training ensures that employees know how to work safely, and it shows that management is committed to safety.

5.   Well-run suggestion programs promote safety by (1) soliciting input from the people who are most likely to know where hazards exist; and (2) involving employees in a way that lets them feel ownership in the safety program.

6.   Safety committees can help promote safety if they are properly structured. The composition of the committee can be a major factor in the committee's success. The most effective committees are composed of a broad cross-section of workers representing all departments.

7.   Union/management agreement is important in promoting safety. Fortunately, safety is an issue on which unions and management can usually agree.

8.   Incentives can promote safety if they are properly applied. To enhance the effectiveness of incentives, the following steps should be followed: (a) define objectives, (b) lead by example, (c) develop specific criteria, (d) make rewards meaningful, (e) keep communications clear, (f) involve employees in planning the incentives, and (g) reward teams.

9.   Competition can promote safety, but it can also get out of hand and do more harm than good. To keep competition positive, involve employees in planning programs of competition and encourage competition between teams rather than individuals.

KEY TERMS AND CONCEPTS

Accountability	Ownership
Commitment	Personal commitment
Competition	Role clarity
Consistency	Safety committee
Employee involvement	Safety policy
Employee participation	Safety rules
Formal response	Safety training
Immediate response	Suggestion programs
Incentives	Union/management participation
Mission	Visual awareness
Objectives	Wellness program
Objectivity	

REVIEW QUESTIONS

1.   What messages should a company's safety policy convey?
2.   Explain why promoting safety by example is so important.
3.   What are the employer's obligations regarding safety rules and regulations?
4.   Explain the concept of negligence as it relates to a company's safety.
5.   What is the significance of objectivity and consistency when enforcing safety rules?
6.   Why is employee participation/involvement so critical in the promotion of safety?

7. If your task was to establish a safety committee, who would you ask to serve on it?
8. List three benefits that companies gain from asking employees to sign a declaration of safety.
9. What are the steps for ensuring that incentives actually promote safety?
10. What are the characteristics of effective teams?

ENDNOTES

1. "Making a Statement About Safety," *Occupational Hazards,* November 1988, pp. 47–48.
2. Payne, M. "Do Employees Know Best?" *Safety & Health,* February 1988, Vol. 137, No. 2, p. 15.
3. Ibid.
4. Ibid.
5. McDermott, B. "Employees Are Best Source of Ideas for Constant Improvement," *Total Quality Newsletter,* July–August 1990, Vol. 1, No. 4, p. 5.
6. Peters, T. *Thriving on Chaos: Handbook for a Management Revolution* (New York: Harper & Row, 1987), p. 92.
7. Fettig, A. "Sign Up for Safety," *Safety & Health,* July 1991, Vol. 144, No. 1, p. 26.
8. Ibid.
9. Ibid., p. 27.
10. Ibid.
11. Puffer, T. "Eight Ways to Construct Effective Service Reward Systems," *Reward & Recognition Supplement, Training,* August 1990, pp. 8–12.
12. Ibid.
13. Myerson, W. A. "Wellness Is the Bottom Line," *Safety & Health,* May 1988, Vol. 137, No. 5, p. 50.
14. Ibid.
15. Ibid., p. 53.
16. National Safety Council. *Today's Supervisor,* August 1991, p. 15.
17. Ibid.

Safety and Health Training

Education and training have been recognized as important components of organized safety programs for many years. In today's rapidly changing high tech workplace, they are more important than ever. Modern safety and health professionals have a key role to play in ensuring that all employees at all levels receive the appropriate types and amounts of training. They must also be prepared to play an active role in preparing, presenting, arranging for the application of, and evaluating safety and health training. This chapter gives prospective and practicing professionals the information they need to play a positive role in providing effective safety and health education.

RATIONALE FOR SAFETY AND HEALTH TRAINING

In 1985, the management team at Will-Burt Co., a steel fabrication and parts manufacturer, was in financial trouble. According to Gregg LaBar,

Low profits, too many product defects, and steep workers' compensation and medical costs were hurting the company. That's when new President and Chief Executive Officer Harry E. Featherstone stepped up and instituted an employee training and education program. . . . Naturally employees get their share of safety training, including lift truck operation, hazard communication, hearing protection, and lockout/tagout, as part of Will-Burt's overall commitment to education.[1]

The employee training program, including the safety components, helped turn Will-Burt's fortunes around. Today, the company is prosperous and competitive. In the words of Jack Bednarowski, director of human resources for Will-Burt, "Our overall attitude toward education, including safety training, is that we can't afford not to do it."[2]

Workers who have not been trained to perform their jobs safely are more likely to have accidents. According to the National Safety Council,

Many studies have been made to determine why people fail to follow safety procedures or to take reasonable precautions on the job. Some of the reasons are that workers have:

- Not been given specific instructions in the operation
- Misunderstood the instructions
- Not listened to the instructions
- Considered the instructions either unimportant or unnecessary
- Disregarded instructions

Any of the above lapses can result in an accident. To prevent such an occurrence, it is essential that safety training work be conducted efficiently.[3]

Besides the commonsense fact that a well-trained employee is more likely to be a safe employee, there are also **legal and ethical reasons** for providing safety and health training.

Legal and Ethical Reasons for Training

The Occupational Safety and Health Act, or OSHAct, mandates that employers provide safety and health training. The OSHAct requires the following:

- Education and training programs for employees
- Establishment and maintenance of proper working conditions and precautions
- Provision of information about all hazards to which employees will be exposed on the job
- Provision of information about the symptoms of exposure to toxic chemicals and other substances that may be present in the workplace
- Provision of information about emergency treatment procedures

In addition, the OSHAct requires that employers make information available to workers concerning the results of medical or biological tests and that workers be given opportunities to observe when activities for monitoring regulated substances are undertaken. For example, if the level of a controlled substance that is used daily must not be allowed to exceed a prescribed level, any employee has the right to observe while readings are taken.

However, the OSHAct did not specifically cover numerous toxic substances in its requirement for employee training. This broad-based exclusion represented a serious and dangerous loophole in the law. Since the passage of the OSHAct, this loophole has been closed. According to Spencer and Simonowitz,

> With the Federal standards promulgated in the early '80s called the **'Right-to-know' laws,** composed of the "Access to Employee Exposure and Medical Records" (29 C.F.R. 1910.20) and the "Hazard Communication" (29 C.F.R. 1910.20), it is now the duty of employers to provide a wide variety of all information.[4]

As a result of the OSHAct and subsequent federal regulations, workers have a right to information about any aspect of the workplace that may affect their safety and health. In addition, the OSHAct and subsequent federal legislation clearly establish the employer's responsibility for providing employees with the information they need to work safely.

One of the specific responsibilities of employers set forth in the Hazard Communication regulation (29 C.F.R. 1910.20) is the provision of **hazard communication** programs. These programs should include such components as warning labels, training, access to records, and the distribution of material safety data sheets.

The Occupational Safety and Health Administration (OSHA) makes grants available for companies to use for programs that will improve the understanding of material safety data sheets (MSDS). "The education program developed . . . may also involve providing training to employees or a combination of both."[5] Information on grants for providing training relating to material safety data sheets is available from the following address:

OSHA Office of Training and Education
Division of Training and Education Programs
1555 Times Drive
Des Plaines, Illinois 60018

Beyond the legal reasons for providing safety and health training, there are also ethical reasons. No amount of legislation can properly regulate every hazardous substance or every potentially hazardous situation. Chemicals and technologies are developed and put in place much more rapidly than is legislation. The federal *Registry of Toxic Effects of Chemical Substances* maintained by the National Institute of Occupational Safety and Health (NIOSH) typically adds several thousand new chemicals to its list each year.

Clearly, the only way to guarantee that employees are well informed about the safety and health aspects of their jobs is for companies to fulfill their moral obligations along these lines. According to Spencer and Simonowitz,

> If the sick person has rights to information and determination regarding his or her body, should the well, healthy employee or working person have any less right to the same consideration when there are health implications of exposure on the job? If the sick person is entitled to know about procedures and prognosis, discomforts and inconveniences, risk, and experience of proposed treatment, surely working persons should have similar rights to information about the nature and toxicity of the substances with which they work, controls and their effectiveness, personal discomforts and inconvenience of both hazards and controls, morbidity and mortality data, and the relative risk."[6]

DISCUSSION CASE

What Is Your Opinion?

"I don't need my employees wasting any more time in training," said the production supervisor. "I need them on the production line operating their machines." The safety director, who had heard this argument many times before, countered, "They won't be able to operate their machines from a hospital bed, and if they don't complete their safety training, the hospital is probably their next stop!"

Both of these managers are interested in the same thing—productivity. The production supervisor is interested in his unit's productivity today. The safety director is interested in the unit's long-term productivity. Who is right in this argument. What is your opinion?

Who Is Responsible for Training?

The OSHAct requires safety and health training, but who is responsible for seeing that training is provided and who is responsible for providing it? Figure 21–1 illustrates the chain of responsibility for safety and health training. The Occupational Safety and Health Administration (OSHA) and the secretary of labor are responsible for both the direct provision of training and for ensuring that industrial firms provide training at the local level.

OSHA's original response to its training mandate was to develop and dispense educational materials. Local companies were to use these materials in the actual provision of training. This continues to be an important component of OSHA's overall training effort. However, OSHA officials soon learned that the printing and distributing of educational materials alone was not sufficient. Consequently, OSHA added three additional components to its training effort:

1. Monetary awards provided on a grant basis to companies, organizations, and educational institutions to finance the provision of safety and health training (the MSDS training grants referred to earlier in this chapter).
2. Training requirements set forth in various standards developed by OSHA.
3. Requirements set forth in the various OSHA regulations known collectively as the "right-to-know" regulations (referred to earlier in this chapter).

Financial awards are typically an effective way to generate activity in a given area. This has been the case with safety and health training, at least partially. The majority of the incentive funds awarded have gone to labor unions whereas a relatively small percentage has gone to individual companies. This is partly because labor unions have been more aggressive in the development of safety and health programs and in seeking funds to cover the start-up costs of those programs. Corporate America has tended to rely primarily on local company programs for providing training.

A major inhibitor in the provision of safety and health training is **mistrust between labor unions and management,** a fact with which safety and health professionals must

Figure 21–1
Who is responsible for pro-viding training?

be prepared to deal. Such professionals are normally part of a company's management team and, as a result, may be viewed with suspicion by workers. Safety and health professionals often feel as if they are caught in the middle with neither side appreciating their work.

When management and labor debate the safety and health issue, both sides usually make many charges and countercharges. In short, management claims labor is not sufficiently sensitive to the bottom line, and labor claims management is not sufficiently sensitive to the safety and health of workers.

The mistrust that is at the heart of this issue runs both ways. Therefore, safety and health professionals have a very important task in convincing both sides that providing a safe and healthy workplace is not only ethically right, but also profitable in the long run.

Safety and health professionals should be able to articulate specifically how employers can meet their legal and ethical obligations concerning workforce training. Spencer and Simonowitz state what employers must do:

> To meet the legal and ethical requirements of educating the workforce, the information given must be sufficiently precise to answer questions about the substance, the kind and degree of exposure, the controls in use, and the degree of their effectiveness, and personal discomfort or inconvenience involved, morbidity and mortality data—both animal and human—and the relative risk. They must be informed not only about emergency procedures for the sudden acute exposure, but also of chronic illness associated with long-term low-level exposure and length of latency periods.[7]

To meet their legal and ethical obligations regarding safety and health training, companies must rely on their safety and health professionals who, in turn, must rely on first-line supervisors. Safety and health professionals and first-line supervisors should work closely in preparing, presenting, applying, and evaluating safety training. The division of labor between safety and health professionals and supervisors is not black and white. However, generally speaking, the following guidelines apply:

- Safety and health professionals train the supervisors and keep them up to date.
- Supervisors provide most of the training to workers.
- Safety and health professionals and supervisors work together to train workers jointly when appropriate.

In general, the less job-specific the information, the more likely it will be provided by safety and health professionals. The more job-specific the information, the more likely it will be provided by supervisors.

EDUCATION AND TRAINING REQUIREMENTS

It is well established that employers are obligated to provide safety and health training, but what are the actual training requirements? The Occupational Safety and Health Administration and the **Mine Safety and Health Administration (MSHA)** have established specific regulations that delineate training requirements by type of industry. For example, **OSHA training requirements** as set forth in Title 29—Labor, *Code of Federal Regulations* cover the following industrial sectors:

General industry	Part 1910
Maritime industry	Part 1915–18
Construction industry	Part 1926
Agriculture industry	Part 1928

Regulations delineating the training requirements for mining workers are set forth in Subsection B of MSHA regulations. This subpart carries the title "Training and Retraining Miners Working at Surface Mines and Surface Areas of Underground Mines." Since the training requirements mandated by OSHA and MSHA differ in accordance with the different industrial sectors for which they were developed, they are best examined separately. The following two subsections summarize the training requirements mandated by OSHA and MSHA.

OSHA's training requirements are set forth in broad occupational categories: general industry, maritime, construction, agriculture, and federal employees. The requirements for general industrial occupations are found in Subparts F–Z of Title 29, Code of Federal Regulations, Part 1910. They are summarized in Figure 21–2. For example, in the area of occupational noise exposure 1910.95(K) paragraph 1 reads as follows:

(1) The employer shall institute a training program for all employees who are exposed to noise at or above an 8-hour time-weighted average of 85 decibels and shall ensure employee participation in such a program.[8]

Subpart F **Powered Platforms, Manlifts, and Vehicle Mounted Work Platforms**
Manlifts

Subpart G **Occupational Health and Environmental Control**
Ventilation, occupational noise exposure, ionizing radiation

Subpart H **Hazardous Materials**
Hydrogen, flammable and combustible agents, explosives and blasting agents, storage and handling of liquefied petroleum gases, storage and handling of anhydrous ammonia, hazardous waste operations and emergency response

Subpart I **Personal Protective Equipment**
Respiratory protection

Subpart J **General Environmental Controls**
Temporary labor camps, specifications for accident prevention signs and tags

Subpart K **Medical and First Aid**
Medical services and first aid

Subpart L **Fire Protection**
Fire brigade training, fixed dry chemical extinguishing systems, local fire alarm signaling systems

Subpart M **Compressed Gas and Compressed Air Equipment**
Safety relief devices for cargo and portable tanks storing compressed gases

Subpart N **Materials Handling and Storage**
Servicing of single piece and multi-piece rim wheels; powered industrial trucks; overhead and gantry cranes; crawler, locomotive, and truck cranes; derricks

Subpart O **Machinery and Machine Guarding**
Woodworking machinery requirements, mechanical power processes, forging machines

Subpart Q **Welding, Cutting, and Brazing**
Welding, cutting, and brazing

Subpart R **Special Industries**
Pulp, paper, and paperboard mills; laundry machinery and operations; sawmills; pulpwood logging; telecommunications

Subpart T **Commercial Diving Operations**
Qualifications of dive team

Subpart Z **Toxic and Hazardous Substances**
Asbestos, 4-nitrobiphenyl, alpha-naphthylamine, methyl chloromethyl ether, 3,3"-dichlorobenzidine (and its salts), bis-chloromethyl ether, beta-naphthylamine, benzidine, 4-aminodiphenyl, ethyleneimine, beta-propiolactone, 2-acetylamineofluorene, 4-dimethylaminoazobenzene, N-nitrosodiummethylamine, vinyl chloride, inorganic arsenic, lead, coke oven emissions, cotton dust, 1,2-dibromo-3-chloropropane, acrylonitrile (vinyl cyanide), ethylene oxide, hazard communication

Figure 21–2
General industry training requirements: 29 C.F.R. Part 1910.

Training requirements in all of the various areas discussed earlier in this section are similarly delineated in Part 1910. Safety and health professionals should be familiar with those that apply in their individual work settings.

MSHA Training Requirements

Subpart B of MSHA regulations contains **MSHA training requirements** for miners. These regulations define a miner as

> Any person working in a surface mine or surface area of an underground mine who is engaged in the extraction and production process or is regularly exposed to mine hazards, or who is a maintenance or service worker (whether employed by operator or contractor) working at the mine for frequent or extended periods.[9]

Since the regulations specify a different set of training requirements for inexperienced miners and newly employed but experienced miners, it is important to understand how MSHA defines *experienced miner*. Paragraph 48.22 of Subpart B defines this term as follows:

> A person currently employed as a miner; or a person who received training acceptable to MSHA from an appropriate state agency within the preceding one month; a person with 12 months experience working in surface operations during the preceding 3 years; a person who received new miner training . . . within the past 12 months.[10]

Paragraph 48.25 of Subpart B sets forth the following training requirements for new or inexperienced miners:

- A minimum of 24 hours of training is required. Typically, this training will take place before the miner begins work. However, with prior approval of MSHA, up to 16 hours of the training may be provided after the new miner begins work. This means that even with a waiver from MSHA, at least eight hours of training must be provided before the miner begins work.
- The first eight hours of a new miner's training must include the following: (1) an introduction to the work environment (orientation); (2) recognition of workplace hazards; and (3) job-specific safety and health measures/concerns.
- All required training beyond the original eight hours must be completed within 60 days. The training program must include at least the following topics: statutory rights of miners and their representatives; authority and responsibility of supervisors; line authority of supervisors and miners' representatives; mine rules; hazard reporting procedures; self-rescue and respiratory devices; transportation controls and communication systems; introduction to the work environment; emergency evacuation and escape procedures; fire warning and firefighting procedures; ground control; personal health; hazard recognition; electrical hazards; MSHA-approved first aid; explosives; and job-specific safety and health procedures.
- Companies must have a training plan that specifies oral, written, or practical demonstration methods will be used to assess whether training has been completed successfully.[11]

Paragraph 48.26 of Subpart B of MSHA regulations lists the following training requirements for newly employed but experienced miners before they begin work: introduction to the work environment (orientation); mandatory safety and health standards, both general and job specific; authority and responsibility of supervisors and miners' representatives; emergency escape and evacuation procedures; fire warning and fire-fighting procedures; ground controls; and hazard recognition.[12]

In addition to these specific training requirements, MSHA requires eight hours per year of refresher training, comprehensive records of each miner's training, and compensation to miners for training time. Training is to be provided during work hours, and employees are to be paid their regular wages. If the training takes place at any location other than the normal worksite, all expenses incurred by miners while participating in training (e.g., lodging, meals, and mileage) must be paid by the employer.[13] Additional information about MSHA training requirements can be obtained by writing to the following address:

Director of Education and Training
MSHA
4015 Wilson Boulevard
Arlington, Virginia 22203

SAFETY AND HEALTH PROFESSIONALS AS TRAINERS

As mentioned earlier, determining whether training should be provided by supervisors or by safety and health professionals is not clear-cut. Generally speaking, supervisors are more likely to provide job- and task-specific training whereas safety and health professionals are more likely to provide more generic training. Regardless of where this distinction is made, it is clear that today's safety and health professionals must be competent at developing, coordinating, and conducting training.

According to the National Safety Council, persons conducting training should have the following characteristics: a thorough knowledge of the topics to be taught; a desire to teach; a positive, helpful, cooperative attitude; strong leadership abilities; a professional attitude and approach; and exemplary behavior that sets a positive example.[14]

In addition to having these characteristics, trainers should be knowledgeable about the fundamental principles of learning and the four-step teaching method. The principles of learning summarize much of what is known about how people learn best. It is important to conduct safety and health training in accordance with these principles. The four-step teaching method is a basic approach to conducting training that has proven effective over many years of use.

Principles of Learning

The **principles of learning** summarize what is known and widely accepted about how people learn. Trainers can do a better job of facilitating learning if they understand the following principles:

1. *People learn best when they are ready to learn.* You cannot *make* employees learn anything. You can only make them *want* to learn. Therefore, time spent motivating employees to want to learn about safety and health is time well spent. Explain why employees need to learn and how they will benefit personally from having done so.

2. *People learn more easily when what they are learning can be related to something they already know.* Build today's learning on what was learned yesterday and tomorrow's learning on what was learned today. Begin each new learning activity with a brief review of the one that preceded it.

3. *People learn best in a step-by-step manner.* This is an extension of the preceding principle. Learning should be organized into logically sequenced steps that proceed from the concrete to the abstract, from the simple to the complex, and from the known to the unknown.

4. *People learn by doing.* This is probably the most important principle for trainers to understand. Inexperienced trainers tend to confuse talking (i.e., lecturing or demonstrating) with teaching. Explanations can be part of the teaching process but are only useful if they are followed by application activities that require the learner to *do* something. To illustrate, consider the example of teaching an employee how to ride a bicycle. You can present a thorough lecture on the principles of pedaling and steering and give a comprehensive demonstration on how to do it. However, until the employee gets on and begins pedaling, he or she will not learn how to ride a bicycle.

5. *The more often people use what they are learning, the better they will remember and understand it.* How many things have you learned in your life that you can no longer remember? People forget what they do not use. Trainers should keep this principle in mind. It means that repetition and application should be built into the learning process.

6. *Success in learning tends to stimulate additional learning.* This principle is a restatement of a fundamental principle in management (i.e., success breeds success). Organize training into long enough segments to allow learners to see progress, but not so long that they become bored.

7. *People need immediate and continual feedback to know if they have learned.* Did you ever take a test and get the results back a week later? If so, that was probably a week later than you wanted them. People who are learning want to know immediately and continually how they are doing. Trainers should keep this principle in mind at all times. Feedback can be as simple as a nod, a pat on the back, or a comment such as "Good job!" It can also be more formal, such as a progress report or a graded activity. Regardless of its form, trainers should concentrate on giving immediate and continual feedback.

Four-Step Teaching Method

Regardless of the setting, teaching is a matter of helping people to learn. One of the most effective approaches for facilitating learning is not new, innovative, gimmicky, or high tech in nature. It is known as the **four-step teaching method,** and it is an effective approach to use for safety and health training. The four steps and a brief description of each follow:

- **Preparation** encompasses all tasks necessary to get participants prepared to learn, trainers prepared to teach, and facilities prepared to accommodate the process. Preparing participants means motivating them to want to learn. Personal preparation involves planning lessons and preparing all of the necessary instructional materials. Preparing the facility involves arranging the room for both function and comfort, checking all equipment to ensure that it works properly, and making sure that all tools and other training aids are in place.

- **Presentation** is a matter of presenting the material that participants are to learn. It may involve giving a demonstration, presenting a lecture, conducting a question/answer session, helping participants to interact with a computer or interactive videodisc system, or assisting participants who are proceeding through self-paced materials. Victor Parachin recommends strategies for giving an effective presentation: Begin dramatically, be brief, be organized, use humor, keep it simple, take charge, be sincere, consider conditions, and tell stories.[15]

- **Application** is a matter of giving learners opportunities to use what they are learning. Application can range from simulation activities in which learners role play to actual hands-on activities in which learners use their new skills in a live format.

- **Evaluation** is a matter of determining the extent to which learning has taken place. In a training setting, evaluation does not need to be a complicated process. If the training objectives were written in measurable, observable terms, evaluation is simple. Employees were supposed to learn how to do X, Y, and Z safely. Have them do X, Y, and Z and then observe the results. In other words, have employees demonstrate proficiency in safely performing a task and observe the results.

PREPARING SAFETY AND HEALTH INSTRUCTION

Preparing instruction involves the following steps: (1) preparing (planning) the instruction; (2) preparing the facility; and (3) preparing the learners. It is important to accomplish all three steps before attempting to present instruction. This section focuses on planning a safety and health course of instruction.

Planning Instruction

The instruction delivered by safety and health professionals will usually be part of a course, workshop, or seminar. In any case, there must be a **course outline** that summarizes the major topics covered by the instruction. The outline should state the expected outcomes of the instruction in broad terms or, in other words, what the learner is supposed to be able to do after completing the course, workshop, or seminar. The outline should also have a brief statement of purpose. More specific instructional objectives are

---

Course Outline
Safety Engineering for Familiarity
(A Course for Supervisors)

Statement of Purpose

This course consists of 15 contact hours of instruction. Its purpose is to familiarize supervisors with the concept of safety engineering and the job of safety engineers so that supervisors and safety engineers can work together more effectively as team members.

Intended Outcomes

Upon completion of this course supervisors should be able to do the following:

- Demonstrate an understanding of safety engineering techniques.
- Analyze workplace environmental hazards safety engineers are concerned with.
- Explain health and safety regulations safety engineers are concerned with.
- Explain the proper relationship between supervisors and safety engineers.

Figure 21–3
Sample course outline.

developed later when preparing lesson plans. Figure 21–3 is a sample outline for a short course on safety engineering for supervisory personnel.

Notice that the course outline contains just two components: a statement of purpose and a list of intended outcomes. Some instructors prefer to add additional components such as a list of equipment and/or training aids needed, but those components shown in Figure 21–3 are sufficient. A good course outline is a broadly stated snapshot of the scope and sequence of the course. Specific details are typically shown in the lesson plans that are developed next.

Lesson plans are an important part of the planning step. They are road maps or blueprints for the actual instruction that is to take place. In addition, they serve to standardize instruction when more than one person may teach the same instruction to different groups. Standardization is particularly important for safety and health training. If even one member of a work team receives less training than the others, the potential for an accident is increased by his or her ignorance.

Lesson plans can vary in format according to the personal preferences of the trainer. However, all lesson plans should include the components discussed in the following paragraphs.

Lesson Title and Number

The lesson title should be as descriptive as possible of the content of the lesson. The number shows where the lesson fits into the sequence of lessons that make up the course.

Statement of Purpose

Like the statement of purpose in the course outline, the **statement of purpose** here consists of a concise description of the lesson's contents, where it fits into the course, and why it is included.

Learning Objectives

Learning objectives are specific statements of what the learner should know or be able to do as a result of completing the lesson. Objectives should be written in behavioral terms that can be measured or easily observed.

Training Aids List

This component serves as a handy checklist to help trainers quickly and conveniently ensure that all the **training aids** needed are present. The list should include every tool, handout, piece of equipment, video, chart, and so on needed to conduct the instruction for that lesson.

Instructional Approach

The **instructional approach** is a brief action plan for carrying out the instruction. It should begin with a short statement describing the instructional methodology to be used (lecture/discussion, demonstration, computer-assisted instruction, and so on). This statement is followed by a step-by-step summary of the trainer's major activities, such as the following: (1) deliver lecture on safety regulations; (2) distribute safety regulations handout, and so on.

Application Assignments

The **application assignments list** details the tasks that learners will be required to complete before they can apply what they are learning.

Evaluation Methodology

The **evaluation methodology** component explains how learning will be evaluated. Will there be a test? Will performance be observed? Will safety and health records be monitored for improvement? Such questions are answered in this section.

Figure 21–4 is an example of a lesson plan developed around one of the intended outcomes in Figure 21–3. This is the typical relationship between the course outline and the lesson plan (i.e., one intended outcome equals one lesson). Occasionally, an outcome may require more than one lesson for adequate coverage; at other times, several outcomes might be covered in one lesson.

PRESENTING SAFETY AND HEALTH INSTRUCTION

Educators hold that learners retain the following percentages from their instruction:

Lesson 1
Safety Engineering Techniques

Statement of Purpose

This is the first lesson in a series of four that make up the course. The purpose of this lesson is to help supervisors understand the job of the safety engineer in the specific area of safety engineering techniques such as noise control, equipment guarding, and dust control.

Learning Objectives

Upon completion of this lesson, learners will be able to do the following:

- Demonstrate how to apply noise control techniques to operating equipment.
- Apply equipment guarding techniques to operating machines.
- Apply dust control techniques in a shop setting.

Training Aids List

The following training aids are needed with this lesson: overhead projector, safety engineering techniques transparencies, dust control measures handout, and sample equipment guarding devices.

Instructional Approach

The lecture/discussion method supplemented with overhead transparencies, handouts, and live examples of equipment guards is used with this lesson. Instruction should proceed as follows: (1) lecture and discussion on noise control with overhead transparencies; (2) learners attach an equipment guarding device to a machine; and (3) learners examine their working area and make recommendations for dust control.

Evaluation

The instructor will observe the performance of learners as they undertake their application activities. Based on these observations they will assign pass or fail assessments to each student.

Figure 21–4
Sample lesson plan.

10 percent of what is read
20 percent of what is heard
30 percent of what is seen
50 percent of what is seen and heard
70 percent of what is seen and spoken
90 percent of what is said while doing what is talked about

Instruction can be presented in several different ways. The most widely used are the lecture/discussion, group instruction, demonstration, conference, and multimedia methods. Regardless of the approach used, trainers should keep in mind the percentages just listed. What they mean is to get the learner actively engaged seeing, saying, listening, and, most importantly, *doing*.

Lecture/Discussion Method

The **lecture/discussion method** of teaching is the oldest, most familiar, most used, and probably most abused. A lecture is a planned, structured, and frequently illustrated (slides, charts, chalkboard, and so on) method of communicating information to a group of people. By itself, the lecture allows for only one-way communication. This serious deficiency is overcome by adding the discussion component. Discussion can be between the instructor and participants or among fellow learners. During discussion, the instructor's job is to keep the discussion on track and moving in the right direction.

The best justification for using the lecture/discussion method is that it is an effective way to communicate information to groups that are too large for individual interaction between instructor and participants. Another reason for using this method is that it allows the instructor to generate enthusiasm among participants about a topic.

The lecture/discussion method, if used properly, can be an effective teaching method. However, it does not work in every situation. Instructors or trainers need to know when to use this method and when to use another. Use the lecture/discussion method when

1. The material to be presented deals strictly with data, theory, or information (no skills development).
2. Participants need to be motivated before beginning a particular lesson.
3. The material to be presented is not available in print.
4. Sharing insight or experience in a particular area will enhance learning.
5. Information must be communicated to a large group in one session.
6. Interaction among participants is desired.

Do not use the lecture/discussion method when

1. The subject matter deals with skills development or how-to information.
2. The participant group is small enough to allow individual learner/teacher interaction.
3. There is no need for interaction among participants.

This is probably the most overlooked step: *Participants must be thoroughly prepared prior to the session.* If they are not, the session will be all lecture and no discussion. Require participants to approach a lecture/discussion session having first prepared themselves. This will ensure that they are active, contributing learners rather than passive spectators. Prepare participants for a lecture/discussion session as follows:

1. Give them a written outline or overview of the lecture the day before so they can familiarize themselves with it.
2. At least a day ahead of time, pass out any handouts that will be used and ask the participants to read them, noting questions that they may have right on the handouts.

3. During the lecture, have group members raise their questions when they think of them. Use their questions to trigger open discussion and other questions.

There is a saying in teaching: "When giving a lecture, tell them what you are going to tell them, then tell them, then tell them what you told them." Although it is said with tongue in cheek, this is actually good advice. A well-planned, properly structured lecture contains three distinct components: the **opening,** in which you "tell them what you are going to tell them"; the **body,** in which you "tell them"; and the **closing,** in which you "tell them what you told them."

Discussion should be interspersed within the body in a workplace-training setting. The best time to discuss an issue is when it is fresh on the minds of the participants. In training language, such instances are known as *Moms,* or moments of maximum opportunity.

The opening, body, and closing of a lecture all contain specific tasks that should be accomplished in order:

1. Opening
   a. Greet the class.
   b. State the title of the lecture.
   c. Explain the purpose of the lecture.
   d. List the objectives so that participants know exactly what they should be learning.
   e. Explain how the current lecture/discussion session relates to past topics studied.
   f. List and define any new terms that will be used during the session.
   g. Present a general overview of the content of the lecture/discussion.
2. Body
   a. Present the information in the order listed in the participants' outline.
   b. Initiate discussion by raising specific questions, calling on participants for comments, and/or soliciting questions from them.
   c. Make frequent reference to all visual aids and supportive materials.
3. Closing
   a. Restate the title, purpose, and objectives.
   b. Briefly summarize major points.
   c. State your conclusions.
   d. Answer remaining questions.
   e. Make follow-up assignments to reinforce the lecture/discussion.

The elements just listed are the fundamental or tangible tasks that should be performed in all lecture/discussion sessions. You should also keep in mind a number of intangibles when conducting lecture/discussion sessions:

1. Make sure that the classroom is arranged to accommodate a lecture/discussion.
2. Be enthusiastic. Enthusiasm is contagious.
3. Call on participants by name. They will appreciate the recognition and feel more at ease.
4. Spread your attention evenly. This will make all participants feel that they are part of the lecture/discussion.

5. Maintain eye contact with all learners in the session.
6. Speak clearly, evenly, and slowly enough to be understood, but not in a monotone.
7. Use facial expressions, body language, and movement to emphasize points.
8. It is all right to use an outline or note cards to keep yourself on track, but never read a lecture to the participants.
9. Use carefully prepared visual aids to reinforce major points.
10. Do not dominate—facilitate. Participation is critical. Remember, this method is called lecture/*discussion*.

Demonstration Method

Demonstration is the process in which the instructor shows participants how to perform certain skills or tasks. While demonstrating, the instructor also explains all operations step by step. The key to giving a good demonstration is preparation. The following checklist contains specific tasks for preparing for a demonstration:

1. Decide exactly what the purpose of the demonstration is, why it will be given, what participants should learn from it, what will be demonstrated and in what order, and how long the demonstration will last.
2. Gather all tools, equipment, and instructional aids. Make sure that everything is available and in working order. Never put yourself in a position of being forced to stop a demonstration unexpectedly because something does not work the way it should or because a necessary part of the demonstration is not on hand.
3. Set up the demonstration so that participants will easily be able to see what is going on and hear what you are saying.
4. Arrange all materials to be used in the demonstration so that they correspond with the order in which the various steps of the demonstration will be presented.
5. Practice the demonstration several times before giving it to work out any bugs.

Just as there are specific tasks to be performed in presenting a lecture, there are specific tasks to be performed in giving a demonstration:

1. Orient participants to the demonstration by explaining its purpose and objectives. Give them a brief overview of the content of the demonstration. Explain how the demonstration ties in with what they already know.
2. Present the demonstration in a slow, deliberate fashion so that participants can easily follow.
3. Pause between stages to determine if all class members are comprehending or to see if they have any questions that should be answered before continuing. Go back over any steps that participants did not seem to grasp.
4. Conclude the demonstration with a brief summary and question/answer session.

Participant activity in skills development is critical. Remember, no matter how well a demonstration is presented, it is really just showing. Showing is important, but participants learn best by doing. Therefore, it is vital to provide hands-on activities after a demonstration. An effective way to follow up demonstrations is to

1. Select several participants and ask them to repeat the demonstration you have just given.
2. Assign several practical application activities in which the participants are required to apply the skills demonstrated.
3. Observe participants on an individual basis as they attempt to perform the practical application activities. Give individual attention and assistance where needed. Be sure to correct misapplication immediately so that the wrong way does not become a habit.
4. Conduct performance evaluations so that skills development can be measured.

Conference Method

The **conference method** is particularly well suited for corporate training settings. It is less formal than a traditional classroom setting and requires that the trainer serve as a facilitator rather than a teacher. It is best used as a problem-solving teaching method. For example, a safety and health professional might use the conference method to make all supervisors aware of a new safety problem while simultaneously soliciting their input on how to solve the problem.

To be effective facilitators, safety and health professionals must become adept at defining the problem, soliciting input from participants, drawing out all of them, summarizing and repeating information, and building consensus. The conference teaching method, when effectively used, should result in both well-informed participants and a plan for solving the problem.

The following example is a scenario in which the conference method might be used. Annette Evans, safety and health manager for Shalimar Machine and Tool Company, has noticed an increase in minor cuts and abrasions on the hands of employees who are handling the new XRT composite material. Through random observation, she has determined that employees are not properly applying the accident prevention techniques they are supposed to have learned for handling XRT. The employees do not appear to understand fully the techniques.

Evans decides to set up a training session for all supervisors of employees who handle XRT. She hopes to accomplish three things: (1) inform supervisors of the problem; (2) reacquaint supervisors with the appropriate accident prevention techniques for handlers of XRT; and (3) develop a plan to get handlers using the techniques properly. According to the National Safety Council, Evans should proceed as follows:

1. State the problem.
2. Break the problem into segments to keep the discussion orderly.
3. Encourage free discussion.
4. Make sure that members have given adequate consideration to all of the significant points raised.
5. Record any conclusions that are reached.
6. State the final conclusion in such a way that it truly represents the findings of the group.[16]

If Annette Evans follows these six steps, all supervisors of workers who handle the new XRT composite material will understand the problem. In addition, they will learn the proper application of the special accident prevention techniques that must be used when handling XRT. Finally, Evans will have solid recommendations that have the full support of the supervisors concerning how to ensure that XRT handlers properly apply the accident prevention techniques.

Other Presentation Methods

In addition to the presentation methods just discussed, several others can be used. The most widely used are simulation, videotapes, and programmed instruction.

Simulation

Simulation, as the name implies, involves structuring a training activity that simulates a line situation. For example, if a safety and health professional is teaching a group of workers how to respond when a fellow worker is electrocuted, he or she might simulate that situation by having a worker role play being electrocuted. Simulation can also be technology-based. Computer simulation activities and those based on interactive laser disc and video technology are becoming more widely used. The military has used technology-based simulation for many years to train pilots.

Videotapes

The use of videotapes for presenting instruction has become common in corporate training settings. In essence, the videotape takes the place of a lecture or demonstration. The pause or stop functions on the video player can be used to allow for discussion, questions and answers, and group interaction. The play-back feature can be used for reviewing material or replaying portions of the tape that are not fully understood by participants. According to John Zeglin of NUS Training Corporation, the greatest advantage that videos offer is consistency of presentation.[17] However, it is important to ensure that videos meet the specific needs of individual companies. Magon suggests that the following features be checked when previewing videos:

1. *Presentation.* What is the quality of the presentation? Look for both content and presentation style.

2. *Focus.* A good video should focus on one subject and give a comprehensive treatment in a time range of 12 to 25 minutes. Multiple subjects should be broken up into more than one video.

3. *Liveliness.* Is the video lively, upbeat, and interesting, or nothing more than a boring presentation by a "talking head"? A lively video will hold the learner's attention.

4. *Accuracy.* Is the content accurate? If not, none of the other features matter. Have subject matter experts preview videos for accuracy before incorporating them into a training program.[18]

Programmed Instruction

Programmed instruction is an approach used to individualize instruction. Traditionally, the programmed medium has been a workbook or text that presents information in segments that proceed as follows: (1) information presentation; (2) information review; (3) questions, problems, or activities based on the information presented for the participant to work on; and (4) a self-test. Before proceeding to the next lesson, the learner must make a specified score on the self-test for the preceding lesson. Increasingly, programmed instruction is becoming computerized. This enhances the interactive nature of the instruction and, with good software, provides almost immediate feedback for the learner.

Interactive Video

Interactive video training combines laser disc and personal computer technologies to create an excellent high tech approach to workplace training. Unlike many other media-based training methods, interactive video is not passive. Rather, it requires the learner to participate actively by making choices, selecting options, and participating in one-on-one simulations of workplace situations. According to Fisher,

> In operation, an employee sits privately at a TV monitor with (optional) headphones. Those being trained need only touch the TV screen for interaction, so no computer literacy is necessary. Employees became actively engaged in the learning process on a one-on-one basis. Dramatizations of workplace situations actually appear on the screen in live action.[19]

Fisher lists the following advantages of interactive video as a training methodology: It individualizes learning, increases retention, is self-paced, evaluates and records progress, reduces training time, simplifies learning, and is dependable, consistent, and flexible.[20]

SAFETY FACT

Safety Training Is a Must

Too many business executives take the MBA (Management-By-Accounting) approach to their jobs. They will accept only dollars on the bottom line as proof of the value of an activity or function. Although the value of training can be shown in dollars and cents, the ubiquitous MBA mentality that persists in business and industry is hard to understand, and even harder to overcome.

The same executives who question the value of safety training have no problem understanding that professional football teams train all week to play just one game. Military units train constantly for the eventuality of a deployment. Training is fundamental to most professions and occupations. It should be fundamental to all.

APPLYING SAFETY AND HEALTH INSTRUCTION

One of the fundamental principles of learning states that *people learn by doing.* To the trainer, this means that learners must be given ample opportunity to apply what they are learning. If the topic of a training session is how to administer CPR, application will involve having the learners actually practice on dummies. While the learners practice, the trainer observes, coaches, and corrects. Regardless of the nature of the material, learners should be given plenty of opportunities to apply what they are learning.

EVALUATING SAFETY AND HEALTH INSTRUCTION

Did the training provided satisfy the training objectives? Safety and health professionals need to know the answer to this question whenever training is given. However, this can be a difficult question to answer.

Evaluating training requires beginning with a clear statement of purpose. What is the overall purpose of the training? The objectives translate this purpose into specific, measurable terms.

The purpose of safety and health training is to improve the individual employee's ability to work safely and to improve the safety and health of the overall work environment.

To determine if training has improved performance, safety and health professionals need to know (1) was the training provided valid? (2) did the employees learn? and (3) has the learning made a difference? Valid training is training that is consistent with the training objectives.

Evaluating training for validity is a two-step process. The first step involves comparing the written documentation for the training, lesson plans, handouts, and so on with the training objectives contained in the course outline. If the training is valid in design and content, the written documentation will match the training objectives. The second step involves determining if the actual training provided is consistent with the documentation. Training that strays from the approved plan will not be valid. Participant evaluations of instruction conducted immediately after completion can provide information on consistency and the quality of instruction.

Determining if employees have learned is a matter of building evaluation into the training. Employees can be tested to determine if they have learned, but be sure that tests are based on the training objectives. If the training is valid and employees have learned, the training should make a difference in their performance. Performance on the job should improve. This means safety and health should be enhanced. Safety and health professionals can make determinations about performance using the same indicators that told them training was needed in the first place.

Can employees perform tasks more safely than they could before the training? Is the accident rate down? Is the amount of time lost to accidents down? Are there fewer health-related complaints? These are the types of questions that safety and health professionals can ask to determine if training has improved performance. Gilda Dangot-Simpkin of Dynamic Development suggests a checklist of questions for evaluating purchased training programs.[21] Here are some of her questions:

- Does the program have specific behavioral objectives?
- Is there a logical sequence for the program?
- Is the training relevant for the trainee?
- Does the program allow trainees to apply the training?
- Does the program accommodate different levels of expertise?
- Does the training include activities that appeal to a variety of learning styles?
- Is the philosophy of the program consistent with that of the organization?
- Is the trainer credible?
- Does the program provide follow-up activities to maintain the training on the job?[22]

TRAINING SUPERVISORS

Supervisors play a key role in the maintenance of a safe and healthy workplace. Consequently, safety and health professionals need to ensure that supervisors have had the training they need to be positive participants in the process. A *Safety & Health* survey revealed that only 53 percent of the companies responding provide safety and health training for their supervisors.[23] This is an unfortunate statistic because supervisors are the safety and health professional's link with employees. In the words of Peter Minetos, "Supervisors . . . are the ones who have to teach employees the safe way to conduct their jobs. With proper training, they can spot and eliminate risks that are waiting to create havoc for their workers."[24]

According to the National Safety Council, the objectives of supervisor safety training are to

1. Involve supervisors in the company's accident prevention programs
2. Establish the supervisor as the key person in preventing accidents
3. Get supervisors to understand their safety responsibilities
4. Provide supervisors with information on causes of accidents and occupational health hazards and methods of prevention
5. Give supervisors an opportunity to consider current problems of accident prevention and develop solutions based on their own and others' experience
6. Help supervisors gain skill in accident prevention activities
7. Help supervisors keep their own departments safe[25]

Responsibility for developing safety training programs for supervisors falls to the safety and health professional. Figure 21–5 is an outline for a basic safety training course for supervisors. This course was developed by the National Safety Council and consists of 14 one-hour lessons.

TRAINING NEW AND TRANSFERRED EMPLOYEES

Perhaps the most important aspect of safety training is orientation for new and transferred employees. Consider the following quote from *Today's Supervisor:*

National Safety Council's
Basic Safety for Supervisors Course

Session 1
Loss Control for Supervisors

Accidents and incidents, areas of responsibility, the cost of accidents, and a better approach to occupational safety and health.

Session 2
Communications

Elements of communication, methods of communication, and effective listening.

Session 3
Human Relations

Human relations concepts, leadership, workers with special problems, and the drug and alcohol problem.

Session 4
Employee Involvement in Safety

Promoting safe-worker attitudes, employee recognition, safety meetings, and off-the-job accident problems.

Session 5
Safety Training

New employee indoctrinations, job safety analysis (JSA), job instruction training (JIT), and other methods of instruction.

Session 6
Industrial Hygiene and Noise Control

General concepts, chemical agents, physical agents, temperature extremes, atmospheric pressures, ergonomics, biological stresses, threshold limit values (TLVs), and controls.

Session 7
Accident Investigation

Finding causes, emergency procedures, effective use of witnesses, and reports.

Figure 21–5
Sample safety training course outline.

Session 8
Safety Inspections

Formal inspections, inspection planning and checklists, inspecting work practices, frequency of inspections, recording hazards, and follow-up actions.

Session 9
Personal Protective Equipment

Controlling hazards, overcoming objections; protecting the head, eyes, and ears; respiratory protective equipment; safety belts and harnesses; protecting against radiation; safe work clothing; and protecting the hands, arms, feet, and legs.

Session 10
Materials Handling and Storage

Materials handling problems; materials handling equipment; ropes, chains, and slings; and material storage.

Session 11
Machine Safeguarding

Principles of guarding, safeguard design, safeguarding mechanisms, and safeguard types and maintenance.

Session 12
Hand Tools and Portable Power Tools

Safe working practices, use of hand tools, use of portable power tools, and maintenance and repair of tools.

Session 13
Electrical Safety

Electrical fundamentals review, branch circuits and grounding concepts, plug- and cord-connected equipment, branch circuit and equipment testing methods, ground fault circuit interrupters, hazardous locations, common electrical deficiencies, safeguards for home appliances, and safety program policy and procedures.

Session 14
Fire Safety

Basic principles, causes of fire, fire-safe housekeeping alarms, equipment, and evacuation; and reviewing the supervisor's fire job.

The confusion and stress that accompany the first day of any job are often the reasons that new employees are more than twice as likely to have accidents as experienced workers. A lack of experience, a strong desire to please and a hesitation to ask for help all cause one in eight new employees to be involved in some type of accident the first year on the job. The first month is the most critical.[26]

The early training provided for new and transferred employees should have at least the following components: orientation, job-specific procedures, and follow-up. These components are discussed in the following paragraphs.

Orientation

Orientation is critical. Too often, companies hand new employees over to an experienced employee who is supposed to "show them the ropes." This is a dangerous practice that can have the effect of minimizing, rather than emphasizing, the importance of safety. Orientation should be structured, and it should involve the new worker's supervisor as well as other personnel as appropriate. A good orientation program should teach the following at a minimum:

1. Management is sincerely interested in preventing accidents.
2. Accidents may occur, but it is possible to prevent them.
3. Safeguarding equipment and the workplace has been done, and management is willing to go further as needs and methods are discovered.
4. Each employee is expected to report to the supervisor any unsafe conditions encountered at work.
5. The supervisor will give job instructions. No employee is expected to undertake a job before learning how to do it and being authorized to do it by a supervisor.
6. The employee should contact the supervisor for guidance before undertaking a job that appears to be unsafe.
7. If an employee suffers an injury, even a slight one, it must be reported at once.

In addition to these points, any safety rules that are a condition of employment, such as wearing eye protection or safety hats, should be understood and enforced from the first day of employment.[27]

Job-Specific Procedures

Before a new employee is allowed to begin work, he or she should be given instruction in safe operation procedures, the use of personal protective equipment, and any other procedures that will promote a safe and healthy work environment.

Follow-Up

After a new employee has worked for three days to a week, a follow-up conference should be called. This conference should be led by the worker's supervisor. It should answer the following questions: Was the initial training effective? Does the new employee have

Orientation Verification Checklist

Please review this checklist and indicate which topics were covered during your orientation by initialing the entry.

Reporting/Authority

☐ Your immediate supervisor
☐ Mid-managers, managers, and executives in your chain of command
☐ Safety and health manager
☐ Human resources manager

Job Requirements

☐ Job description
☐ Responsibility
☐ Authority
☐ Performance standards
☐ Evaluation system

Pay/Benefits

☐ Wages
☐ Pay days
☐ Payroll deduction options
☐ Medical benefits
☐ Life insurance
☐ Disability coverage
☐ Annuity opportunities
☐ Saving deductions
☐ Retirement

Rules/Regulations

☐ Company rules/regulations
☐ Department rules/regulations
☐ Unit rules/regulations
☐ General safety rules
☐ Job-specific safety rules
☐ Consequences of breaking rules

Tools/Equipment

☐ Check-out/check-in procedures
☐ Maintenance procedures
☐ Orientation to use
☐ Emergency shut-down
☐ Personal protective equipment

Work Hours

☐ Work days
☐ Work hours
☐ Overtime compensation
☐ Holidays/down days
☐ Vacation
☐ Sick leave
☐ Nontraditional scheduling opportunities

Employee signature: _____ Date: _____

Figure 21–6
Sample employee orientation checklist.

questions or concerns? Does the new employee have suggestions for improving the safety and health of his or her work environment? It can also be a good idea to ask the employee to verify that he or she has been adequately and properly oriented by signing a checklist such as the one in Figure 21–6.

JOB SAFETY ANALYSIS AS A TRAINING TECHNIQUE

Job safety analysis (JSA) is a process through which all of the various steps in a job are identified and listed in order. Each step is then analyzed to identify any potential hazards associated with it. The final step involves developing procedures for reducing the hazard potential associated with each respective step. For example, say, in performing a job safety analysis for a solid ceramic molding process, you identified a step in which the technician could be seriously burned. With the setup and the potential hazard clearly identified, a procedure can be developed for reducing or completely eliminating the hazard.

Figure 21–7 is an example of a form that can be used to expedite the process. The form should be completed as described in the following paragraphs.

Break Down the Job into Steps

Go through the job slowly, taking note of the steps involved. Be especially cognizant of changes in direction, activity, or movement. Such changes typically signify the end of one step and the beginning of another. List all steps required to do the job even if a given step is not required every time.

Identify Potential Hazards

The most effective way to identify the potential hazards that may be associated with each step is to observe as another worker performs the job. While this is taking place, ask, "What could go wrong here? Is there danger of a back injury, burn, slip, being caught between objects, a fall, or muscle strain? Is there danger of exposure to dust, radiation, toxic fumes, or chemicals?" Record your concerns on the lines that correspond to the step(s) in question.

Develop Accident Prevention Procedures

Having identified potential hazards, ask the question, "How can this hazard be eliminated or reduced to the maximum extent possible?" Will redesigning the job eliminate the hazard? Is personal protective equipment needed? Should worker-equipment interaction be changed ergonomically? Will better housekeeping solve the problem? State the accident prevention procedures decided on in behavioral (action) terms. Do not use vague generalities. Be specific as to what must be done to prevent an accident (e.g., tighten the clamp firmly).

Kelton Electronics

Industrial Park Causeway
Fort Walton Beach, FL 32548
904-729-5218

Job Safety Analysis

Job title: _____

Analysis by: _____

Approved by: _____

Date: _____

Step-by-Step Sequence	Potential Hazard	Accident Prevention Procedure

Figure 21–7
Sample job safety analysis form.

Using the JSA as a Training Technique

Conducting a JSA can be a valuable learning experience for both new and experienced employees. Not only does it help them understand their job better, it familiarizes them with potential hazards and involves them in developing accident prevention procedures. Workers are more likely to follow procedures that they had a voice in planning. Finally, the JSA process causes employees to think about safety and how it relates to their job.

NSC SAFETY TRAINING INSTITUTE[28]

Prospective and practicing safety and health professionals should be familiar with the National Safety Council's Safety Training Institute. Courses may be taken at the NSC's main headquarters in Chicago or at the western regional office in Foster City, California. Customized on-site courses can also be arranged for companies that want the training brought to them.

Safety and health professionals can earn an Advanced Safety Certificate from the institute by completing the following three courses within a span of five years: principles of occupational safety and health, safety training methods, and safety management techniques. The content of these courses is summarized in the following paragraphs.

Principles of Occupational Safety and Health

This course is designed to give safety and health professionals the skills they need to develop and implement an effective safety program for their company. Major topics covered include safety organization, record keeping, communications for safety, accident investigation, electrical hazards, fire prevention and control, industrial hygiene accident reporting, human factors engineering, safety inspections, new employee indoctrination, machine safeguarding, personal protective equipment, job safety analysis, job instruction training, federal legislation, materials handling, safety meetings, and office safety.

Safety Training Methods

This course is designed to expose participants to proper techniques for planning, presenting, and evaluating instruction for adult learners. Major topics covered include methods of teaching, learning processes of adults, visual aid techniques, developing and using lesson plans, practice training, group involvement techniques, training needs of supervisors, computing training costs, course development, and extemporaneous speaking workshops.

Safety Management Techniques

This course is designed to equip participants with communications and management skills. Major topics covered include safety management, meeting leadership, conference leading practice, managing change (communication/listening), human relations, safety management by objectives, problem solving, decision making, accident investigation, and safety program audits.

Other Courses Offered

The courses offered by the institute vary each year. However, the following courses are typically available at a minimum: principles of occupational safety and health, safety training methods, safety management techniques, advanced safety concepts, practical aspects of industrial hygiene, safety in chemical operations, product safety management, OSHA hazard communication standard, motor fleet accident investigation, fundamentals of occupational ergonomics, back power master trainer, and defensive driving. For additional information about the NSC's Safety Training Institute, call or write:

NSC Safety Training Institute
444 N. Michigan Avenue
Chicago, Illinois 60611
(312) 527–4800, ext. 5105

ILLITERACY AND SAFETY

In recent years, industry has been forced to face a tragic and potentially devastating problem: *adult illiteracy,* which is having a major impact on the productivity of industry in the United States. It is estimated that more than 60 million people, or approximately one-third of the adult population in this country, are marginally to functionally illiterate.[29] The problem is compounded by non-English-speaking immigrants who may also be illiterate in their native language.

Illiteracy in the broadest sense is the inability to read, write, compute, solve problems, and communicate. Functionally illiterate people read at the fourth grade level or below. Marginally literate people read at the fifth through eighth grade levels. Functionally literate people read at the ninth grade level or above. This is the literacy continuum as it now exists. However, technological developments are having the effect of shifting these levels up the continuum. As a result, what constitutes functional or marginal literacy now will have to be redefined periodically.

People are sometimes shocked to learn that the number of illiterate adults is so high. Fields lists several reasons why this number has been obscured in the past and why there is now a growing awareness of the adult illiteracy problem.[30]

1. Traditionally, the number of low-skilled jobs available has been sufficient to accommodate illiterate adults.
2. Faulty research methods for collecting data on illiterate adults have obscured the reality of the situation.
3. Reticence on the part of illiterate adults to admit that they have a problem and to seek help has further obscured the facts.[31]

The reasons we are now becoming more aware of the adult illiteracy problem, according to Fields, are

1. Basic skill requirements are being increased by technological advances and the need to compete in the international marketplace.
2. Broader definitions of literacy that go beyond just reading and writing now include speaking, listening, and mathematics abilities.
3. We are realizing that old views of what constitutes literacy no longer apply.[32]

Impact of Illiteracy on Industry

The basic skills necessary to be productive in a modern industrial setting are increasing steadily. At the same time, the high school dropout rate nationwide continues to increase as does the number of high school graduates who are illiterate in spite of their diploma. This means that while the number of high-skill jobs in modern industry is increasing, the number of people able to fill them is on the decline. The impact this will have on industry in the United States can be summarized as follows:

1. Difficulty in filling high-skill jobs.
2. Lower levels of productivity and, as a result, a lower level of competitiveness.
3. Higher levels of waste.
4. Higher potential for damage to sophisticated technological systems.
5. Greater number of dissatisfied employees in the workplace.
6. Greater potential for safety and health problems in the workplace.

This last entry is of grave concern to prospective and practicing safety and health professionals. According to Frank Frodyma of OSHA,

> Intuitively, the greater degree of illiteracy there is, the more dangerous the workplace becomes. When you are working with a lot of technical equipment, chemicals, or hazardous materials, there's more of a reliance on written information, and there's a potential for problems.[33]

With written notification of potential hazards and the precautions that should be taken to avoid them so much a part of the typical safety program, illiteracy poses a difficult challenge. In Bruening's words,

> The issue of basic literacy skills in the workplace has become even more critical following the extension of OSHA's hazard communication standard to all industries. . . . The standard has required virtually all employers to provide written programs to transmit information on hazardous chemicals and materials to workers via container labels and material safety data sheets (MSDSs).[34]

OSHA and the Illiteracy Problem

Complaints that OSHA's **material safety data sheets** (MSDSs) are too technical and too complex are common. In fact, OSHA awards grants aimed at simplifying the MSDSs. The American Subcontractors Association has noted a gap between the reading level of MSDSs and the average worker.[35] According to Mike O'Brien of the association,

> MSDSs are written at a college level. That is an extreme problem, not only for people who are illiterate, but for people who have the most basic education. A lot of our contractors can't

even read MSDSs, and that really negates any of the effectiveness that MSDSs could have in the workplace.[36]

OSHA responds to complaints relating to illiteracy in the workplace by taking the stand that it is not their problem. They accept no responsibility since safety training does not have to be in written form. As the situation evolves, however, OSHA and other governmental agencies are becoming more sensitive to the issue of illiteracy. The Department of Labor has increased its efforts to deal with the problem by sponsoring research and demonstration projects designed to develop literacy training program models, test and demonstrate new and innovative instructional techniques, develop instruments for assessing literacy levels, and encourage the development of state-level policies that promote workforce literacy.[37]

Industry's Role in Fighting Illiteracy

Industry in the United States has found it necessary to confront the illiteracy problem head on. This is being done by providing remedial education for employees in the workplace. Some companies contract with private training firms, others provide the remediation themselves, and others form partnerships with community colleges or vocational schools.

The National Center for Research in Vocational Education conducted a study of industry-based adult literacy training programs in several industrial firms.[38] According to this study, Texas Instruments requires math, verbal and written communication, and basic physics of its employees. Physics skills have not traditionally been viewed as required for functional literacy. However, employees must have these skills to function in this high tech company. The approach used by Texas Instruments to provide literacy training can be one of the least expensive. They work with public community colleges or vocational schools to provide literacy training at little or, in some cases, no cost. Safety and health professionals should establish a working relationship with representatives of local community colleges and vocational schools.[39]

Rockwell International also defines literacy more stringently than has been typical in the past. To function effectively at Rockwell International, employees must be skilled in chemistry and physics. Rockwell hires its own certified teachers and provides its own literacy program.[40]

Polaroid takes an aggressive approach in defining functional literacy. The skills taught in Polaroid's program are also indicative of the trend toward higher levels of knowledge. Polaroid requires statistics, problem solving, and computer literacy. The need for these skills is technology driven. Statistics is needed in order to use statistical process control (SPC), which is becoming a widely used quality control method in automated manufacturing settings. Few employees in a modern industrial firm get by without using a computer on the job, hence the need for computer literacy training.

Problem-solving skills are becoming critical as companies implement total quality management (TQM) programs. Such programs involve all employees in identifying and correcting problems that negatively affect quality or do not add value to the company's products. This is the rationale for including problem-solving skills in the definition of functional literacy.

The Role of Safety and Health Professionals in Literacy Training

In their study of industry-based literacy training, Fields et al. made recommendations to assist in the planning and implementation of such programs.[41] Several of these recommendations have relevance for safety and health professionals:

1. The definitions should be driven by the needs of the company.
2. Companies should establish an environment in which employees feel comfortable having their literacy skills assessed and in seeking help to raise those skills.
3. Whenever possible, companies should establish programs to raise the skills of existing employees rather than laying them off and hiring new employees.
4. Whenever possible, companies should collaborate with educational institutions or education professionals in providing literacy training.

As industrial companies continue to enhance their technological capabilities and safety and health concerns increase correspondingly, the literacy levels of the workforce will also have to increase. Because the number of people in the labor force who are highly literate is not increasing, the need for workforce literacy training is a fact of life that safety and health professionals will have to confront for some time to come.

SUMMARY

1. The rationale for safety and health training is that workers who know how to do their jobs properly are less likely to have accidents.
2. The OSHAct of 1970 established a legal foundation for safety and health training. In addition to producing and distributing training materials, OSHA also promotes safety and health training by doing the following: providing grants, publishing training requirements, and writing training-oriented regulations.
3. The division of labor between safety and health professionals and supervisors is not clear-cut. However, generally speaking, the following guidelines apply: Safety and health professionals help train supervisors and keep them up to date; supervisors provide most of the training to workers; and safety and health professionals and supervisors work together as appropriate to train workers.
4. The Mine Safety and Health Administration (MSHA) sets forth training requirements for miners in Subpart B of its regulations. Paragraph 48.25 contains the regulations that pertain to new or inexperienced miners.
5. Any person who is going to conduct safety and health training should have the following characteristics: thorough knowledge of the subject matter, desire to teach, positive attitude, leadership ability, professional approach, and exemplary behavior.
6. The principles of learning summarize what is known about how people learn. Trainers should be familiar with these principles.
7. The four-step teaching method is as follows: preparation, presentation, application, and evaluation.

8. Widely used presentation methods include lecture/discussion, demonstration, conference, simulation, videotapes, and programmed instruction.

9. People learn by doing. Therefore, it is critical that people be given plenty of opportunities to apply what they are learning.

10. In evaluating training, safety and health professionals need to know: (a) was the training valid? (b) did the employees learn? and (c) has the learning made a difference?

11. Supervisors play a key role in the maintenance of a safe and healthy workplace. Consequently, safety and health professionals need to ensure that supervisors have had the training they need to be positive participants in the process.

12. One of the most important aspects of a safety program is the orientation of new employees. It should consist of a general orientation, job-specific procedures, and follow-up.

13. Job safety analysis, or JSA, can be an excellent way to teach safety. It consists of three steps: (a) breaking down a job into a chronological sequence of steps; (b) identifying potential hazards for each step; and (c) developing accident procedures to eliminate or reduce potential hazards.

KEY TERMS AND CONCEPTS

Application

Application assignments list

Conference method

Course outline

Demonstration method

Ethical reasons

Evaluation

Evaluation methodology

Four-step teaching method

Hazard communication

Immediate/continual feedback

Instructional approach

Interactive video training

Job safety analysis (JSA)

Labor/management mistrust

Learning objectives

Lecture/discussion method

Legal reasons

Lesson plan

Material safety data sheets

Mine Safety and Health Administration (MSHA)

MSHA training requirements

Opening/body/closing of a lecture

Orientation

OSHA training requirements

Planning instruction

Preparation

Presentation

Principles of learning

Programmed instruction

Right-to-know laws

Simulation

Statement of purpose

Training aids

REVIEW QUESTIONS

1. Explain briefly the rationale for providing safety training.
2. Establish the legal framework for providing safety and health training.
3. What does the hazard communication regulation (29 C.F.R. 1910) require of employers?
4. Beyond the legal requirements of the OSHAct, how does OSHA promote safety and health training?
5. Summarize the mistrust that sometimes exists between labor and management concerning safety and health training.
6. What document could you use as a guide to the OSHA training requirements?
7. How does MSHA define the term *experienced miner?*
8. Summarize the MSHA training requirements for newly employed but experienced miners.
9. What are the characteristics that should be present in persons who plan to conduct safety training?
10. List five principles of learning.
11. Explain briefly each of the steps in the four-step teaching method.
12. What are the essential components of a lesson plan?
13. Give three examples of when the lecture/discussion method is best used.
14. Give three examples of when the lecture/discussion method is not appropriate.
15. What are the three components of a lecture?
16. Give an example of when the demonstration method might be used.
17. Briefly summarize the National Safety Council's recommendations for using the conference teaching method.
18. List five questions that should be asked when evaluating training.
19. What are the objectives of supervisor safety training, according to the National Safety Council?
20. List the minimum content of a good orientation program.
21. Explain how to conduct a job safety analysis.
22. What is the safety training value of a job safety analysis?

ENDNOTES

1. LaBar, G. "Worker Training: An Investment in Safety," *Occupational Hazards,* August 1991, p. 25.
2. Ibid.
3. Laing, P. M. (ed.). *Supervisor's Safety Manual,* 7th ed. (Chicago: National Safety Council, 1991), p. 35.
4. Spencer, J. A., and Simonowitz, J. A. "Employee Education," in Joseph LaDon, ed., *Introduction to Occupational Health and Safety* (Chicago: National Safety Council, 1986), p. 276.
5. Morris, B. K. "Grants Available for Training Workers on MSDS," *Occupational Health & Safety Letter,* December 26, 1990, Vol. 20, No. 26, p. 210.

6. Spencer and Simonowitz, "Employee Education," pp. 277–78.
7. Ibid., p. 280.
8. Title 29, Code of Federal Regulations, Part 1910.95(k)(1).
9. Paragraph 48.22, Subpart B, MSHA Regulations as published in the *Federal Register,* October 13, 1978, Vol. 43, No. 199.
10. Ibid.
11. Paragraph 48.25, Subpart B, MSHA Regulations as published in the *Federal Register,* October 13, 1978, Vol. 43, No. 199.
12. Ibid.
13. Ibid.
14. Laing, *Supervisor's Safety Manual,* p. 36.
15. Parachin, V. "10 Tips for Powerful Presentations," *Training,* July–August 1990, pp. 71–83.
16. National Safety Council. *Accident Prevention Manual for Industrial Operations: Administration and Programs,* 9th ed. (Chicago: National Safety Council, 1988), p. 195.
17. Magon, K. M. "Videos Add New Dimension to Safety Programs," *Safety & Health,* January 1991, Vol. 143, No. 1, p. 34.
18. Ibid., p. 35.
19. Fisher, R. "Interactive Training, Wave of the Future," *Safety & Health,* January 1984, Vol. 138, No. 1, p. 34.
20. Ibid., pp. 36–37.
21. Dangot-Simpkin, G. "How to Get What You Pay For," *Training,* July–August 1990, pp. 53–54.
22. Ibid.
23. Minetos, P. "Supervisors: Teach Them Well," *Safety & Health,* October 1989, Vol. 140, No. 4, p. 64.
24. Ibid.
25. National Safety Council, *Accident Prevention Manual for Industrial Operations,* p. 186.
26. Knowles, K. (ed.). "Start Safety Training Early," *Today's Supervisor,* July 1991, Vol. 55, No. 7, p. 4.
27. National Safety Council, *Accident Prevention Manual for Industrial Operations,* p. 199.
28. Piepho, C. "Safety Training Institute," *Safety & Health,* May 1990, Vol. 141, No. 5, pp. 80 85.
29. Fields, E. L., Hull, W. L., and Sechler, J. A. *Adult Literacy: Industry-Based Training Programs* (Columbus, OH: The National Center for Research in Vocational Education, 1987), p. vii.
30. Ibid.
31. Ibid.
32. Ibid.
33. Bruening, J. C. "Workplace Illiteracy: The Threat to Worker Safety," *Occupational Hazards,* October 1989, p. 120.

34.   Ibid., p. 212.
35.   Ibid., p. 121.
36.   Ibid.
37.   Ibid.
38.   Fields et al. *Adult Literacy,* p. vii.
39.   Ibid.
40.   Ibid.
41.   Ibid.

Computers, Automation, and Robots

Automation of the workplace has changed and continues to change how work is done. The introduction of automated processes that involve computers and robots has changed the environment of the modern workplace and what is needed in order to succeed in it. According to Rinzo Ebukuro of NEC Corporation,

> Developments . . . have changed the concepts of the factory and product, created knowledge-intensive jobs, and encouraged the development of management technology. Innovative technology calls for intelligent manpower and increases the severity of work.[1]

IMPACT OF AUTOMATION ON THE WORKPLACE

The advent of **automation** in the workplace was the next logical step on a continuum of developments intended to enhance productivity, quality, and competitiveness. This continuum began when humans first developed simple tools to assist them in doing work. This was the age of hand tools and **manual work.** It was eventually superseded by the age of **mechanization** during the Industrial Revolution. During the age of mechanization, machines were developed to do work previously done by humans using hand tools.

503

The 1960s saw the beginnings of broad-based efforts at automating mechanical processes and systems.

These early attempts at automation resulted in **islands of automation**, or individual automated systems lacking electronic communication with other related systems. Examples of islands of automation are a stand-alone computer numerical control milling machine or a personal computer-based word processing system, neither of which is connected to other related systems. Local area networks (LANs) for integrating personal computers are an example of integration in the office. Computer-integrated manufacturing is an example in the factory.

These developments are having an impact on the workplace. According to Ebukuro, automation and integration are having the following effects on workers:

- Reducing the amount of physical labor workers must perform
- Increasing the amount of mental work required
- **Polarizing work** into mental jobs and labor-intensive jobs
- Increasing the stress levels of managers
- Decreasing the need for traditional blue-collar workers
- Decreasing the feelings of loyalty that workers feel toward employers
- Increasing workers' feelings of powerlessness and helplessness[2]

These various effects of automation are resulting in a marked increase in the amount of stress experienced by workers. Two factors in particular lead to increased levels of stress: rapid, continual change and an accompanying feeling of **helplessness.** With automation, the rate of change has increased. As a result, workers must continually learn and relearn their jobs with little or no relief. In addition, automated machines do more and more of the work that used to be done by humans. This can leave workers feeling as if they might be replaced by a machine and powerless to do anything about it. According to Ostberg,

> The development of a special curriculum known as robot medicine in Japan acknowledges the connection between emerging technology and stress. This course of study is offered to industrial doctors by the University of Occupational and Environmental Health, Japan. It is defined as the branch of medicine studying stress and other mental and physical problems of workers in automated and roboticized production.[3]

Work stress is a complex concept involving physiological, psychological, and social factors. People become stressed when there is an imbalance between the demands placed on them and their ability to respond.[4] Automation appears to be increasing the instances in which such an imbalance occurs. This chapter focuses on the safety and health concerns associated with computers, robots, and automation and appropriate measures for dealing with these concerns.

VDTs IN OFFICES AND FACTORIES

A safety and health concern brought about by the advent of computers has to do with the impact of **video display terminals (VDTs).** Does prolonged use of VDTs cause safety and health problems? Are pregnant women who work at VDTs more likely to miscarry? Are such problems as eye fatigue, muscle stiffness, and mental fatigue caused by VDT use?

In 1991, the National Institute for Occupational Safety and Health (NIOSH) published a study showing that "women who work with VDTs have no greater risk of miscarriage than those who do not."[5] The study began in 1984 in response to reports of a perceived high incident rate of miscarriages among women whose jobs involved regular use of VDTs. According to Castelli,

> NIOSH researchers studied telephone operators in eight southeastern states. One group did not use computer terminals; the other group, directory-assistance operators, used them seven hours a day. The researchers interviewed 2,430 operators and gathered information on 882 pregnancies that occurred among the workers between 1983 and 1986. About half of the pregnant workers used VDTs.[6]

The miscarriage rate among VDT users was actually lower than that for nonusers (14.8 percent for VDT users and 15.9 percent among nonusers). Both figures are close to the national average of 15 percent. As a result of this study, NIOSH plans to conduct research into two additional areas of concern: (1) whether VDT use can be tied to birth defects, premature births, or low birthweight; and (2) potential risks of extremely low frequency (ELF) radiation, a type of electromagnetic energy emitted by VDTs, power lines, electrical wires, and appliances.[7]

Not everyone agreed with the findings of the NIOSH study, however. The Service Employees' International Union (SEIU) challenged the findings as inconclusive because the effects of stress and VDT use on pregnancy were not considered. SEIU claimed that the study did not give VDTs a "clean bill of health" and that "employers should likewise not conclude from this study that they have no obligation to develop safety guidelines and better designed equipment."[8]

Regardless of which side of the miscarriage issue one believes, there is ample evidence that such problems as eye fatigue, blurred vision, eye strain, and nervousness are associated with VDT use. As far back as 1976, a study by Ostberg suggested that VDT users frequently complained of eye symptoms including dryness, heaviness, discomfort, pain, blurring, and double vision.[9] Further research by Smith in 1981 supported Ostberg's findings, as have numerous subsequent studies.[10]

The **eyestrain** caused by prolonged VDT use poses safety and health problems from two different perspectives. First, the visual health of the operator experiencing the strain is impaired. Second, there is an increased likelihood of accidents caused by impaired work performance and increased psychological stress.

According to Kurimoto and associates of the Department of Ophthalmology at Japan University of Occupational and Environmental Health, the eye functions most noticeably affected by VDT use are accommodation, convergence, and lacrimation.[11] **Accommodation** is the ability of the eye to become adjusted after viewing the VDT so as to be able to focus on other objects, particularly objects at a distance. **Convergence** is the coordinated turning of the eyes inward so as to focus on a nearby point or object. **Lacrimation** is the process of excreting tears. The Kurimoto research confirms in nonmedical terms that prolonged VDT use can render operators unable to focus on either distant or near objects. It can also impair the tearing function, leading to dry eyes.[12]

Kurimoto recommends the following strategies for reducing the physiological and psychological problems associated with VDT use:[13]

- Faster computer response time. This is a matter of upgrading the computer's processing capability or replacing it with one that has more processing power.

- More frequent breaks from VDT use or a work rotation schedule that allows users to intersperse non-VDT work in their daily routine.

- **Work design** that recognizes and accommodates the need to break up continual VDT use.

- Arranging the keyboard properly so it is located in front of the user, not to the side. Body posture and the angle formed by the arms are critical factors.

- Adjusting the height of the desk. Taller employees often have trouble working at average height desks.

- Adjusting the tilt of the keyboard. The rear portion of the keyboard should be lower than the front.

- Encouraging employees to use a soft touch on the keyboard and when clicking a mouse. A hard touch increases the likelihood of injury.

- Encouraging employees to avoid wrist resting. Resting the wrist on any type of edge can increase pressure on the wrist.

- Placing the mouse within easy reach. Extending the arm to its full reach increases the likelihood of injury.

- Removing dust from the mouse ball cavity. Dust can collect, making it difficult to move the mouse. Blowing out accumulated dust once a week will keep the mouse easy to manipulate.

- Locating the VDT at a proper height and distance. The height of the VDT should be such that the top line on the screen is slightly below eye level. The optimum distance between the VDT and user will vary from employee to employee, but will usually be between 16 and 32 inches.

- Minimizing glare. Glare from a VDT can cause employees to adopt harmful postures. Glare can be minimized by changing the location of the VDT, using a screen hood, and closing or adjusting blinds and shades.

- Reducing lighting levels. Vision strain can be eliminated by reducing the lighting level in the area immediately around the VDT.

- Dusting the VDT screen. VDT screens are magnets to dust. Built-up dust can make the screen difficult to read, contributing to eyestrain.

- Eliminating telephone cradling. Cradling a telephone receiver between an uplifted shoulder and the neck while typing can cause a painful disorder called cervical radiculopathy (compression of the cervical vertebrae in the neck). Employees who need to talk on the telephone while typing should wear a headphone.

HUMAN/ROBOT INTERACTION

Every new tool developed to enhance the ability of humans to work efficiently and effectively has brought with it a new safety and health hazard. This is particularly the case with industrial robots. What makes robots more potentially dangerous than other machines can be summarized as follows: (1) their ability to acquire intelligence through programming; (2) their flexibility and range of motion; (3) their speed of movement; and (4) their power.

The often-discussed **peopleless factory** is still far in the future. However, robots are so widely used now that they are no longer the oddity they once were. Consequently, there is plenty of **human/robot interaction** in modern industry. According to Yamashita, "At ordinary factories . . . human workers and robots coexist, creating such problems as cooperation and competition between man and machines and safety."[14]

How does human/robot interaction differ from human interaction with other machines? This is an important question for safety and health professionals. According to Lena Martensson of the Royal Institute of Technology in Stockholm, Sweden, the modern factory has or is moving toward having the following characteristics:

■ Workers will supervise machine systems rather than interact with individual pieces of production equipment.

■ Workers will communicate with machines via video display terminals on which complex information processed by a computer will be displayed.

■ Workers will be supported by expert systems for fault identification, diagnosis, and repair.[15]

Robots and other intelligent computer-controlled machines will play an increasingly important role in modern industry. As this happens, safety and health professionals must be concerned about the new workplace hazards that will be created.

SAFETY AND HEALTH PROBLEMS ASSOCIATED WITH ROBOTS

Robots are being used in industry for such applications as arc welding, spot welding, spray painting, material handling and assembly, and loading/unloading of machines. Figures 22–1 and 22–2 are examples of modern industrial robots being used in typical applications. According to the National Safety Council the principal hazards associated with robots are as follows:

■ Being struck by a moving robot while inside the work envelope. The **work envelope** of a robot is the total area within which the moving parts of the robot actually move. Figure 22–3 is an example of a robot's work envelope.

■ Being trapped between a moving part of a robot and another machine, object, or surface.

■ Being struck by a workpiece, tool, or other object dropped or ejected by a robot.[16]

Figure 22–1
Industrial robot in use.
Courtesy of Cincinnati Milcron.

Until a worker enters the work envelope of a robot, there is little probability of an accident. However, any time a worker enters a functioning robot's work envelope, the probability becomes very high. The only logical reason for a worker to enter the work envelope of an engaged robot is to teach it a new motion. According to the National Safety Council,

> Many applications of robots use a mode of operation called the **teach mode** that makes them unique among industrial equipment as regards safeguarding them. In the teach mode, an operator may be required to place himself within the operating range of the robot in order to program its movements within very close tolerance parameters. The teaching of a robot is done at a greatly reduced speed of operation; however, a hazard may arise if the robot goes out of control and moves in an unpredictable fashion at a high rate of speed. It is in the teach mode where the highest degree of hazard exists. A Japanese survey . . . showed that the greatest risk of accidents involving robots occurs during programming, teaching, and maintenance; all times when a person may be within the operating envelope of the robot.[17]

Minimizing the Safety and Health Problems of Robots

If human workers never had to enter a work envelope, the safety and health problems associated with robots would be minimal. However, workers must occasionally do so. Therefore safety and health professionals must be concerned with ensuring safe human/robot interaction.

The National Safety Council recommends several strategies for minimizing the hazards associated with robots. The general strategies are summarized as follows:

- Ensure a glare-free, well-lighted robot site. The recommended light intensity is 50–100 foot-candles.
- Keep the floors in and around the robot site carefully maintained, clean, and free of obstructions so that workers do not trip or slip into the work envelope.
- Keep the robot site free of associated hazards such as blinding light from welding machines or vapors from a paint booth.
- Equip electrical and pneumatic components of the robot with fixed covers and guards.
- Clear the work envelope of all nonessential objects and make sure all safeguards are in place before starting the robot.
- Apply lockout and proper test procedures before entering the work envelope.
- Remove and account for all tools and equipment used to maintain the robot before starting it.[18]

SAFETY AND HEALTH IN OFFICE AUTOMATION

According to Nishiyama, the objectives of office automation are increased efficiency, personnel reductions, economy of personnel expenditures, improved service to customers, improved planning and estimating, increased processing speed, and improved working conditions.[19] These goals are being achieved with varying degrees of success. However, automation also introduced a new set of safety and health problems into the office environment.

Figure 22-2
Industrial robot in use.
Courtesy of Unimation, Inc.

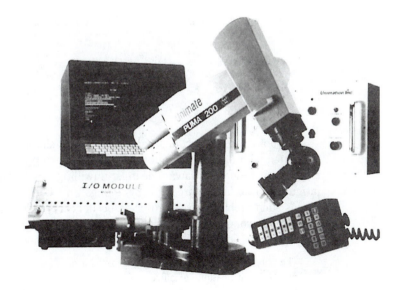

Figure 22–3
A robot's work envelope.
Courtesy of Unimation, Inc.

Model 762

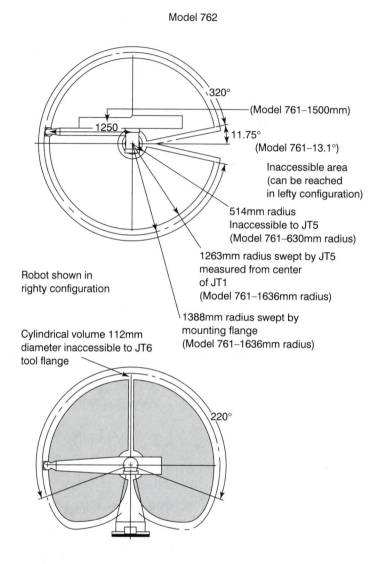

320°

(Model 761–1500mm)

1250

11.75°

(Model 761–13.1°)

Inaccessible area
(can be reached
in lefty configuration)

514mm radius
Inaccessible to JT5
(Model 761–630mm radius)

1263mm radius swept by JT5
measured from center
of JT1
(Model 761–1636mm radius)

Robot shown in
righty configuration

1388mm radius swept by
mounting flange
(Model 761–1636mm radius)

Cylindrical volume 112mm
diameter inaccessible to JT6
tool flange

220°

Morooka and Yamoda of Japan's Tokai University identified the following problems associated with office automation: eye fatigue, seeing double images and complementary colors, headache, yawny feelings, unwillingness to talk, **shoulder fatigue, neck fatigue,** dryness in the throat, sleepy feelings, and whole body tiredness.[20]

As such problems began to be associated with office automation, interest grew in establishing workplace and technology standards to minimize them. Benjamin C. Amick III of Congress' Office of Technology Assessment described this situation as follows:

The public policy issues developing as a result of office automation range from issues of labor management relations to office, work-station, and human-computer interface standards. The current scientific research suffers from a lack of integration and clear definition of what is causing what. . . . Clearly, the current upswing in the purchasing of office automation equipment provides a unique opportunity to effect changes in the quality of worklife on a national basis. The question facing policy-makers is how best to create the policy to maintain the current level of creativity and innovation in the marketplace while not sacrificing the quality of worklife.[21]

By 1991, states that had introduced legislation establishing standards relating to office automation were Oregon, Washington, California, Florida, Colorado, Missouri, Iowa, Minnesota, Wisconsin, Illinois, Indiana, Ohio, Pennsylvania, West Virginia, Maryland, New Jersey, Connecticut, Rhode Island, Massachusetts, New Hampshire, and Maine. New Mexico established standards by executive order.

Most legislation dealing with office automation concerns standards for VDT interaction. For example, legislation introduced in Maryland requires that VDT users (1) have an eye examination every year; (2) have an adjustable chair with adjustable backrest height and tension; and (3) take a 15-minute break from the VDT every hour. Legislation introduced in other states is similar to the Maryland proposal.[22]

Public policy debate is increasing in the United States over the safety and health concerns inherent in office automation. Unfortunately, the debate has outpaced the research on this important issue. Consequently, the probability is high that policies being adopted are based on insufficient information. Regarding research that needs to be conducted, Amick recommends the following:

- Testing of the biological plausibility of health hazards from VDT work
- Ergonomic intervention research to determine the contribution of workstation design and office design to the worker's health
- Examination of the interaction of physical and **psychosocial stressors** in the high-technology workplace
- Development of prospective case-control studies to determine the temporal relationships
- Establishment of a national high-technology surveillance system for worker safety and health
- Development of organizational and job-intervention programs
- Multidisciplinary studies examining the relative contribution of various working conditions to the health and well being of the office worker
- Programs that bring system designers and/or building designers into the total automation process[23]

The Japanese Association of Industrial Health developed the following set of principles upon which future safety and health measures relating to office automation should be based:

- More attention should be focused on bringing employers and employees together for the purpose of improving working conditions in automated offices.
- Since VDT use is becoming so common in so many different occupations, it should no longer be considered specialized. Consequently, jobs should be designed to accommodate VDT use.

■ The amount of time spent doing VDT work exclusively should be kept short, and employees should be able to perform the work in their own way.

■ Working conditions should be established that prevent safety and health problems so that management is acting instead of reacting.

■ Special emphasis should be placed on education and training as the best way to prevent adverse effects from office automation.

■ VDT work should not be done part-time at home or contracted out since working conditions cannot be properly controlled or supervised under these circumstances.[24]

INDUSTRIAL MEDICINE AND ROBOTS

Industrial medicine is a specialized field that is concerned with work-related safety and health issues. Practitioners of **industrial medicine** are becoming increasingly concerned with the interaction of humans and automated machines, particularly robots and computers. They are concerned about maladaptation to an automated environment.

According to Masamitsu Oshima of Japan's Medical Information System Development Center, **maladaptation** can manifest itself as an urge to quit work; **fatigue;** problems with human relations; a drop in work performance; **social pathological phenomena** such as drug use or crime; mood swings; a loss of motivation; and accidents.[25] Practitioners of industrial medicine are concerned with improving the relationship between humans and automated machines by establishing methods whereby humans can work more adequately with the machines.[26]

Oshima makes the following recommendations for improving the interaction between humans and robots, computers, and other automated machines in the high tech workplace:

■ Match the human system and the computer system.

■ Position machine systems as human-supportive systems.

DISCUSSION CASE

What Is Your Opinion?

Mountain Trek, Inc., is one of the most successful catalog ordering houses in the western United States. The company employs 115 VDT operators who work in eight-hour shifts. Customers may place orders by telephone or Internet 24 hours a day, seven days a week. Mountain Trek has taken no steps to prevent VDT-related problems (psychological or physiological). What types of problems could the company begin to experience. How could these problems be prevented? What is your opinion?

- Adapt human-computer interaction to human use.
- Initiate job-changing opportunities.
- Allow suitable rest periods for users of automated equipment.
- Vitalize the workplace.
- Encourage recreation.
- Promote the effective use of nonworking hours.
- Increase the contact with nature.
- Free people from hazardous, dirty, and harmful jobs.
- Shorten working hours and promote work sharing.
- Expand human contact.
- Harmonize people, things, and the environment ergonomically.[27]

It should be obvious from reviewing this list that it was developed with the Japanese culture in mind. However, these recommendations can also be applied in the American workplace. As human workers continually increase the amount of their interaction with automated machines, the potential for maladaptation also increases. One key to preventing maladaptation is to design automated systems around the needs of humans rather than designing them in a vacuum and then expecting humans to adapt. Another is to pay special attention to establishing and maintaining human contact for workers who interact primarily with automated machines and systems. These are areas in which attentive safety and health professionals can have a positive impact.

SAFETY FACT

There Is More to RSI than Carpal Tunnel Syndrome

The personal computer has become an all-pervasive and universal tool. Jobs from the shop floor to the executive office now involve frequent, repetitive computer use. This means that people in the workplace are typing and clicking at an unprecedented pace. Frequent, and for some, constant computer use has led to an explosion of injuries heretofore seen mostly in the meat-packing industry. Collectively, these injuries are known as repetitive strain injuries, or RSIs.

RSI is an umbrella term that covers a number of cumulative trauma disorders (CTDs) caused by forceful or awkward hand movements repeated frequently over time. Other aggravating factors include poor posture, an improperly designed workstation, and job stress. RSIs occur to the muscles, nerves, and tendons of the hands, arms, shoulders, and neck. For years, RSI has been incorrectly referred to as carpal tunnel syndrome, which in reality is, itself, one type of RSI. This is like referring to all trees as oaks.

TECHNOLOGICAL ALIENATION IN THE AUTOMATED WORKPLACE

As technology has become more widely used in the workplace, particularly automated technology, some workers have come to resent its impact on their lives. This concept is known as **technological alienation.** According to Gary Benson of the University of Wyoming, this concept has several meanings, all of which encompass one or more of the following:

- **Powerlessness** is the feeling that workers have when they are not able to control the work environment. Powerless workers may feel that they are less important than the technology with which they work and that they are expendable.

- **Meaninglessness** is the feeling that workers get when their jobs become so specialized and so technology-dependent that they cannot see the meaning in their work as it relates to the finished product or service.

- **Normlessness** is the phenomenon in which people working in a highly automated environment can become estranged from society. Normless people lose sight of societies, norms, rules, and mores.[28]

Benson investigated what he considers to be the most devastating form of technological alienation—mindlessness.[29] **Mindlessness** is the result of the process of "**dumbing down**" the workplace. This is a concept that accompanied automation. In the past, machines have been used to do physical work previously done by human workers. With the advent of computers, robots, and automation, machines began doing mental work. According to Benson,

> The net result is jobs and work environments where people do not have to use their minds or think to do their work—an environment where computers, robots, and other forms of high technology do the thinking.[30]

Mindlessness on the job should be of interest to safety and health professionals because of the other problems it can create. According to Benson, these problems can include an increase in alcoholism, drug abuse, employee theft, work-related accidents, absenteeism, sick leave abuse, turnover rates, and employee personal problems. Mindlessness can also lead to a decrease in job performance, productivity, and work quality.[31]

Each of these problems can, in turn, increase the potential for safety and health problems on the job. Employees who abuse alcohol and/or drugs represent a serious threat in the workplace. Absent employees force their fellow workers to double up. High turnover results in a steady influx of inexperienced workers. Employees with personal problems may not be properly focused on accident prevention measures. Finally, when productivity and quality fall, supervisors can feel so much pressure to improve performance that they overlook or put aside safety precautions. These are the potentially negative safety and health effects of mindlessness in the automated workplace.

According to Benson,

> What is needed now are more employees who are aware of and willing to do something about the problem and more research into the phenomenon and causes of the solutions for mindlessness in technological alienation. This is surely a phenomenon of modern-day worklife that must be dealt with effectively—and immediately.[32]

MINIMIZING THE PROBLEMS OF AUTOMATION

The infusion of technology into the workplace has presented safety and health professionals with an entirely new set of challenges. Among the most pressing of these is the need to identify and minimize the new safety and health problems specifically associated with automation. Behavioral scientist A. B. Cherns developed a **sociotechnical system theory** for doing so that consists of the following components: variance control, boundary location, work group organization, management support, design process, and quality of work life.[33]

Although the sociotechnical system theory was developed in 1977, it has even more relevance now than it did then. According to Yoshio Hayashi of Japan's Keio University,

> The safety and health of workers in this high technology age cannot be discussed within the conventional framework of one worker assigned to one machine. . . . The socio-technical system may be roughly understood if the *man* and *machine* in the man-machine system are replaced by *socio* and *technical* respectively. It refers to a system composed of a work group and high technology.[34]

The various components in the sociotechnical system theory explain what must happen if humans and technological systems are going to work together harmoniously and safely. Safety and health professionals can apply the theory as they work to minimize the potential problems associated with automation in the modern workplace. These components can be summarized as follows:

- **Variance control** involves controlling the unexpected events that can be introduced by new technologies. For example, a runaway, out-of-control industrial robot introduces unexpected safety hazards at variance with the expectations of workers and management. Variance control involves bringing the situation under control and establishing appropriate preventive measures for the future.

- The concept of **boundary location** involves the classification of work. What specific tasks are included in an employee's job description? Does a robot technician just operate the robot, or is she also required to teach and maintain the robot? The accident prevention measures learned by an employee should cover all tasks in his or her job description.

- The concept of **work group organization** involves identifying the tasks that a work group is to perform and how these tasks are to be performed. The key is to make sure that all work group members have the training needed to accomplish effectively and safely all tasks assigned to them.

- **Management support** is perhaps the most important of the components of the sociotechnical system theory. It states that, in the age of high technology, managers must be willing to accept occasional temporary declines in productivity without resorting to shortcuts or improvement efforts that might be unsafe or unhealthy. Management must be willing to emphasize safety in spite of temporary declines in productivity.

- The **design process** component refers to the ability of an organization to design itself in ways that promote productivity, quality, competitiveness, safety, and health.

It also involves the ability to continually redesign as technological advances and other circumstances dictate.

■ **Quality of work life** involves determining ways to promote the morale and best interests of workers. The key is to ensure that technology extends the abilities of humans and that technological systems are *human-centered*. In other words, it is important to ensure that people control systems rather than vice versa.[35]

If the sociotechnical system theory is fully applied, the safety and health hazards of the automated workplace can be minimized. Safety and health professionals can play a key role in making sure the theory is applied. To play such a role, these professionals must be technicians, diplomats, trainers, and lobbyists. They must work with the technical aspects of variance control, boundary location, work group organization, and the design process. They must be diplomats in working with supervisors and employees in promoting adherence to safe work practices. They must be trainers in order to ensure that all employees know how to apply safe work practices and appropriate accident prevention techniques. Finally, they must be lobbyists as they continually interact with management to establish and maintain management support for safety, health, and quality of life issues.

Safety Measures for Automated Systems

The sociotechnical system theory discussed in the previous section is broad and conceptual in nature. Modern safety and health professionals also need to know specific measures that can be taken to minimize the hazards associated with robots and other automated systems. Minoru Goto of the Nissan Motor Company's safety department developed specific safety measures in the categories of technological systems, auxiliary equipment, and training.[36]

Examples of safety measures that can be used at the technological systems level include the following:

■ Construction of a safety fence around the system that defines the work envelope of the system

■ Control of the speed of movement of system components when working inside the work fence

■ Installation of an emergency stop device colored red and placed in an easily accessible location

■ Location of the control panel for the system outside of the safety fence

■ Establishment of automatic shutdown switches that activate any time a system component goes beyond its predetermined operational range[37]

Safety measures relating to training include training system operators to work safely within the work envelope and to work together as a team when interacting with the system. Maintenance workers should be trained on the technical aspects of maintaining all machines and equipment that make up the system. This is important because the safety level of the system is the sum of the safety levels of its individual components. A system with four properly operating components and just one faulty component is an unsafe system.[38]

CHALLENGE FOR THE FUTURE

Much more effort has gone into developing automated systems to improve productivity than has gone into the appropriate matching of people and technology. Now, with the speed of technical development being what it is, the safety and health problems associated with automation, particularly stress-related problems, are likely to increase. According to Kensaburo Tsuchiya of Japan's University of Occupational Health and Safety, the challenge for the future is "to create jobs which are free from stress and musculoskeletal overloads while at the same time being challenging and interesting for the individual."[39]

The future holds a number of problems that will have to be addressed to meet Tsuchiya's challenge. The most prominent of these are as follows:

■ Increasingly intense international competition may magnify the tendency for companies to neglect safety and health precautions in favor of short-term productivity gains.

■ The level of mental stress is likely to increase as the automated manipulation of information forces workers to try continually to handle too much information that is poorly understood.

■ Automation and competition are likely to increase the level of anxiety as workers are required to make split-second decisions while knowing that their actions or inactions may have dire consequences.

■ New **occupational diseases** relating to mental, visual, and **musculoskeletal problems** may arise whose remedies must be sought through a combination of ergonomics, psychology, occupational medicine, and design.

■ There is likely to be increased introduction of robots into the workplace with even less foresight that will, in turn, introduce more unexpected safety and health risks.

■ Ignorance may lead to the introduction of automation in an office or factory in forms that do not require workers to think, reason, or make judgments, giving rise to alienation and frustration.[40]

Since they are likely to face these inhibitors, the safety and health professionals of the future must be prepared to deal with them. They need to know what has to happen if the inhibitors are to be overcome. Tsuchiya suggests the following strategies for enhancing the safety and health of tomorrow's automated workplace:

■ Technological systems and processes must be designed to take into account the physical, mental, and emotional needs of human workers.

■ Workers will need training and continual retraining so that they can effectively and efficiently operate technological systems and interact with them from the perspective of mastery rather than inadequacy.

■ Safety and health professionals, management, workers, psychologists, ergonomists, and practitioners of occupational medicine will have to work together as a team in all aspects of the safety and health program.

- The quality of work life and safety and health considerations will have to receive as much attention in the design and implementation of automated systems and processes as do economic and technological concerns.
- Additional research will have to be conducted to determine more clearly the psychological and physiological effects of human interaction with automated technologies.
- Much more comprehensive accident reporting will be needed. Implementation of the "critical incident" reporting system used in commercial aviation might be considered by companies for collecting safety and health data.
- **Ergonomists** should become involved in accident prevention. They should focus their accident prevention activities on accident/error analysis and simulation of accidents for training purposes.[41]

Tsuchiya sums up his thoughts regarding the future of the automated workplace as it relates to safety and health as follows:

> A motivating, satisfying and good quality job should be consistent with a safe, healthy and efficient automated workplace. Intermediate stages of new technologies sometimes lead to repetitive and monotonous tasks and these must be replanned to minimize adverse reactions to them. Where technology reduces the number of operators, isolation should be avoided for safety and to increase social contacts to reduce stress. When people can work or act together in small groups, even for a short period of the working day, human and productivity advantages can arise (e.g., group discussion of production activities). Where technologies or products are changing rapidly a workforce with good intercommunications can be even more important in keeping productivity high.[42]

SUMMARY

1. The introduction of automation in the workplace has had several different effects on workers including the following: reduced the amount of physical labor required, increased the amount of mental work required (in some cases), increased stress levels, and increased feelings of powerlessness and helplessness.
2. There is growing concern but little solid evidence over the potential negative effects of prolonged VDT use. Safety and health professionals are concerned about various musculoskeletal and visual problems that may be associated with VDT use.
3. Strategies for reducing the physiological and psychological problems associated with VDT use include the following: faster computer response time, more frequent breaks from VDT use or a work rotation schedule that allows non-VDT work to be interspersed in the daily routine, and work design that breaks up continual VDT use.
4. What makes robots potentially dangerous to humans are the following factors: their ability to acquire intelligence through programming, their flexibility and range of motion, their speed of movement, and their power.
5. Specific safety and health risks associated with robots include the following: being struck by a moving robot while inside the work envelope, entrapment between a moving robot and another machine, and being struck by a workpiece, tool, or other object dropped or ejected by a robot.

6. There are numerous strategies for minimizing the safety and health hazards of robots and other automated machines. They include the following: ensuring a well-lighted, glare-free robot site, maintaining good housekeeping around the robot site, keeping the robot site free of associated hazards, having fixed covers over the electrical and pneumatic components of the robot, keeping the work envelope clear of all nonessential objects, using appropriate lockout and test procedures, and removing maintenance tools and supplies from the work envelope before starting the robot.

7. Safety and health problems associated with office automation include eye fatigue, double images, complementary colors, headaches, shoulder fatigue, neck fatigue, dryness in the throat, sleepy feelings, whole body tiredness, and an unwillingness to talk.

8. Maladaptation to automated technologies manifests itself as an urge to quit work, fatigue, problems with human relations, a drop in work performance, social pathological phenomena such as drug use and crime, mood swings, a loss of motivation, and accidents.

9. Technological alienation is the state of mind that exists when workers resent the impact of new technologies on their lives. It is characterized by feelings of powerlessness, meaninglessness, and normlessness.

10. Mindlessness is the result of the "dumbing down" of the workplace so that workers are not required to use their minds in their work.

11. Problems associated with mindlessness include an increase in alcoholism, drug abuse, employee theft, work-related accidents, absenteeism, sick leave abuse, turnover rates, and employee personal problems.

12. The sociotechnical system theory consists of the following components: variance control, boundary location, work group organization, management support, design process, and quality of work life.

13. Problems that are likely to be associated with automation in the future include the following: the tendency to overlook safety and health precautions for short-term productivity gains may be exacerbated by increasingly intense international competition; the level of mental stress to which workers are subjected is likely to increase; new occupational diseases are likely to be introduced; and the tendency to "dumb down" the workplace is likely to continue.

14. Strategies for overcoming anticipated future problems include the following: better design of technological systems; training and continual retraining; teaming safety and health professionals with management, workers, psychologists, ergonomists, and practitioners of occupational medicine; more research; better accident reporting; and the involvement of ergonomists in accident prevention.

KEY TERMS AND CONCEPTS

Accommodation	Convergence
Automation	Design process
Boundary location	Dumbing down

Ergonomists
Eye strain
Fatigue
Helplessness
Human/robot interaction
Industrial medicine
Islands of automation
Lacrimation
Maladaptation
Management support
Manual work
Meaninglessness
Mechanization
Mindlessness
Musculoskeletal problems
Neck fatigue
Normlessness

Occupational diseases
Peopleless factory
Polarizing work
Powerlessness
Psychosocial stressors
Quality of work life
Shoulder fatigue
Social pathological phenomena
Sociotechnical system theory
Teach mode
Technological alienation
Variance control
Video display terminal (VDT)
Work design
Work envelope
Work group organization
Work stress

REVIEW QUESTIONS

1. Briefly summarize how automation has changed the workplace.
2. List five effects that automation of the workplace has had on workers.
3. What are the safety and health problems most widely associated with VDT use?
4. List three strategies for reducing the psychological and physiological problems associated with VDT use.
5. Explain the four factors that make robots more potentially dangerous than other machines.
6. There are specific hazards associated with human/robot interaction. List and explain them.
7. Explain four specific strategies for minimizing the hazards associated with interacting with a robot.
8. Name six safety and health problems widely associated with office automation.
9. Explain three of the principles set forth by the Japanese Association of Industrial Health for developing safety and health measures for office automation.
10. Define the term *maladaptation*. Explain how it manifests itself in workers.
11. List five strategies for minimizing the potential for occurrences of maladaptation.
12. Define the following automation-related terms: technological alienation, powerlessness, meaninglessness, normlessness, and mindlessness.
13. Mindlessness in the workplace can lead to a number of other problems. Name five of them.

14. Define the term *sociotechnical system theory.* Explain each of its six components.
15. Explain how you would ensure that a new robot system was safe for its operators and other workers.
16. What effect may increasingly intense international competition and the need to improve productivity have on workplace safety and health in the future?
17. What role may you play as a new safety and health professional in meeting the future challenges of an automated workplace?

=== ENDNOTES ===

1. Ebukuro, R. "Alleviation of the Impact of Microelectronics on Labour," in *Occupational Health and Safety in Automation and Robotics,* ed. K. Noro ((Chicago: National Safety Council, 1987), p. 11.
2. Ibid., p. 5.
3. Ostberg, O. "Emerging Technology and Trends in Blue-Color Stress," in *Occupational Health and Safety in Automation and Robotics,* ed. K. Noro (Chicago: National Safety Council, 1987), p. 15.
4. Ibid., p. 17.
5. Castelli, J. "NIOSH Releases Results of VDT Study," *Safety & Health,* June 1991, Vol. 143, No. 6, p. 69.
6. Ibid.
7. Ibid.
8. Business Publisher, Inc. *Occupational Health & Safety Letter,* March 20, 1991, Vol. 21, No. 6, p. 49.
9. Ostberg, O. "CRTs Pose Health Problems for Operators," *International Journal of Occupational Health and Safety,* January 1975, pp. 24–52.
10. Smith, M. J. "An Investigation of Health Complaints and Job Stress in Video Display Operations," *Human Factors,* Vol. 23, 1981, pp. 387–400.
11. Kurimoto, S., Tsuneto, I., Kageyu, N., Yamamoto, S., and Komatsubara, A. "Eye Strain in VDT Work from the Standpoint of Ergophthalmology," in *Occupational Health and Safety in Automation and Robotics,* ed. K. Noro (Chicago: National Safety Council, 1987), p. 112.
12. Ibid., pp. 112–33.
13. Ibid., p. 133.
14. Yamashita, T. "The Interaction Between Man and Robot in High Technology Industries," in *Occupational Health and Safety in Automation and Robotics,* ed. K. Noro (Chicago: National Safety Council, 1987), p. 140.
15. Martensson, L. "Interaction Between Man and Robots with Some Emphasis on 'Intelligent' Robots," in *Occupational Health and Safety in Automation and Robotics,* ed. K. Noro (Chicago: National Safety Council, 1987), p. 144.
16. National Safety Council. *Robots,* Data Sheet 1–717–85 (Chicago: National Safety Council, 1991), p. 1.
17. Ibid.
18. Ibid., pp. 2–3.

19. Nishiyama, K. "Introduction and Spread of VDT Work and Its Occupational Health Problems in Japan," in *Occupational Health and Safety in Automation and Robotics,* ed. K. Noro (Chicago: National Safety Council, 1987), p. 251.

20. Morooka, K., and Yamoda, S. "Multivariate Analysis of Fatigue on VDT Work," in *Occupational Health and Safety in Automation and Robotics,* ed. K. Noro (Chicago: National Safety Council, 1987), p. 236.

21. Amick, B, III. "The Impacts of Office Automation on the Quality of Worklife: Considerations for United States Policy," in *Occupational Health and Safety in Automation and Robotics,* ed. K. Noro (Chicago: National Safety Council, 1987), p. 232.

22. Ibid., p. 229.

23. Ibid., p. 223.

24. Ibid., pp. 261–62.

25. Oshima, M. "The Role of Industrial Medicine at the Man-Robot Interface," in *Occupational Health and Safety in Automation and Robotics,* ed. K. Noro (Chicago: National Safety Council, 1987), pp. 284–85.

26. Ibid., p. 284.

27. Ibid., p. 285.

28. Benson, G. "Mindlessness: A New Dimension of Technological Alienation—Implications for the Man-Machine Interface in High Technology Work Environments," in *Occupational Health and Safety in Automation and Robotics,* ed. K. Noro (Chicago: National Safety Council, 1987), pp. 326–27.

29. Ibid., p. 328.

30. Ibid.

31. Ibid., p. 332.

32. Ibid., p. 336.

33. Cherns, A. B. "Can Behavioral Science Help Design Organizations?" *Organizational Dynamics,* Spring 1977, pp. 44–64.

34. Hayashi, Y. "Measures for Improving the Occupational Health and Safety of People Working with VDTs or Robots—Small-Group Activities and Safety and Health Education," in *Occupational Health and Safety in Automation and Robotics,* ed. K. Noro (Chicago: National Safety Council, 1987), p. 383.

35. Ibid., p. 384.

36. Goto, M. "Occupational Safety and Health Measures Taken for the Introduction of Robots in the Automobile Industry," in *Occupational Health and Safety in Automation and Robotics,* ed. K. Noro (Chicago: National Safety Council, 1987), pp. 399–417.

37. Ibid., pp. 404–408.

38. Ibid., pp. 411–13.

39. Tsuchiya, K. "Summary Report on the Fifth University of Occupational and Environmental Health International Symposium," in *Occupational Health and Safety in Automation and Robotics,* ed. K. Noro (Chicago: National Safety Council, 1987), p. 422.

40. Ibid., pp. 422–26.

41. Ibid.

42. Ibid., pp. 425–26.

CHAPTER TWENTY-THREE

Ethics and Safety

Practically everyone agrees that business practices of industrial firms should be above reproach with regard to ethical standards. Few people are willing to defend unethical behavior. For the most part, industry in the United States operates within the scope of accepted legal and ethical standards. According to Peter Drucker,

> Business ethics is rapidly becoming the 'in' subject.... there are countless seminars on it, speeches, articles, conferences and books, not to mention the many earnest attempts to write "business ethics" into the law.[1]

However, unethical behavior does occur frequently enough that modern safety and health professionals should be aware of the types of ethical dilemmas that they may occasionally face and should know how to deal with such issues. How to deal successfully and effectively with ethics on the job is the subject of this chapter.

AN ETHICAL DILEMMA

According to Stead, Worrell, and Stead,

> Managing ethical behavior is one of the most pervasive and complex problems facing business organizations today. Employee's decisions to behave ethically or unethically are influenced by a myriad of individual and situational factors. Background, personality, decision history, man-

523

agerial philosophy, and reinforcement are but a few of the factors which have been identified by researchers as determinants of employee's behavior when faced with ethical dilemmas.[2]

Consider the following example of an ethical dilemma:

Mil-Tech Manufacturing Company is a Department of Defense contractor that produces air- and watertight aluminum containers for shipping nonnuclear munitions such as missiles, bombs, and torpedoes. Business has been good and Mil-Tech is prospering. However, the company's management team has a problem.

Mil-Tech has been awarded a contract to produce 10,000 boxes in six months. The company's maximum capacity is currently 1,000 boxes per month. Unless Mil-Tech can find a way to increase its capacity, the company will be forced to add new facilities, equipment, and personnel—an expensive undertaking that will quickly eat up the projected profits of the new contract.

The most time-consuming bottleneck in the production of the boxes is the painting process, the last step. The problem is with the paint that Mil-Tech uses. It poses no health, safety, or environmental hazards, but it is difficult to apply and requires at least two hours to dry. Clearly, the most expeditious way to increase productivity is to find a paint that is easier to apply and takes less time to dry.

The production manager has been searching frantically for a substitute paint for two weeks and has finally found one. The new paint is easy to apply, and it dries almost on contact. However, it is extremely toxic and can be dangerous to anyone exposed to it at any time before it dries. Personal protective equipment and other hazard-prevention techniques can minimize the health problems, but they must be used properly with absolutely no shortcuts. In addition, it is recommended that every employee who will work with the paint complete three full days of training.

Mil-Tech's management team is convinced that the union will not consent to the use of this paint even if the personal protective equipment is purchased and the training is provided. To complicate matters, the supplier of the paint cannot provide the training within a timeframe that meets Mil-Tech's needs. In a secret meeting, top management officials decide to purchase personal protective equipment, use the new paint, and forego the training. More importantly, the management team decides to withhold all information about the hazards associated with the new paint.

Camillo Garcia, Mil-Tech's safety and health manager, was not invited to the secret meeting. However, the decisions made during the meeting were slipped to him anonymously. Garcia now faces an ethical dilemma. What should he do? If he chooses to do nothing, Mil-Tech employees may be inappropriately exposed to an extremely hazardous substance. If he confronts the management team with what he knows, he could fall into disfavor or even lose his job. If he shares what he knows with union leaders, he might be called on to testify about what he knows. This is an example of the type of ethical dilemma that safety and health professionals face on the job.

ETHICS DEFINED

There are many definitions of the term *ethics*. However, no one definition has emerged as universally accepted. According to Paul Taylor, the concept can be defined as "inquiry into the nature and grounds of morality where morality is taken to mean moral judgments, standards, and rules of conduct."[3] According to Arlow and Ulrich, ethical dilemmas in the workplace are more complex than ethical situations in general.[4] They involve societal expectations, competition, and social responsibility as well as the potential consequences of an employee's behavior on customers, fellow workers, competitors, and the public at large. The result of the often-conflicting and contradictory interests of workers, customers, competitors, and the general public is a natural tendency for ethical dilemmas to occur frequently in the workplace.

Any time that ethics is the topic of discussion, such terms as **conscience, morality,** and **legality** are frequently heard. Although these terms are closely associated with ethics, they do not, by themselves, define it. For the purpose of this book, **ethics** is defined as follows:

> *Ethics is the study of morality.*

Morality refers to the **values** that are subscribed to and fostered by society in general and individuals within society. Ethics attempts to apply reason in determining rules of human conduct that translate morality into everyday behavior. **Ethical behavior** is that which falls within the limits prescribed by morality.

How, then, does a safety and health professional know if someone's behavior is ethical? Ethical questions are rarely black and white. They typically fall into a **gray area** between the two extremes of right and wrong. This gray area is often clouded further by personal experience, self-interest, point of view, and external pressure.

Guidelines for Determining Ethical Behavior

Guidelines are needed for safety and health professionals to use when trying to sort out matters that are not clearly right or wrong. First, however, it is necessary to distinguish between the concepts of *legal* and *ethical*. They are not the same thing. Just because an option is legal does not necessarily mean it is ethical.

In fact, it is not uncommon for people caught in the practice of questionable behavior to use the "I didn't do anything illegal" defense. A person's behavior can be well within the scope of the law and still be unethical. The following guidelines for determining ethical behavior assume that the behavior in question is legal (Figure 23–1):

Figure 23–1
Guidelines to determine what is ethical.

Guidelines for Ethical Choices

1. Apply the morning-after test
2. Apply the front-page test
3. Apply the mirror test
4. Apply the role-reversal test
5. Apply the common-sense test

- Apply the **morning-after test**. This test asks, "If you make this choice, how will you feel about it tomorrow morning?"
- Apply the **front-page test**. This test encourages you to make a decision that would not embarrass you if printed as a story on the front page of your hometown newspaper.
- Apply the **mirror test**. This test asks, "If you make this decision, how will you feel about yourself when you look in the mirror?"
- Apply the **role-reversal test**. This test requires you to trade places with the people affected by your decision and view the decision through their eyes.
- Apply the **common-sense test**. This test requires you to listen to what your instincts and common sense are telling you. If it feels wrong, it probably is.

Blanchard and Peale suggest their own test for deciding what the ethical choice is in a given situation.[5] Their test consists of the following three questions:

1. Is it legal?
2. Is it balanced?
3. How will it make me feel about myself?

If a potential course of action is not legal, no further consideration is in order. If an action is not legal, it is also not ethical. If an action is balanced, it is fair to all involved. This means that safety and health professionals and their team members have responsibilities that extend well beyond the walls of their unit, organization, and company. If a course of action is in keeping with your own moral structure, it will make you feel good about yourself. Blanchard and Peale also list the following "Five P's of Ethical Power":

- *Purpose.* Individuals see themselves as ethical people who let their conscience be their guide and in all cases want to feel good about themselves.
- *Pride.* Individuals apply internal guidelines and have sufficient self-esteem to make decisions that may not be popular with others.
- *Patience.* Individuals believe right will prevail in the long run, and they are willing to wait when necessary.
- *Persistence.* Individuals are willing to stay with an ethical course of action once it has been chosen and see it through to a positive conclusion.
- *Perspective.* Individuals take the time to reflect and are guided by their own internal barometer when making ethical decisions.[6]

These tests and guidelines will help safety and health professionals make ethical choices in the workplace. In addition to internalizing the guidelines themselves, safety and health professionals may want to share these values with all employees with whom they interact.

ETHICAL BEHAVIOR IN ORGANIZATIONS

Research by Trevino suggests that ethical behavior in organizations is influenced by both individual and social factors.[7] Trevino identified three personality measures that

---

SAFETY FACT

Models for Determining Ethical Behavior

In addition to the various tests that can be used for determining ethical behavior, there are also numerous models:

- Categorical imperative (black and white)
- Conventionalistic ethic (anything legal is ethical)
- Disclosure rule (explain actions to a wide audience)
- Doctrine of the mean (virtue through moderation)
- Golden Rule (do unto others . . .)
- Intuition rule (what is right is just known)
- Market ethic (whatever makes a profit is right)
- Means-end ethic (end justifies the means)
- Might-equals-right ethic (self explanatory)
- Organizational ethic (loyalty to the organization)
- Practical imperative (treat people as ends, not means)
- Equal freedom (full freedom unless it deprives another)
- Proportionality ethic (good outweighs the bad)
- Professional ethic (do only what can be explained to your peers)
- Revelation ethic (answers revealed by prayer)
- Rights ethic (protect rights of others)
- Theory of justice (impartial, even-handed)

---

can influence an employee's ethical behavior: (1) ego strength, (2) Machiavellianism, and (3) locus of control.

An employee's **ego strength** is his or her ability to undertake self-directed tasks and to cope with tense situations. A measure of a worker's **Machiavellianism** is the extent to which he or she will attempt to deceive and confuse others. **Locus of control** is the perspective of workers concerning who or what controls their behavior. Employees with an internal locus of control feel that they control their own behavior. Employees with an external locus of control feel that their behavior is controlled by external factors (e.g., rules, regulations, their safety and health professional, and so on).

Preble and Miesing suggest that social factors also influence ethical behavior in organizations.[8] These factors include gender, role differences, religion, age, work experience, nationality, and the influence of other people who are significant in an individual's life. Luthans and Kreitner feel that people learn appropriate behavior by observing the behavior of significant role models (parents, teachers, public officials, and so on).[9] Since safety and health professionals represent a significant role model for their team members, it is critical that they exhibit ethical behavior that is beyond reproach in all situations.

THE SAFETY AND HEALTH PROFESSIONAL'S ROLE IN ETHICS

Using the guidelines set forth in the previous section, safety and health professionals should be able to make responsible decisions concerning ethical choices. Unfortunately, deciding what is ethical is much easier than actually doing what is ethical. In this regard, trying to practice ethics is like trying to diet. It is not so much a matter of knowing you should cut down eating, it is a matter of following through and actually doing it.

It is this fact that defines the role of safety and health professionals with regard to ethics. Their role has three parts. First, they are responsible for setting an example of ethical behavior. Second, they are responsible for helping fellow employees make the right decision when facing ethical questions. Finally, safety and health professionals are responsible for helping employees follow through and actually undertake the ethical option once the appropriate choice has been identified. In carrying out their roles, safety and health professionals can adopt one of the following approaches (Figure 23–2): the best-ratio approach, the black-and-white approach, or the full-potential approach.

Best-Ratio Approach

The **best-ratio approach** is the pragmatic approach. Its philosophy is that people are basically good and under the right circumstances will behave ethically. However, under certain conditions, they can be driven to unethical behavior. Therefore, the safety and health professional should do everything possible to create conditions that promote ethical behavior and try to maintain the best possible ratio of good choices to bad. When hard decisions must be made, the appropriate choice is the one that will do the most good for the most people. This is sometimes referred to as *situational ethics*.

Figure 23–2
Three basic approaches to handling ethical problems.

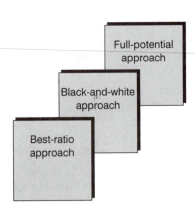

Full-potential approach

Black-and-white approach

Best-ratio approach

Black-and-White Approach

Using the **black-and-white approach**, right is right, wrong is wrong, and circumstances are irrelevant. The safety and health professional's job is to make ethical decisions and carry them out. It is also to help employees choose the ethical route. When difficult decisions must be made, safety and health professionals should make fair and impartial choices regardless of the outcome.

Full-Potential Approach

Safety and health professionals who use the **full-potential approach** make decisions based on how the outcomes will affect the ability of those involved to achieve their full potential. The underlying philosophy is that people are responsible for realizing their full potential within the confines of morality. Choices that can achieve this goal without infringing on the rights of others are considered ethical.

Decisions made might differ, depending on the approach selected. For example, consider the ethical dilemma presented at the beginning of this chapter. If the safety and health manager, Camillo Garcia, applies the best-ratio approach, he may decide to keep quiet, encourage the proper use of personal protective equipment, and hope for the best. On the other hand, if he takes the black-and-white approach, he will be compelled to confront the Mil-Tech management team with what he knows.

THE COMPANY'S ROLE IN ETHICS

Industrial firms have a critical role to play in promoting ethical behavior among their employees. Safety and health professionals cannot set ethical examples alone or expect

---

DISCUSSION CASE

What Is Your Opinion?

The city council members are in a real quandary about the request from International Plastics Corporation (IPC) to open a new plant on a 100-acre plot owned by the city. On the one hand, the city needs the new jobs that the IPC plant would bring—badly. High unemployment is the city's most serious problem, and every member of the city council ran on a job-creation platform in the last election. On the other hand, the council members have learned that IPC is not always a good corporate citizen in spite of its claims to the contrary.

Several cities with IPC plants have had problems enforcing their safety and health regulations. The consensus among other cities is that IPC officials say all of the right things until contracts are signed. Then, all of a sudden, they begin to procrastinate, stonewall, and break promises. Should the city council allow IPC to build the new plant? What is your opinion?

employees to behave ethically in a vacuum. A company's role in ethics can be summarized as (1) creating an internal environment that promotes, expects, and rewards ethical behavior; and (2) setting an example of ethical behavior in all external dealings (Figure 23–3).

Creating an Ethical Environment

A company creates an **ethical environment** by establishing policies and practices that ensure all employees are treated ethically and then enforcing these policies. Do employees have the right of due process? Do employees have access to an objective grievance procedure? Are there appropriate safety and health measures to protect employees? Are hiring practices fair and impartial? Are promotion practices fair and objective? Are employees protected from harassment based on race, sex, or other reasons? A company that establishes an environment that promotes, expects, and rewards ethical behavior can answer yes to all these questions.

One effective way to create an ethical environment is to develop an **ethics philosophy** and specific, written guidelines for implementing that philosophy that are shared with all employees. Martin Marietta Corporation of Orlando, Florida, has a *Code of Ethics and Standards of Conduct* that is shared with all employees. The code begins with the following statement of philosophy:

> Martin Marietta Corporation will conduct its business in strict compliance with applicable laws, rules, regulations, and corporate and operating unit policies, procedures and guidelines, with honesty and integrity, and with a strong commitment to the highest standards of ethics. We have a duty to conduct our business affairs within both the letter and the spirit of the law.[10]

This statement sets the tone for all employees at Martin Marietta. It lets them know that higher management not only supports ethical behavior, but also expects it. This approach makes it less difficult for safety and health professionals when they find themselves caught in the middle between the pressures of productivity and the maintenance of safe work practices.

In addition to its corporate ethics philosophy, Martin Marietta publishes a more specific *Credo and Code of Conduct*. The Martin Marietta credo is summarized as follows:

> In our daily activities we bear important obligations to our country, our customers, our owners, our communities, and to one another. We carry out these obligations guided by certain unifying principles:

Figure 23–3
Ethics cannot be practiced in a vacuum. The company has a critical role to play.

- Our foundation is INTEGRITY
- Our strength is our PEOPLE
- Our style is TEAMWORK
- Our goal is EXCELLENCE[11]

This **ethics credo** shows employees that they have obligations extending well beyond their work units and that how they perform their work can have an impact, negative or positive, on fellow employees, the company, customers, and the country. Key concepts set forth in the credo are *integrity, people, teamwork,* and *excellence.* Safety and health professionals who stress, promote, and model these concepts will make a major contribution to ethical behavior in the workplace.

Although the emphasis on ethics in the workplace is relatively new, the concept is not. According to Shanks, Robert Wood Johnson, the leader who built Johnson & Johnson into a major international corporation, developed an ethics credo for his company as early as the mid-1940s.[12] Johnson's credo read as follows:

- To customers and users: quality and service at reasonable prices
- To suppliers: a fair opportunity
- To employees: respect, equal opportunity, and a sense of job security
- To communities: a civic responsibility
- To the environment: protection
- To shareholders: a fair return[13]

Written philosophies and guidelines such as those developed by Martin Marietta Corporation and Johnson & Johnson are the first step in creating an ethical environment in the workplace. Safety and health professionals can play a key role in promoting ethical behavior on the job by encouraging higher management to develop written ethics philosophies/credos/guidelines and then by modeling the behavior that they encourage.

Setting an Ethical Example

Companies that take the "Do as I say, not as I do" approach to ethics will not succeed. Employees must be able to trust their company to conduct all external and internal dealings in an ethical manner. Companies that do not pay their bills on time; companies that pollute; companies that place short-term profits ahead of employee safety and health; companies that do not live up to advertised quality standards; companies that do not stand behind their guarantees; and companies that are not good neighbors in their communities are not setting a good ethical example. Such companies can expect employees to mimic their unethical behavior.

A final word on the company's role in ethics is in order. In addition to creating an ethical internal environment and handling external dealings in an ethical manner, companies must support safety and health professionals who make ethically correct decisions. This support must be given not just when such decisions are profitable, but in all cases. For example, in the ethical dilemma presented earlier in this chapter, say, Camillo Garcia decided that his ethical choice was to confront the management team with his knowledge of the hazards associated with the new paint. Management gave the order to withhold critical information. This is obviously the profitable choice in the short run. But is it the ethical choice? If Camillo Garcia does not think so, will Mil-Tech stand behind him? If not, everything else that the company does to promote ethics will fail.

HANDLING ETHICAL DILEMMAS

No person will serve long as a safety and health professional without confronting an ethical dilemma. How, then, should one proceed when confronting an ethical dilemma? There are three steps (Figure 23–4):

1. Apply the various guidelines for determining what is ethical presented earlier in this chapter.
2. Select one of the three basic approaches to handling ethical questions.
3. Proceed in accordance with the approach selected, and proceed with consistency.

Apply the Guidelines

In this step, as a safety and health professional, you should apply as many of the tests set forth in Figure 23–1 as necessary to determine the ethically correct decision. In applying these guidelines, attempt to block out all mitigating circumstances and other factors that tend to cloud the issue. At this point, the goal is only to identify the ethical choice. Deciding whether to implement the ethical choice comes in the next step.

Select the Approach

When deciding how to proceed after step 1, you have three basic approaches. These approaches, as set forth in Figure 23–2, are the best-ratio, black-and-white, and full-potential approaches. These approaches and their ramifications can be debated ad infinitum; however, selecting an approach to ethical questions is a matter of personal choice. Factors that will affect the ultimate decision include your personal makeup, the expectations of the company, and the degree of company support.

Proceeding with the Decision

The approach selected in step 2 will dictate how you should proceed as a safety and health professional. Two things are important in this final step. The first is to pro-

Figure 23–4
Handling ethical dilemmas.

Steps for Handling Ethical Dilemmas
1. Apply the guidelines
2. Select the approach
3. Proceed accordingly and consistently

ceed in strict accordance with the approach selected. The second is to proceed consistently. **Consistency** is critical when handling ethical dilemmas. Fairness is a large part of ethics, and consistency is a large part of fairness. The grapevine will ensure that all employees know how a safety and health professional handles an ethical dilemma. Some will agree, and some will disagree, regardless of the decision. Such is the nature of human interaction. However, regardless of the differing perceptions of the problem, employees will respect consistency. Conversely, even if the decision is universally popular, you may lose respect if the decision is not consistent with past decisions.

QUESTIONS TO ASK WHEN MAKING DECISIONS

Safety and health professionals often must make decisions that have ethical dimensions. A constant state of tension often exists between meeting production schedules and maintaining employee safety. Safety and health professionals usually are right in the middle of these issues. Following are some questions that can and should be asked by managers when making decisions about issues that have ethical dimensions. Safety and health professionals should ask these questions themselves, and they should encourage other decision makers within their organizations to do the same.

1. Has the issue or problem been thoroughly and accurately defined?
2. Have all dimensions of the problem (productivity, quality, cost, safety, health, and so on) been identified?
3. Would other stakeholders (employees, customers) agree with your definition of the problem?
4. What is your real motivation in making this decision? Meeting a deadline? Outperforming another organizational unit, or a competitor? Self-promoting? Getting the job done right? Protecting the safety and health of employees? Some combination of these?
5. What is the probable short-term result of your decision? What is the probable long-term result?
6. Who will be affected by your decision and in what way? In the short term? In the long term?
7. Did you discuss the decision with all stakeholders (or all possible stakeholders) before making it?
8. Would your decision withstand the scrutiny of employees, customers, colleagues, and the general public?

Safety professionals should ask themselves these questions, but equally important, they should insist that other managers do so. The manager responsible for meeting this month's production quota may be so focused on the numbers that he or she overlooks safety. The manager who is feeling the pressure to cut production costs may make decisions that work in the short term but have disastrous consequences in the long term. Questions such as those posed herein can help managers broaden their focus and consider the long-range impact when making decisions.

DISCUSSION CASE

What Is Your Opinion?

Following are two ethical dilemmas that a safety and health professional may face. What is the right thing to do in each case? What is your opinion?

- You are the safety director for West Coast Power Company. Your son is a lineman for a branch of the company that is located in another city. He is visiting you while recuperating from a back injury for which he is collecting workers' compensation. While visiting, he jogs, lifts weights, and plays softball with friends. You finally realize that he is not really injured. What should you do?

- The manufacturing director and the union representative have just had a chin-to-chin argument about removing the new machine guards from the milling machines. They have asked you—the company's safety director—to mediate the dispute. According to the manufacturing director, "We ran these machines for five years without an accident. The only reason we put them on was because some OSHA inspector suggested it. They're fine when we are not in a hurry, but they slow us down when the rush is on. Unless we remove the guards, this job will not be shipped on time." The union representative counters by saying, "These machines are dangerous." What is the right thing to do here? What is your opinion?

SUMMARY

1. Ethics is the study of morality. Morality refers to the values that are subscribed to and fostered by society. Ethics attempts to apply reason in determining rules of human conduct that translate morality into everyday behavior.
2. Ethical behavior is that which falls within the limits prescribed by morality.
3. *Legal* and *ethical* are not the same. If something is illegal, it is also unethical. However, just because something is legal does not mean that it is ethical. An act can be legal but unethical.
4. To determine if a choice is ethical, you can apply the following tests: morning after, front page, mirror, role reversal, and common sense.
5. Safety and health professionals have a three-pronged role with regard to ethics. They are responsible for setting an ethical example, helping employees to identify the ethical choices when facing ethical questions, and helping employees to follow through and actually undertake the ethical option.
6. Safety and health professionals have three approaches available in handling ethical dilemmas: best ratio, black and white, and full potential.
7. The company's role in ethics is to create an ethical environment and to set an ethical example. An effective way is to develop a written ethics philosophy and share it with all employees.

8. Three personality characteristics that can influence an employee's ethical behavior are ego strength, Machiavellianism, and locus of control.
9. People facing ethical dilemmas should apply the tests for determining what is ethical, select one of the three basic approaches, and proceed consistently.

========== KEY TERMS AND CONCEPTS ==========

Best-ratio approach   Legality
Black-and-white approach   Locus of control
Common-sense test   Machiavellianism
Conscience   Mirror test
Consistency   Morality
Ego strength   Morning-after test
Ethical behavior   Patience
Ethical environment   Persistence
Ethics   Perspective
Ethics credo   Pride
Ethics philosophy   Purpose
Front-page test   Role-reversal test
Full-potential approach   Values
Gray area

========== REVIEW QUESTIONS ==========

1. Define the term *morality*.
2. Define the term *ethics*.
3. Briefly explain each of the following ethics tests: morning-after test, front-page test, mirror test, role-reversal test, and common-sense test.
4. What is the safety and health professional's role with regard to ethics?
5. Briefly explain the following approaches to handling ethical behavior: best-ratio approach, black-and-white approach, and full-potential approach.
6. Briefly explain a company's role with regard to ethics.
7. Explain how one should proceed when facing an ethical dilemma.
8. Write a brief ethics philosophy for a chemical company.
9. List the individual and social factors that may influence an employee's ethical behavior.
10. List and briefly describe the "Five P's of Ethical Power" as set forth by Blanchard and Peale.
11. What questions should safety and health professionals ask when making decisions that have an ethical component?

════════ ENDNOTES ════════

1. Drucker, P. F. "What Is Business Ethics?" *Across the Board*, October 1981, pp. 22–32.
2. Stead, E. W, Worrell, D. L., and Stead, J. G. "An Integrative Model for Understanding and Managing Ethical Behavior in Business Organizations," *Journal of Business Ethics*, Vol. 5, No. 9, September 1990, p. 233.
3. Taylor, P. *Principles of Ethics: An Introduction* (Encino, CA: Dickson Publishing Company, 1975), p. 96.
4. Arlow, P. and Ulrich, T. A. "Business Ethics, Social Responsibility, and Business Students: An Empirical Comparison of Clark's Study," *Akron Business and Economic Review,* Vol. 4, No. 3, March 1980, pp. 17–23.
5. Blanchard, K., and Peale, N. V. *The Power of Ethical Management* (New York: Ballantine Books, 1988), pp. 10–17.
6. Ibid., p. 79.
7. Trevino, L. K. "Ethical Decision Making in Organizations: A Person-Situation Interactionist Model," *Academy of Management Review,* Vol. 11, No. 3, 1986, pp. 601–17.
8. Preble, J. F., and Miesing, P. "Do Adult MBA and Undergraduate Business Students Have Different Business Philosophies?" *Proceedings of the National Meeting of the American Institute for the Decision Sciences,* November 1984, pp. 346–48.
9. Luthans, F., and Kreitner, R. *Organizational Behavior Modification and Beyond: An Operant and Social Learning Approach* (Glenview, IL: Scott Foresman, 1985).
10. Martin Marietta Corporation. *Code of Ethics and Standards of Conduct* (Orlando, FL: Corporate Ethics Office, 1990), inside front cover.
11. Ibid., p. 2.
12. Shanks, D. C. "The Role of Leadership in Strategy Development," *Journal of Business Strategy*, January/February 1989, p. 32.
13. Ibid., p. 33.

Bloodborne Pathogens in the Workplace

Acquired immunodeficiency syndrome, or **AIDS,** has become one of the most difficult issues that safety and health professionals are likely to face today. It is critical that they know how to deal properly and appropriately with this controversial disease. The major concerns of safety and health professionals with regard to AIDS are knowing the facts about AIDS; knowing the legal concerns associated with AIDS; knowing their role in AIDS education and related employee counseling; and knowing how to ease unfounded fears concerning the disease while simultaneously taking the appropriate steps to protect employees from infection. In addition to AIDS, the modern safety and health manager must be concerned with other bloodborne pathogens including human immunodeficiency virus (HIV) and hepatitis B (HBV).

FACTS ABOUT AIDS[1]

Consider the following facts about HIV and AIDS:

- As of the fall of 1993, almost 340,000 Americans had been diagnosed with AIDS, and almost 205,000 of them had died. By the fall of 1994, the number diagnosed had risen to over 400,000, and the number of fatalities had risen to more than 217,000.

■ Worldwide, almost 990,000 cases of AIDS had been reported to the World Health Organization (WHO) by June 1994. However, WHO officials estimate that the actual number of people stricken with AIDS was more than 4 million in 1994, and that this number increases by 37 percent every year.[2]

AIDS is feared, misunderstood, and controversial. Modern safety and health professionals need to know the facts about AIDS and be prepared to use these facts to make the workplace safer.

SYMPTOMS OF AIDS

AIDS and various related conditions are caused when humans become infected with the **human immunodeficiency virus**, or **HIV**. This virus attacks the human immune system, rendering the body incapable of repelling disease-causing microorganisms. Symptoms of the onset of AIDS are as follows:

■ Enlarged lymph nodes that persist

■ Persistent fevers

■ Involuntary weight loss

■ Fatigue

■ Diarrhea that does not respond to standard medications

■ Purplish spots or blotches on the skin or in the mouth

■ White, cheesy coating on the tongue

■ Night sweats

■ Forgetfulness

How AIDS Is Transmitted

The HIV virus is transmitted in any of the following three ways: (1) **sexual contact**, (2) blood contact, and (3) mother-to-child during pregnancy or childbirth. Any act in which **body fluids** are exchanged can result in infection if either partner is infected. The following groups of people are at the highest level of risk with regard to AIDS: (1) homosexual men who do not take appropriate precautions; (2) **IV drug users** who share needles; (3) people with a history of multiple **blood transfusions** or blood-product transfusions, such as hemophiliacs; and (4) sexually promiscuous people who do not take appropriate precautions.

How AIDS Is Not Transmitted

There is a great deal of misunderstanding about how AIDS is transmitted. This can cause inordinate fear among fellow employees of **HIV-positive** workers. Safety and health professionals should know enough about AIDS transmission so that they can reduce employees' fears about being infected through casual contact with an HIV-positive person.

Occupational Health and Safety magazine provides the following clarifications concerning how AIDS is *not* transmitted:

> AIDS is a blood-borne, primarily sexually transmitted disease. It is not spread by casual social contact in schools, workplaces, public washrooms, or restaurants. It is not spread via handshakes, social kissing, coughs, sneezes, drinking fountains, swimming pools, toilet facilities, eating utensils, office equipment, or by being next to an infected person.
>
> No cases of AIDS have been reported from food being either handled or served by an infected person in an eating establishment.
>
> AIDS is not spread by giving blood. New needles and transfusion equipment are used for every donor.
>
> AIDS is not spread by mosquitoes or other insects.
>
> AIDS is not spread by sexual contact between uninfected individuals—whether homosexual or heterosexual—if an exclusive sexual relation has been maintained.[3]

AIDS IN THE WORKPLACE

The first step in dealing with AIDS at the company level is to develop a comprehensive **AIDS policy**. Safety and health professionals should be part of the team that drafts the initial policy and updates an existing policy. If a company has no AIDS policy, the safety and health professional should encourage the company to develop one. In all likelihood, most companies won't take much convincing.

According to Peter Minetos,

> Industry is doing its part to eliminate any unnecessary fear: Nearly half of companies surveyed offer their employees literature or other materials to keep them informed on the disease; more than half have an Employee Assistance Program (EAP) to deal with emotional problems concerning AIDS; two-thirds of those who have not yet addressed AIDS with employees plan to do so in the future.[4]

AIDS is having a widely felt impact in the workplace, particularly on employers. According to Minetos, employers are feeling the impact of AIDS in increased insurance premiums and health-care costs, time-on-the-job losses, decreased productivity, AIDS-related lawsuits, increased stress, and related problems that result from misconceptions about AIDS.[5]

The starting point for dealing with AIDS in the workplace is the development of a company policy that covers AIDS and other bloodborne pathogens. The policy should cover the following areas at a minimum: employee rights, testing, and education (Figure 24–1).

Employee Rights

An AIDS policy should begin by spelling out the rights of employees who have tested positive for the disease. The report *National Academy of Sciences Confronting AIDS:*

Figure 24–1

Components of a corporate policy for bloodborne pathogens.

- Employee rights
- Testing
- Education

Update, 1988 makes the following recommendations for developing the **employee rights** aspects of an AIDS policy:

- Treat HIV-positive employees compassionately, allowing them to work as long as they are able to perform their jobs.

- Develop your company's AIDS policy and accompanying program before learning that an employee is HIV positive. This will allow the company to act instead of having to react.

- Make reasonable allowances to accommodate the HIV-positive employee. The U.S. Supreme Court has recognized AIDS as a handicapping condition. Consequently, reasonable allowances must include modified work schedules and special adaptations to the work environment.

- Ensure that HIV-positive employees have access to private health insurance that covers the effects of AIDS. Also, work with state and federal government insurance providers to gain their support in helping to cover the costs of health care for HIV-positive employees.

- Include provisions for evaluating the work skills of employees to determine if there has been any degradation of ability caused by the disease.[6]

Testing

According to the Centers for Disease Control, there is no single test that can reliably diagnose AIDS.[7] However, there is a test that can detect antibodies produced in the blood to fight the virus that causes AIDS. The presence of these antibodies does not necessarily mean that a person has AIDS. According to the Centers for Disease Control,

> Presence of HTLV-III antibodies [now called HIV antibodies] means that a person has been infected with that virus . . . The antibody test is used to screen donated blood and plasma and assist in preventing cases of AIDS resulting from blood transfusions or use of blood products, such as Factor VIII, needed by patients with hemophilia. For people who think they may be infected and want to know their health status, the test is available through private physicians, most state or local health departments and at other sites. Anyone who tests positive should be considered potentially capable of spreading the virus to others.[8]

Whether a company can, or even should, require AIDS tests of employees or potential employees is widely debated. The issue is contentious and controversial. However, there is a growing body of support for mandatory testing. The legal and ethical concerns surrounding this issue are covered in the next section. The testing component of a company's AIDS policy should take these concerns into account.

Education

The general public is becoming more sophisticated about AIDS. People are beginning to learn how AIDS is transmitted. However, research into the causes, diagnosis, treatment, and prevention of this disease is ongoing. The body of knowledge changes continually. Consequently, it is important to have an ongoing education program to keep employees up to date and knowledgeable. According to Minetos,

> AIDS education campaigns can be conducted in many forms. Literature, slide shows, and video presentations are all communication vehicles. Presentations by health professionals are one of the most popular and effective methods of communicating information on AIDS. The primary purpose of each is to convey basic knowledge and, subsequently, eliminate unnecessary fear among co-workers.[9]

Once a comprehensive AIDS policy has been developed and shared with all employees, a company has taken the appropriate and rational approach for dealing with this deadly and controversial disease. If an employer has not yet taken this critical step, safety and health professionals should encourage such action immediately. It is likely that most companies either employ now, or will employ in the future, HIV-positive employees. A poll conducted by *U.S. News and World Report* found that 48 percent of the companies responding indicated that AIDS was a concern.[10]

Minetos writes that "AIDS is having an effect on the workplace. Yet . . . only 5 percent of all employers have a written corporate policy on AIDS. For a disease that according to some estimates is reaching epidemic proportions, this is an extremely low percentage."[11] Clearly, one of the major challenges facing safety and health professionals is convincing their employers to develop an AIDS policy.

LEGAL CONCERNS

There are legal considerations relating to AIDS in the workplace with which safety and health professionals should be familiar. They grow out of several pieces of federal legislation, including the Rehabilitation Act of 1973, the Occupational Safety and Health Act of 1970, and the Employee Retirement Income System Act of 1974.

The **Rehabilitation Act of 1973** was enacted to give protection to people, including workers, with handicaps. Section 504 of the act makes discrimination on the basis of a handicap unlawful. Any agency, organization, or company that receives federal funding falls within the purview of the act. Such entities may not discriminate against individuals who are handicapped but **otherwise qualified**. Through various court actions, this concept has been well defined. A person with a handicap is "otherwise qualified" when he or she can perform what the courts have described as the **essential functions** of the job.

When the handicap that a worker has is a contagious disease such as AIDS, it must be shown that there is no significant risk of the disease being transmitted in the workplace. If there is a significant risk, the infected worker is not considered otherwise qualified. Employers and the courts must make these determinations on a case-by-case basis.

Another concept associated with the Rehabilitation Act is the concept of **reasonable accommodation**. In determining if a worker with a handicap can perform the essential functions of a job, employers are required to make reasonable accommodations to help the worker. This concept applies to workers with any type of handicapping condition including a communicable disease such as AIDS. What constitutes reasonable accommodation, just as what constitutes otherwise qualified, must be determined on a case-by-case basis.

The concepts growing out of the Rehabilitation Act of 1973 give the supervisor added importance when dealing with AIDS-infected employees. The supervisor's knowledge of the various jobs in his or her unit will be essential in helping company officials make an otherwise qualified decision. The supervisor's knowledge that AIDS is transmitted only by exchange of body fluids coupled with his or her knowledge of the job tasks in question will be helpful in determining the likelihood that AIDS may be transmitted to other employees. Finally, the supervisor's knowledge of the job tasks in question will be essential in determining what constitutes reasonable accommodation and what the actual accommodations should be. Therefore, it is critical that safety and health professionals work closely with supervisors and educate them in dealing with AIDS in the workplace.

In arriving at what constitutes reasonable accommodation, employers are not required to make fundamental changes that alter the nature of the job or result in undue costs or administrative burdens. Clearly, good judgment and a thorough knowledge of the job are required when attempting to make reasonable accommodations for an AIDS-infected employee. Safety and health professionals should involve supervisors in making such judgments.

The Occupational Safety and Health Act of 1970 (**OSHAct**) requires that employers provide a safe workplace free of hazards. The act also prohibits employers from retaliating against employees who refuse to work in an environment they believe may be unhealthy (Section 654). This poses a special problem for employers of AIDS-infected employees. Other employees may attempt to use Section 654 of the OSHAct as the basis for refusing to work with such employees. For this reason, it is important that companies educate their employees about AIDS and how it is transmitted. If employees know how AIDS is transmitted, they will be less likely to exhibit an irrational fear of working with an infected colleague.

Even when a comprehensive AIDS education program is provided, employers should not automatically assume that an employee's fear of working with the infected individual is irrational. Employers have an obligation to treat each case individually. Does the complaining employee have a physical condition that puts him or her at greater risk of contracting AIDS than other employees? If so, that employee's fears may not be irrational. However, a fear of working with an AIDS-infected co-worker is usually irrational, making it unlikely that Section 654 of the OSHAct could be used successfully as the basis of a refusal to work.

The Employee Retirement Income Security Act (**ERISA**) of 1974 protects the benefits of employees by prohibiting actions taken against them based on their eligibility for benefits. This means that employers covered by ERISA cannot terminate an employee with AIDS or who is suspected of having AIDS as a way of avoiding expensive medical costs. With ERISA, it is irrelevant whether the employee's condition is considered a handicap since the act applies to all employees regardless of condition.

The Testing Issue

Perhaps the most contentious legal concern growing out of the AIDS controversy is the issue of testing. Writing in the *AAOHN Journal*, Beatrice Crofts Yorker says,

> Few topics have generated the amount of controversy that currently exists in the area of testing for Acquired Immune Deficiency Syndrome (AIDS). Proponents and opponents have strong arguments, often based on emotional reactions to this deadly epidemic. In the workplace, the issues of AIDS testing are very specific and have implications for health policies in occupational settings. Few clear laws or statutes specifically regulate AIDS testing in the workplace.[12]

Issues regarding testing for AIDS and other diseases with which safety and health professionals should be familiar are summarized as follows (see Figure 24–2):

1. *State laws.* Control of communicable diseases is typically considered to be the province of the individual state. In response to the AIDS epidemic, several states have

Figure 24–2
Disease testing issues.

Health and safety professionals should be familiar with how the following factors might affect the issue of AIDS testing at their companies:

- Applicable state laws
- Applicable federal laws and regulations
- Case law from civil suits
- Company policy

passed legislation. Some states prohibit the use of pre-employment AIDS tests to deny employment to infected individuals. Because of the differences among states with regard to AIDS-related legislation, safety and health professionals should familiarize themselves with the laws of the state in which their company is located.

2. *Federal laws and regulations.* The laws protecting an individual's right to privacy and due process apply to AIDS testing. These laws fall within the realm of constitutional law. They represent the primary federal contribution to the testing issue.

3. *Civil suits.* **Case law** serves the purpose of establishing precedents that can guide future decisions. One precedent-setting case was taken all the way to the Supreme Court (*School Board of Nassau County v. Arline,* 1987), where it was decided that an employer cannot discriminate against an employee who has a communicable disease.

4. *Company policy.* It was stated earlier that companies should have an AIDS policy that contains a testing component. This component should include at least the following: a strong rationale, procedures to be followed, employee groups to be tested, the use and dissemination of results, and the circumstances under which testing will be done. Safety and health professionals should be knowledgeable about their company's policy and act in strict accordance with it.[13]

On one side of the testing controversy are the issues of fairness, accuracy, and **confidentiality** or, in short, the rights of the individual. On the other side of the controversy are the issues of workplace safety and public health. Individual rights' proponents ask such questions as: What tests will be used? How do test results relate to the maintenance of workplace safety? How will test results be used and who will see them? Workplace safety proponents ask such questions as: What is the danger of transmitting the disease to other employees? Can the safety of other workers be guaranteed?

Bayer, Levine, and Wolf recommend the following guidelines for establishing testing programs that satisfy the concerns of both sides of the issue:

- The purpose of screening must be ethically acceptable.
- The means to be used in the screening program and the intended use of the information must be appropriate for accomplishing the purpose.
- High-quality laboratory services must be used.
- Individuals must be notified that the screening will take place.
- Individuals who are screened must have a right to be informed of the results.
- Sensitive and supportive counseling programs must be available before and after screening to interpret the results, whether they be positive or negative.
- The confidentiality of screened individuals must be protected.[14]

Facts About Testing for AIDS and Other Diseases

In addition to understanding the legal concerns associated with disease testing, safety and health professionals should also be familiar with the latest facts about AIDS tests and testing. A concerned employee might ask for recommendations concerning AIDS testing.

Ensuring the accuracy of an **HIV antibody test** (there is no such thing as an AIDS test) requires two different tests, one for initial screening and one for confirmation. The **screening test** currently used is the enzyme linked immunosorbent assay, or **ELISA test**. The **confirmation test** is the immuno-florescent (**IFA**), or the **Western Blot test**. The ELISA test is relatively accurate, but it is susceptible to both false positive and false negative results. A **false positive** test is one that shows the presence of HIV antibodies when in reality no such antibodies exist. A **false negative** result is one that shows no HIV antibodies in people who in reality are infected. A negative result indicates that no infection exists at the time of the test. A confirmed positive result indicates that HIV antibodies are present in the blood.[15]

The American College Health Association makes the following recommendations concerning the HIV antibody test:

1. The test is not a test for AIDS, but a test for antibodies to HIV, the virus that can cause AIDS.
2. Talk to a trained, experienced, sensitive counselor before deciding whether to be tested.
3. If you decide to be tested, do so *only* at a center that provides both pre- and post-test counseling.
4. If possible, use an *anonymous* testing center.
5. Be sure that the testing center uses two ELISA tests and a Western Blot or IFA test to confirm a positive result.
6. A positive test result is *not* a diagnosis of AIDS. It does mean you have HIV infection and that you should seek medical evaluation and early treatment.
7. A positive test result *does* mean that you can infect others and that you should avoid risky or unsafe sexual contact and IV needle sharing.
8. It can take six months (and—although rarely—sometimes even longer) after infection to develop antibodies, so the test result may not indicate whether you have been infected during that period.
9. A negative test result *does not* mean that you are immune to HIV or AIDS, or that you cannot be infected in the future.[16]

Safety and health professionals need to share this type of information with employees who ask about AIDS tests and testing. Employees who need more detailed information should be referred to a health-care professional.

AIDS EDUCATION

The public is becoming more knowledgeable about AIDS and how the disease is spread. However, this is a slow process, and AIDS is a complex and controversial disease. Unfortunately, many people still respond to the disease out of ignorance and inaccurate information. For this reason, it is imperative that a company's safety and health program include an AIDS education program.

A well-planned **AIDS education program** can serve several purposes: (1) It can give management the facts needed to develop policy and make informed decisions with regard to AIDS; (2) it can result in changes in behavior that will make employees less likely to contract or spread the disease; (3) it can prepare management and employees to respond appropriately when a worker falls victim to the disease; and (4) it can decrease the likelihood of legal problems resulting from an inappropriate response to an AIDS-related issue. Consequently, safety and health professionals should be prepared to participate in developing AIDS education programs.

Planning an AIDS Education Program

The first step in planning an AIDS education program is to decide its purpose. A statement of purpose for an AIDS education program should be a broad conceptual declaration that captures the company's reason for providing the education program. Following is an example:

> The purpose of this AIDS education program is to deal with the disease in a positive proactive manner that is in the best interests of the company and its employees.

The next step in the planning process involves developing goals that translate the statement of purpose into more specific terms. The goals should tell specifically what the AIDS education program will do. Sample goals are as follows:

■ The program will change employee behaviors that might otherwise promote the spread of AIDS.

■ The program will help the company's management team to develop a rational, appropriate AIDS policy.

■ The program will help managers to make responsible decisions concerning AIDS issues.

■ The program will help employees to protect themselves from the transmission of AIDS.

■ The program will alleviate the fears of employees concerning working with an AIDS-infected co-worker.

■ The program will help managers to respond appropriately and humanely to the needs of AIDS-infected workers.

Once goals have been set, a program is developed to meet the goals. The various components of the program must be determined. These components may include confidential one-on-one counseling, referral, posters, a newsletter, classroom instruction, self-paced multimedia instruction, group discussion sessions, printed materials, or a number of other approaches. Figure 24–3 is a suggested outline for a course on AIDS.

COUNSELING INFECTED EMPLOYEES

The employee who learns that he or she has AIDS will be angry, frightened, and confused. Safety and health professionals who are confronted by such an employee should proceed as follows:

■ Listen.

■ Maintain a nonjudgmental attitude.

Figure 24–3
Course outline for AIDS education course.

> **Statement of Purpose**
>
> The purpose of this course is to give employees the knowledge they need to deal with AIDS in a positive, proactive manner.
>
> **Major Topics**
>
> - What is AIDS?
> - What causes AIDS?
> - How is AIDS transmitted?
> - Who is most likely to get AIDS?
> - What are the symptoms of AIDS?
> - How is AIDS diagnosed?
> - Who should be tested for AIDS?
> - Where can I get an AIDS test?
> - How can I reduce my chances of contracting AIDS?
> - How is AIDS treated?
> - Can AIDS be prevented?
> - What are common myths about AIDS?

- Make the employee aware of the company's policy on AIDS.
- Respond in accordance with company policy.

Listen carefully as you would with an employee who comes to you with any problem. If you must ask a question for clarification, do so, but be objective, professional, and nonjudgmental. Make the employee aware of the company's policy on AIDS and respond in strict accordance with the policy.

The U.S. Public Health Service recommends the following steps for persons who have determined they are HIV positive.

- Seek regular medical evaluation and follow-up.
- Either avoid sexual activity or inform your prospective partner of your antibody test results and protect him or her from contact with your body fluids during sex. (Body fluids include blood, semen, urine, feces, saliva, and women's genital secretions.) Use a condom, and avoid practices that may injure body tissues (for example, anal intercourse). Avoid oral-genital contact and open-mouthed intimate kissing.
- Inform your present and previous sex partners, and any persons with whom needles may have been shared, of their potential exposure to HTLV-III [HIV] and encourage them to seek counseling and antibody testing from their physician or at appropriate health clinics.
- Don't share toothbrushes, razors, or other items that could become contaminated with blood.
- If you use drugs, enroll in a drug treatment program. Needles and other drug equipment must never be shared.

- Don't donate blood, plasma, body organs, other body tissue, or sperm.
- Clean blood or other body fluid spills on household or other surfaces with freshly diluted household bleach—1 part bleach to 10 parts water. (Don't use bleach on wounds.)
- Inform your doctor, dentist, and eye doctor of your positive HTLV-III status so that proper precautions can be taken to protect you and others.
- Women with a positive antibody test should avoid pregnancy until more is known about the risks of transmitting HTLV-III from mother to infant.[17]

Safety and health professionals can pass these recommendations along to employees who have contracted the virus. Any information that is requested beyond this should be provided by a qualified professional. Safety and health professionals should be prepared to make appropriate referrals. The Public Health Service's AIDS hotline number is: 1-800-342-AIDS.

Employee Assistance Programs

Company-sponsored **employee assistance programs (EAPs)** should have an AIDS component so that employees can seek confidential advice and **counseling** about the disease. EAPs may provide on-site services or contract for them through a private organization or agency. In either case, confidential counseling, referral, the provision of AIDS-related information, seminars, and other forms of assistance should be provided. From an employee assistance perspective, AIDS should be treated like other health problems such as stress, depression, substance abuse, and so on.[18]

EASING EMPLOYEES' FEARS ABOUT AIDS

During his first inaugural address on March 4, 1933, in the depths of the Great Depression, President Franklin D. Roosevelt said, "So first let me assert my firm belief that the only thing we have to fear is fear itself."[19] This is a message that safety and health professionals should be prepared to spread among employees. Although the level of sophistication concerning AIDS is increasing, people tend to react to the disease from an emotional perspective. Typically, their fears are unfounded and are more likely to cause them problems than does AIDS.

Fear, panic, and even hysteria are all common reactions to AIDS. Fellow employees are likely to respond this way when they discover that a co-worker has AIDS. Consequently, safety and health professionals need to know how to ease the fears and misconceptions associated with AIDS. According to Brown and Turner, the following strategies will help:

- Work with higher management to establish an AIDS education/awareness program that covers the following topics at a minimum: (1) how HIV is transmitted, (2) precautions that workers can take, and (3) concerns about AIDS testing.
- Conduct group round-table discussions that allow employees to express their concerns.
- Correct inaccuracies, rumors, and misinformation about AIDS as soon as they occur.[20]

PROTECTING EMPLOYEES FROM AIDS

Safety and health professionals should be familiar with the precautions that will protect employees from HIV infection on and off the job. OSHA's guidelines for preventing exposure to HIV infection identify three categories of work-related tasks: Categories I, II, and III. Jobs that fall into Category I involve routine exposure to blood, body fluids, or tissues that might be HIV-infected. Category II jobs do not involve routine exposure to blood, body fluids, or tissues, but some aspects of the job may involve occasionally performing Category I tasks. Category III jobs do not normally involve exposure to blood, body fluids, or tissues.

Most industrial occupations fall into Category III, meaning there is very little risk of contracting AIDS on the job. However, regardless of the category of their job, employees should know how to protect themselves, and safety and health professionals should be prepared to tell them how.

The U.S. Public Health Service recommends the following precautions for reducing the chances of contracting AIDS:

- Abstain from sex or have a **mutually monogamous marriage/relationship** with an infection-free partner.
- Refrain from having sex with multiple partners or with a person who has multiple partners. The more partners one has, the greater the risk of infection.
- Avoid sex with a person who has AIDS or who you think might be infected. However, if you choose not to take this recommendation, the next logical course is to take precautions against contact with the infected person's body fluids (blood, semen, urine, feces, saliva, and female genital secretions).
- Do not use **intravenous drugs** or, if you do, do not share needles.[21]

Safety and health professionals should make sure that all employees are aware of these common precautions. They should be included in the company-sponsored AIDS education program, available through the company's EAP, and posted conspicuously for employee reading.

CPR and AIDS

It is not uncommon for an employee to be injured in a way that will require resuscitation. Consequently, many companies provide employees with CPR training. But what about AIDS? Is CPR training safe? Writing for *Safety and Health*, Martin Eastman said,

> In the early 1960s when CPR training procedures were being developed, some thought was given to the cleaning and disinfection of manikins. This was before acquired immune deficiency syndrome (AIDS) and hepatitis-B became headline diseases. In the fearful climate that has followed the spread of these diseases, practices that once appeared merely unsanitary now seem truly life-threatening.[22]

The HIV virus has been found in human saliva. Because CPR involves using your fingers to clear the airway and placing your mouth over the victim's, there is concern about contracting AIDS while trying to resuscitate someone. Although there is no hard evidence that HIV can be transmitted through saliva, there is some legitimacy to the concern.

Because of the concern, **disposable face masks** and various other types of personal protective devices are now being manufactured. Typical of these devices is the rescue key ring produced by Ambu, Inc., of Hanover, Maryland.[23] The ring contains a face mask made of a transparent film material that has a one-way valve. The valve prevents the passage of potentially contaminated body fluids from the victim to the rescuer.

Safety and health professionals should ensure that such devices are used in both training and live situations involving CPR. These devices should be readily available in many easily accessible locations throughout the company.

HEPATITIS B (HBV) IN THE WORKPLACE

Although the spread of HIV receives more attention, a greater risk is from the spread of **Hepatitis B.** This bloodborne virus averages approximately 300,000 new cases per year compared with AIDS which averages approximately 35,000 new cases. The Hepatitis B virus is extremely strong as compared with HIV. For example, it can live on surfaces for up to a week if it is exposed to air. Hepatitis B is also much more concentrated than HIV.

Hepatitis B is caused by a double-shelled virus **(HBV).** It can be transmitted in the workplace in the following ways:

- Contact with blood
- Contact with bodily fluids including tears, saliva, and semen

The Hepatitis B virus can live in bodily fluids for years. Carriers of the virus are at risk themselves, and they place others at risk. Persons infected with HBV may contract chronic hepatitis, cirrhosis of the liver, and/or primary heptocellular carcinoma. An HBV-infected individual is more than 300 times more likely to develop primary liver cancer than is a noninfected individual from the same environment. Unfortunately, it is possible to be infected and not know it because the symptoms can vary so widely from person to person.

The symptoms of Hepatitis B are varied but include the following:

- Jaundice
- Joint pain

SAFETY FACT

Employees at Greatest Risk from Viral Hepatitis

Employees who are at the greatest risk of necrosis of liver cells from viral hepatitis are those who are elderly and those who have diabetes mellitus, cancer, or any other illness severe enough to require surgery and transfusions.

- Rash
- Internal bleeding

The next section explains OSHA's standard on occupational exposure to bloodborne pathogens. This standard applies to all bloodborne pathogens including HBV and HIV.

OSHA'S STANDARD ON OCCUPATIONAL EXPOSURE TO BLOODBORNE PATHOGENS

OSHA's standard on occupational exposure to bloodborne pathogens is contained in 29 C.F.R. Part 1910.1030. The purpose of the standard is to limit the exposure of personnel to blood and to serve as a precaution against bloodborne pathogens that may cause diseases.

Scope of Application

This standard applies to all employees whose job duties may bring them in contact with blood or other potentially infectious material. There is no attempt on OSHA's part to list all occupations to which 1910.1030 applies. The deciding factor is the *reasonably anticipated* theory. If it can be reasonably anticipated that employees may come in contact with blood in the normal course of performing their job duties, the standard applies. It does not apply in instances of *good Samaritan* type acts in which one employee attempts to assist another employee who is bleeding.

The standard speaks to blood and other infectious materials. These other materials include the following:

- Semen
- Vaginal secretions
- Cerebrospinal fluid
- Synovial fluid
- Pleural fluid
- Peritoneal fluid
- Amniotic fluid
- Saliva
- Miscellaneous body fluids mixed with blood

In addition to these fluids, other potentially infectious materials include the following:

- Unfixed human tissue, or organs other than intact skin
- Cell or tissue cultures
- Organ cultures
- Any medium contaminated by Human Immunodeficiency Virus (HIV) or Hepatitis B (HBV)

Exposure Control Plan

OSHA's 1910.1030 requires employers to have an *Exposure Control Plan* to protect employees from exposure to bloodborne pathogens. It is recommended that such plans have at least the following major components:

Part 1	Administration
Part 2	Methodology
Part 3	Vaccinations
Part 4	Post-Exposure Investigation and Follow-Up
Part 5	Labels and Signs
Part 6	Information and Training

Administration

This component of the plan should clearly define the responsibilities of employees, supervisors, and managers regarding exposure control. It should also designate an exposure control officer (usually the organization's safety and health manager or a person who reports to this manager). The organization's Exposure Control Plan must be readily available to all employees, and the administration component of it must contain a list of locations where copies of the plan can be examined by employees. It should also describe the responsibilities of applicable constituent groups/individuals. These responsibilities are summarized in the following sections.

Employees   All employees are responsible for the following: knowing which of their individual and specific job tasks may expose them to bloodborne pathogens; participating in training provided concerning bloodborne pathogens; carrying out all job duties in accordance with the organization's various control procedures; and practicing good personal hygiene.

Supervisors and Managers   Supervisors and managers are responsible for coordinating with the exposure control officer to implement and monitor exposure control procedures in their areas of responsibility.

Exposure Control Officer   This individual is assigned overall responsibility for carrying out the organization's exposure control plan. In addition to the administrative duties associated with it, this position is also responsible for employee training. In small organizations, exposure control officers may double as the training coordinator. In larger organizations, a separate training coordinator may be assigned exposure control as one more training responsibility.

In either case, the exposure control officer is responsible for the following duties: overall development and implementation of the exposure control plan; working with management to develop other exposure-related policies; monitoring and updating the plan; keeping up-to-date with the latest legal requirements relating to exposure; acting as liaison with OSHA inspectors; maintaining up-to-date training files and documentation showing training; developing the needed training program; working with other managers to establish appropriate control procedures; establishing a Hepatitis B vaccination program as appropriate; establishing a post-exposure evaluation and follow-up system; displaying labels and signs as appropriate; and maintaining up-to-date, confidential medical records of exposed employees.

Methodology

This section of the plan describes the procedures established to protect employees from exposure. These procedures fall into one of the following five categories:

- General precautions
- Engineering controls
- Work practice controls
- Personal protection equipment
- Housekeeping controls

General precautions include such procedures as assuming that all body fluids are contaminated, and acting accordingly. Engineering controls are design and technological precautions that protect employees from exposure. Examples of engineering controls include self-sheathing needles, readily accessible hand-washing stations equipped with antiseptic hand cleaners, leakproof specimen containers, and puncture-proof containers for sharp tools that are reusable. Work practice controls are precautions that individual employees take, such as washing their hands immediately after removing potentially contaminated gloves or refraining from eating or drinking in areas where bloodborne pathogens may be present. Personal protection equipment includes any device designed to protect an employee from exposure. Widely used devices include the following:

- Gloves (disposable and reusable)
- Goggles and face shields
- Respirators
- Aprons, coats, and jackets

Examples of housekeeping controls include the use, disposal, and changing of protective coverings; decontamination of equipment; and regular cleaning of potentially contaminated areas.

Vaccinations

The OSHA standard requires that employers make Hepatitis B vaccinations available to all employees for whom the *reasonable anticipation* rule applies. The vaccination procedure must meet the following criteria: available at no cost to employees; administered at a reasonable time and place within ten days of assignment to a job with exposure potential; and administered under the supervision of an appropriately licensed health-care professional, according to the latest recommendations of the U.S. Public Health Service.

Employees may decline the vaccination, but those who do must sign a *declination form* stating that they understand the risk to which they are subjecting themselves.

Employees who decline are allowed to change their minds. Employees who do must be allowed to receive the vaccination.

Part of the exposure control officer's job is to keep accurate, up-to-date records showing vaccinated employees, vaccination dates, employees who declined, and signed declination forms.

Post-Exposure Investigation and Follow-Up

When an employee is exposed to bloodborne pathogens, it is important to evaluate the circumstances and follow up appropriately. How did it happen? Why did it happen? What should be done to prevent future occurrences? The post-exposure investigation should determine at least the following:

- When did the incident occur?
- Where did the incident occur?
- What type of contaminated substances or materials were involved?
- What is the source of the contaminated materials?
- What type of work tasks were being performed when the incident happened?
- What was the cause of the incident?
- Were the prescribed precautions being observed when the incident occurred?
- What immediate action was taken in response to the incident?

Using the information collected during the investigation, an incident report is written. This report is just like any other accident report. Keeping the exposed employee fully informed is important. The employee should be informed of the likely avenue of exposure and the source of the contaminated material, even if the source is another employee. If the source is another employee, that individual's blood should be tested for HBV or HIV, and the results should be shared with the exposed employee. Once the exposed employee is fully informed, he or she should be referred to an appropriately certified medical professional to discuss the issue. The medical professional should provide a written report to the employer containing all pertinent information and recommendations. The exposed employee should also receive a copy.

Labels and Signs

This section of the plan describes the procedures established for labeling potential biohazards. Organizations may also use warning signs as appropriate and color-coded containers. It is important to label or designate with signs the following:

- Biohazard areas
- Contaminated equipment
- Containers of potentially contaminated waste
- Containers of potentially contaminated material (i.e., a refrigerator containing blood)
- Containers used to transport potentially contaminated material

Information and Training

This section of the plan describes the procedures for keeping employees knowledgeable, fully informed, and up-to-date regarding the hazards of bloodborne pathogens. The key to satisfying this requirement is training. Training provided should cover at least the following:

- OSHA Standard 1910.1030
- The exposure control plan
- Fundamentals of bloodborne pathogens (e.g., epidemiology, symptoms, and transmission)
- Hazard identification
- Hazard prevention methods
- Proper selection and use of personal protection equipment
- Recognition of warning signs and labels
- Emergency response techniques
- Incident investigation and reporting
- Follow-up techniques
- Medical consultation

It is important to document training and keep accurate up-to-date training records on all employees. These records should be available to employees and their designated representatives (e.g., family members, attorneys, and physicians), and to OSHA personnel. They must be kept for at least three years and must include the following:

- Dates of all training
- Contents of all training
- Trainers' names and qualifications
- Names and job titles of all participants

Record Keeping

OSHA Standard 1910.1030 requires that medical records be kept by employers on all employees who are exposed to bloodborne pathogens. These records must be confidential and should contain at least the following information:

- Employee's name and social security number
- Hepatitis B vaccination status
- Results of medical examinations and tests
- Results of incident follow-up procedures
- A copy of the written opinion(s) of health-care professionals
- A copy of all information provided to health-care professionals following an exposure incident

SUMMARY

1. Symptoms of the onset of AIDS include the following: enlarged lymph nodes that persist, persistent fevers, involuntary weight loss, fatigue, diarrhea that does not respond to standard medications, purplish spots or blotches on the skin or in the mouth, white cheesy coating on the tongue, night sweats, and forgetfulness.

2. AIDS is known to be transmitted in three ways: sexual contact, blood contact, and mother-to-child during pregnancy or childbirth. AIDS is not spread through casual contact such as handshakes, toilet facilities, eating utensils, or coughing.

3. High-risk groups with regard to AIDS are as follows: homosexual men who do not take appropriate precautions, IV drug users, people with a history of multiple blood transfusions (including hemophiliacs), and sexually promiscuous people who do not take appropriate precautions.

4. AIDS is having an impact on the workplace in the form of higher insurance premiums, time-on-the-job losses, decreased productivity, cost of AIDS-related lawsuits, and increased stress.

5. A corporate AIDS policy should have at least three components: employee rights, testing, and education.

6. Federal legislation relating to AIDS includes the Rehabilitation Act of 1973, Occupational Safety and Health Act of 1970, and Employee Retirement Income Security Act of 1974. Other legal concerns include state laws, case law, testing, and company policy.

7. Tests for AIDS are actually tests to identify the presence of HIV antibodies. For initial screening, the ELISA test is widely used. For verification, the IFA and Western Blot tests are used.

8. A well-planned AIDS education program can serve several purposes including the following: give management the facts needed to develop policy and make informed decisions; and result in positive behavioral changes among employees.

9. An AIDS education program may consist of any or all of the following components: one-on-one counseling, referral, posters, a newsletter, classroom instruction, self-paced multimedia instruction, group discussion sessions, and/or printed materials.

10. When counseling AIDS-infected employees, safety and health professionals should proceed as follows: listen, maintain a nonjudgmental attitude, make the employee aware of the company's AIDS policy, and respond in accordance with company policy.

11. Safety and health professionals can use the following strategies for helping quell fears about AIDS: help establish an AIDS education program; conduct group round-table discussions; and correct inaccuracies, rumors, and misinformation about AIDS as soon as they occur.

12. The following precautions will help to reduce the chances of contracting AIDS: abstinence; mutually monogamous relationship/marriage with an infection-free partner; no multiple sexual partners; avoidance of sex with a high-risk person without taking proper precautions; no use of intravenous drugs or, if you do, no sharing of needles; and use of a protective device while practicing live or simulated CPR.

13. Hepatitis B poses an even greater threat than HIV. It is caused by a double-shelled virus (HBV). HBV is transmitted through blood, tears, saliva, and semen. It can stay alive for years in bodily fluids.

KEY TERMS AND CONCEPTS

Acquired immunodeficiency syndrome (AIDS)

AIDS education program

AIDS policy

Blood transfusions

Body fluids

Case law

Confidentiality

Confirmation test

Counseling

Disposable face mask

ELISA test

Employee assistance programs (EAPs)

Employee rights

ERISA of 1974

Essential functions

False negative

False positive

HBV

Hepatitis B

HIV antibody test

HIV-positive

Human immunodeficiency virus (HIV)

IFA

Intravenous drugs

IV drug users

Mutually monogamous relationship

OSHAct of 1970

Otherwise qualified

Reasonable accommodation

Rehabilitation Act of 1973

Screening test

Sexual contact

Western Blot test

REVIEW QUESTIONS

1. What does the acronym AIDS mean?
2. List five symptoms of AIDS.
3. What are the three known ways that AIDS is transmitted?
4. List the groups of people who are considered high risk with regard to AIDS.
5. List five ways that AIDS is *not* transmitted.
6. Describe the ways in which AIDS is having an impact in the workplace.
7. Briefly explain the minimum components of a corporate AIDS policy.
8. Explain how the following legal concepts relate to AIDS: otherwise qualified, essential functions, and reasonable accommodation.
9. Briefly explain both sides of the AIDS testing controversy.
10. What are the most widely used HIV antibody tests for screening and confirming AIDS?
11. Name four purposes of a well-planned AIDS education program.

12. How should a safety and health professional proceed when confronted by an employee who thinks he or she has AIDS?
13. Briefly explain the steps that a company can take to alleviate the fears of employees about AIDS.
14. Explain how OSHA categorizes work tasks relative to exposure to HIV infection.
15. List four ways to guard against contracting AIDS.
16. How can CPR be administered safely?
17. Explain why HBV poses more of a problem for safety personnel than HIV does.

ENDNOTES

1. Brown, K. C., and Turner, J. G. *AIDS: Policies and Programs for the Workplace* (New York: Van Nostrand Reinhold, 1989), pp. 1–8, 15, 16.
2. "AIDS Cases Soar," GLOBAL AIDSNEWS, The Newsletter of the World Health Organization Global Programme on AIDS, No. 3, 1994, p. 2.
3. "AIDS—The Basic Facts," *Occupational Health and Safety*, Winter 1987, Vol. 10, No. 3, p. 6.
4. Minetos, P. "Corporate America vs. AIDS," *Safety & Health,* December 1988, Vol. 138, No. 6, p. 34.
5. Ibid.
6. Ibid., p. 36.
7. Centers for Disease Control. U.S. Department of Health and Human Services brochure, *Facts About AIDS*, Spring 1987, p. 4.
8. Ibid.
9. Minetos, "Corporate America vs. AIDS," p. 36.
10. Ibid.
11. Ibid.
12. Yorker, B. C. "AIDS Testing," *AAOHN Journal*, May 1988, Vol. 36, No. 5, p. 231.
13. Ibid.
14. Bayer, R., Levine, C., and Wolf, S. M., "HIV Antibody Screening: An Ethical Framework for Evaluating Proposed Programs," in Yorker, B. C. "AIDS Testing," *AAOHN Journal,* May 1988, Vol. 36, No. 5, p. 232.
15. American College Health Association brochure, *The HIV Antibody Test* (Rockville, MD: 1989).
16. Ibid.
17. Centers for Disease Control, *Facts About AIDS*, p. 8.
18. National Safety Council. "Is the AIDS Fear a Threat at Work?" *Safety & Health,* June 1989, Vol. 139, No. 6, p. 52.
19. Seldes, G. *The Great Quotations* (Secaucus, NJ: Castle Books, 1966), p. 590.
20. Brown and Turner, *AIDS: Policies and Programs for the Workplace*, pp. 106–107, 116–17.
21. Centers for Disease Control, *Facts About AIDS*, p. 6.
22. Eastman, M. "CPR Training: Is It Still Safe?" *Safety & Health*, November 1990, Vol. 142, No. 5, p. 36.
23. Ibid. p. 39.

Environmental Safety and ISO 14000

American industry and government are currently dealing with such issues as acid rain, ground-level ozone, lack of **stratospheric ozone,** radon gas, the need for additional clean-up of PCBs, polluted ground water, higher-than-acceptable levels of lead in drinking water, and the potential for disastrous levels of environmental damage from oil exploration in the Arctic National Wildlife Refuge. A subtitle of an environmental article printed in *Fortune* magazine read, "Stunned by evidence of ecological damage, Americans are again declaring war on pollution."[1] In the words of Jeremy Main,

> Heightened concern about the environment is justified. Scientists have detected increasing evidence that chlorofluorocarbons and other chemicals are rising to the stratosphere and eat-

ing the ozone layer, which protects life from cancer-causing ultraviolet rays. Carbon dioxide and other gases have accumulated over decades of industrial-age combustion and threatened to raise Earth's temperature through the greenhouse effect."[2]

SAFETY, HEALTH, AND THE ENVIRONMENT

Much progress has been made in cleaning up and protecting the **environment** since Congress initially became interested in the 1960s. During the 1960s and 1970s, Congress passed a continual proliferation of environmental regulations aimed primarily at air, water, and ground pollution. According to the **Environmental Protection Agency (EPA),** the following progress was made in cleaning up air and water pollution between 1970 and 1986:

- Carbon monoxide emissions were down by 38 percent.
- Volatile organic compounds were down by 29 percent.
- Sulfur oxides were down by 25 percent.
- Ocean dumping of industrial wastes was down by 94 percent.
- Cities without adequate sewage treatment were down by 80 percent.
- Miles of river unfit for swimming were down by 44 percent.[3]

The progress made in cleaning up air and water pollution is encouraging. However, not all of the news is good. According to the EPA, from 1970 to 1986, nitrogen oxide emissions increased by 6 percent and ocean dumping of sewage sludge increased by 60 percent.[4]

Much work remains to be done about environmental safety and health. Jeremy Main recommends the following actions along these lines:

- *Acid rain.* Gradually reduce sulfur oxide emissions while concurrently avoiding expensive new regulation-induced clampdowns until sufficient evidence is available to justify such actions.
- *Ground-level ozone.* Develop new regulations that focus on gas stations and other small sources of ground-level ozone emissions while concurrently avoiding any new drastic restrictions on automobile emissions.
- *Global warming.* Focus more on the use of nuclear power while continuing efforts to use fossil fuels more efficiently.
- *Water pollution.* Build more sewage treatment plants throughout the nation as quickly as possible.
- *Toxic wastes.* Develop and implement incentive programs to encourage a reduction in the volume of toxic waste.
- *Garbage.* Increase the use of recycling while simultaneously reducing the overall waste stream. Burn or bury what cannot be recycled, but under strict controls.[5]

Clearly, a great deal of work remains to be done regarding safety and the environment, and business and industry have a major role to play. The DuPont Corporation is setting an example of how industrial firms can move responsibly to do their part in maintaining a safe and healthy environment. DuPont has adopted a corporate policy to ensure that all of its facilities operate in accordance with an identical set of safety, health, and environmental standards.

According to Donald G. Windsor, manager of the Safety and Occupational Health Division of DuPont,

> Our worldwide capital investment in environmental control facilities totals nearly $900 million—not an insignificant percent of our total capital investment. We spend more than $350 million a year to operate our pollution control facilities and support environmental research. And, we have the equivalent of more than 3,000 employees worldwide working to ensure that our environmental safeguards are adequate.[6]

Safety and health professionals need to be familiar with environmental legislation and regulations; the different types of environments; the various hazards in the environment; the role of waste reduction in environmental safety; and the future of the environment as it relates to industrial safety. This chapter provides practicing and prospective safety and health professionals with a fundamental understanding of these key topics.

LEGISLATION AND REGULATION

A fairly clear-cut division of authority exists for legislation and regulations concerning the environment. The Occupational Safety and Health Administration (OSHA) is responsible for regulating the work environment within an individual company or plant facility. Environmental issues that go beyond the boundaries of the individual plant facility are the responsibility of the EPA. Of course, some environmental issues and concerns do not fall clearly within the scope of either OSHA or the EPA. Therefore, these two agencies have begun to cooperate closely in dealing with environmental matters.

OSHA/EPA Partnership

Writing for the *Occupational Health & Safety Letter*, Gershon Fishbein had this to say about the ability of OSHA and the EPA to work together:

> It has taken 20 years to do it, but OSHA and EPA have finally consummated a marriage, sort of. In a Memorandum of Understanding (MOU), the two agencies have agreed to conduct joint inspections, refer to each other any possible violations of occupational or environmental statutes discovered during separate inspections, exchange data and conduct joint training programs for inspectors and coordinate compliance and enforcement activities.[7]

The OSHA/EPA partnership has ramifications for safety and health professionals because most companies have traditionally separated environmental and workplace safety concerns into two different departments. Environmental health professionals are usually engineers or scientists who specialize in environmental matters. The workplace safety professional may be an occupational physician, industrial hygienist, or safety manager. To respond effectively and efficiently to joint efforts of OSHA and the EPA, companies may be forced to combine their environmental health and workplace safety departments.[8]

The MOU originally signed by both agencies in 1990 requires that OSHA and the EPA cooperate in developing a plan for joint enforcement and information sharing. Such a plan is intended to eliminate the bureaucratic overlaps and inefficiencies sometimes associated with the governmental agencies. Joint training and data maintenance also will result.

Clean Air Act

One of the most important pieces of federal environmental legislation has been the **Clean Air Act**. Originally passed in 1970, the act was amended in 1990. Writing in the magazine *Occupational Hazards*, Greg LaBar had this to say about the amended Clean Air Act:

> Relying on technology-driven, market-based strategies, the law is designed to reduce air pollution—in the form of hazardous air pollutants, acid rain, and smog—by 56 billion lb. per year. This includes a 75 percent reduction in air toxins, a 50 percent cut in acid rain, and 40 percent decrease in smog over the next 20 years or so. Like the original Clean Air Act (CAA) of 1970, the new law does not on its own guarantee clean air, or even cleaner air. It's a framework from which the Environmental Protection Agency (EPA) and state agencies will develop the implementing regulations.[9]

The EPA and various state regulatory agencies have a great deal of latitude in developing regulations to implement fully the intentions of Congress. According to LaBar, the EPA may have to draft over 250 regulations to implement fully all of the titles set forth in the Clean Air Act as amended in 1990.[10]

The new Clean Air Act contains provisions that require companies to take whatever actions are necessary to prevent or minimize the potential consequences of the **accidental release** of pollutants into the air. It also establishes an independent chemical safety and hazard investigation board that will investigate accidental releases of pollutants that result in death, serious injury, or substantial property damage.

The Clean Air Act originally signed into law in 1970 was revised in 1977. The revised law was still a relatively small statute. However, the Clean Air Act as amended in 1990 contains approximately 350 pages detailing the requirements of seven titles:

Title I:	Urban Air Quality
Title II:	Mobile Sources
Title III:	Hazardous Air Pollutants
Title IV:	Acid Rain Control
Title V:	Permits
Title VI:	Stratospheric Ozone Provisions
Title VII:	Enforcement

Figure 25–1 gives a brief summary of each title in the act.

- **Title I: Urban Air Quality**
  Establishes five classes for noncompliance with ambient air quality standards: marginal, moderate, serious, severe, and extreme.
- **Title II: Mobile Sources**
  Increases the emission standards for automobiles in two steps (1998 and 2003). Also requires the sale of reformulated gasoline in selected high-pollution cities.
- **Title III: Hazardous Air Pollutants**
  Requires EPA to use maximum control technologies to regulate the use of 189 toxic substances.
- **Title IV: Acid Rain Control**
  Allows EPA to issue acid rain allowances to existing companies. Allowances can be used, saved, or sold to other companies.
- **Title V: Permits**
  Requires major pollution emitters to obtain a special permit. Contains provisions that allow concerned citizens to petition for the revocation of a company's permit.
- **Title VI: Stratospheric Ozone Provisions**
  Bans the manufacture of methyl chloroform and requires that the manufacture of the five most ozone-destructive chemicals be phased out completely by 2000.
- **Title VII: Enforcement**
  Increases the ability of EPA and corresponding state agencies to impose both criminal and civil penalties against violators of the Clean Air Act.

Figure 25–1
Clean Air Act.

Economics of Environmental Regulation

Since Earth Day 1970, a proliferation of legislation and regulations have been developed in the nation's capital in an effort to clean up and protect the environment. However, there has been very little work done to determine the costs vs. benefits of this legislation and corresponding regulations. According to Paul R. Portney, a senior fellow at Resources for the Future,

> One way of measuring the success of environmental policies is benefit-cost analysis. Yet no such analyses exist with respect to the vast majority of environmental laws and regulations implemented since Earth Day 1970. As the United States enters the third decade of environmental regulation, it may be prudent to supply decision-makers with benefit and cost evaluations which will help them weigh the pros and cons of legislative initiatives from both environmental and broader social perspectives.[11]

There is ample evidence to suggest that legislation and regulations passed since 1970 have resulted in substantial improvements. Carbon monoxide emissions are down, volatile organic compounds in the environment are down, the amount of sulfur oxide is down, the extent of ocean dumping of industrial wastes is down, the number of cities without adequate sewage treatment is down, and the number of miles of rivers unfit for swimming is down. However, what no one has yet been able to produce is accurate information on the cost of benefits derived from these improvements. In Portney's words,

> What is the dollar valuation of the health improvements, reduced damage to exposed materials, increased agricultural output, improved visibility, and other physical changes that have accompanied reduced air pollution concentrations?[12]

There are acceptable estimates of the annual costs of implementing federal environmental regulations. It costs approximately $85 billion per year to carry out the mandates of the Clean Air Act, the Clean Water Act, and more than 20 other major federal laws regulating such things as pesticides, herbicides, drinking water contaminants, solid and hazardous wastes, and new chemicals.[13]

Each time a state or federal agency gets involved in an environmental issue, the question of the monetary benefits of proposed solutions is raised. Since it is possible to quantify the costs of federally mandated solutions, it is critical that a way be found to quantify benefits. A cleaner, safer, healthier environment has the potential to yield a variety of benefits. We need accurate data on the effect of the country's annual investment in a clean environment on the following:

- The inflation rate
- The unemployment rate
- The rate of growth of the gross national product
- The international trade balance
- Human health
- Productivity in the workplace
- Reduced damage to exposed materials
- Agricultural output
- Industrial modernization
- Research and innovation in the chemical and pharmaceutical industries

The economics of safety and health is an issue that gains importance every year. Prospective safety and health professionals need to be aware of the importance of being able to provide both costs and benefits data when safety and health measures are proposed. This does not mean that only measures that are economical should be undertaken. According to Portney,

> Those who oppose the use of benefit-cost analysis in environmental policy making often do so on the grounds that there are important values and effects that defy quantification. They are absolutely right, and for this reason benefit-cost analysis can never—and should never—be the only basis for making regulatory or other decisions.[14]

However, **cost/benefit analysis** information should be part of the process. This is particularly necessary since opposition to environmental safety measures, just as with any other safety and health measure, is likely to come in the form of economic argument. By including the cost/benefit information in the decision-making process, the process is improved along with the quality of the decision.

TYPES OF ENVIRONMENTS

What do you think of when you hear the term *environment*? The air, forests, lakes, rivers, oceans, and other natural resources? These are all part of our natural environment. However, the natural environment is not the only type of environment with which we interact. According to Hammer, "Environments can be divided into either natural or induced, and then into those that might be controlled or artificial, or free or closed."[15] These classifications are explained in the following paragraphs:

1.   The **natural environment** is not man-made. It is the environment we typically think of as the Earth and all of its natural components, including the ground, the water, flora and fauna, and the air. The natural environment is full of both beauty and hazards. Some of the hazards of nature can be controlled or at least reduced. However, there are hazards over which we have little or no control. These include hurricanes, tornadoes, lightning, floods, and other natural disasters. **Induced environments** are those that have been affected in some way by human action. For example, the highly polluted air that results from the exhaust emissions of automobiles in heavily populated cities is part of an induced environment.[16]

2.   A **controlled environment** is a natural or induced environment that has been changed in some way to reduce or eliminate potential environmental hazards. An example of a controlled environment is a home or workplace that is heated to reduce the potential hazards associated with cold. An **artificial environment** is one that is fully created to prevent definite hazardous conditions from affecting people or material. For example, the environment within the space shuttle is an artificial environment.[17]

3.   A **closed environment** is one that is completely or almost completely shut off from the natural environment. Both a controlled and an artificial environment must be closed. A **free environment** is one that does not interfere with the free movement of air.[18]

Safety and health professionals need to be prepared to confront potential hazards in all of these types of environments. Dealing with hazards of the environment involves protecting workers within a plant from dangerous environmental conditions and protecting the general public from unsafe conditions that may be created by the operations and/or products of the plant. This is a critical area for modern safety and health professionals and one in which their level of sophistication will have to increase continually.

ROLE OF SAFETY AND HEALTH PROFESSIONALS

Writing for *Occupational Hazards* magazine, Sandy Moretz uses the example of David Arthur of Smith's Tool Company. She shows three consecutive printings of Arthur's business card. On the first business card, his title is safety director. On the next card, his title is safety and health manager. In the latest version of his business card, David Arthur's title is director—safety, health, and environment.[19]

In the 1980s, workplace safety and health and environmental compliance were typically separated into two different departments. However, with the signing of the memorandum of understanding between OSHA and EPA discussed earlier in this chapter, the lines between safety and health in the workplace and environmental compliance are no longer clear-cut. According to Moretz,

> Once distinct, the boundary between plant safety and environmental issues has become increasingly blurred. As companies meet a growing list of federal, state, and local environmental regulations, safety and health professionals are finding themselves called upon to deal with a burgeoning environmental compliance workload. At some companies, they're being given direct responsibility for environmental clients—in addition to their current safety and health duties. At other companies, they are assuming supervisory responsibilities for the work of environmental specialists.[20]

In some companies, the safety and health professional will be asked to take on environmental responsibilities in addition to the traditional safety and health responsibilities. Other companies will employ one person or persons to be responsible for workplace safety and health and a separate person or persons to be responsible for environmental concerns, both positions reporting to an overall health, safety, and environmental issues manager. Such factors as the size of the company, the type of product produced, and the perceived impact of governmental regulations are also factors that determine how health, safety, and environmental departments will be structured.

A safety and health professional should be prepared to undertake the increased workload associated with environmental safety. Safety and health students should be prepared to study environmental courses or portions of courses in their college curricula. Practicing safety and health professionals should be prepared to undertake continuing education in environmental issues. In the words of Roy Barresi, director of safety and environmental health for a Houston-based oil refining company,

> Safety and health professionals no longer have the luxury of passing off environmental issues to other members of the organization. Because legislation overlaps the safety, health, and environmental areas, the safety and health professional himself has to be aware of the environmental regulations, even if he does not have responsibility for them.[21]

HAZARDS OF THE ENVIRONMENT

Being aware of the various hazards in the environment and the negative impact that these hazards can have on humans is another requirement for safety and health professionals. This is true whether the environment is the natural environment or that of the workplace. Hammer lists some of the most common **environmental hazards:** "High humidity conditions, low relative humidity, sunlight, high temperature conditions, low temperature conditions, airborne salts, dust, sand, dirt . . . meteorological and micrometeorological conditions, lightning, high and low pressure, radiation, vibration and sound."[22]

Each of these potential hazards can have an effect on people, machines, systems, and other inhabitors of a given environment. For example, high humidity conditions can cause condensation that in turn results in short circuits and inadvertent activations or disruptions of electrical systems.[23] For this reason, manufacturers of computers, consumer electric products, and other microelectronics-based products pay particular attention to humidity control in their plant. However, low relative humidity can also cause problems such as drying out and cracking of organic materials, an increased tendency for the creation of static electricity, and easier ignition of accidents and fires.[24]

Indoor Air Pollution

According to Norman Beddows of the EPA,

> The evidence for **indoor air quality** affecting health is plentiful. The most dramatic is that of Legionnaires' disease that was universally confounding initially. This is caused by a bacterium (*Legionella pneumophila*) carried in aerosol form and conveyed through ventilation systems. The disease continued to arise because of poor engineering or hygiene practices.[25]

Noise, lighting, work stress, and glare are some of the many safety and health concerns in the indoor work environment; however, indoor air quality is one of the most important. Safety and health professionals should be attentive to ensuring high-quality indoor air in both new and old buildings. However, it is particularly important to be attentive to air quality in old buildings.

Writing for *Safety & Health*, Beddows has this to say about indoor air quality:

> Air quality in public and commercial, multi-story buildings is of increasing concern to occupants and building owners alike. It is a major current topic in industrial hygiene, and one that is expected to assume even more importance in the 1990s because of increasing public awareness of the so-called "sick building" or "tight building" syndrome and anticipated involvement of state and local government and office workers in indoor quality programs.[26]

Safety and health professionals should be particularly attentive to the following factors relating to the quality of indoor air:

- Heating, air conditioning, and ventilation systems
- Air filtration
- Temperature and humidity
- Volatile and/or toxic organic compounds and gases in the air
- Mineral and/or organic fibers and particles in the air
- Molds, bacteria, and other biological matter in the air[27]

Of particular concern in older buildings is the issue of asbestos and asbestos-containing material (ACM). A major concern for safety and health professionals is that the inhalation of fibers from ACM will lead to cancer. This concern is compounded because there is no universally accepted safe intake level for the inhalation of asbestos fibers. The EPA reported in 1988 that 733,000 public and commercial buildings contained asbestos.[28] Areas of particular concern in old buildings are spaces above the ceiling and below the roof, spaces behind walls, peripheral heat exchangers, air vents, lighting fixtures, and other forms of duct work. All of these spaces or system components may contain or be covered with various types of asbestos-containing materials. Another major concern for safety and health professionals is the so-called **sick building syndrome**. "Sick" buildings are those that contain unhealthy levels of biological organisms in the air as a result of faulty ventilation and air filtration systems. A common cause of sick building syndrome is the introduction of unhealthy outdoor air that is brought in and circulated through the cooling system. According to Beddows,

> On certain days at different times of the year, the use of make-up air may be minimal or maximal. This will occur to obtain operational economy, using the prevailing outdoor conditions, reducing electrical load or minimizing pollutant levels. When outdoor air is rated unhealthy such as when levels of ozone, peroxyacyl nitrates (PANs), sulphur dioxide or particulate matter are elevated above their respective, relevant OA standards, treatment of incoming air, when practical, is preferable to total indoor air recirculation for the duration of the episode.[29]

Peroxyacyl nitrates (PANs) are organic nitrates (oxidants) formed from hydrocarbons, nitrogen oxides, and ozone under the influence of ultraviolet light.[30]

HAZARDOUS WASTE REDUCTION

One of the most effective ways to ensure a safe and healthy environment is to reduce the amount of **hazardous waste** being introduced into it. In 1976, the EPA developed a policy statement designed to encourage industrial firms to reduce their hazardous waste output. The policy contains three broad provisions. First, companies must certify on the transportation manifest that they have a program in operation to reduce the volume and toxicity of hazardous waste each time hazardous waste is transported off-site. Second, in order to qualify for permits to treat, store, and/or dispose of hazardous waste, companies must implement and operate a hazardous waste reduction program. Finally, companies must submit biennial plans to the EPA describing actions taken to reduce the volume and toxicity of their hazardous waste.

In 1976, the **Resource Conservation and Recovery Act (RCRA)** became the first piece of federal legislation to encourage the reduction of hazardous waste. It was followed in 1984 by passage of the **Hazardous and Solid Waste Amendments (HSWA)**. However, the most significant federal regulation pertaining to hazardous waste reduction is the OSHA Hazardous Waste Standard, which took effect in 1990.

OSHA Hazardous Waste Standard

On March 6, 1990, a new OSHA standard (1910.120) went into effect setting standards for dealing with hazardous materials. According to Moretz,

> It is expected to protect some 1.75 million potentially exposed employees. The standard is not just a hazardous waste worker standard, however. Although it protects workers at cleanup operations at uncontrolled hazardous waste dump sites and those working at waste treatment, storage, and disposal (TSD) facilities licensed by EPA, it also protects firefighters, police officers, and ambulance and response team personnel who respond to hazardous materials spills.[31]

Figure 25–2 summarizes the requirements of the **OSHA Hazardous Waste Standard**. These standards require that companies identify, evaluate, and control hazardous substances; train employees in the proper accident prevention procedures and emergency response procedures; reduce exposure to hazardous substances; provide medical monitoring; keep employees informed concerning hazardous substances on the job; implement decontamination procedures; and develop both on- and off-site emergency response plans.

The goal of these and other hazardous materials regulations is to (1) encourage companies to minimize the amount and toxicity of hazardous substances that they use; (2) ensure that all remaining hazardous materials are used safely; and (3) ensure that companies are prepared to respond promptly and appropriately when accidents occur.

According to Philips,

Approximately 300 million tons of hazardous waste are deposited annually by land burial in the United States. This estimate has forced government into passing legislation to control waste disposal. Waste reduction requirements for industry have been on the books for many years. Although legislation has been a strong motivating factor, it has not insured 100% compliance. Industry has generally attacked the problem of waste management from the wrong end of the pipe. The preferred method in the past has been off-site recycling, treatment, and

---

- Develop a safety and health program to control hazards and provide for emergency response.

- Conduct a preliminary site evaluation to identify potential hazards and select appropriate employee protection strategies.

- Implement a site control program to prevent contamination of employees.

- Train employees before allowing them to undertake hazardous waste operations or emergency response activities that could expose them to hazards.

- Provide medical surveillance at least annually and/or at the end of employment for employees exposed to greater than acceptable levels of specific substances.

- Implement measures to reduce exposure to hazardous substances to below established acceptable levels.

- Monitor air quality to identify and record the levels of hazardous substances in the air.

- Implement a program to inform employees of the names of people responsible for health and safety and of the requirements of the OSHA standard.

- Implement a decontamination process that is used each time an employee or piece of equipment leaves a hazardous area.

- Develop an emergency response plan to handle on-site emergencies.

- Develop an emergency response plan to coordinate off-site services.

Figure 25–2
General requirements of the OSHA Hazardous Waste Standard.

disposal. These techniques pose more harm to the environment and are often more costly than simply reducing the amount of waste generated. This expense is expected to increase as regulations concerning waste become more stringent. In meeting new regulations, companies will have to place strong emphasis on source reduction, on-site recycling and substitution of less hazardous materials.[32]

Consequently, safety and health professionals need to be familiar with the concept of hazardous waste reduction, how to organize a reduction program, and how to conduct a reduction audit. Hazardous waste reduction can be defined as follows:

> *Hazardous waste reduction is reducing the amount of hazardous waste generated and, in turn, the amount introduced into the waste stream through the processes of source reduction and recycling.*

The two key elements of this definition are **source reduction** and **recycling**. Figure 25–3 breaks down these two elements into more specific subelements. Philips gives the following examples of source reduction and recycling strategies.[33]

Management Improvements in Operating Efficiency

A large consumer product company in California achieved a 75 percent reduction in hazardous waste by forming advisory teams of employees and supervisors from all departments that purchase and store hazardous materials. The teams attacked the problem from a materials handling perspective. Their most effective strategies were as follows: (1) reducing the inventory of perishable materials; (2) segregating hazardous and nonhazardous materials by storing them in different locations; and (3) reducing equipment leaks and spills.

An adhesive manufacturer reduced its hazardous waste through better production scheduling. Production machines have to be broken down and cleaned thoroughly each time that a different type of adhesive material is produced. This creates a great deal of

Figure 25–3
Key elements in hazardous waste reduction.

Source Reduction
- Management improvements in operating efficiency
- Better use of modern technology and processes
- Better selection of the materials used in production processes
- Product revisions

Recycling
- Reclamation of useable materials from hazardous waste materials
- Reconstituting of waste for reuse as an original product

hazardous waste left over from the cleaning process. By more carefully scheduling production runs to coincide with actual orders, the number of cleanings required and, in turn, the amount of hazardous waste generated was reduced.

Better Use of Technology

By purchasing a mechanical cleaning system to replace its chemical cleaning units, a company completely eliminated an entire category of hazardous waste. The company manufactures metal products, including nickel and titanium wire. The wire has to be cleaned before it can be used. The traditional cleaning process consisted of an alkaline chemical bath that produced large quantities of hazardous waste. The company substituted a mechanical cleaning system that uses silk, carbide pads, and pressure to clean the wire. The mechanical system produces no hazardous wastes.

Better Materials Selection

Printed circuit boards are often cleaned with solvents that become hazardous waste. One electronics manufacturer found that it could clean its boards using a water-based process, thereby eliminating the use of hazardous solvents.

Reclamation

Silver is a hazardous by-product of photo-processing operations. Wastewater containing silver cannot be discharged into a sewer system without expensive pretreatment. One photo-processing company solved this problem by investing in a silver **reclamation** system. The system uses an electrolytic cell to recover silver from the rinse water used in processing film.

Organizing a Waste Reduction Program

There are four steps that must be accomplished in establishing a waste reduction program. The first step is to convince top-level managers that the program is not just environmentally and ethically necessary, but also cost effective. The cost/benefit data associated with waste reduction should be enough to convince the most cost-conscious manager of the feasibility of such a program. The potential for reduced costs in the areas of regulatory compliance, legal liability, and workers' compensation should be included in all briefings given to managers. The other steps involved in organizing a waste reduction program are discussed in the following paragraphs (see Figure 25–4).

Figure 25–4
Steps in establishing a hazardous waste reduction program.

1. Gain a full commitment from the executive level of management.
2. Form the waste reduction team.
3. Develop a comprehensive waste reduction plan.
4. Implement, monitor, and adjust as necessary.

Form the Waste Reduction Team

The **waste reduction team** should have representatives from every department that purchases, stores, and/or uses hazardous waste. In every case possible, these representatives should be supervisors to ensure maximum involvement of, and commitment by, all employees. The team should be chaired by an executive-level manager to show that management supports the effort and to ensure the team access to the highest level of decision makers when recommendations are made. The appropriate role of the safety and health professional on the team is that of ex-officio staff member. The person who staffs the committee works closely with the chairperson to arrange meetings, record proceedings, facilitate communication, and take care of the logistical and administrative aspects of committee operation.

Develop a Comprehensive Plan

Figure 25–5 contains a basic outline for a comprehensive **waste reduction plan**. It is important that the final plan contain all of these components. The statement of purpose explains briefly and succinctly the overall purpose of the waste reduction program. The goals translate the statement of purpose into specific action items with timetables for their accomplishment. Each goal should be accompanied by proposed strategies for its accomplishment. This forces the waste reduction team to go beyond just the question of *what* to a consideration of *how*. By identifying potential inhibitors at the outset, the team can avoid wasted time, effort, and resources. Inhibitors should be identified for each individual goal. The measures of success answer the question, "How do we know when we've accomplished this goal?" For example, if a goal is to reduce the average monthly inventory of a given hazardous substance, the measure of success may be the monthly inventory printout. The actual measures used are less important than ensuring from the outset that everyone agrees on the measures. This will prevent disagreements over whether a goal has been achieved. To apply the measures of success, the team must be able to track and monitor hazardous material purchases, storage, use, and waste. The process that will be used for ongoing audits of the program should also be explained in the plan. The program audit process is explained later in this chapter.

Figure 25–5
Outline of waste reduction plan.

- Statement of purpose
- Goals with timetables
- Strategies for accomplishing each goal
- Potential inhibitors associated with each goal
- Measures of success for each goal
- Tracking system
- Audit process

Implement, Monitor, and Adjust

When the plan is complete, the program can be implemented. For best results, select an implementation date and have a high-profile ceremony involving the company's top officials. The perception of high-level support among employees will be critical to the success of the program. Such a ceremony preceded or followed by an intercompany memorandum of support from a top executive to all employees can establish the desired perception at the outset. At this point, the waste reduction audit begins, results are monitored by a system of continual tracking, and adjustments are made as necessary. The waste reduction audit is the most critical component of the program. The next section is devoted to this subject.

Waste Reduction Audit

Dan Philips is coordinator of the widely acclaimed Hazardous Materials Management Program at Pensacola Junior College in Florida. He has this to say about hazardous waste audits:

> A waste reduction audit is a procedure for identifying and assessing waste reduction opportunities within a plant site. The audit consists of a careful review of the plant's operations and waste streams and the selection of specific streams and/or operations to assess. The audit is a very useful tool in analyzing how a facility can reduce wastes destined for treatment and/or disposal sites. A waste reduction audit is basically a search for waste reduction opportunities without great concern for legal implications. However, in a waste reduction program, legal and regulatory compliance should be considered in order to be within present and future regulations.[34]

The most important steps in a **waste reduction audit** are as follows: (1) target processes; (2) analyze processes; (3) identify reduction alternatives; (4) consider the cost/benefit ratio for each alternative; and (5) select the best options (Figure 25–6).

Target Processes

In the first step, the waste reduction team identifies all processes that use hazardous materials and/or generate hazardous waste. These processes are targeted for analysis, and a targeted process list is developed.

Figure 25–6
Major steps in a waste reduction audit.

1. Target processes for analysis and create the target list.
2. Analyze processes to identify sources of waste generation.
3. Identify reduction options for each waste generation source.
4. Consider the cost/benefit of each reduction goal.
5. Select the best options.

Analyze Processes

Each process on the list should be observed from start to finish. It is important to analyze processes in person rather than relying on flowcharts or operating manuals. Members of the waste reduction team should actually see every step in the process. A list of points in the process where waste is generated should be compiled and the types of waste recorded. In addition to the production aspects of a process, the waste reduction team should also analyze all related processes (e.g., maintenance, storage of materials, materials handling, and waste treatment).

Identify Reduction Options

For every type of waste generated, **reduction options** should be identified to eliminate or reduce the waste. This is both an art and a science. Options can be identified by reviewing professional literature, talking with specialists, and drawing on the expertise of professional organizations and environmental agencies. Team members should be encouraged to think creatively. Both product and process changes may be considered as options.

Consider the Cost/Benefit of Each Option

There will be both costs and benefits associated with all proposed reduction options. The first step is to identify the costs of continuing the current approach. These include the cost of materials, regulatory compliance, handling and use precautions, storage, personnel, treatment, training, and disposal. The impact of each reduction option on these and all other costs should be considered.

Select the Best Option

Based on the results of a cost/benefit analysis, the best option is selected and implemented. The procedure is then repeated for each process on the target list.

ENVIRONMENTAL PRIORITIES FOR THE FUTURE

In September 1990, the Environmental Protection Agency released a report entitled *Reducing Risk: Setting Priorities and Strategies for Environmental Protection*. According to this report, the top environmental priorities and problems for the future are

- Habitat alteration and destruction
- Species extinction
- Loss of biological diversity
- Stratospheric ozone depletion
- Global climate change
- Outdoor/indoor air pollution
- Exposure to industrial and agricultural chemicals
- Contaminated drinking water[35]

Hazardous waste-generating industries have an impact on all of these areas of concern. Consequently, safety and health professionals need to be familiar with future trends regarding how these concerns will be handled. There is a growing consensus that, although regulation is critical, regulation alone will not bring the desired results. The consensus has given birth to several trends for the future. The most prominent of these are as follows:

1. *Pollution prevention will get more attention than after-the-fact cleanup.* The old adage "an ounce of prevention is worth a pound of cure" is particularly appropriate when it comes to environmental safety and health. Hazardous waste reduction by the means explained earlier in this chapter will receive more attention and have more resources devoted to it in the future.

2. *Intra-agency cooperation will increase.* As was explained earlier in this chapter, OSHA and EPA have signed a memorandum of understanding to work together on data gathering, inspections, and training. Now that these two large federal agencies have taken the lead, other agencies and organizations at all levels of government can be expected to follow their example. This should eventually simplify regulatory compliance for hazardous waste generators.

3. *There will be more use of economic incentives to reduce the levels of hazardous waste introduced into the environment.* The future will see more innovative attempts to use market forces to improve environmental conditions. Such incentives will supplement rather than supplant governmental regulations.

4. *Particularly hazardous substances will be targeted for phaseouts.* In February 1991, the EPA introduced a program that encouraged companies to reduce voluntarily their use of 17 highly toxic chemicals in the workplace.[36] Such programs are likely to be used increasingly in an effort to reduce the use of targeted substances. If the voluntary approach does not bring the desired result, reductions of selected chemicals may become mandatory.

ISO 14000 INTRODUCED

Globalization of the marketplace has created a competitive environment that requires peak performance and continual improvement. The unrelenting demands of the modern marketplace have given rise to new philosophies for doing business, most of which fall under the broad umbrella of Total Quality Management, or TQM. One of the initiatives under the TQM umbrella is the ISO 9000 family of quality standards.

These standards contain criteria for promoting effective quality management systems. The International Organization for Standardization **(ISO)**—the same organization that developed the ISO 9000 quality standards—has now developed the ISO 14000 family of standards to promote effective environmental management systems.

Just as the decision to adopt the TQM philosophy or the ISO 9000 standards is voluntary, adoption of ISO 14000 is based on voluntary organizational commitment to environmental protection rather than government coercion. For safety and health managers accustomed to complying with government mandates, ISO 14000 certification is a novel approach.

Shifting attitudes and greater public awareness are making it essential that business firms be good neighbors in their communities. The marketplace demands that businesses produce high-quality products at competitive prices, without harming the environment. ISO 14000 provides the framework for making effective environmental management part of the organization's overall management system.

WHAT IS ISO?

ISO is the acronym for International Organization for Standardization, a worldwide organization of national standards bodies (Figure 25-7). The complete membership roster for ISO contains the standards bodies of 118 countries. The overall goal of ISO is as follows:

> . . . to promote the development of standardization and related activities in the world with a view to facilitating the international exchange of goods and services and to developing cooperation in the sphere of intellectual, scientific, technological, and economic activity.[37]

ENVIRONMENTAL MANAGEMENT SYSTEM (EMS)

A management system, regardless of its application, is the component of an organization responsible for leading, planning, organizing, and controlling (Figure 25-8). An **environmental management system,** or EMS, is the component of an organization with pri-

---

Australia
Standards Australia
1 The Crescent
Homebush—N.S.W. 2140
P. O. Box 1055
Strathfield—N.S.W. 21335
Tel: +61 2 746-4700

France
Association francaise de normalisation
Tour Europe
F-92049 Paris La DeFeuse Cedex
Tel: +33 1 42 91 55 55

United Kingdom
British Standards Institution
389 Chiswick High Road
GB-London W4 4AL
Tel: +44 181 996 90 00

Canada
Standards Council of Canada
45 O'Connor Street
Suite 1200
Ottawa, Ontario KIP 6N7
Tel: 1 613 238-3222

Germany
Deutsches Institut fur Normung
Burggrafen strasse 6
D-10772 Berlin
Tel: +49 30 26 01 23 44

United States
Institute ANSI
American National Standards
11 West 42nd Street
13th Floor
New York, N.Y. 10036
Tel: +1 212 642-4900

Figure 25–7
Addresses of selected ISO members.

Figure 25–8
The principal functions of
management.

> **Leading**
> - Establishing a vision
> - Maintaining effective communication
> - Setting an example of commitment
> - Inspiring and motivating
> - Providing adequate support/resources
>
> **Planning**
> - Planning strategy
> - Planning operation
> - Developing policy
> - Developing procedure
>
> **Organizing**
> - Establishing structure
> - Defining staff and line functions
> - Delegating authority
> - Establishing span of control
>
> **Controlling**
> - Establishing benchmarks
> - Monitoring performance
> - Adjusting as necessary

mary responsibility for these functions as they relate specifically to the impact of an organization's processes, products, and/or services on the environment.

An organization's EMS may be a subset of its safety and health management system (Figure 25-9) or a separate component of the organization's overall management system. Regardless of where and how it fits into an organization, the EMS should do the following:[38]

- Establish a comprehensive environmental-protection policy (planning).
- Identify all government regulations and requirements that apply to the organization's processes, products, and/or services (controlling).
- Establish organization-wide commitment to environmental protection (leading).
- Establish responsibility and accountability relating to environmental protection (organizing).
- Incorporate environmental concerns in all levels of organizational planning including strategic, operational, and procedural (planning).

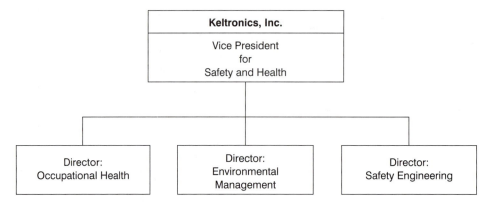

Figure 25–9
Sample organizational structure.

■ Establish management processes for achieving performance benchmarks (controlling).
■ Provide sufficient resources to ensure that performance benchmarks can be achieved on a continual basis (leading).
■ Establish and maintain an effective emergency preparedness program (leading, planning, organizing, and controlling).
■ Assess the organization's environmental performance against all applicable benchmarks and adjust as necessary (controlling).
■ Establish a review process for auditing the EMS and identifying opportunities for improvement.
■ Establish and maintain communications linkages with all stakeholders, internal and external.
■ Promote the establishment of an EMS in contractors and suppliers.

RATIONALE FOR THE EMS MOVEMENT

Different organizations have become interested in better environmental management for different reasons (Figure 25-10). Some organizations are responding to pressure brought by environmental advocacy groups and watchdog organizations. Such groups have become increasingly proficient in winning public support for their individual causes. This support, if effectively focused, can be translated into market pressure. When this happens, better environmental management becomes a market imperative.

Environmental groups have learned how to use the political process to gain support among elected officials and in key government agencies. Some elected officials and government employees are, themselves, environmental advocates. In most industrialized countries, a number of so-called "green parties" even run candidates for elective offices on environmental platforms. These factors, taken together, are giving environmental advocates steadily increasing influence. Business and industry are responding to this influence with increasing interest in environmental management.

One of the key drivers behind the EMS movement is the concept of competitive advantage. Some forward-looking organizations, particularly those that must compete in the global marketplace, have begun to view having an effective EMS as a competitive advantage. The advantage is the result of the following factors:

■ Minimization of funds siphoned off into nonproductive activities such as litigation and crisis management.

■ Ease of compliance with government regulations in the foreign countries that make up the global marketplace.

■ Better public image in countries (especially European countries) where interest in environmental protection is high.

Improved public relations is an incentive to improve environmental management. The mood of the general public relating to the environment has become strongly pro-environment over the past several decades. Companies that are perceived as being harmful to the environment have begun to feel the public's displeasure in the marketplace. This is one of the reasons why so many have begun recycling and voluntary clean-up programs.

An environmentally sensitive public often manifests itself in the form of environmentally sensitive customers. Such customers demand effective environmental management as a condition of their long-term retention. Positive environmental impact is rapidly becoming an attribute that customers consider when deciding whether to select and/or stay with a given vendor/supplier. Direct pressure from the marketplace (customers) is one of the most potent motivators behind the EMS movement.

Another powerful motivator is the potential threat of legal liability and the economic threat that it represents. In today's litigious business environment, a company can see its financial solvency evaporate overnight as the result of class-action lawsuits. Increasingly, individuals and advocacy groups are using the courts as their first response in disagreements over environmental issues, and the economic stakes are high. Juries are

Figure 25–10
Forces driving the interest in EMS.

✓	Public pressure from environmental advocacy groups
✓	Influence of environmental advocates within elected bodies and governmental organizations
✓	Potential to gain a competitive advantage in the global marketplace
✓	Improved public relations
✓	Customer demand
✓	Fear of liability and risk
✓	Pre-empt new and additional government oversight
✓	Reduced duplication
✓	De facto requirement to adopt a standard
✓	Get ahead of government adoption

showing an increasing tendency to award not only damages, but also huge punitive fines against companies that are careless with the environment.

Another motivating factor driving the EMS movement is the desire of companies to pre-empt additional government oversight. The idea is to adopt voluntary standards before the government establishes its own additional standards and makes them mandatory. Supporters of this approach believe that a form of voluntary peer review is inherently more effective than government oversight which, in their opinion, is often unwieldy and bureaucratic.

Reducing duplication promotes efficiency which, in turn, promotes competitiveness. Companies that do business on a global basis are accustomed to dealing with regulations that can vary markedly from country to country. Differences in regulations often cause differences in production processes. Whenever a process must be modified to accommodate regulatory variations—or for any other reason—the price of the product is driven up. Consequently, globally competitive companies are interested in globally uniform standards—standards that help to level the playing field in such key areas as environmental management.

The potential emergence of a de facto requirement to adopt an international standard in order to do business in a given market has also created interest in EMS. Although ISO 14000 is a voluntary standard, potential customers may decide unilaterally to do business only with companies that adopt the standard. In such cases, the marketplace makes adoption mandatory even though the government of the country in question has no such requirement.

The corollary to a de facto requirement is an actual government mandate to adopt an international EMS standard. The potential for a government mandate has gotten many companies interested in getting a headstart in the adoption of an EMS.

Potential Benefits of an EMS

Regardless of why a company becomes interested, there are many benefits that can be realized from adopting an EMS. Joseph Cascio, chairman of the Technical Advisory Group that represented the United States in developing ISO 14000, lists the following potential benefits:[39]

1. *Ease of trade.* Uniform international standards knock down the barriers created by country-to-country variations.

2. *Improved compliance with regulations.* When all players must obey the same rules, regulatory compliance is improved by the resulting application of uniform accountability criteria.

3. *Credibility.* Third-party certification takes politics out of the process, thereby enhancing its credibility.

4. *Reduction in liability/risk.* Certification forces companies to focus on the issue of environmental impact. This focus on being environmentally friendly, in turn, reduces the likelihood of environmentally hazardous behavior that might lead to expensive, nonproductive litigation.

5. *Regulatory incentives.* Companies that show initiative in establishing an effective EMS can take advantage of incentives that reward organizations for showing leadership in protecting the environment.

6. *Sentencing mitigation.* Adoption of a comprehensive EMS may serve as a mitigating factor when fines are assessed for failing to comply with regulations.

7. *Pollution prevention and waste reduction.* Better environmental management results in less waste and less pollution. These, in turn, result in attendant savings. This is the ounce-of-prevention/pound-of-cure concept.

8. *Profit.* Better management of any kind—quality, human resources, time, or environmental—translates into better profits.

9. *Improved internal management.* ISO 14000 certification requires the use of management methods that can improve all aspects of internal management. This, in turn, will lead to a better overall profit ratio.

10. *Community good will.* Good environmental management makes companies good corporate citizens in their neighborhoods which, in turn, leads to community good will.

11. *Retention of a high-quality workforce.* Employees live in the communities where their companies are located. Companies that are good corporate citizens will find it easier to retain their best and brightest employees than will companies whose environmental practices are embarrassing or irresponsible.

12. *Insurance.* Companies with established, effective environmental management systems will have fewer problems finding insurance policies that can be written at reasonable prices.

13. *Preference in lending.* One of the key factors that lending institutions consider before making a loan is the company's ability to pay. Since costly environmental litigation can bankrupt even the most solvent of companies, lending institutions may begin to give preference to those with established environmental management systems.

ISO 14000 SERIES OF STANDARDS

The term **ISO 14000 Series** refers to a family of environmental management standards that cover the five disciplines shown in Figure 25-11. All of the standards are in a constant state of evolution. The ISO 14000 Series contains two types of standards: specification standards and guidance standards.

Specification vs. Guidance Standards

A **specification standard** contains only the specific criteria that can be audited internally or externally by a third party. A **guidance standard** explains how to develop and implement environmental management systems and principles. Guidance standards are descriptive standards that also explain how to coordinate among various quality management systems.

Figure 25–11
Environmental disciplines in the ISO 14000 series of standards.

> ✓ Environmental management system
> ✓ Environmental auditor criteria (These criteria may be used by internal auditors and external third-party auditors.)
> ✓ Environmental performance evaluation criteria
> ✓ Environmental labeling criteria
> ✓ Life-cycle assessment methods

Guides and Technical Reports

ISO also develops **guides** and **technical reports.** A guide (not a guidance standard) is a tool to assist organizations in the improvement of environmental management. Guides are used voluntarily and do not contain criteria for certification. There is just one guide pertaining to the ISO 14000 Series: *ISO Guide 64—Guide for the Inclusion of Environmental Aspects in Product Standards.*[40]

Technical reports are written only when one of the following circumstances exist:

- When a technical committee of ISO cannot reach consensus on an issue. In such cases, the technical report explains why consensus was not possible.
- When an issue is still under development or when there is reason to believe that an undecided issue that cannot be immediately resolved will be resolved in the future.
- When a technical committee that is working on a standard collects information that is different from the kind normally published as a standard. This information may be published as a technical report.

Classification of ISO 14000 Standards

Figure 25-12 illustrates the two broad classifications of standards in the ISO 14000 family and their respective subclassifications. The two broad classifications are as follows:

- Process- or organization-oriented standards
- Product-oriented standards

Process-oriented standards cover the following three broad areas: environmental management system, environmental performance evaluation, and environmental auditing. Product-oriented standards cover three different broad areas of concern as follows: life-cycle assessment, environmental labeling, environmental aspects in product standards.

STANDARDS IN THE ISO 14000 SERIES

The six subclassifications shown in Figure 25-12 represent six distinct, but interrelated, areas of concern. There are actually 20 separate documents in the ISO 14000 series, all related to

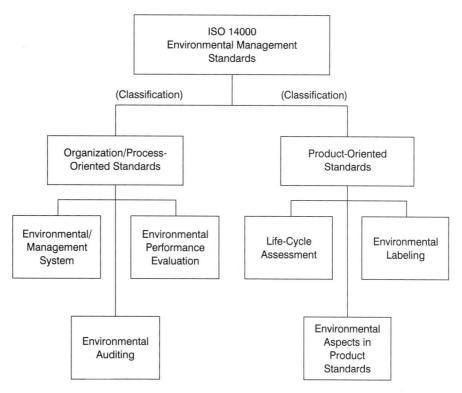

Figure 25–12
Environmental management standards.

one of these six subclassifications. Figure 25-13 is a checklist of the process-oriented documents in the ISO 14000 series. Figure 25-14 is a checklist of the product-oriented documents. Of these 20 documents, only ISO 14000 is a standard against which a company is audited. The rest are guidance documents. This is an important point to remember.

ISO 14001 STANDARD[41]

The ISO 14001 Standard represents an approach to protecting the environment and, in turn, a company whose processes and products may affect the environment. The standard is unique because it relies on voluntary motivation instead of mandatory compliance. To understand the ISO 14001 standard, one must first understand the following concepts: environmental management system (EMS), EMS audit, environmental aspect, continual improvement, and plan-do-check-adjust model.

1. *Environmental Management System (EMS).* An organization's EMS provides the structure for implementing its environmental policy. The management system consists of the organization's structure, personnel, processes, and procedures relating to environmental management.

Figure 25–13
ISO 14000 process-oriented
documents.

- ISO 14001
  Environmental Management Systems—Specification with
  Guidance for Use
- ISO 14004
  Environmental Management Systems—General Guidelines
  on Principles, Systems, and Supporting Techniques
- ISO 14010
  Guidelines for Environmental Auditing—General Principles on
  Environmental Auditing
- ISO 14011/1
  Guidelines for Environmental Auditing—Audit Procedures—
  Auditing of Environmental Management Systems
- ISO 14012
  Guidelines for Environmental Auditing—Qualification Criteria
  for Environmental Auditors
- ISO 14014
  Initial Reviews
- ISO 14015
  Environmental Site Assessments
- ISO 14031
  Evaluation of Environmental Performance
- ISO 14020
  Goals and Principles of All Environmental Labeling
- ISO 14021
  Environmental Labels and Declarations—Self-Declaration
  Environmental Claims—Terms and Definitions

2. *EMS audit.* The EMS audit is the process used to verify that an EMS actually does what an organization says it will do. As part of the audit, results are reported to the organization's top management team.

3. *Environmental aspect.* Any aspect of an organization's processes, products, or services that can potentially affect the environment.

4. *Continual improvement.* In the global marketplace, good enough is never good enough. Performance that is competitive today may not be tomorrow. Consequently, an EMS must be improved continually, forever.

5. *Plan-do-check-adjust model.* The plan-do-check-adjust (PDCA) model comes from the work of J. Edwards Deming in the area of Total Quality Management. The model works well in any management system that must be continually improved. Figure 25-15 lists some of the types of activities—relating to each of the model's four components—that are associated with continually improving an EMS.

Figure 25–14
ISO 14000 product-oriented
documents.

- ISO 14022
  Environmental Labels and Declarations—Symbols
- ISO 14023
  Environmental Labels and Declarations—Testing and
  Verifications
- ISO 14024
  Environmental Labels and Declarations—Environmental
  Labeling
  Type 1—Guiding Principles and Procedures
- ISO 1402X
  Type III Labeling
- ISO 14040
  Life Cycle Assessment—Principles and Framework
- ISO 14041
  Life Cycle Assessment—Life Cycle Inventory Analysis
- ISO 14042
  Life Cycle Assessment—Impact Assessment
- ISO 14043
  Life Cycle Assessment—Interpretation
- ISO 14050
  Terms and Definitions—Guide on the Principles for ISO/TC
  207/SC6 Terminology Work
- ISO Guide 64
  Guide for the Inclusion of Environmental Aspects of Product
  Standards

6. *Structure and application of the ISO 14001 standards.* The ISO 14001 Standard is subdivided into five broad components as shown in Figure 25-16. It is important to understand what the ISO 14001 standard does and does not do. The standard *does* allow a company voluntarily and proactively to establish a framework for involving all employees in improving environmental management. The standard does *not* establish product or performance standards, set pollutant levels, specify test methods, mandate zero emissions, nor expand on governmental regulations.

REQUIREMENTS OF THE ISO 14001 STANDARD[42]

The ISO 14001 Standard contains criteria subdivided into the following broad categories: General environmental policy; planning; implementation and operations; checking and corrective action; and management review. These broad categories are in most cases, subdivided further.

Figure 25–15
Plan-do-check-adjust (PDCA)
model applied to an EMS.

Plan

- Identify environmental aspects of processes, products, and services
- Evaluate all environmental aspects identified
- Identify legal requirements associated with the EMS
- Develop an environmental policy for the organization
- Develop performance criteria for internal audits of the EMS
- Establish EMS objectives, benchmarks, and performance targets

Do

- Provide the resources needed to support the EMS
- Assign responsibility for all aspects of the EMS
- Clarify accountability as it relates to the performance of EMS
- Provide training to ensure that all parties associated with the EMS have the knowledge and skills needed
- Communicate continually
- Develop and disseminate reports
- Document all aspects of the EMS and its performance
- Establish operational control

Check

- Monitor the performance of the EMS by making appropriate measurements
- Apply preventive action wherever possible
- Monitor EMS records continually
- Conduct audits of the EMS

Act

- Take corrective action when necessary
- Implement adjustments as needed continually

Following is a checklist of criteria that translates ISO 14001 into a format that can be used for conducting internal audits. The checklist ties into each category and subcategory of the standard.

4.0 General*

1. Has the organization established an EMS that meets the requirements of ISO 14001?
2. Does the organization properly maintain the EMS?

4.0.1 Environmental Policy

3. Has the organization developed an appropriate environmental policy?
4. Has the organization ensured that its environmental policy meets the following criteria:
   a. Appropriate in terms of the nature, scale, and potential environmental impact of the organization's processes, products, and/or services?
   b. Contains a statement of commitment to continual improving and to preventing pollution?
   c. Contains a statement of commitment to comply with all applicable and relevant regulations (federal, state, and local)?
   d. Establishes a framework for setting environmental objectives and for monitoring progress in achieving the objectives?
   e. Is effectively deployed:
      - Documented?
      - Implemented?
      - Maintained?
      - Co-communicated to all employees?
   f. Made available to all stakeholders including the general public?

4.2 Planning

4.2.1 Environmental Aspects

5. Has the organization established a procedure to identify the environmental aspects of its processes, products, activities, and/or services?
6. Does the organization properly maintain all aspects of its procedure?
7. Does the organization consider all instances of environmental impact when setting environmental objectives?
8. Does the organization keep its information relating to environmental impact up-to-date?

4.2.2 Legal and Other Requirements

9. Has the organization established a procedure to identify legal requirements that apply to its processes, products, activities, and/or services?

---

* Complete the remainder of the checklist before answering the questions in this section.

Figure 25–16
Structure of the ISO 14001
Standard.

4.0	**General**
4.0.1	Policy
4.2	**Planning**
4.2.1	Environmental aspects
4.2.2	Legal and other requirements
4.2.3	Objectives and targets
4.2.4	Environmental Management programs
4.3	**Implementation and Operations**
4.3.1	Structure and Responsibility
4.3.2	Training, Awareness, and Competence
4.3.3	Communication
4.3.4	Environmental management system documentation
4.3.5	Document control
4.3.6	Operational control
4.3.7	Emergency preparedness and response
4.4	**Checking and Corrective Action**
4.4.1	Monitoring and measurement
4.4.2	Nonconformance and corrective/preventive action
4.4.3	Records
4.4.4	Environmental management system audit
4.5	**Management Review**
Annex A:	Guidance on use of the specification
Annex B:	Bibliography
Annex C:	Links between ISO 14001 and ISO 9000

10. Has the organization established a procedure for gaining access to all legal requirements identified?
11. Does the organization properly maintain these procedures?

4.2.3 Objectives and Targets

12. Has the organization established documented environmental objectives at all relevant functional levels?
13. Does the organization properly maintain these documented environmental objectives? (For example, are both the objectives and the documents related to them kept up-to-date?)

14. When establishing/updating its environmental objectives, does the organization consider the following factors:
    a. Legal requirements?
    b. Environmental aspects?
    c. Technological options?
    d. Financial requirements?
    e. Operational requirements?
    f. Business requirements?
    g. Views of all stakeholders?
15. Are the organization's objectives consistent with the environmental policy?

4.2.4 Environmental Management Program(s)

16. Has the organization established programs for achieving its environmental objectives?
17. Do the organization's programs include the following:
    a. Designation of the responsible party for environmental objectives at each functional level?
    b. Means by which the objectives will be accomplished?
    c. Timeframe within which each objective is to be accomplished?
18. Does the organization amend its programs to include new products, processes, activities, and/or services as they are introduced?

4.3 Implementation and Operation

4.3.1 Structure and Responsibility

19. Has the organization clearly defined roles, responsibilities, and authority relating to environmental management?
20. Has the organization documented roles, responsibilities, and authority?
21. Has the organization communicated roles, responsibilities, and authority to all employees and all other stakeholders?
22. Does the organization commit the resources necessary to implement, operate, and maintain the EMS? (For example, human, financial, and technological resources)
23. Has the organization appointed a specific management representative with responsibility and authority for the following:
    a. Ensuring that all EMS requirements are met in accordance with the standard?
    b. Keeping top management informed concerning the performance of the EMS so that performance can be improved continually?

4.3.2 Training, Awareness, and Competence

24. Has the organization identified the training needs at all of its functional levels?
25. Does the organization ensure that all employees whose work may affect the environment receive the appropriate training?
26. Has the organization established procedures to make all employees at all levels in all relevant functions aware of the following:
    a. Importance of positive voluntary compliance with the environmental policy and procedures and all requirements of the EMS?

    b. The environmental aspects of their work activities and the environmental benefits of improving their individual performance?

    c. Their roles and responsibilities in helping the organization comply with its environmental policy and EMS?

    d. The potential negative consequences of failing to follow specified operating procedures?

27. Does the organization properly maintain these procedures?

28. Does the organization ensure that all personnel performing tasks that are potentially hazardous to the environment have the appropriate education, training, and/or experience?

4.3.3   Communication

29. Has the organization established procedures for the following:

    a. Ensuring effective communication among its various levels and functions?

    b. Ensuring effective communication with external stakeholders about environmental concerns?

30. Has the organization recorded its decision concerning how to handle external communication about its significant environmental aspects?

4.3.4   Environmental Management System Documentation

31. Has the organization developed information in an electronic or hardcopy format that does the following?

    a. Describes the core elements of its EMS and how these elements interact?

    b. Gives direction to other documentation related to the EMS?

32. Is this information properly maintained?

4.3.5   Document Control

33. Has the organization established procedures for controlling documents required by ISO 14001?

34. Does the organization ensure that all documents meet the following criteria:

    a. Can documents be located?

    b. Are documents regularly reviewed, updated, and approved by the appropriate authority?

    c. Are up-to-date documents available at all locations where essential functions are performed?

    d. Are outdated documents promptly removed to ensure against their use?

    e. Are outdated documents that must be retained for legal or historical purposes clearly marked for ease of identification?

35. Does the organization ensure that its documents are:

    a. Legible?

    b. Regularly reviewed, updated, and approved by the appropriate authority?

    c. Readily identifiable?

    d. Maintained in an orderly manner?

    e. Retained for a specified period?

36. Has the organization established procedures for the creation and modification of its documents?
37. Has the organization assigned responsibility for carrying out the procedures for creation and modification of its documents?
38. Does the organization properly maintain all of its procedures relating to its documents?

4.3.6 Operational Control

39. Has the organization identified its operations and activities that are associated with its environmental aspects and tied them to policy and objectives?
40. Has the organization planned in such a way that operational control activities are carried out under specified conditions by doing the following:
    a. Establishing/maintaining operational controls that ensure against deviations from the environmental policy and/or objectives?
    b. Building operational criteria into the procedures?
    c. Establishing/maintaining operational procedures relating to significant environmental impacts and communicating them to suppliers and contractors?

4.3.7 Emergency Preparedness and Response

41. Has the organization established procedures for identifying potential accidents, incidents, and emergency situations?
42. Has the organization established procedures for responding to accidents, incidents, and emergency situations?
43. Has the organization established procedures for preventing/mitigating the environmental effects of accidents, incidents, and emergency situations?
44. Does the organization properly maintain its procedures relating to the environmental aspects of accidents, incidents, and emergency situations?
45. Does the organization periodically review/revise its procedures relating to emergency preparedness and response?
46. Does the organization periodically test (wherever possible) its procedures relating to emergency preparedness and response?

4.4 Checking and Corrective Action

4.4.1 Monitoring and Measurement

47. Has the organization established procedures for monitoring/measuring its operations that may affect the environment?
48. Do the monitoring/measurement procedures include the following:
    a. Recording of information for tracking performance?
    b. Relevant operational controls?
    c. Compliance with environmental objectives?
49. Does the organization properly calibrate its monitoring equipment?
50. Does the organization properly maintain its monitoring equipment?
51. Does the organization make/retain records of calibration and maintenance activities?

52. Has the organization established documented procedures for periodically assessing its compliance with all applicable environmental legislation/regulations?
53. Are the documented procedures properly maintained?

4.4.2   Nonconformance and Corrective/Preventive Action

54. Has the organization established documented procedures for defining responsibility/authority for the following:
    a. Handling/investigating nonconformance?
    b. Taking action to mitigate the negative effects of nonconformance?
    c. Initiating and taking corrective/preventive action?
55. Does the organization ensure that all corrective/preventive action taken is:
    a. Appropriate to the magnitude of the problem?
    b. Commensurate with the environmental problem in question?
56. Does the organization implement changes in documented procedures that result from corrective/preventive action?
57. Does the organization record changes in documented procedures so that the procedures are up-to-date?

4.4.3   Records

58. Has the organization established documented procedures for the following:
    a. Identification of environmental records?
    b. Maintenance of environmental records?
    c. Disposition of environmental records?
59. Are the organization's documented procedures properly maintained?
60. Do the organization's records include the following:
    a. Training records?
    b. Audits?
    c. Reviews?
61. Does the organization ensure that its environmental records:
    a. Are legible?
    b. Are identifiable and traceable to the applicable activity, product, or service?
62. Are the organization's environmental records stored in such a way that they are:
    a. Readily retrievable?
    b. Protected against damage, deterioration, or loss?
63. Has the organization established and recorded retention times for its environmental records?
64. Does the organization maintain its environmental records in a way that:
    a. Is appropriate to the system?
    b. Is appropriate to the organization?
    c. Conforms to the requirements of ISO 14000?

4.4.4   Environmental Management System Audit

65. Has the organization established procedures for periodically auditing its EMS?

DISCUSSION CASE

What Is Your Opinion?

The executive managers of Petroleum Processing, Inc. are debating the relative merits of seeking ISO 14000 certification. The managers have polarized around two opinions. The first opinion is that the company should proceed with preparations for certification immediately. In the words of one ISO proponent, "The sooner we get started, the sooner we will realize the benefits of ISO 14000." The second opinion is that any certification that is not required should be ignored. In the words of one ISO opponent, "Certification is time-consuming and expensive. Why bother if we don't have to? We already have to comply with OSHA. Isn't that enough?" Where do you stand on this issue? What is your opinion?

---

66. Do the audit procedures do the following:
    a. Determine whether the EMS conforms to the ISO 14000 Standard and is properly maintained?
    b. Provide audit results to the organization's executive management team?
67. Are the audit procedures/schedule based on the environmental importance of the activities in question and on the results of previous audits?
68. Are the audit procedures sufficiently comprehensive that they cover:
    a. The audit scope?
    b. Frequency and method?
    c. Responsibilities/requirements for conducting audits?
    d. Responsibilities/requirements for reporting results?

4.5 Management Review

69. Does the organization's executive management team periodically review the EMS?
70. Does the organization's management review process ensure that the information needed to properly evaluate the EMS is collected?
71. Are all management reviews properly documented?

═══ SUMMARY ═══

1. Environmental problems currently facing the United States include acid rain, ground-level ozone, global warming, water pollution, toxic wastes, and the proliferation of garbage.
2. The federal agencies most involved in environmental safety and health are OSHA and the EPA. In 1990, these two organizations signed an historic memorandum of understanding in which each agreed to work with the other in collecting data, conducting inspections, and providing training.

3. One of the most important pieces of federal environmental legislation is the Clean Air Act as amended in 1990. It is designed to reduce hazardous air pollutants, acid rain, and smog by significant percentages over the next 20 years. The act deals with urban air quality, mobile sources, hazardous air pollutants, acid rain control, permits, stratospheric ozone provisions, and enforcement.

4. Measures taken to clean up and protect the environment are ethically right. They may also be cost-effective. However, research into the economics of a clean environment is limited at best. Research is needed to determine the effect of the country's investment in a clean environment not only on people's health, but also on inflation, unemployment, growth of the GNP, the international trade balance, productivity, research, innovation, and other economic factors.

5. There are different kinds of environments. These include natural, induced, controlled, artificial, closed, and free environments.

6. Safety and health professionals should be prepared to take responsibility for both workplace safety and environmental safety.

7. Some of the most common environmental hazards include high and low humidity, sunlight, high and low temperatures, airborne substances, meteorological and micrometeorological conditions, lightning, high and low air pressure, radiation, vibration, and sound.

8. Indoor air quality is as important as outdoor air quality. The following factors affect indoor air quality: heating, ventilation, and air conditioning; air filtration; temperature; humidity; airborne compounds and gases; and airborne molds, bacteria, and other biological matter.

9. The goal of the OSHA Hazardous Waste Standard and other regulations is to (a) encourage companies to reduce the amount and toxicity of the hazardous substances that they use; (b) ensure that remaining hazardous materials are used safely, and (c) ensure that companies are prepared to respond promptly and appropriately when accidents occur.

10. Hazardous waste reduction involves reducing the amount of hazardous waste generated and, in turn, the amount introduced into the waste stream through the processes of source reduction and recycling.

11. Source reduction and recycling strategies include (a) management improvements in operating efficiencies, (b) better use of technology, (c) better material selection, and (d) reclamation.

12. Organizing a waste reduction program involves accomplishing the following steps: (a) gaining a commitment from executive management; (b) forming a waste reduction team; (c) developing a comprehensive plan; (d) implementing, monitoring, and adjusting the plan.

13. The waste reduction audit is the most important part of the waste reduction program. It involves targeting processes, analyzing processes, identifying reduction options, considering cost/benefit data, and selecting the best option.

14. Trends for the future include more emphasis on prevention than cleanup, more interagency cooperation, use of economic incentives to supplement regulations, and targeting/phasing out of toxic chemicals.

15. ISO is the acronym for International Standards Organization, a worldwide consortium of national standards bodies. ISO is the organization that developed the ISO 14000 family of standards.

16. A management system is that component of an organization that is responsible for leading, planning, organizing, and controlling. An environmental management system, or EMS, is that component of an organization with primary responsibility for these functions as they relate specifically to an organization's impact on the environment.

17. The potential benefits of an EMS include the following: ease of international trade, improved compliance, credibility, reduction in liability/risk regulatory incentives, sentencing mitigation, pollution prevention, waste reduction, profits, improved internal management, community good will, employee retention, insurance, and lending preferences.

18. The ISO 14001 Standard is structured as follows: general, planning, implementation and operation, checking and corrective action, and management review.

=========== KEY TERMS AND CONCEPTS ===========

Accidental release

Acid rain

Artificial environment

Clean Air Act

Closed environment

Continual improvement

Controlled environment

Cost/benefit analysis

Economic incentives

EMS audit

Environment

Environmental aspect

Environmental hazards

Environmental Management System

Environmental Protection Agency (EPA)

Free environment

Garbage

Global warming

Ground-level ozone

Guidance standard

Guide

Hazardous and Solid Waste Amendments (HSWA)

Hazardous waste reduction

Indoor air quality

Induced environment

ISO

ISO 14000 Series

Natural environment

OSHA Hazardous Waste Standard

Peroxyacyl nitrates (PANs)

Plan-do-check-adjust model

Reclamation

Recycling

Reduction options

Resource Conservation and Recovery Act (RCRA)

"Sick-building" syndrome

Source reduction

Specification standard

Stratospheric ozone

Technical report

Toxic wastes

Waste reduction audit

Waste reduction plan

Waste reduction team

Water pollution

REVIEW QUESTIONS

1. Briefly describe the progress made in cleaning up the environment between 1970 and 1986.
2. Describe the OSHA/EPA memorandum of understanding signed in 1990.
3. What are the major goals of the Clean Air Act as amended in 1990?
4. Briefly explain the following titles of the Clean Air Act: Title I, Title II, and Title III.
5. Why is it important to undertake more research into the economics of a clean environment?
6. Distinguish between the following types of environments: natural and induced; controlled and artificial.
7. How are environmental concerns changing the role of safety and health professionals?
8. List five of the most common environmental hazards.
9. What are the factors that most affect indoor air quality?
10. Explain the term *sick building syndrome*.
11. Summarize briefly the requirements of the OSHA Hazardous Waste Standard.
12. What are the goals of hazardous materials regulations in general?
13. Define the term *hazardous waste reduction*.
14. List and explain three hazardous waste reduction strategies.
15. What are the steps involved in organizing a waste reduction program?
16. List the minimum contents of a hazardous waste reduction plan.
17. Define the term *waste reduction audit*.
18. What are the most important steps in a waste reduction audit?
19. What are EPA's top environmental priorities for the future?
20. List four trends for the future regarding environmental protection.
21. What is an environmental management system?
22. What factors are behind the competitive advantage that can result from implementation of an EMS?
23. List five potential benefits of an EMS.
24. Distinguish between process- and product-oriented ISO 14000 Standards.

ENDNOTES

1. Main, J. "Here Comes the Big New Clean-Up," *Fortune,* November 21, 1988, Vol. 118, No. 12, p. 102.
2. Ibid.
3. Ibid.
4. Ibid.

5. Ibid.
6. Windsor, D. G. "DuPont's Environmental Program," *Safety & Health*, November 1988, Vol. 138, No. 5, p. 39.
7. Fishbein, G. "Twenty-Year Friendship Culminates in Wedding Bells for OSHA and EPA," *Occupational Health & Safety Letter*, December 12, 1990, p. 203.
8. Ibid., pp. 203–204.
9. LaBar, G. "Seeing Past the Clean Air Act," *Occupational Hazards*, March 1991, p. 29.
10. Ibid.
11. Portney, P. R. "Taking the Measure of Environmental Regulation," *Resources*, Spring 1990, p. 2.
12. Ibid.
13. Ibid., p. 3.
14. Ibid., p. 4.
15. Hammer, W. *Occupational Safety Management and Engineering* (Upper Saddle River, NJ: Prentice Hall, 1989), p. 126.
16. Ibid., p. 129.
17. Ibid., pp. 129–30.
18. Ibid., pp. 130–31.
19. Moretz, S. "The Greening of Safety and Health," *Occupational Hazards*, April 1990, p. 61.
20. Ibid.
21. Ibid., p. 63.
22. Hammer, *Occupational Safety Management and Engineering*, pp. 127–28.
23. Ibid., p. 127.
24. Ibid.
25. Beddows, N. A. "Indoor Air: A New Problem for Old Buildings," *Safety & Health*, February 1989, Vol. 139, No. 2, p. 61.
26. Ibid. p. 58.
27. Ibid.
28. Ibid.
29. Ibid., p.60.
30. Ibid.
31. Moretz, S. "Industry Prepares for OSHA's Hazardous Waste Rule," *Occupational Hazards*, November 1989, p. 39.
32. Philips, D. *Waste Reduction in Industrial Processes* (Pensacola: Pensacola Junior College/Florida Department of Environmental Regulation, 1991), p. I–5.
33. Ibid., pp. I-7 to I-12.
34. Ibid., p. I-15.
35. National Safety Council. "Future Holds Hope for America's Environment," *Safety & Health*, June 1991, Vol. 143, No. 6, p. 29.
36. Ibid., p. 31.
37. Cascio, Joseph (ed.). *The ISO 14000 Handbook* (Fairfax, VA: CEEM Information Services, 1996), p. 4.

38. Ibid., pp. 9–10.
39. Ibid., pp. 10–11.
40. Tibor, Tom. *ISO 14000: A Guide to the New Environmental Management Standards* (Chicago: IRWIN Professional Publishing, 1996), pp. 171–76.
41. ISO DIS 14001—*Environmental Management Systems—Specification with Guidance for Use* (Chicago: American Society for Quality Control, 1996).
42. Ibid.

CHAPTER TWENTY-SIX

Product Safety and Liability

The United States has become the most litigious industrialized country in the world. With approximately 70 percent of the world's attorneys doing business in the United States, a heavy volume of litigation is not surprising. One of the fastest growing areas of the law is **product liability**. Product manufacturers are being sued in large numbers by users, misusers, and even abusers of their products.

In the words of the National Safety Council:

> Injuries resulting from the use (or often misuse) of products are the basis for an increasing number of product liability lawsuits. These suits cost industry millions of dollars each year. The best way the manufacturer can prevent or defend such claims is by manufacturing a reasonably safe and reliable product, and, where necessary, by providing instructions for its proper use. The key to achieving a reasonably safe and reliable product and, at the same time, reducing the product liability exposure is to build in product safety.[1]

An example of a widely used consumer product that has risks associated with it is the microwave oven. Since they were first introduced, microwave ovens have been carefully scrutinized for potential radiation leak hazards. As a result, radiation leakage from microwave ovens is no longer of major concern as a public safety and health issue.

However, the number of burns associated with microwave ovens is on the rise. According to the Consumer Product Safety Commission, "There were 2,612 hospital emergency room admissions for microwave-related burns in 1989, up from 2,352 in 1988."[2] Scalding is the most common form of burn associated with microwave ovens. There were 2,060 scalding cases in 1989, most of which occurred when people spilled heated liquids or were exposed to steam as it escaped from food containers.[3] Burns and scaldings together accounted for almost 5,000 microwave-related injuries in 1989. Every one of these incidents represents a potential lawsuit against a microwave manufacturer. Clearly, these manufacturers are seeking ways to decrease their potential exposure to product safety litigation.

PRODUCT LIABILITY AND THE LAW

Prior to 1916, consumers and employees used products and machines at their own risk. If injured while using a manufacturer's product, a person had no legal recourse. Consequently, there was no vehicle through which damages could be sought. Finally, in 1916, the concept of negligent manufacture was established in the law. Since then, the concepts of breach of warranty and strict liability in tort have also been added to the body of law relating to product safety/product liability.

Lawsuit Process

What happens when an individual or organization decides to bring a product liability case against a manufacturer? Before dealing with this question, it is important to understand the nature of product liability law. Typically, product liability lawsuits fall within the realm of civil law. This means that such suits do not involve criminal charges. Rather, they involve one party seeking redress from another party in the form of monetary damages.

Figure 26–1 illustrates the various steps that take place when a product liability lawsuit is filed. The **discovery period** (shown as step 3) can be lengthy. The collection of evidence can involve taking depositions from expert witnesses and examining the product in question. Product examination may involve running a variety of tests and observing simulations. The length of time involved in completing this step often leads both parties to seek out-of-court settlements.[4]

Figure 26-1

Product liability lawsuit process.

Source: D. A Colling, *Industrial Safety: Management and Technology* (Upper Saddle River, NJ: Prentice Hall, 1990), p. 263.

1. An injured party decides to seek redress against the manufacturer of a product and engages an attorney.
2. The attorney files a complaint in state or federal court.
3. The discovery period is undertaken during which evidence is collected, depositions are taken, and the product is examined.
4. A trial date is set.
5. The trial takes place or a settlement is reached out of court.

History of Product Liability Law[5]

Product liability law is relatively new. In fact, until 1960, manufacturers would not be held liable unless they produced flagrantly dangerous products. The concept of nonliability, established in the courts of England in 1842, persisted in the law in this country until the turn of the century. Since then, four landmark cases have established what is still the foundation of product liability law in the United States. These cases are discussed in the following paragraphs.

MacPherson vs. Buick Motor Company

This case established the concept of **negligent manufacture**, which means that the maker of a product can be held liable for its performance from a safety and health perspective. While driving a Buick Motor Company car, MacPherson lost control and was injured in the resulting accident. The cause of the accident was a defective wheel. The courts ruled in MacPherson's favor.

Henningson vs. Bloomfield Motors, Inc.

This case established the concept of **breach of warranty**, broadened the manufacturer's liability to include people without contractual agreements with the manufacturer, and limited the protection that manufacturers could derive from using disclaimers with their products. In this case, Henningson had purchased a new car for his wife, who was subsequently involved in an accident while driving it. The cause of the accident was a defective steering system. Two issues made this case different from other negligent manufacture cases: (1) the car was damaged to such an extent that negligent manufacture could not be proven; and (2) Mrs. Henningson had no contractual agreement with the manufacturer because Mr. Henningson had actually made the purchase. The court ruled that an implied warranty existed for the performance of the car and that it extended to persons other than those with an actual contractual agreement with the manufacturer.

Greenman vs. Yuba Products, Inc.

This case established the concepts of **strict liability in tort** and **negligent design**. Greenman sued after being injured while using a lathe in the prescribed manner. His attorneys argued that the lathe's design was defective and that a breach of warranty had occurred. In finding in Greenman's favor, the court broadened the concept of negligent manufacture to include design of the product.

Van der Mark vs. Ford Motor Company

This case confirmed the concepts established in *Greenman vs. Yuba Products, Inc.* Van der Mark sued when his new car was destroyed because of a defective braking system. In finding for Van der Mark, the Supreme Court overturned a lower court decision against him, thereby confirming the concept of negligent design.

Strict Liability in Tort

Paragraph 402A of the Second Restatement of Torts (American Law Institute) reads as follows:

1. One who sells any product in a defective condition unreasonably dangerous to the user or consumer or to his property is subject to liability for physical harm thereby caused to the ultimate consumer or user, or to his property if:
   a. the seller is engaged in the business of selling such a product, and
   b. it is expected to and does reach the user or consumer without substantial change in the condition in which it is sold.
2. The rule stated in Subsection (1) applies although:
   a. the seller has exercised all possible care in preparation and sale of his product, and
   b. the user or consumer has not bought the product from or entered into any contractual relation with the seller.[6]

This definition of strict liability has become the standard and is used in the courts of most states.

Tort law also includes the concept of **duty to warn**. This is why there are warning labels on so many consumer products, particularly those made for children. Paragraph 388 of the Second Restatement of Torts established the following three criteria for determining whether a duty to warn exists:

1. The potential for an accident when the product is used without a warning, provided the use to which it is put is reasonably predictable
2. Probable seriousness of injuries if an accident does occur
3. Potential positive effectiveness and feasibility of a warning[7]

Statutory Product Liability Laws

In addition to the common and tort law concepts discussed in the previous section, a number of statutory laws relate to product liability. Those that have the most significant impact on product liability are as follows:

- Consumer Product Safety Act
- Flammable Fabrics Act
- Food, Drug, and Cosmetics Act
- Hazardous Substances Act
- Mine Health and Safety Act
- National Traffic and Motor Vehicle Safety Act
- Occupational Health and Safety Act
- Poison Prevention Packaging Act
- Refrigerator Safety Act
- Toxic Substances Control Act

■ Workers' Compensation Act

Of these various statutes, the **Consumer Product Safety Act** has the most direct application to product liability. This is the statute with which Ralph Nader is closely associated. The act has four basic purposes:

1. To protect the public from the risk of injuries incurred while using consumer products
2. To help consumers make objective evaluations of the risks associated with using consumer products
3. To encourage uniformity in standards and regulations and to minimize conflicts among regulations at the various levels of government
4. To encourage research into the causes of product-related injuries, health problems, and deaths and how these things can be prevented[8]

The act is administered by the Consumer Product Safety Commission (CPSC). Since its inception in 1972, the CPSC has been successful in facilitating changes to widely used consumer products that have enhanced their safety. These changes include safety covers on matchbooks and childproof caps on medicines and other potentially dangerous substances. The CPSC also collects information on the safety of consumer products and maintains a database through the National Electronic Information Surveillance System (NEISS). NEISS is a nationally coordinated computer system containing hospital records of injuries resulting from the use of consumer products.[9]

Applying Product Liability Laws

To recover damages in a product liability case, a person must satisfy the burden of proof by meeting certain criteria, listed in Figure 26–2. Several important legal concepts are used to apply these criteria, including the following: patent defect, latent defect, prudent man concept, reasonable risk, and unreasonable risk.

■ The product is defective or unreasonably unsafe as produced by the manufacturer.

■ The condition that causes the product to be defective or unreasonably unsafe existed when the product left the care of the original manufacturer.

■ The defective or unreasonably unsafe condition actually caused the injury/damage/loss that is the subject of the case.

■ The nature of the complainant's injury/damage/loss is related to the defective or unreasonably unsafe condition.

Figure 26–2
Burden-of-proof criteria for product liability cases.
Source: D. A Colling, *Industrial Safety: Management and Technology* (Upper Saddle River, NJ: Prentice Hall, 1990), p. 266.

A **patent defect** is one that occurs in all items in a manufactured batch. For example, if a manufacturer produces a batch of 10,000 copies of a given product and the same defect occurs in all copies, it is classified as a patent defect. Patent defects are the kind that sometimes results in product recalls. **Latent defects** occur in only one or a limited number of copies in a batch.

A key concept in product liability cases is the concept of unreasonable risk. In determining whether a risk is reasonable or unreasonable, the **prudent man concept** is applied. A **reasonable risk** exists when consumers (1) understand risk, (2) evaluate the level of risk, (3) know how to deal with the risk, and (4) accept the risk based on reasonable risk/benefit considerations. In other words, they behave prudently. An **unreasonable risk** exists when (1) consumers are not aware that a risk exists, (2) consumers are not able to adequately judge the degree of risk even when they are aware of it, (3) consumers are not able to deal with the risk, and (4) risk could be eliminated at a cost that would not price the product out of the market.[10]

Community Right-to-Know Act

An important piece of federal legislation relating to product liability is Title III of the Superfund Amendments and Reauthorization Act (SARA). The law is commonly referred to as the **Community Right-to-Know Act**. The act gives people the right to obtain information about hazardous chemicals being used in their communities. It applies to all companies that make, transport, store, distribute, or use chemicals.

The act has four main components: (1) emergency planning, (2) emergency notification, (3) reporting requirements, and (4) toxic chemical release reporting. These four components are explained in the following paragraphs.

Emergency Planning

This component of the act requires the establishment of local emergency planning committees (**LEPCs**) and state emergency response commissions (**SERCs**). LEPCs are required to do the following: (1) develop an emergency response plan for the local community, (2) provide information through public hearings, and (3) designate a community coordinator. SERCs must (1) supervise all LEPCs in the state, and (2) review the emergency response plans for all LEPCs.

Emergency Notification

Companies must report immediately the release of hazardous chemicals that exceed limits specified by the LEPC and SERC. When this occurs, companies must report on the following: (1) chemical name or names; (2) date, time, place, and amount of the release; (3) potential hazards to safety and health; (4) recommended precautions to limit the potential hazards of the release; and (5) name of a contact person who can provide additional information.

Reporting Requirements

Companies are required to maintain accurate, up-to-date information on chemicals that they produce, store, use, transport, and so on. They are required to provide material safety data sheets (MSDSs) or other communication devices that contain at least the following information: name, formula, and other technical information about the chemical; potential safety and health hazards; handling precautions; and emergency procedures. Finally, companies are required to give LEPCs and SERCs information about how much of a given chemical is typically present on their site or sites and where it is stored.

Toxic Chemical Release Reporting

Companies must report their annual level of toxic chemical emissions to the Environmental Protection Agency and to state-level regulatory agencies.

For information about the act, SERCs, and LEPCs, prospective and practicing safety and health professionals should use the Environmental Protection Agency's toll-free hot line telephone number: 1-800-535-0202.

DEVELOPING A PSM PROGRAM

The purpose of a **product safety management (PSM) program** is to limit as much as possible a company's exposure to product liability litigation and related problems. The key to limiting liability exposure is to develop and maintain a comprehensive PSM program. At a minimum, the program should have three functional components: PSM coordinator, PSM committee, and PSM auditor.

PSM Coordinator

A successful PSM program must involve all departments within the company (design, manufacturing, marketing, sales, installation/service, accounting, and so on). Therefore, it is important to have a **PSM coordinator** whose role is to coordinate and facilitate this involvement.

The factor that will contribute most to the success or failure of a PSM program is the coordinator's level of authority. The higher the level of authority, the more likely the program is to succeed. Giving coordination responsibility to a person who lacks executive-level access and decision-making powers can contribute to the failure of a PSM program. This is why it is sometimes advantageous to add PSM responsibilities to those of an existing executive-level decision-maker who (1) has line authority over a major component of the company, (2) has access to other executive-level managers, and (3) has authority equal to fellow executives.

According to the National Safety Council, the PSM coordinator should have the authority to undertake the following actions:

- Assist in setting PSM program policy
- Recommend product recalls, field modifications, product redesign, and special analysis
- Conduct complaint, incident, or accident analysis
- Coordinate all program documents
- Facilitate communication among all parties involved in the program

- Develop a base of product safety and liability information for use by all parties involved in the program
- Establish and maintain relationships with agencies and organizations that have missions relating to product safety and liability
- Conduct PSM program audits[11]

PSM Committee

The PSM coordinator is just that—a coordinator. Effective product safety management is the responsibility of all departments. For this reason, many companies find it advantageous to form a **PSM committee** with a representative from all major departments. This approach gives the PSM coordinator a broad base of expertise to call on and encourages broad-based support among all departments.

PSM Auditor

Auditing is an important component of the PSM program, since the effectiveness of the program is determined through audits. According to the National Safety Council,

> The program auditor's main duty is to evaluate the adequacy of the organization's PSM program activities in relation to actual and potential exposures. This evaluation determines what the organization should do to prevent product-related losses by comparing what is being done against what should be done.[12]

The **PSM auditor** is responsible for evaluating the overall organization and individual departments within it. Specific duties of the PSM auditor include the following:

- Identify evidence of a lack of commitment on management's part
- Observe the action taken when a product deficiency is identified (i.e., does management move immediately to determine the cause and correct the problem?)
- Bring deficiencies to the attention of management and make corresponding recommendations
- Review documentation of actions taken to correct product deficiencies (Were design changes made? Were manufacturing changes made? Was a product recall initiated?)

The National Safety Council summarizes the duties of the PSM program auditor as follows: "[T]o interview all key management personnel, to observe the actual manufacturing operation, and to investigate, question, and verify performance."[13] Obviously, the auditor needs a certain amount of autonomy to be able to perform these duties. Can an internal employee have the autonomy needed to do the job? Can the program coordinator? These are important questions to ask when establishing a PSM program.

Opinions vary on these questions. Some think that the PSM coordinator should also conduct PSM program audits. Others think that this is an inappropriate mix of duties with built-in conflict of interest. Some think that an external person or agency should conduct audits to provide complete objectivity. Others think an internal person who is familiar with the company, its processes, and its personnel can perform a better audit. In reality, what works best depends on the company, the individual situation, and the personnel involved. If management is truly committed to PSM, it won't matter whether the auditor is internal or external, nor will it matter whether the auditor is the PSM program coordinator or a different individual. The key to the success of a PSM program is commitment.

EVALUATING THE PSM PROGRAM

The job of the PSM auditor is to evaluate the program continually to identify and recommend corrections to the causes of product liability problems. Figure 26–3 lists some of the more common causes of product liability exposure. The PSM auditor should constantly search for these and other causes.

Preparing to Evaluate the Program

Before beginning an evaluation, the auditor should allow time to prepare for the task. Proper preparation can improve both the level of cooperation and the quality of information obtained. The following preparatory activities can improve the quality of the audit:

- Meet with top management officials and review the purpose of the audit
- Work with all managers involved to coordinate schedules so that audit activities occur at mutually acceptable times
- Let managers know how the audit will affect their organizations and what will be expected of them
- Review all documentation relating to the PSM program (e.g., minutes of meetings, memoranda, and letters)
- Review product-related literature (e.g., related processes, procedures, standards, and technical data)
- Review copies of all warning labels and other printed precautionary material relating to the product[14]

Figure 26–3
Common causes of product liability exposure.

- Insufficient research during product development
- Faulty product design
- Insufficient testing of product prototypes
- Faulty manufacturing
- Insufficient quality control
- Poorly written instructions
- Insufficient or unclear warnings
- Unethical representation of the product (recommending it for use in situations outside the scope of its design and intended use)

Conducting the Evaluation

Having accomplished the preparation activities just described, the auditor is ready to conduct the actual evaluation. Conducting the evaluation is a matter of going to all pertinent departments and looking for evidence of the types of factors set forth in Figure 26–3. The following departments should be audited at a minimum: engineering/design, manufacturing/production, marketing/sales, service, and purchasing.

ROLE OF THE SAFETY AND HEALTH PROFESSIONAL

There is no set rule for the role played by safety and health professionals in the operation of a PSM program. In some cases, the safety and health professional is a member of the PSM committee, typically in large companies. In other cases, the safety and health professional doubles as the PSM program coordinator, more often in smaller companies. In yet other cases, the safety and health professional may be designated as the PSM auditor, which may happen in both large and small companies.

Regardless of how their role is structured, safety and health professionals can make several contributions to the program. According to the National Safety Council, these contributions include the following:

■ Safety professionals, because of their knowledge of plant operations and because of their general expertise in safety-related matters, can evaluate and offer comments on the company's PSM program.

■ Safety professionals, because of their experience in safety training, can evaluate product safety-related training programs developed as part of the PSM program.

■ Safety professionals, because of their knowledge of accident investigation techniques, can assist in conducting product-related accident investigations.

■ Safety professionals can provide product safety surveillance in production areas. This can help to prevent errors and product-related accidents. All product safety-related complaints or problems uncovered as a result of surveillance should be discussed

---

DISCUSSION CASE

What Is Your Opinion?

"Product liability is not my job," said Mark Conners, safety director for Richfield Toys, Inc. "The safety of the toys we manufacture is a design problem. Leave it to the engineers." "Engineers have a role to play, of course," said Amanda Garner, CEO of Richfield Toys. "But they are more function-oriented than safety-oriented. We need all of our products to be scrutinized from a safety perspective." Does the safety director have a role to play here? If so, what is it? If not, why? What is your opinion?

with the production manager and formally documented, with copies sent to the PSM program coordinator.

■ Because of their past experience in developing and implementing employee safety programs, safety professionals are aware not only of potential product hazards, but also of ways in which customers misuse products. Therefore, a knowledgeable safety professional should be used as a consultant to the PSM program coordinator and the auditor.[15]

TOTAL QUALITY MANAGEMENT AND PRODUCT SAFETY

It is widely accepted that the best way to limit liability exposure is to produce a quality product. But what is a **quality product?** For the purposes of this book, the following definition is used:

> *A quality product is one that meets or exceeds customer standards and expectations.*

An inherent expectation is that the product, if used properly, will cause no health or safety problems.

An approach to management with a large number of proponents is **total quality management (TQM)**. TQM can be an effective way to ensure that a company's products consistently meet or exceed customer standards and expectations, thereby reducing the company's exposure to product liability. Managing a company using TQM means expecting a total and willing commitment to quality by all personnel at all levels. With TQM, every employee is (1) responsible for quality and its continual improvement, and (2) empowered to make decisions/recommendations to improve quality continually. With TQM, quality is the overriding factor in all decisions at all levels.

A key element in this concept is the empowerment of employees. Empowered employees are able to think creatively, act independently, and pursue innovative solutions to problems. This approach is easy enough to talk about, but not so easy to implement. Figure 26–4 lists some of the main elements of TQM. These key points make it clear that TQM is not just another superficial quick-fix program puffed up by catchy slogans but lacking in substance. TQM requires a complete change in a company's culture and management philosophy.

Another key element of TQM is the elimination of the **vacuum mentality**. Workers often think that they work in a vacuum. They don't realize that their work affects that of other employees and vice versa. Products are produced through a sequentially arranged series of processes. Every task performed in one step affects all of those that follow. This is why another key element of TQM is teamwork.

In a TQM setting, **teamwork** is engendered through five key strategies (Figure 26–5):

■ **Involvement** of all personnel who must implement decisions in making those decisions.

- Requires a secure work environment
- Requires leadership at all levels
- Defines quality in terms of customer standards/expectations
- Focuses on the continual improvement of products and processes
- Requires long-term commitment of both management and employees
- Uses the teamwork approach in producing products
- Emphasizes continual training
- Requires total employee involvement

Figure 26–4
Major elements of total quality management (TQM).

- **Empowerment** of all personnel to take the action necessary to bring about product and/or process improvements in their areas of responsibility and to recommend action outside of their areas.
- **Communication**, both vertically and horizontally, on a continual basis. Communication must be a two-way activity. This means that supervisors and managers must *listen*.
- **Reinforcement** of teamwork-oriented behavior and product/process improvements. Any rewards given, regardless of whether they involve money, recognition, or any other reinforcer, should be given to teams, not individuals.
- **Respect** for the dignity and worth of all team members, regardless of status. This is critical. A team is like a family in that every member is important.

Figure 26–5
Team-building strategies.

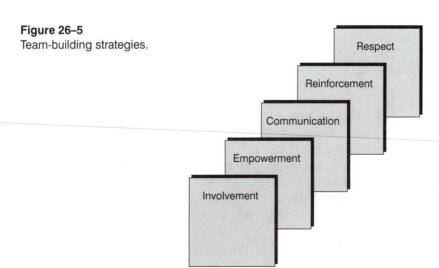

Respect

Reinforcement

Communication

Empowerment

Involvement

Training and continual retraining are fundamental to TQM. Employees must know how to do their job before they can be expected to do it well. Doing well means approaching all tasks in a manner that results in continual productivity and quality improvements while simultaneously ensuring a safe and healthy workplace. Therefore, training should be broad-based. It should encompass specific job skills training, team-building training, and safety/health training.

PSM PROGRAM RECORD KEEPING

Despite a company's best efforts, it may still have to defend itself in a product safety lawsuit. When this happens, it is critical that the company be able to produce records to support its defense. In the words of the National Safety Council,

> Complete, accurate records can be extremely convincing in a court of law. Consequently, records should be retained that are pertinent to all phases of a company's manufacturing, distributing, and importing activities, from the procurement of raw materials and components, through production and testing, to the marketing and distribution of the finished products. In addition to their usefulness in demonstrating to courts and juries that a company manufactures safe, reliable products, comprehensive PSM program records also permit a company to readily identify and locate products that its data collection and analysis system indicates may have reached the customer in a defective condition. Should a product recall or field-modification program become necessary, comprehensive PSM program records can result in successful implementation.[16]

In addition to the litigation-related reasons, there are other compelling reasons for keeping accurate, up-to-date records. One key reason is compliance. Comprehensive product safety records are necessary to satisfy compliance requirements of many federal regulations. The most prominent of these are listed in Figure 26–6. Additional reasons for maintaining accurate product safety records are as follows:

■ Comprehensive, up-to-date records are evidence of a company's commitment to producing quality products that pose no health or safety problems.

SAFETY FACT

Teams Must Have a Charter

A PSM team, like any other team, needs a charter. Teams need to know what their responsibilities are and what is expected of them. A team charter should contain at least the following elements:

- Clearly stated mission
- Broad goals
- Specific activities with schedules

Figure 26–6
Federal regulations requiring
product safety record keeping.

- Consumer Product Safety Act (PL 92-573)
- Federal Hazardous Substances Act (15 USC 1261)
- Federal Food, Drug, and Cosmetic Act (21 USC 321)
- Poison Prevention Packaging Act (PL 91-601)
- Occupational Safety and Health Act (PL 91-596)
- Child Protection and Toy Act (PL 91-113)
- Magnuson-Moss Warranty—Federal Trade Commission Improvement Act (PL 93-637)

- Records document the amount of care required to produce, market, and distribute a high-quality product that is safe and reliable and poses no health risks.
- Records allow company officials to track products from the point of production, through the distribution system, and to customers.
- Records give company officials a database of information needed for determining and comparing insurance costs, identifying sources of supply, and determining the costs of field modifications or product recalls.[17]

An issue that has traditionally caused problems for PSM program personnel is record storage and retention. In most companies, either corporate policy or the requirements of outside agencies dictate how long records of various types should be retained. This is not the case with product safety records. Since there is no way to know when the records might be needed as evidence in a lawsuit, the only way to be sure that they are retained long enough is to retain them forever.

Fortunately, record storage is no longer the problem that it was in the past. Technologies such as ROM (read-only-memory) and WORM (write once, read many) have made it easy to store huge quantities of records in an extremely small amount of space forever. Both ROM and WORM are computer technologies that were originally used by libraries for storing periodicals, newspapers, encyclopedias, and other research materials that rapidly absorb shelf space. These technologies are now available for any type of record storage.

USER FEEDBACK COLLECTION AND ANALYSIS

It is critical to the success of a PSM program that feedback from users of the company's products be collected and analyzed. **User feedback** can come in the form of compliments, testimonials, complaints, problems, or accident reports. Such feedback can help identify modifications that should be made in a product's design, problems with manufacturing processes, the need for a product recall, and potential lawsuits.

Customer feedback can be collected in many different ways. Companies can publish a toll-free "hot line" telephone number that customers can call. Customers can be tracked and periodically surveyed. The Mazda Corporation does both. After purchasing a new Mazda vehicle, the customer receives a wallet-size card containing a toll-free hot line to call with complaints or problems of any kind. The customer also receives periodic surveys and telephone calls from Mazda representatives.

Regardless of the collection methods used, it is important to have one central location into which feedback flows. It is also important that the PSM program coordinator have open access to this information and that it be shared regularly with the PSM committee for analysis.

The National Safety Council says the following about consumer feedback collection and analysis:

> Based on information acquired from complaints, incidents, and accident reports from customers, distributors or dealers, and state (or provincial) or federal agencies, a company must be capable of determining immediately if a substantial hazard exists. Techniques for making the determination include: on-site investigation of the complaint, incident, or accident; hazard and failure analyses of the unit involved in the complaint, incident, or accident by the company or an independent laboratory; analysis of other units of the same product batch; evaluation of in-house test, or other records. If a company finds that a substantial product hazard does exist, the appropriate product recall or field-modification plan should be implemented.[18]

PRODUCT LITERATURE AND SAFETY

A key component in a product safety management program is a company's **product literature**. Such literature includes assembly instructions, warning labels, technical manuals, and operating instructions. It is not uncommon for product liability lawsuits to include charges relating to poorly written product literature. In fact, from a legal perspective, the quality of product literature is as important as the quality of the product itself.

Darzinskis describes the purpose of product literature as follows:

> [the] proper role of safety instructions and warnings is to tell users about hazards that cannot be removed by design or controlled by guards and safety devices. A secondary role is to disclose a product's intended use, if this is not self-evident, as well as to recount the product's safety features and guards with the hope that understanding these features will encourage users not to remove, bypass, or alter them.[19]

Figure 26–7 contains a list of subjects that may be included in product literature. This list should be viewed as a menu from which subjects are selected. Once the subjects to be covered in a given piece of product literature have been identified, the literature can be developed. Darzinskis recommends the strategies discussed in the following paragraphs for producing high-quality, effective product literature.[20]

1. *Minimize and simplify narrative text.* Long, rambling paragraphs tend to turn off the hurried reader. Therefore, the use of narrative writing should be minimized. The text should consist of short, simple words and sentences written at a reading level below that of the intended audience.

Figure 26–7
Potential subjects for product literature.
Source: K. Darzinskis, "How to Put Safety into Product Literature," *Safety & Health,* June 1991, Vol. 143, No. 6, p. 74.

- General information
- Safety information
- How to remove shipping crates from around the product
- Product components
- Handling instructions/precautions
- Assembly instructions
- Installation instructions
- Tools, equipment, special clothing, and other materials needed in conjunction with the product
- Instructions for proper use
- Routine servicing instructions
- Maintenance information
- Troubleshooting procedures
- Overhaul requirements/instructions
- Disposal procedures

2. *Use illustrations whenever possible.* The old saying that a picture is worth a thousand words applies when developing product literature. Tables, charts, graphs, flowcharts, photographs, line drawings, and so on should be used instead of narrative text whenever possible. Illustrations supplemented by the written word can be an effective approach.

3. *Consider the eye appeal of the layout.* The **eye appeal** of product literature can be enhanced by applying the following tips: (a) do not justify (line up) right-hand margins of text material; (b) use boldface print or underlining for short passages only; (c) use a typeface that has large lowercase letters as compared with its uppercase letters; (d) avoid the use of reverse print except for short passages; (e) use blue, black, green, or purple ink for single-color printing on packages.

4. *Maximize drawing power.* It is important to produce product literature that compels the reader to read it. The use of headlines, widely recognized symbols, and color will increase the **drawing power** of the literature. Other strategies include increasing the amount of white space and using boxes around high-priority messages.

In addition to producing literature that is readable, compelling, and appealing to the eye, it is important to produce literature that is accurate. Product literature should never exaggerate or mislead users. The best design is rendered useless if the content of the readable, compelling, appealing information is incorrect.

=== SUMMARY ===

1. Prior to 1916, consumers and employees used products and machines at their own risk. In 1916, the concept of negligent manufacture was established in the law.

Since then, the concepts of breach of warranty and strict liability in tort have also been established.

2. Typically, product liability lawsuits fall within the realm of civil rather than criminal law. Redress comes in the form of a monetary award or settlement.

3. The concept of strict liability in tort makes manufacturers of products liable for physical harm that results from their use as long as the product is not altered by a party other than the manufacturer while it is en route to the customer.

4. Tort law gives manufacturers a duty to warn consumers of potential hazards associated with products.

5. The most significant product liability statutes include the following: Consumer Product Safety Act; Flammable Fabrics Act; Food, Drug, and Cosmetics Act; Hazardous Substances Act; Mine Health and Safety Act; National Traffic and Motor Vehicle Safety Act; Occupational Health and Safety Act; Poison Prevention Packaging Act; Refrigerator Safety Act; and Workers' Compensation Act.

6. The Consumer Product Safety Act of 1972 is designed to protect the public from injuries, help consumers evaluate the risk associated with products, encourage uniform standards, and encourage research.

7. Important legal concepts associated with product liability law are patent defect, latent defect, prudent man concept, reasonable risk, and unreasonable risk.

8. Title III of the Superfund Amendments and Reauthorization Act (SARA) is more commonly known as the Community Right-to-Know Act. The act has four components: emergency planning, emergency notification, reporting requirements, and toxic chemicals release reporting.

9. The three fundamental components of a PSM program are the PSM coordinator, PSM committee, and PSM auditor.

10. The role of the safety and health professional in the operation of a PSM program can vary depending on the size of the company and local circumstances. In some cases, a safety and health professional may serve as the program coordinator; in others, as the program auditor, and in others, as neither. In yet other cases, the safety and health professional may serve as an ex-officio in-house consultant to the process.

11. Total quality management, or TQM, is an approach to management that makes quality the responsibility of all employees while at the same time empowering them to make decisions/recommendations to improve quality. The goal is to meet or exceed customer expectations by building in quality. Key concepts in TQM are involvement, empowerment, communication, reinforcement, and respect.

12. It is important to collect customer feedback and use it to make product improvements. It can be collected by mail surveys, telephone contact, and a variety of other methods. Regardless of how it is collected, user feedback should be shared with the PSM coordinator and committee.

13. From a legal perspective, the quality of product literature can be as important as the quality of the actual product. The purpose of product literature is to define the product's intended use and protect consumers from hazards that cannot be eliminated in the design or manufacturing processes.

=========== KEY TERMS AND CONCEPTS ===========

Breach of warranty

Communication

Community Right-to-Know Act

Consumer Product Safety Act

Discovery period

Drawing power

Duty to warn

Empowerment

Eye appeal

Involvement

Latent defect

LEPC

Negligent design

Negligent manufacture

Patent defect

Product liability

Product literature

Product safety management (PSM) program

Prudent man concept

PSM auditor

PSM committee

PSM coordinator

Quality product

Reasonable risk

Reinforcement

Respect

SERC

Strict liability in tort

Teamwork

Total quality management (TQM)

Unreasonable risk

User feedback

Vacuum mentality

=========== REVIEW QUESTIONS ===========

1. Explain the best way for a manufacturer to limit its exposure to product liability lawsuits.
2. What are the steps to follow in filing a product liability lawsuit?
3. What is the discovery period in a product liability lawsuit?
4. Briefly explain the concept of strict liability in tort.
5. List the three criteria governing a company's duty to warn.
6. What are two purposes of the Consumer Product Safety Act?
7. Define the following concepts associated with product liability law: patent defect, latent defect, and prudent man concept.
8. Explain the reporting requirements component of the Community Right-to-Know Act.
9. List and briefly explain the three components of a comprehensive PSM program.
10. How may assigning PSM coordination duties to an executive of the company affect the success of the program?
11. What role do you think safety and health professionals should play in a PSM program?
12. Define the term *total quality management*. Explain the five key strategies of TQM.
13. Why is record keeping so important with PSM?
14. List four strategies for producing high-quality, effective product literature.

ENDNOTES

1. National Safety Council. *Accident Prevention Manual for Industrial Operations,* 9th ed. (Chicago: National Safety Council, 1988), p. 425.
2. Castelli, J. "Are Microwave Ovens Safer Than Ever?" *Family Safety and Health,* Summer 1991, Vol. 50, No. 2, p. 28.
3. Ibid.
4. Ibid.
5. Ibid., p. 264.
6. American Law Institute. Second Restatement of Torts, 1965, paragraph 402A.
7. Ibid., paragraph 388.
8. Colling, *Industrial Safety: Management and Technology,* p. 267.
9. Ibid.
10. Ibid., p. 266.
11. Ibid., p. 426.
12. National Safety Council, *Accident Prevention Manual for Industrial Operations,* p. 426.
13. Ibid.
14. Ibid., p. 427.
15. Ibid.
16. Ibid., p. 428.
17. Ibid., p. 434.
18. Ibid., p. 436.
19. Darzinskis, K. "How to Put Safety into Product Literature," *Safety & Health,* June 1991, Vol. 143, No. 6, p. 74.
20. Ibid., pp. 76–78.

CHAPTER TWENTY-SEVEN

Roles of Safety and Health Personnel

This book was designed for use by prospective and practicing safety and health managers. People with such titles are typically responsible to higher management for the safety and health of a company's workforce. Modern safety and health managers seldom work alone. Rather, they usually head a team of specialists that may include engineers, physicists, industrial hygienists, occupational physicians, and occupational health nurses.

It is important for safety and health managers today to understand not just their roles, but the roles of all members of the safety and health team. This chapter provides the information that prospective and practicing safety and health managers need to know about the roles of safety personnel in the age of high technology.

MODERN SAFETY AND HEALTH TEAMS

The issues that concern modern safety and health managers are multifaceted and complex. They include such diverse issues as stress; explosives; laws, standards, and codes; radiation; AIDS; product safety and liability; ergonomics; ethics; automation; workers' compensation; and an ever-changing multitude of others.

It would be unreasonable to expect one person to be expert regarding all of the many complex and diverse issues faced in the modern workplace. For this reason, the practice of safety and health management in the age of high technology has become a team sport. Figure 27–1 illustrates the types of positions that may comprise a safety and health team. In the remaining sections of this chapter, the roles, duties, responsibilities, and relationships of members of the safety and health team are described.

SAFETY AND HEALTH MANAGER

The most important member of the safety and health team is its manager. Companies that are committed to providing a safe and healthy workplace employ a **safety and health manager** at an appropriate level in the corporate hierarchy. The manager's position in the hierarchy is an indication of the company's commitment and priorities. This, more than anything else, sets the tone for a company's safety and health program.

In times past, companies with a highly placed safety and health manager were rare. However, passage of the OSHAct in 1970 (see Chapter 4) began to change this. The OSHAct, more than any other single factor, put teeth in the job descriptions of safety and health professionals. OSHA standards, on-site inspections, and penalties have encouraged a greater commitment to safety and health than was evident in the past. Environmental, liability, and workers' compensation issues have also had an impact, as has the growing awareness that providing a safe and healthy workplace is the right thing to do from both an ethical and a business perspective.

Job of the Safety and Health Manager

The job of the safety and health manager is complex and diverse. Figure 27–2 is an example of a job description for such a position. The description attests to the diverse

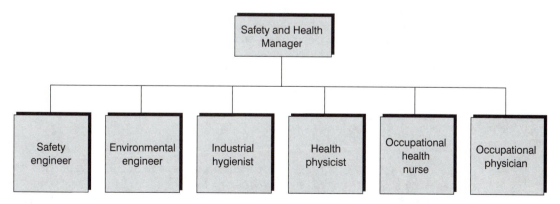

Figure 27–1
A modern safety and health team.

POULTRY PROCESSING, INC.

Highway 90 West
Crestview, Florida 32536

Vacancy Announcement

Position Title: Safety and Health Manager

Position Description: The Safety and Health Manager for PPI is responsible for establishing, implementing, and managing the company's overall safety and health program. The position reports to the local plant manager. Specific duties include the following:

- Establish and maintain a comprehensive company-wide safety and health program.
- Assess and analyze all departments, processes, and materials for potential hazards.
- Work with appropriate personnel to develop, implement, monitor, and evaluate accident prevention/hazard control strategies.
- Ensure company-wide compliance with all applicable laws, standards, and codes.
- Coordinate the activities of all members of the company's safety and health program.
- Plan, implement, and broker, as appropriate, safety and health-related training.
- Maintain all required safety and health-related records and reports.
- Conduct accident investigations as necessary.
- Develop and maintain a company-wide Emergency Action Plan (EAP).
- Establish and maintain an ongoing safety promotion effort.
- Analyze the company's products from the perspectives of safety, health, and liability.

Qualifications Required: The following qualifications have been established by the PPI management team with input from all levels and all departments.

- *Minimum Education.* Applicants must have at least a one-year community college certificate or an associate of science or applied science degree in industrial safety or a closely related degree (A.S. holders will begin work at a salary 15% higher than a certificate graduate).
- *Preferred Education.* Applicants with a baccalaureate degree in any of the following major fields of study will be given first priority: industrial safety and health, industrial technology, industrial management, manufacturing technology, engineering technology, and related. Degree programs in these fields must include at least one three-semester or five-quarter-hour course in industrial or occupational safety and health.

Figure 27–2
Safety and health manager job description.

nature of the job. Duties range from hazard analysis to accident reporting to standards/compliance to record keeping to training to emergency planning and so on.

The minimum educational requirement set by Poultry Processing, Inc. (PPI) is an occupational certificate from a community college with the full associate of science or applied science degree preferred. Preference is given to applicants who hold a baccalaureate degree in specifically identified fields.

Role in the Company Hierarchy

The safety and health manager described in Figure 27–2 reports to PPI's local plant manager and has line authority over all other members of the safety and health team. This and the duties set forth in the job description are evidence of the company's commitment to safety and health. They are also evidence that PPI is large enough to have a dedicated safety and health manager.

In some companies, the safety and health manager may also have other duties such as production manager or personnel manager. In such cases, the other members of the safety and health team, such as those shown in Figure 27–1, are not normally company employees. Rather, they are available to the company on a part-time or consultative basis as needed. The role that safety and health managers are able to play in a company will depend in part on whether their safety and health duties are full-time or added on to other duties.

Another role determinant is the issue of authority. Does the safety and health manager have line or staff authority? **Line authority** means that the safety and health manager has authority over and supervises certain employees (i.e., other safety and health personnel). **Staff authority** means that the safety and health manager is the staff person responsible for a certain function, but he or she has no line authority over others involved with that function.

Those occupying staff positions operate like internal consultants—that is, they may recommend, suggest, and promote, but they do not have the authority to order or mandate. This is typically the case with safety and health managers. Even managers with line authority over other safety and health personnel typically have a staff relationship with other functional managers (e.g., personnel, production, or purchasing). For example, consider the following safety and health-related situations:

1. A machine operator continually creates unsafe conditions by refusing to practice good housekeeping.
2. A certain process is associated with an inordinately high number of accidents.
3. A new machine is being purchased that has proven itself to be unsafe at other companies that have purchased it.

In the first example, the safety and health manager could recommend that the employee be disciplined but could not normally undertake or administer disciplinary measures. In the second example, the safety and health manager could recommend that the process be shut down until a thorough analysis could be conducted, hazards identified, and corrective measures taken. However, the manager would rarely have the authority to order the process to be shut down. In the final example, the safety and health manager could recommend that an alternative machine be purchased, but he or she would not normally have the authority to stop the purchase.

Maintaining a safe and healthy workplace while playing the role of internal consultant is often the greatest challenge of safety and health managers. It requires managers to be resourceful, clever, astute with regard to corporate politics, good at building relationships, persuasive, adept at trading for favors, credible, and talented in the development and use of influence.

Problems Faced by Safety and Health Managers

As if the diversity and complexity of the job were not enough, there are a number of predictable problems that safety and health managers are likely to face. These problems are discussed in the following paragraphs.

Lack of Commitment

Top management may go along with having a company-wide safety and health program because they see it as a necessary evil. The less enthusiastic may even see safety and health as a collection of government regulations that interfere with profits. Although this is less often true now than it has been in the past, safety and health professionals should be prepared to confront a less than wholehearted commitment in some companies.

Production vs. Safety

Industrial firms are in business to make a **profit.** They do this by producing or processing products. Therefore, anything that interferes with production or processing is likely to be looked on unfavorably. At times, a health or safety measure will be viewed by some as interfering with productivity. A common example is removal of safety devices from machines as a way to speed production. Another is running machines until the last possible moment before a shift change rather than shutting down with enough time left to perform routine maintenance and housekeeping tasks.

The modern marketplace has expanded globally and, therefore, become intensely competitive. To survive and succeed, today's industrial firm must continually improve its productivity, quality, cost, image, response time, and service. This sometimes puts professionals who are responsible for safety and health at odds with others who are responsible for productivity, quality, cost, and response time.

Sometimes, this cannot be avoided. At other times, it is the fault of a management team that is less than fully committed to safety and health. However, sometimes the fault rests squarely on the shoulders of the safety and health manager. This is because one of the most important responsibilities of this person is to convince higher management, middle management, supervisors, and employees that, in the long run, the safe and healthy way of doing business is also the competitive, profitable way of doing business. The next section explains several strategies for making this point.

Gaining a Commitment to Safety and Health

In many cases, safety and health managers have been their own worst enemy when it comes to gaining a company-wide commitment. The most successful are those who understand the goals of improved productivity, quality, cost, image, service, and response time and are able to convey the message that a safe and healthy workplace is the best way to accomplish these goals. The least successful are those who earn a reputation for being grumpy in-house bureaucrats who quote government regulations chapter and verse but know little and care even less about profits. Unfortunately, in the past, there have been too many safety and health managers who fall into the latter category. This is not the way to gain a company-wide commitment to safety and health, but it is a sure way to engender resentment.

Today's safety and health manager must understand the bottom-line concerns of management, supervisors, and employees and be able to use these concerns to gain a commitment to safety and health. Figure 27–3 illustrates the essential message that **competitiveness** comes from continually improving a company's productivity, quality, cost, image, service, and response time. These continual improvements can be achieved and maintained best in a safe and healthy work environment.

Safety and health managers should use this message to gain a commitment from management and employees. Following are some strategies that can be used to get the point across.

Productivity, Quality, Cost, and Response Time

These four factors, taken together, are the key to productivity in the age of high technology and global competitiveness. The most **productive** company is the one that generates the most output with the least input. Output is the company's product. Input is any resource—time, talent, money, technology, and so on—needed to produce the product. **Quality** is a measure of reliability and customer satisfaction. **Cost** is the amount of money required to purchase the item. If all other factors are equal, customers will select the product that costs less. **Response time** is the amount of time that elapses between an order being placed and the product being delivered.

Figure 27–3
Factors that produce competitiveness.

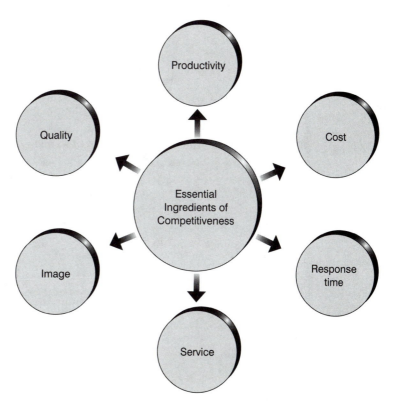

To compete in the global marketplace, industrial companies must continually improve these four factors. At the most fundamental level, successfully competing in the global marketplace means having the best people and the best technology and getting the most out of both by applying the best management strategies.

Safety and health managers who understand this can use their knowledge to gain a commitment to their programs. In attempting to do so, the following five points are helpful:

1. If it is important to attract and keep the best people, a safe and healthy workplace will help.
2. If it is important to get the most out of talented people, it must be important to keep them safe and healthy so that they are functioning at peak performance levels (for example, the best technician in the world can't help when he or she is out of action as the result of an injury or illness).
3. Employees cannot concentrate fully on quality when they are concerned for their safety and health.
4. Keeping industrial technologies up-to-date requires the continual investment of funds. Profits that are siphoned off to pay the costs associated with accidents, emergencies, and health problems cannot be reinvested in the latest technologies needed to stay competitive.
5. With the skyrocketing costs of medical care, workers' compensation, and litigation, it costs less to prevent accidents than to pay for them.

Image and Service

Image and **service** are also important factors in the competitiveness equation. Of the two, image relates more directly to safety and health.

In today's intensely competitive marketplace, a company's image, internal and external, can be a deciding factor in its ability to succeed. Companies that establish a solid internal image in terms of safety and health will find it easier to attract and keep the best employees. Companies that establish a solid external image with regard to environmental and product safety issues will find it easier to attract and retain customers.

Gaining a full and real commitment to safety and health in the workplace is one of the most important roles of the safety and health manager. Traditionally, safety and health managers have argued their cases from the perspectives of ethics or government mandates. The ethical argument is as valid now as it has always been and should continue to be used.

However, in today's competition-driven workplace, managers responsible for the bottom line may resent arguments that are based on government mandates and regulations. On the other hand, these same managers may respond positively if they can be shown that resources invested in safety and health can actually improve a company's competitiveness. Using the points made earlier about productivity, quality, cost, image, and response time can go a long way in helping to gain management commitment to safety and health.

Education and Training for Safety and Health Managers

Advances in technology, new federal legislation, the potential for costly litigation, and a proliferation of standards have combined to make the job of safety and health professionals more complex than ever before. These factors have correspondingly increased the importance of education and training for safety and health managers. The ideal formula for safety and health professionals is **formal education** prior to entering the profession supplemented by **in-service training** on a lifelong basis afterward.

Universities, colleges, and community colleges across the country have responded to the need for formal education for safety and health managers as well as other safety and health personnel. Many community colleges offer occupational certificates and **associate of applied science** and/or **associate of science degrees** with such program titles as industrial safety, occupational safety, environmental technology, safety and health management, and industrial hygiene.

Universities have responded to the need for formal education by making safety and health-related courses either optional or required parts of such **baccalaureate degree** programs as industrial technology, manufacturing technology, engineering technology, industrial engineering technology, industrial management, and industrial engineering. Some universities offer baccalaureate degrees in industrial safety and health, occupational safety management, and industrial hygiene.

Formal education provides the foundation of knowledge needed to enter the profession. Having begun a career as a safety and health manager, the next challenge is keeping up as the laws, regulations, standards, and overall body of knowledge relating to safety and health grow, change, and evolve.

In-service training, ongoing interaction with professional colleagues, and continued reading of professional literature are effective ways to stay current. New safety and health managers should move immediately to get themselves "plugged into" the profession. This means joining the appropriate professional organizations, becoming familiar with related government agencies, and establishing links with relevant standards organizations. The next section covers agencies and organizations that can be particularly helpful to safety and health managers.

Helpful Agencies and Organizations

Numerous agencies and organizations are available to help the safety and health manager keep up-to-date. These agencies and organizations provide databases, training, and/or professional literature. There are professional societies, trade associations, scientific organizations, certification boards, service organizations, and emergency service organizations. Those listed here represent only a portion of those available.

■ *Certification boards.* Professional certification is an excellent way to establish one's status in the field of safety and health. To qualify to take a certification examination, safety and health managers must have the required education and experience and submit letters of recommendation as specified by the certification board. Figure

27–4 contains the names and addresses of **certification boards** of interest to safety and health managers.

■ *Professional societies.* **Professional societies** are typically formed for the purpose of promoting professionalism, adding to the body of knowledge, and forming networks among colleagues in a given field. Numerous professional societies focus on various safety and health issues. Figure 27–5 summarizes some of these.

■ *Scientific standards/testing organizations.* Scientific **standards/testing organizations** conduct research, run tests, and establish standards that identify the acceptable levels for materials, substances, conditions, and mechanisms to which people might be exposed in the modern workplace. Figure 27–6 summarizes those of critical importance to safety and health managers.

■ *Government agencies.* Many **government agencies** are concerned with various aspects of workplace-related safety and health. Some of the most helpful agencies for the safety and health manager are listed in Figure 27–7.

■ *Trade associations.* The purpose of a **trade association** is to promote the trade that it represents. Consequently, material produced by trade associations can be somewhat self-serving. Even so, trade associations can be valuable sources of information and training for safety and health managers. Figure 27–8 lists some of the trade associations that can be particularly helpful.

ENGINEERS AND SAFETY

Engineers can make a significant contribution to safety. Correspondingly, they can cause, inadvertently or through incompetence, accidents that result in serious injury and property damage. The engineer has more potential to affect safety in the workplace than any other person does. The following example illustrates this:

> A car-pooler transports himself and three fellow employees to work each day. He is not a particularly safe driver and does not insist that his passengers use seatbelts. After running a red light, he crashes into the side of a building while swerving to avoid another vehicle. The two passengers wearing seatbelts are not hurt, but the driver and one other passenger, neither of whom were wearing seatbelts, are critically injured.

Figure 27–4
Professional certification boards.

Safety Related:

Board of Certified Safety Professionals of America
208 Burwash
Savoy, Illinois 61874

Health Related:

American Board of Industrial Hygiene
4600 W. Saginaw, Suite 101
Lansing, Michigan 48917

Figure 27–5

Professional societies.
* The National Safety Council is the most important professional organization for safety and health managers.

American Academy of Industrial Hygiene
302 S. Waverly Road
Lansing, Michigan 48917

American Industrial Hygiene Association
475 Wolf Ledges Parkway
Akron, Ohio 44311

American Occupational Medical Association
2340 S. Arlington Heights Road
Arlington Heights, Illinois 60005

American Society of Safety Engineers
1900 E. Oakton Street
Des Plaines, Illinois 60016

National Safety Council*
444 North Michigan Avenue
Chicago, Illinois 60611

Society of Toxicology
1133 I Street, N.W., Suite 800
Washington, D.C. 20005

Figure 27–6

Scientific standards/testing organizations.

American National Standards Institute (ANSI)
1430 Broadway
New York, New York 10018

International Atomic Energy Agency
Wagramstrasse 5
A01400 Vienna, Austria

National Fire Protection Association (NFPA)
Batterymarch Park
Quincy, Massachusetts 02269

Underwriters Laboratories, Inc. (UL)
333 Pfingsten Road
New York, New York 10017

American Society of Mechanical Engineers (ASME)
345 East 47th Street
New York, New York 10017

American Society for Testing and Materials (ASTM)
1916 Race Street
Philadelphia, Pennsylvania 19103

Bureau of Mines Department of the Interior 2401 E. Street, N.W. Washington, D.C. 20241	National Center for Toxicological Research Food and Drug Administration 5600 Fishers Lane Rockville, MD 20857

Bureau of Mines
Department of the Interior
2401 E. Street, N.W.
Washington, D.C. 20241

National Institute for Occupational Safety and
  Health
Department of Health and Human Services
Parklawn Building
5600 Fishers Lane, Road 1401
Rockville, MD 20857

Center for Devices and Radiological Health
Food and Drug Administration
8757 Georgia Avenue
Silver Spring, MD 20910

Consumer Product Safety Commission
1111 Eighteenth Street, N.W.
Washington, D.C. 20207

Department of Labor
200 Constitution Avenue, N.W.
Washington, D.C. 20210

Environmental Protection Agency
401 M Street, S.W.
Washington, D.C. 20460

Mine Safety and Health Administration
Department of Labor
4015 Wilson Boulevard
Arlington, VA 22203

National Bureau of Standards Headquarters
Route I-270 and Quince Orchard Road
Gaithersburg, MD 20899

National Center for Toxicological Research
Food and Drug Administration
5600 Fishers Lane
Rockville, MD 20857

National Technical Information Service
  Department of Commerce
5285 Port Royal Road
Springfield, VA 22161

Nuclear Regulatory Commission
1717 H Street, N.W.
Washington, D.C. 20555

Occupational Safety and Health Commission
Department of Labor
200 Constitution Avenue, N.W.
Washington, D.C. 20210

Occupational Safety and Health Review
  Committee
1825 K Street N.W.
Washington, D.C. 20006

Office of Energy Research
Department of Energy
1000 Independence Avenue, S.W.
Washington, D.C. 20585

Office of Hazardous Materials Transportation
400 Seventh Street, S.W.
Washington, D.C. 20590

U.S. Fire Administration
Federal Emergency Management Agency
16825 S. Seton Avenue
Emmitsburg, MD 21727

Figure 27–7
Government agencies.

This brief story illustrates an accident that has two things in common with many workplace accidents. The first is a careless worker—in this case, the driver. The second is other workers who do not follow prescribed safety rules—in this case, failing to use seatbelts. Employees such as these can and do cause many workplace accidents, but even the most careless employee cannot cause a fraction of the problems caused by a careless engineer. The following example illustrates this point:

An engineer is charged with the responsibility for designing a new seatbelt that is comfortable, functional, inexpensive, and easy for factory workers to install. He designs a belt that meets all of these requirements, and it is installed in 10,000 new cars. As the cars are bought and accidents begin to occur, it becomes apparent that the new seatbelt fails in crashes involving speeds over 36 miles per hour. The engineer who designed the belt took all factors into consideration except one: *safety*.

This brief story illustrates how far-reaching an engineer's impact can be. With a poorly designed seatbelt installed in 10,000 automobiles, the engineer has inadvertently endangered the lives of as many as 40,000 people (estimating a maximum of four passengers per automobile).

The engineer's opportunity for both good and bad comes during the design process. The process is basically the same regardless of whether the product being designed is a small toy, an industrial machine, an automobile, a nuclear power plant, a ship, a jumbo jetliner, or a space vehicle. Safety and health professionals should be familiar with the design process so that they can more fully understand the role of engineers concerning workplace safety.

Not all engineers are **design engineers**. However, engineers involved in design are usually in the aerospace, electrical, mechanical, and nuclear fields. The following paragraphs give an overview of these design-oriented engineering fields as seen in the course descriptions of a college catalog.

Figure 27–8
Trade associations.*
* Complete addresses are available in the reference section of most college and university libraries.

Alliance of American Insurers
American Foundrymen's Association
American Insurance Association
American Iron and Steel Institute
American Metal Stamping Association
American Petroleum Institute
American Welding Society
Associated General Contractors of America
Compressed Gas Association
Industrial Safety Equipment Association
Institute of Makers of Explosives
Lead Industrial Association
National Electrical Manufacturers Association
National LP-Gas Association
National Machine Tool Builders Association
Scaffolding, Shoring, and Forming Institute
Soap and Detergent Association
Technical Association of the Pulp and Paper Industry

SAFETY FACT

The Calculus Controversy in Safety and Health Programs

Many of the programs in colleges and universities that prepare safety and health professionals require calculus. Some practicing professionals in the field think that college programs require too much math and too little management, business, and international safety. These professionals argue that calculus has no value to them on the job, but that more business and management courses would. The opposing faction in the field continues to defend calculus as necessary and important. A third faction supports calculus and business/management courses, claiming that the safety and health professional needs both.

Aerospace Engineering

The Bachelor of Science in Engineering (Aerospace Engineering) program incorporates a solid foundation of physical and mathematical fundamentals which provides the basis for the development of the engineering principles essential to the understanding of both atmospheric and extra-atmospheric flight. Aerodynamics, lightweight structures, flight propulsion, and related subjects typical of aeronautical engineering are included. Other courses introduce problems associated with space flight and its requirements. Integration of fundamental principles with useful applications is made in design work in the junior and senior years. Thus, the program prepares the student to contribute to future technological growth which promises exciting and demanding careers in aerospace engineering. . . . Examples of concentration areas are: aerodynamics; design; flight propulsion; flight structures; space technology; stability, control, and guidance.[1]

Electrical Engineering

Electrical Engineering is a science-oriented branch of engineering primarily concerned with all phases and development of the transmission and utilization of electric power and intelligence. The study of electrical engineering can be conveniently divided into the academic areas of circuits, electronics, electromagnetics, electric energy systems, communications, control, and computer engineering.[2]

Mechanical Engineering

Mechanical Engineering is the professional field that is concerned with motion and the processes whereby other energy forms are converted into motion. Mechanical engineers are the people who are responsible for conceiving, designing, manufacturing, testing, and marketing devices and systems that alter, transfer, transform, and utilize the energy forms that ultimately cause motion. Thus mechanical engineers . . . are the people who make the engines that power ships, trains, automobiles, and spacecraft; they design the power plants which convert the energy in fuels, atoms, waterfalls, and sunlight into useful mechanical forms; and they construct intelligent machines and robots as well as the gears, cams, bearings, and couplings that facilitate and control all kinds of mechanical motion.[3]

Nuclear Engineering

Nuclear Engineering Sciences comprises those fields of engineering and science directly concerned with the release, control, and safe utilization of nuclear energy. Applications range over such broad topics as the design, development, and operation of nuclear reactor power systems to the applications of radiation in medicine, space, industry, and other related areas. The nuclear engineer, by virtue of his/her engineering and science-based training, is in a unique position to contribute to the many diverse aspects of this major component of the energy radiation field.[4]

The Design Process

Professor William S. Chalk describes the **design process** as follows:

> The design process is a plan of action for reaching a goal. The plan, sometimes labeled problem-solving strategy, is used by engineers, designers, drafters, scientists, technologists, and a multitude of professionals.[5]

The design process proceeds in five sequential steps:

1. *Problem identification.* Engineers draft a description of the problem. This involves gathering information, considering constraints, reviewing specifications, and combining all of these into a clear and concise description of the problem.

2. *Synthesis.* Engineers combine or synthesize systematic, scientific procedures with creative techniques to develop initial solutions to the problem identified in step 1. At this point, several possible solutions might be considered.

3. *Analysis and evaluation.* All potential solutions developed in the previous step are subjected to scientific analysis and careful evaluation. Such questions as the following are asked: Will the proposed solution satisfy the functional requirements? Will it meet all specifications? Can it be produced quickly and economically?

4. *Document and communicate.* Engineering drawings, detailed calculations, and written specifications are prepared. These document the design and communicate its various components to interested parties. It is common to revise the design at this point based on feedback from different reviewers.

5. *Produce and deliver.* Shop or detail drawings are developed and the design is produced, usually as a prototype. The prototype is analyzed and tested. Design changes are made if necessary. The product is then produced and delivered.

The design process gives engineers unparalleled opportunities to contribute significantly to safety in the workplace and in the marketplace by producing products with safety built into them. However, in too many cases, the design process does not serve this purpose. There are two primary reasons for this:

1. In analyzing and evaluating designs, engineers consider such factors as function, cost, life span, and manufacturability. All too often safety is not even considered or is only a secondary consideration.
2. Even when engineers do consider safety in analyzing and evaluating designs, many are insufficiently prepared to do so effectively.

Engineers who design products may complete their entire college curriculum without taking even one safety course. Safety courses, when available to engineering students in design-oriented disciplines, often tend to be electives. This limits the contributions that design engineers can make to both product and workplace safety. Figure 27–9 is a typical core-course listing for a mechanical engineering student.

Safety Engineer

The title **safety engineer** is often a misnomer in the modern workplace. It implies that the person filling the position is a degreed engineer with formal education and/or special training in workplace safety. Although this is sometimes the case, typically the title is given to the person who has overall responsibility for the company's safety program (the safety manager) or to a member of the company's safety team. This person is responsible for the traditional aspects of the safety program, such as preventing mechanical injuries; falls, impact, and acceleration injuries; heat and temperature injuries; electrical accidents; fire-related accidents, and so on.

In the former case, the person should be given a title that includes the term *manager*. In the latter case, the title *safety engineer* is appropriate. However, persons with academic credentials in areas other than engineering should be encouraged to seek such positions since they are likely to be at least as well prepared and possibly even better prepared than persons with engineering degrees. These other educational disciplines include industrial technology, industrial engineering technology, manufacturing technology, engineering technology, industrial management, and industrial safety technology (baccalaureate or associate degree).

There are signs that engineering schools are becoming more sensitive to safety and health issues. Graduate degrees in such areas as nuclear physics and nuclear engineering now often require safety courses. The federal government sponsors postgraduate studies

Figure 27–9
Typical required core courses for the mechanical engineering degree.

Engineering mechanics: Statics

Engineering mechanics: Dynamics

Mechanics of materials

Kinematics and dynamics of machinery

Manufacturing processes

Control of mechanical engineering systems

Mechanical vibrations

Machine analysis and design

Thermodynamics

Heat transfer

Fluid dynamics

in safety. However, the following quote summarizes clearly and succinctly the current status of safety engineering:

> Four states now have registration of professional engineers in a Safety Engineer discipline. Registration gives the registrant the right to use the title "Safety Engineer," but the enabling law has no other requirement that the services of such an engineer be used. The contracting office of one military service responsible for development of advanced high-tech systems does require certain hazard analyses and documents to be signed off and approved at specific points in the designs. Such approval will be valid only under the signature of a registered safety engineer or other engineer shown to have had extensive experience in safety programs. The principal problem is that in a new advanced design project, there may be 400 engineers with no training in accident avoidance who may make critical errors and only one or two safety engineers to find them.[6]

Industrial Engineers and Safety

Industrial engineers are the most likely candidates from among the various engineering disciplines to work as safety engineers. Their knowledge of industrial systems, both manual and automated, can make them valuable members of a design team, particularly one that designs industrial systems and technologies. They can also contribute after the fact as a member of a company's safety team by helping to design job and plant layouts for both efficiency and safety.

The industrial engineering discipline can be described as follows:

> Industrial growth has created unusual opportunities for the industrial and systems engineer. Automation and the emphasis on increased productivity coupled with higher levels of systems sophistication are providing impetus to the demand for engineering graduates with a broad interdisciplinary background. The industrial engineering option prepares the student for industrial practice in such areas as product design, process design, plant operation, production control, quality control, facilities planning, work system analysis and evaluation, and economic analysis of operational systems.[7]

Although industrial engineers are more likely to work as safety engineers than are engineers from other disciplines, they are not much more likely to have safety courses as a required part of their program of study. However, their focus on industrial systems and the integration of people and technology does give industrial engineers a solid foundation for additional learning through either in-service training or graduate work.

Environmental Engineers and Safety

A relatively new discipline (when compared with more traditional disciplines such as mechanical engineering) is environmental engineering. This discipline may be described as follows:

> Environmental Engineering Sciences is a field in which the application of engineering and scientific principles is used to protect and preserve human health and the well-being of the environment. It embraces the broad field of the general environment including air and water quality, solid and hazardous wastes, water resources and management, radiological health,

environmental biology and chemistry, systems ecology, and water and waste-water treatment.[8]

With the addition of health concerns to the more traditional safety concerns, **environmental engineers** will be sought as members of corporate safety and health teams. The coursework they take is particularly relevant since all of it relates either directly or indirectly to health. Figure 27–10 shows the types of courses typically required of environmental engineering students.

A person with the type of formal education shown in this figure would be a valuable addition to the safety and health team of any modern industrial firm. Environmental engineers typically report to the overall safety and health manager and are responsible for those elements of the program relating to hazardous waste management, atmospheric pollution, indoor air pollution, water pollution, and wastewater management.

Chemical Engineers and Safety

Increasingly, industrial companies are seeking **chemical engineers** to fill the industrial hygiene role on the safety and health team. Their formal education makes people in this discipline well equipped to serve in this capacity. Chemical engineering may be described as follows:

> Although chemical engineering has existed as a field of engineering for only about 80 years, its name is no longer completely descriptive of this dynamic, growing profession. The work of the chemical engineer is neither restricted to the chemical industry nor limited to chemical changes or chemistry. Instead, modern chemical engineers, who are also called process engineers, are concerned with all the physical and chemical changes of matter to produce economically a product or result that is useful to mankind. Such a broad background has made the chemical engineer extremely versatile and capable of working in a wide variety of industries: chemical, petroleum, aerospace, nuclear, materials, microelectronics, sanitation, food processing, and computer technology. The chemical industry alone provides an opportunity for the chemical engineer to participate in the research, development, design, or operation of

Figure 27–10
Typical required core courses for environmental engineering students.

> Environmental biology
> Environmental chemistry
> Water chemistry
> Atmospheric pollution
> Solid waste management
> Water and wastewater
> Hazardous waste control
> Environmental resources management
> Air pollution control design
> Hydraulic systems design

plants for the production of new synthetic fibers, plastics, chemical fertilizers, vitamins, antibiotics, rocket fuels, nuclear fuels, paper pulp, photographic products, paints, fuel cells, transistors, and the thousands of chemicals that are used as intermediates in the manufacture of the above products.[9]

INDUSTRIAL HYGIENIST

Industrial hygiene is defined by the American Industrial Hygiene Association as the "science and art devoted to the recognition, evaluation, and control of those environmental factors or stresses, arising in and from the workplace, which may cause sickness, impaired health and well-being, or significant discomfort and inefficiency among workers or among citizens of the community."[10] The National Safety Council describes the job of the **industrial hygienist** as follows:

> An industrial hygienist is a person having a college or university degree or degrees in engineering, chemistry, physics, medicine, or related physical and biological sciences who, by virtue of special studies and training, has acquired competence in industrial hygiene. Such special studies and training must have been sufficient in all of the above cognate sciences to provide the abilities: (a) to recognize environmental factors and to understand their effect on humans and their well-being; (b) to evaluate, on the basis of experience and with the aid of quantitative measurement techniques, the magnitude of these stresses in terms of ability to impair human health and well-being; and (c) to prescribe methods to eliminate, control, or reduce such stresses when necessary to alleviate their effects.[11]

Industrial hygienists are primarily concerned about the following types of hazards: solvents, particulates, noise, dermatoses, radiation, temperature, ergonomics, toxic substances, biological substances, ventilation, gas, and vapors. In a safety and health team, the industrial hygienist typically reports to the safety and health manager.

HEALTH PHYSICIST

Health physicists are concerned primarily with radiation in the workplace. Consequently, they are employed by companies that generate and/or use nuclear power. Their primary duties include the following: monitoring radiation inside and outside the facility, measuring the radioactivity levels of biological samples, developing the radiation components of the company's emergency action plan, and supervising the decontamination of workers and the workplace when necessary.

Nuclear engineering and nuclear physics are the two most widely pursued fields of study for health physicists. A study conducted by Moeller and Eliassen gave the following breakdown of the academic preparation of practicing health physicists:[12]

Associate degree   5.41%

Baccalaureate degree   28.38%

Master's degree   42.43%

Doctorate degree   19.46%

The remaining practitioners are nondegreed personnel who have completed various types of noncollege credit training. This breakdown shows that graduate study is particularly important for health physicists. Professionals in this field may be certified by the American Board of Health Physics (ABHP).

OCCUPATIONAL PHYSICIAN

Occupational medicine as a specialized field dates back to World War II, when the United States experienced unprecedented industrial expansion. Production of manufactured goods skyrocketed, and workplace-related medical needs followed suit. Occupational medicine was not classified as a medical specialty by the American Board of Preventive Medicine until 1955, however. The National Safety Council describes the main concerns of the **occupational physician** as follows:

- Appraisal, maintenance, restoration, and improvement of the workers' health through application of the principles of preventive medicine, emergency medical care, rehabilitation, and environmental medicine.
- Promotion of a productive and fulfilling interaction of the worker and the job, via application of principles of human behavior.
- Active appreciation of the social, economic, and administrative needs and responsibilities of both the worker and work community.
- Team approach to safety and health, involving cooperation of the physician with occupational or industrial hygienists, occupational health nurses, safety personnel, and other specialties.[13]

Occupational physicians are fully degreed and licensed medical doctors. In addition they must have completed postgraduate work in the following areas of safety:

> . . . biostatistics and epidemiology, industrial toxicology, work physiology, radiation (ionizing and nonionizing), noise and hearing conservation, effects of certain environmental conditions such as high altitude and high pressures (hyperbaric and hypobaric factors), principles of occupational safety, fundamentals of industrial hygiene, occupational aspects of dermatology, psychiatric and psychological factors, occupational respiratory diseases, biological monitoring, ergonomics, basic personnel management functions, record and data collection, governmental regulations, general environmental health (air, water, ground pollution, and waste management control).[14]

Bernardino Ramazzini is widely thought of as being the first occupational physician. This is primarily as a result of his study of the work-related problems of workers in Modena, Italy, and a subsequent book he authored entitled *The Diseases of Workers* (1700).[15] The first leading occupational physician in the United States was Alice Hamilton, M.D. According to the National Safety Council,

> In 1910 Dr. Hamilton became managing director of the Illinois Occupational Disease Commission in the United States. Members of this commission were given one year to study and report on health hazards in Illinois industries. Dr. Hamilton investigated lead poisoning in workers who manufactured white lead carbonate, ceramics, railway carriages, automobiles, batteries, printing and mining, and smelting of the metal itself. She fought with company

officials for changes in their plants; she lectured to the public; she worked with the Hull House community; she wrote books and magazine articles on occupational disease; she was appointed to Harvard Medical School in 1921 as assistant professor of medicine (the first woman to hold a teaching position there); and she was the only woman member of the League of Nations Health Committee in the 1920s.[16]

Pioneers such as Ramazzini and Hamilton paved the way for the approximately 3,000 occupational physicians practicing today. Whereas in the past the primary role of the occupational physician was treatment, today the primary role is prevention. This means more analytical, diagnostic, and intervention-related work than in the past and a much more proactive rather than passive approach.

One major difference in the role of modern occupational physicians when compared with those in the past is in their relationship with employers. In the past, the occupational physician tended to be an in-house physician. Over 80 percent of the Fortune 100 companies still have at least one in-house occupational physician. However, the trend is away from this approach to the contracted approach in all but the largest of companies. Companies contract with a private physician, clinic, or hospital to provide specific medical services. The health-care provider is typically placed on a retainer in much the same manner as an attorney.

This contracted approach presents a new challenge for the safety and health manager. Including the occupational physician in planning, analyzing, assessing, monitoring, and other prevention-related duties can be more difficult when the physician is not in-house and readily available. For this reason, the safety and health manager must work with higher management to develop a contract that builds in time for prevention activities and compensates the health-care provider appropriately.

The National Safety Council makes the following recommendations concerning occupational physicians (OPs):

> Whether your company uses an in-house OP, an OP that works as a consultant or a clinic that provides all health care, remember:
>
> - There should be a written medical program available to all management and employees.
> - The OP should understand the workplace and the chemicals used and produced.
> - Periodic tours of all facilities are necessary for an understanding of possible work-related injuries, and also to aid in job accommodation.
> - The OP should be familiar with OSHA and NIOSH health mandates.
> - The OP should be the leader of other medical personnel.
> - And, the OP should understand what the company expects and what the OP expects from the company.[17]

OCCUPATIONAL HEALTH NURSE

Occupational health nurses have long been important members of corporate safety and health programs. According to the American Association of Occupational Health Nurses,

> Occupational health nursing is the application of nursing principles in conserving the health of workers in all occupations. It involves prevention, recognition, and treatment of illness and injury, and requires special skills and knowledge in the areas of health education and counselling, environmental health, rehabilitation, and human relations.[18]

DISCUSSION CASE

What Is Your Opinion?

"What do you mean, we will need a *team* of safety and health professionals? Can't we just hire a safety manager or a safety engineer?" asked the frustrated CEO. The consultant smiled understandingly, but stood firm. "Look," said the consultant, "you have just acquired two new companies. One is a chemical processing plant. The other is a manufacturer that produces hazardous by-products. Both have enormous profit potential, but right now those profits are being drained off by legal and medical expenses. The little you invest in the safety and health personnel that I'm recommending will pay off many times over in the future." "But I don't like to hire support personnel," countered the CEO. "If they don't make a direct contribution, why hire them?" Who is right in this discussion? What is your opinion?

Like occupational physicians, today's occupational health nurses have seen their profession evolve over the years. The shift in emphasis is away from after-the-fact treatment to prevention-related activities such as analysis, monitoring, counseling, and education.

The American Association of Occupational Nurses defines the objectives of occupational nurses as follows:

■ To adopt the nursing program to meet the specific needs of the individual company

■ To give competent nursing care for all employees

■ To ensure that adequate resources are available to support the nursing program

■ To seek out competent medical direction if it is not available on-site

■ To establish and maintain an adequate system of records relating to workplace health care

■ To plan, prepare, promote, present, and broker educational activities for employees

■ To establish and maintain positive working relationships with all departments within the company

■ To maintain positive working relationships with all components of the local health-care community

■ To monitor and evaluate the nursing program on a continual basis and adjust accordingly[19]

Like all members of the safety and health team, occupational nurses are concerned about cost containment. Cost containment is the driving force behind the emphasis on activities promoting health and accident prevention.

Occupational nurses typically report to an occupational physician when he or she is part of the on-site safety and health team. In companies that contract for off-site physician services, occupational nurses report to the overall safety and health manager.

RISK MANAGER

Organizations are at risk every time they open their doors for business. On any given day, an employee may be injured, a customer may have an accident, or a consumer may be injured using the organization's product. *Risk* is defined as a specific contingency or peril. Because the situations that put organizations at risk can be so expensive when they do occur, many organizations employ risk managers.

Risk management consists of the various activities and strategies that an organization can use to protect itself from situations, circumstances, or events that may undermine its security. You are, yourself, a risk manager. You take action every day to protect your personal and economic security. Take, for instance, the act of driving an automobile—a risky undertaking. Every time you drive you put yourself at risk of incurring injuries, medical bills, lawsuits, and property damage.

You manage the risk associated with driving using two broad strategies: **reduction** and **transference.** The risk associated with driving an automobile can be reduced by wearing a seatbelt, driving defensively, and obeying traffic laws. The remaining risk is managed by transferring it to an insurance company by purchasing a policy that covers both collision and liability.

The same approach—managing risk by using reduction and transference strategies—applies in the workplace. Risk managers work closely with safety and health personnel to reduce the risk of accidents and injuries on the job. They also work closely with insurance companies to achieve the most effective transference possible.

SUMMARY

1. The modern safety and health team is headed by a safety and health manager. Depending on the size of the company and the commitment of its management, the team may include people in the following additional positions: safety engineer, industrial hygienist, environmental engineer, health physicist, occupational physician, and occupational health nurse.

2. The job of the safety and health manager is complex and diverse, focusing on analysis, prevention, planning, evaluation, promotion, and compliance. Educational requirements range from technical certificates to graduate degrees. Typical college majors held by practitioners include industrial safety and health technology, industrial technology, industrial engineering technology, manufacturing technology, industrial management, and engineering technology.

3. Engineers can have a significant impact on safety in the workplace and the marketplace by designing safety into products. The engineering disciplines geared most closely to design are aerospace, electrical, nuclear, and mechanical. Not all safety engineers are engineering majors. In fact, graduates of various technology degrees may have more formal education in safety than engineering graduates do.

4. Industrial hygienists are concerned with the following: (a) recognizing the impact of environmental factors on people, (b) evaluating the potential hazards of environmental stressors, and (c) prescribing methods to eliminate stressors.

5. Health physicists are concerned primarily with radiation in the workplace. Their duties include monitoring radiation in the air, measuring radioactivity levels in biological samples, developing the radiation components of a company's emergency action plan, and supervising decontamination activities.

6. Occupational physicians are medical doctors who specialize in workplace-related health problems and injuries. In the past, occupational physicians treated injuries and illnesses as they occurred. Today's occupational physicians focus more attention on anticipating and preventing problems.

7. Occupational health nurses are concerned with conserving the health of workers through prevention, recognition, and treatment. Like occupational physicians, occupational health nurses focus more attention on anticipation and prevention than they did in the past.

8. Risk management involves the application of risk reduction strategies and transferring remaining risk to insurance companies.

KEY TERMS AND CONCEPTS

Aerospace engineering	Mechanical engineering
Analysis and evaluation	Nuclear engineering
Associate degree	Occupational health nurse
Baccalaureate degree	Occupational physician
Certification boards	Problem identification
Chemical engineer	Produce and deliver
Competitiveness	Production vs. safety
Cost	Productivity
Design engineers	Professional societies
Design process	Profits
Document and communicate	Quality
Electrical engineering	Reduction
Environmental engineer	Response time
Formal education	Risk management
Government agencies	Safety and health manager
Health physicist	Safety engineer
Image	Service
In-service training	Staff authority
Industrial engineer	Standards/testing organizations
Industrial hygienist	Synthesis
Lack of commitment	Trade associations
Line authority	Transference

=========== REVIEW QUESTIONS ===========

1. What types of positions may be included in a modern safety and health team?
2. Briefly explain the impact that such issues as workers' compensation and the environment have had on the commitment of corporate management to safety and health.
3. What is the difference between a staff and a line position?
4. Explain the types of problems that safety and health managers can expect to confront in attempting to implement their programs.
5. Briefly explain what a company must do to succeed in today's competitive global marketplace.
6. How can safety and health managers use the competitiveness issue to gain a commitment to their programs?
7. List five different college majors that can lead to a career as a safety and health manager.
8. Explain the importance of ongoing in-service training for modern safety and health managers and how to get it.
9. How can safety and health managers become certified in their profession?
10. Name three professional societies that a modern safety and health manager may join.
11. What is meant by the following statement? "If a physician makes an error, he may harm one person, but an engineer who errs may harm hundreds."
12. Explain how the design process can affect safety.
13. What types of engineers are most likely to work as design engineers?
14. Why is the title *safety engineer* sometimes a misnomer?
15. What specific strengths may industrial engineers bring to bear as safety engineers?
16. What specific strengths may environmental engineers bring to the safety and health team?
17. What specific strengths may chemical engineers bring to the safety and health team?
18. Describe the job of the industrial hygienist.
19. What is a health physicist?
20. Describe the job of the occupational physician.
21. Describe the job of the occupational health nurse.
22. Explain the concept of risk management.

=========== ENDNOTES ===========

1. *The University Record.* Undergraduate Catalog of the University of Florida, 1991–92, p. 67.
2. Ibid., p. 70.
3. Ibid., p. 74.
4. Ibid., p. 75.
5. Goetsch, D. L., Nelson, J., and Chalk, William S. *Technical Drawing,* 2nd ed. (Albany: Delmar Publishers, 1989), p. 791.

6. Hammer, W. *Occupational Safety Management and Engineering*, 4th ed. (Upper Saddle River, NJ: Prentice Hall, 1989), p. 791.

7. *The University Record*, p. 73.

8. Ibid., p. 72.

9. Ibid., p 68.

10. American Industrial Hygiene Association (AIHA). *Engineering Field Reference Manual* (Akron: AIHA, 1984), p. 6.

11. Berry, C. (revised by Barbara A. Plog). "The Industrial Hygienist," Chapter 24 in *Fundamentals of Industrial Hygiene* (Chicago: National Safety Council, 1988), p. 571.

12. Ibid., p. 572.

13. Zenz, C. "The Occupational Physician," Chapter 26 in *Fundamentals of Industrial Hygiene* (Chicago: National Safety Council, 1988), p. 607.

14. Ibid.

15. National Safety Council. "Occupational Physicians: What Is Their New Role?" *Safety and Health*, April 1990, Vol. 141, No. 4, p. 37.

16. Ibid.

17. Ibid., p. 38.

18. American Association of Occupational Health Nurses (AAOHN). *A Guide for Establishing an Occupational Health Nursing Service* (New York: AAOHN, 1977), p. 78.

19. Ibid., p. 79.

TSM: Safety Management in a TQM Setting

TQM is the acronym for **Total Quality Management**, an approach to doing business that began to gain wide acceptance in the United States in the late 1980s and early 1990s. Various individual components of this concept had been used by forward-looking organizations for years. However, not until the 1990s were these components pulled together into a cohesive philosophy of how to do business in a competitive global environment.

The safety and health professional of today is likely to work in an organization that practices the TQM philosophy. Consequently, students of safety and health should understand TQM and how to apply its principles to the management of an organization's safety and health programs. After providing an overview of TQM, this chapter presents a comprehensive explanation of **Total Safety Management,** or **TSM,** the management of workplace safety and health according to the principles of TQM. The TSM material contained herein is adapted from *Implementing Total Safety Management* (Prentice Hall, 1997) by the author.

WHAT IS TQM?

The TQM concept goes by several different names. It is sometimes called TQL for *Total Quality Leadership*, TQC for *Total Quality Control*, or just TQ for *Total Quality*. Regardless of the name used, the concept can be defined as follows:

> *Total Quality Management (TQM) is an approach to doing business that maximizes the competitiveness of an organization through continuous improvement of its products, services, people processes, and environments.*[1]

Other approaches to doing business could lay claim to this or a similar definition. What distinguishes TQM from the other philosophies is the following list of characteristics. These characteristics describe *how* TQM achieves its purpose.

1. *Customer focus.* Total Quality organizations continually solicit input (up front) and feedback (after the fact) from both external and internal customers.

2. *Obsession with quality.* In a Total Quality organization, quality is every employee's job, from top management down through line employees. In such organizations, all employees learn to focus on what they can do to improve quality and competition every day.

3. *Scientific approach.* Total Quality organizations apply scientific decision making. This means that they operate on facts rather than assumptions and apply logic rather than so-called *gut feelings*. Employees in such organizations learn to use decision-making tools such as Pareto charts, fishbone diagrams, histograms, and control charts to identify clearly root causes before solving problems. This ensures that they solve problems, rather than just attacking symptoms.

4. *Long-term commitment.* The executive management team of a Total Quality organization is committed to this approach for the long term. TQM is not a quick-fix approach that will produce overnight miracles. It requires a long-term investment of both time and effort.

5. *Teamwork.* Total Quality organizations are teamwork oriented. Work is done by teams, and performance is measured against team, rather than individual, benchmarks.

6. *Continual process improvements.* Total Quality organizations continually improve the processes that make their products or deliver their services. *Good enough* is never good enough. Today's improvements are just the starting point for tomorrow's improvement.

7. *Education and training.* In a Total Quality organization, education and training are a normal part of doing business. They are used to equip all employees to be knowledgeable, skilled participants in the continual improvement of quality.

8. *Freedom through control.* A Total Quality organization frees its employees to improve their work processes continually without relinquishing appropriate management controls. This is done by establishing the parameters within which employees must work and then empowering them to make improvements within those parameters.

9. *Unity of purpose.* Along with a long-term commitment from top management, Total Quality organizations have unity of purpose from top to bottom. The organization's purpose is represented by its *vision statement.* All employees understand the vision, believe in it, and unite around it.

10. *Empowerment.* Total Quality organizations tap the enormous well of knowledge, experience, and potential that their employees represent. They do this by involving employees in decisions that affect their work and by empowering them to do what is necessary to ensure and improve quality continually.

HOW DOES TQM RELATE TO SAFETY?

TQM has proven itself to be an effective way to maximize an organization's long-term competitiveness. It is also an excellent approach for maximizing the effectiveness of an organization's safety and health programs. TQM can solve the same problem for safety managers that it solves for quality managers—the problem of *isolation.*

Often in a traditionally managed organization, quality is viewed as the sole responsibility of the quality department or the quality manager. The weaknesses of this approach became evident when organizations wedded to it began losing market share to foreign competition in the 1970s and 1980s. TQM solved this problem by making quality everybody's job and casting the quality manager in the role of facilitator and catalyst.

The same type of isolation often occurs with safety and health managers. Management and employees sometimes view safety as the responsibility of the safety department or the safety manager. In the case of quality, it is a prescription for failure. In the case of safety, it is a prescription for disaster. TQM principles can solve the problem of isolation by making safety everybody's job and casting the safety manager in the role of facilitator and catalyst.

The isolation issue is not the only tie between TQM and safety management. To fully understand how TQM relates to safety, one must examine each of the distinguishing characteristics of TQM from a safety and health perspective. These characteristics, as viewed from this perspective, are as follows:

1. *Customer focus.* In addition to soliciting input and feedback about the quality of its products, organizations should also collect customer data about the safety and health features of the products. After all, these features have a significant bearing on the perceived quality of a product. An unsafe product is not a quality product regardless of the attractiveness of its other characteristics. In the same way, it is important to collect input and feedback from internal customers (i.e., the employees) concerning the safety of work processes and environments. Internal customer data are the most effective data available for continually improving safety and health procedures.

2. *Obsession with quality.* In addition to being obsessed with quality, organizations should become obsessed with safety. The two concepts are actually inseparable. An organization with unsafe processes and/or products will eventually pay too high a price to be able to keep up in a competitive marketplace.

3. *Scientific approach.* When dealing with accidents and incidents, it is critical to identify root causes rather than symptoms. Future accidents cannot be prevented when the action taken in response to a current accident is based on symptoms rather than the root cause. The same scientific tools used to identify root causes of quality problems can be used to identify root causes of safety and health problems.

4. *Long-term commitment.* The long-term commitment of top management is as essential to safety as it is to quality. Executive-level managers show their commitment to safety and health in the same way that they show it to quality—by making safety an organizational priority.

5. *Teamwork.* Part of making safety everybody's responsibility is taking a teamwork approach to the continuous improvement of work processes and the work environment from a safety perspective. Just as cross-functional teams are assigned specific quality improvement projects, similar teams should be formed to tackle specific safety and health issues.

6. *Continual process improvements.* Total Quality organizations structure themselves to improve work processes continually from the perspective of quality. This can and should be done simultaneously from a safety perspective. The continual improvement of safety can be achieved in the same way as the continual improvement of quality.

7. *Education and training.* Quality and safety don't just happen. In both cases, employees at all levels develop the knowledge, skills, and attitudes needed to make a difference. This is accomplished in both cases through ongoing education and training for employees at all levels.

8. *Freedom through control.* Too often, safety precautions are viewed as unwanted inhibitors of productivity. Actually, the safe way is not always the fastest way in the short run. However, over the long run, when factoring in the downtime and expenses associated with injured employees and damaged machines, the safe way is typically the most productive way. Maximum productivity can be achieved by giving employees the freedom to apply their own initiatives in improving both quality and safety within the guidelines and parameters prescribed by management.

9. *Unity of purpose.* Unity of purpose is just as important to safety as it is to quality. The guiding principles that accompany an organization's vision statement should clearly establish safety and health as organizational priorities. Organization-wide unity concerning safety and health will go a long way toward ensuring the effectiveness of a safety program.

10. *Empowerment.* In a Total Quality organization, some of the best ideas for quality improvements come from employees who are empowered to think, suggest, and act. The same employees, equally empowered, could put their creative energy into making safety and health improvements.

TQM has excellent potential as an approach to safety and health management. In fact, other ties between quality and safety are so close that it can be argued that safety and quality must be improved simultaneously.

DISCUSSION CASE

What Is Your Opinion?

"TQM is a quality initiative. It has nothing to do with safety," said the CEO with a laugh. "On the contrary," replied the vice president for safety and health, "TQM and safety management have a great deal in common. In fact, the two concepts are complementary." Is there a relationship between TQM and safety management? If so, what is it? If not, why not? What is your opinion?

SAFETY MANAGEMENT IN A TQM SETTING

The concept of Total Safety Management (TSM) was introduced by the author in his book, *Implementing Total Safety Management*. It grew out of a need to transform safety and health management from a strictly compliance-orientation to a performance-orientation in which compliance is an important issue, but not the only issue. A safe and healthy work environment should do more than keep employers out of trouble with regulatory agencies. Pressure from state and federal agencies in the area of workplace safety fluctuates in accordance with the prevailing political climate. However, the need to maximize the performance of employees and organizations is constant. Safety and health should be a key element in an organization's strategic plan for gaining a competitive advantage in the global marketplace. TSM is an approach to safety and health management that is rooted in organizational performance and global competitiveness. Its purpose is to give organizations the sustainable competitive advantage of a safe and healthy work environment that allows employees to achieve consistent peak performance.

WHAT IS TSM?

TSM is to safety and health management what TQM is to quality management. It is safety management according to the principles of TQM. TQM has revolutionized the way in which organizations in the United States do business. As a result, the United States is slowly but steadily reclaiming marketshare lost to Japan and other countries in such critical market sectors as automobiles, consumer electronics, and computers. TSM can produce similar results for occupational safety and health, thereby making organizations that adopt it better able to gain the competitive advantage needed to compete in the global marketplace.

Just as TQM involves the total organization in continually improving quality, TSM involves the total organization in establishing and maintaining a work environment that is safe and conducive to quality and productivity. Both concepts are rooted firmly in the pressures of the global marketplace.

TSM Defined

The origin of TSM can be traced back to the globalization of the marketplace that began after World War II, but really took hold in the 1970s. The need for TSM was created by the need for organizations to be competitive globally. Consequently, TSM is defined as follows:

> *A performance-oriented approach to safety and health management that gives organizations a sustainable competitive advantage in the global marketplace by establishing a safe and healthy work environment that is conducive to consistent peak performance and that is improved continually forever. It involves applying the principles of TQM to the management of safety and health.*

This definition contains several key elements that must be understood if one is to comprehend TSM fully. These elements are as follows:

1. *Sustainable competitive advantage.* Every organization that competes at any level, but especially those that compete at the global level, must have competitive advantages. These are capabilities or characteristics that allow them to outperform the competition. For example, if the organization in question is a baseball team, it may have such competitive advantages as an excellent pitching staff, several speedy baserunners, two or three power hitters, and/or outstanding fielders in key positions. These advantages, if exploited wisely, will help make the baseball team a winner. If these advantages can be sustained over time, they will help make the team a consistent winner. This same concept applies to organizations that compete in the global marketplace. To survive and prosper, companies need as many competitive advantages as possible. Traditionally, competitive advantages have been sought in the key areas of quality, productivity, service, and distribution. However, peak-performing organizations have learned that a safe and healthy work environment is essential to gaining competitive advantages in all of these critical areas. In fact, a safe and healthy work environment is itself a competitive advantage. In today's competitive marketplace, high-performance employers are adding one more critical area to the list of those in which competitive advantages are sought. This new addition is the work environment.

Peak performance organizations are learning that a safe and healthy work environment gives them a doubly-effective competitive advantage. First, it ensures that employees work in an environment that allows them to focus all of their attention, energy, and creativity on continually improving performance. Second, it prevents an organization's limited resources from being drained off by the non-value-added costs associated with accidents and injuries.

2. *Peak performance.* The primary driver behind TSM performance is organizational, team, and individual. An organization's ability to survive and prosper in the global marketplace is determined largely by the collective performance of individuals and teams. Consistent peak performance by all individuals and teams in an organization is essential

---

SAFETY FACT

Cost of Accidents

The cost of accidents that occur on or off the job each year in the United States is approximately $150 billion.

---

to long-term success in the global marketplace. The quality of the work environment is a major determinant of the performance levels that individuals, teams, and organizations are able to achieve. A better work environment promotes better performance.

3. *Continual improvement forever.* People work in an environment, and the quality of that environment affects the quality of their work. The work environment is a major determinant of the quality of an organization's processes, products, and services. In the age of global competition, quality is an ever-changing phenomenon. Quality that is competitive today may not be tomorrow. Consequently, continual improvement is essential. If quality must be improved continually, it follows that the work environment must also be improved continually. Quality and safety are more than complementary; they are inseparable.

TRANSLATING TSM INTO ACTION

There are three fundamental components through which the TSM philosophy is translated into action on a daily basis. These three components are the **TSM steering committee, improvement project teams (IPTs),** and the **TSM facilitator.**

TSM Steering Committee

The TSM steering committee oversees the organization's safety and health program. It is responsible for the formulation of safety and health policies, the approval of internal regulations and work procedures relating to safety and health, the allocation of resources, and the approval of recommendations made by IPTs.

Improvement Project Teams

Improvement project teams (IPTs) are ad hoc, or temporary, teams formed by the TSM steering committee for the purpose of pursuing specific improvement projects relating to the work environment. Through the on-going process of hazard identification, in which all employees in a TSM organization participate, potential safety and health problems are identified. When a potential problem is identified, an IPT is formed to analyze the situation and make recommendations for improvements. These recommendations are presented to the steering committee by the TSM facilitator. For example, if an organization's workers' compensation costs have increased sharply, or if the number of accidents in a given department is on the rise, IPTs may be formed to analyze these situations and make recommendations. IPTs do not supplant the regular tasks and duties of safety and health professionals. Rather, they supplement these activities.

Typically, IPTs are cross-functional. This means that all stakeholders concerned about the problem in question are represented on the IPT that will look into it. A stakeholder is any individual, team, or department that is affected by a given problem, directly or indirectly. Consequently, IPTs usually have members from several different departments. Hence, the term *cross-functional.*

However, IPTs can also be natural work teams. These are teams made up of people who work together on a daily basis in performing their jobs. How an IPT is constituted is determined by the nature and scope of the problem. If its effects are confined to a single department, the IPT should be a natural work team. If its effects cross over into several different departments, the IPT should be cross-functional.

Once an IPT fulfills the charter it is given by the steering committee, it is disbanded. While one IPT is working on an assigned project, others are formed to pursue other projects. This is an ongoing process that never stops because a fundamental aspect of TSM is continual improvement of the work environment forever.

TSM Facilitator

The TSM facilitator must be a safety and health professional from within the organization. Ideally, he or she should be the organization's chief safety and health officer. The TSM facilitator serves as the steering committee's resident expert on the technical and compliance aspects of safety and health. He or she is responsible for the overall implementation and operation of the organization's TSM program.

FUNDAMENTAL ELEMENTS OF TSM

TSM differs in several ways from traditional safety and health management. To appreciate these differences, one must understand the following fundamental elements of TSM:

- Strategically based
- Performance oriented
- Dependent on executive-level commitment
- Teamwork oriented
- Committed to employee empowerment and enlistment
- Based on scientific decision making
- Committed to continual improvement
- Involves comprehensive, ongoing training
- Promotes unity of purpose

SAFETY FACT

Ranking of Industries by Death Rates

On the basis of the number of deaths per 100,000 employees, various industries can be ranked as follows (from the highest death rates to the lowest):

- Mining/quarrying
- Agriculture
- Construction
- Transportation/publications
- Government
- Manufacturing
- Services
- Trades

Strategic Basis

The fact that TSM is being **strategically based** means that an organization views a safe and healthy workplace as giving it a competitive advantage in the marketplace. Consequently, the organization makes maintaining a safe and healthy workplace a part of its strategic plan. Traditional organizations, on the other hand, tend to view safety and health more from a compliance perspective. TSM is strategically based because it is considered part of an organization's overall strategy for competing globally.

To compete in the global marketplace, organizations must plan for and establish sustainable competitive advantages. This is known as *strategic planning*. Once the planned advantages have been established, they must be maintained over time and exploited to the fullest possible extent. Strategic planning is a matter of answering the following questions:

■ *As an organization, what would we like to be? What is our dream?* This is the vision component of the strategic plan.

■ *As an organization, what is our purpose?* This is the mission component of the strategic plan.

■ *As an organization, what values are most important to us?* This is the guiding-principles component of the strategic plan. A guiding principle is a high-priority organizational value.

■ *As an organization, what do we want to accomplish?* The is the broad-objectives component of the strategic plan. Figure 28-1 contains an excerpt from the strategic plan for a hypothetical company, Allied Manufacturing.

Allied Manufacturing Company

Vision

Allied Manufacturing Company will be the preferred supplier of ceiling fans in the United States.

Mission

The mission of Allied Manufacturing Company is to design and manufacture high-quality ceiling fans for residential and commercial use.

Guiding Principles

The highest-priority corporate values of Allied Manufacturing Company are as follows:

- ► Customer satisfaction
- ► Product and service quality
- ► Cost leadership
- ► Employee safety and health
- ► Continual improvement
- ► Ethical business practices

Figure 28-1
Core of the strategic plan for Allied Manufacturing Company.

With TSM, a safe and healthy work environment shows up in the strategic plan as a competitive advantage that will be actively pursued and, having been established, will be maintained and continually improved over time. Safety and health concerns appear either in the guiding principles section of the strategic plan, the broad objectives section, or both. For example, in Figure 28-1, employee safety/health is a guiding principle value of Allied Manufacturing Company.

Employee safety and health may also appear in the strategic plan as a broad objective. For example, the following broad objective may appear in an organization's strategic plan:

> *To maintain a safe and healthy work environment that is conducive to consistent peak performance on the part of all employees.*

Performance Oriented

In any organization that is subject to federal and state regulations, compliance is an important issue. Failure to comply with applicable regulations can lead to fines and other non-value-added expenses. Companies in such sectors as manufacturing, processing, construction, transportation, and maritime services must comply with numerous government safety regulations.

In a compliance-driven setting, the safety department has a regulation orientation. Such an orientation can breed a big-brother-is-watching-you mentality among employees and managers that can cause them to resent the safety department's efforts. The resentment is magnified when employees and managers are pressed to meet deadlines, and they see safety regulations as slowing them down.

With TSM, the safety department is **performance oriented,** and safety and health personnel are viewed as part of the equation for continually improving performance. Compliance is still important, of course, but compliance and productivity are no longer viewed as being mutually exclusive entities. Compliance is viewed as a natural extension of an organization's performance-improvement strategies and continues to be an important responsibility of the TSM facilitator.

Executive Commitment

The TSM approach can be implemented successfully and fully only if an organization's executive management team is committed to providing a safe and healthy work environment as a performance-improvement strategy. For this reason, the safety and health professional must be able to articulate the TSM philosophy convincingly. If **executive commitment** does not exist, it will have to be generated. Naturally, responsibility for generating the necessary commitment falls to safety and health professionals.

Teamwork Oriented

In a traditional setting, workplace safety and health tend to be viewed as the responsibility of the safety department. In a **teamwork-oriented** TSM setting, although the safety department plays the key facilitating role, ensuring a safe and healthy workplace is everybody's responsibility. IPTs are formed by the TSM steering committee and given charters to improve specific aspects of the work environment. When a team satisfies its charter, it is disbanded. As the need to improve the work environment continues, new teams are chartered, and the process continues forever.

Employee Empowerment and Enlistment

Empowerment means involving employees in ways that give them a real voice. With empowerment, employees are allowed to give input concerning workplace issues, problems, and challenges. If the issue is how to make the work environment safer, empowered employees are allowed and encouraged to make suggestions, and their suggestions are given serious consideration. **Enlistment** takes the concept of empowerment one step further. With enlistment, rather than allowing employees to give input, the organization expects them to do so.

In a TSM setting, employees are viewed as invaluable sources of information, knowledge, insight, and experience that should be tapped continually. Consequently, employee input is not just wanted, it is needed; it is not just sought, it is expected. A criteria that should be included on the performance-appraisal form of modern organizations reads as follows:

■   To what extent does this employee provide input that is useful in making continual improvements in the workplace?

Frequently   ____
Sometimes   ____
Seldom   ____
Never   ____

Scientific Decision Making

It is no exaggeration to say that decisions about workplace safety and health can, on occasion, be life-or-death decisions. Whether they are or not, all such decisions are too important to be based on guesswork. With TSM, **scientific decision making** is the norm. This means that employees in organizations that adopt TSM must learn to use such decision making tools as Pareto charts, scatter diagrams, histograms, and fishbone diagrams. Such tools can take the guesswork out of decision making and problem solving and, as a result, lead to better decisions and better solutions.

Continual Improvement

With TSM, the status quo is never good enough, no matter how good it is. In a competitive marketplace, it must be assumed that the competition is improving continually. Consequently, today's competitive performance may not be tomorrow's. The state of the work environment can always be improved, and it should be. In an organization that adopts TSM, IPTs are always working to recommend accident prevention strategies that will make the workplace safer and more productive.

Comprehensive, Ongoing Training

In a TSM setting, training is an ongoing part of the job. If all employees are going to be involved in hazard identification and accident prevention, they must be trained in the basic techniques. If employees are going to be members of IPTs, they must be trained in the fundamentals of teamwork. If employees are going to use scientific tools in making recommendations for solving problems, they must be given instructions in the proper use of these tools.

Training should be **comprehensive and ongoing.** Figure 28-2 contains a partial list of courses, seminars, and/or workshops to which employees in a TSM setting should

Figure 28-2
TSM-related training opportunities.

- TSM: What It Is and How to Do It
- Teamwork Fundamentals
- Scientific Decision-Making/Problem-Solving
- Stress and Safety
- Mechanical Hazards and Body Guarding
- Lifting Hazards
- Falling, Impact, and Acceleration Hazards
- Heat and Temperature Hazards
- Pressure Hazards
- Electrical Hazards
- Fire Hazards
- Toxic Substances
- Explosive Hazards
- Radiation Hazards
- Noise and Vibration Hazards
- Hazard Analysis
- Accident Prevention
- Emergency Preparation/Response
- Facility Assessment/Hazard Identification
- Accident Investigation and Reporting
- Industrial Hygiene
- Automation-Related Hazards
- Bloodborne Pathogens and Related Hazards
- Ergonomics in the Workplace

have access. The specific training opportunities that should be provided are determined by the nature of the organization, its products, processes, and employees.

Unity of Purpose

In a TSM setting, all employees at all levels understand that safety and health are the responsibilities of everyone. The TSM facilitator and other safety and health professionals are responsible for planning, facilitating, coordinating, advising, and monitoring. In addition, employees are responsible for doing their part to make the workplace safe and healthy. Employees satisfy their responsibilities in this regard by doing the following:

- Setting a positive example of working safely
- Encouraging fellow employees to work safely

- Practicing hazard identification techniques constantly
- Recommending accident prevention strategies
- Serving effectively on IPTs

Gaining **unity of purpose** begins with the commitment of executive management. An organization's executive management team must commit to safety and health, make it a priority issue in the organization's strategic plan, stress its importance through both words and example, and expect it to be a priority of all employees.

RATIONALE FOR TSM

The correlation between work environment and job performance is strong. It is this correlation that forms the basis of the rationale for TSM. Globalization has increased the level of competition in the marketplace exponentially. For many organizations, adjusting to globalization has been like jumping from high school athletic competition to the Olympics.

Organizations that find themselves in these circumstances need every competitive advantage that they can muster. They need to do everything possible to improve performance continually. What every organization can, and must, do is provide employees with a safe and healthy work environment that is conducive to consistent peak performance.

IMPLEMENTING TSM: THE MODEL

Figure 28-3 contains a three-phase, 15-step model that can be used for successfully implementing TSM in any organization. The three broad phases of activity are **planning** and **preparation, identification** and **assessment,** and **execution.** Phase One—planning and preparation—encompasses steps 1 to 7. Each step is completed in order. Pursuing steps out of order can throw the implementation process into disarray. For example, step 1 is a critical preparation activity. Failure to complete step 1 successfully can jeopardize all subsequent steps. This is because the organization's attitude toward TSM will be determined primarily by the attitudes of executive managers.

Step 2 is the activity in which the TSM steering committee is established. The ideal TSM steering committee is the organization's executive management team, possibly augmented as necessary to ensure that all functional departments are well represented. Steps 3 and 4 are preparation activities that involve training. In step 3, the TSM steering committee undergoes teamwork training and team-building activities. In step 4, the steering committee undergoes safety and health awareness training.

Steps 5 and 6 are planning activities in which the steering committee develops a mini-strategic plan for safety and health. This plan consists of a vision, guiding principles, mission, and broad objectives, all of which focus solely on safety and health in the workplace. This mini-strategic plan becomes a subset of the organization's overall strategic plan. In step 7, the plan is communicated to all employees.

Figure 28-3
Model for implementation of
TSM.

Planning and Preparation

1. Gain executive-level commitment
2. Establish the TSM steering committee
3. Mold the steering committee into a team
4. Give the steering committee safety and health awareness training
5. Develop the organization's safety and health vision and guiding principles
6. Develop the organization's safety and health mission and objectives
7. Communicate and inform

Identification and Assessment

8. Identify the organization's safety and health strengths and weaknesses
9. Identify safety and health advocates and resistors
10. Benchmark initial employee perceptions concerning the work environment
11. Tailor implementation to the organization
12. Identify specific improvement projects

Execution

13. Establish, train, and activate improvement project teams
14. Activate the feedback loop
15. Establish a TSM culture

Phase Two of the implementation model—identification and assessment—encompasses steps 8-12. The steps in this phase allow the steering committee to identify strengths that may work in favor of the implementation and inhibitors that may work against it. Having identified strengths and inhibitors, the steering committee can tailor the implementation so as to exploit strengths while minimizing inhibitors.

Phase Three of the implementation model—execution—encompasses steps 13 to 15. The steps in this phase involve actually assigning teams to specific improvement projects. As soon as a given improvement has been made, that IPT is disbanded, and a new team is formed to pursue other improvements. This process repeats itself continually forever. An outgrowth of the repetitive nature of the model is that TSM becomes ingrained in the organization's culture.

===== SUMMARY =====

1. TQM is the acronym for Total Quality Management, which is an approach to doing business that maximizes the competitiveness of an organization through continuous improvement of its products, services, people, processes, and environments.

2. TQM achieves its purpose by developing the following characteristics in an organization: customer focus, obsession with quality, scientific approach, long-term commitment, teamwork, continual process improvement, education and training, freedom through control, unity of purpose, and empowerment.

3. TQM can solve the same problems for safety managers that it solves for managers concerned about quality. TQM makes quality everybody's responsibility, rather than limiting responsibility to quality personnel. It can do the same for safety managers. In addition, a quality product is a safe product, and the best environment in which to produce quality products is a safe and healthy environment.

4. TQM is implemented using a 20-step process. These steps are divided into three distinct phases: preparation, planning, and execution. Safety and health should figure prominently in each of these phases.

5. TSM is an approach to safety and health management that gives organizations a sustainable competitive advantage in the global marketplace. This is accomplished by involving all employees in establishing, maintaining, and continually improving the work environment so that it is conducive to consistent peak performance.

6. The fundamental elements of TSM are its strategic basis, performance orientation, executive commitment, teamwork orientation, employee empowerment and enlistment, scientific decision making, continual improvement, comprehensive and ongoing training, and unity of purpose.

7. The rationale of TSM can be found in the connection between job performance and the work environment. To compete in the global marketplace, organizations need all employees performing at peak levels on a consistent basis. A safe and healthy workplace promotes peak performance.

8. The model for implementing TSM consists of 15 steps arranged in three phases. These phases are planning and preparation, identification and assessment, and execution.

===== KEY TERMS AND CONCEPTS =====

Assessment	Enlistment
Comprehensive, ongoing training	Execution
Continual improvement	Executive commitment
Continual process improvement	Freedom through control
Customer focus	Identification
Education and training	Improvement project teams (IPTs)
Employee empowerment and enlistment	Long-term commitment
Empowerment	Obsession with quality

Peak performance	Teamwork
Performance oriented	Teamwork oriented
Planning	Total Quality Management
Preparation	TQM
Scientific approach	Total Safety Management
Scientific decision making	TSM
Steering committee	TSM facilitator
Strategically based	TSM steering committee
Sustainable competitive advantage	Unity of purpose

REVIEW QUESTIONS

1. Define *TQM*.
2. What are the critical characteristics of TQM?
3. How does TQM relate to safety?
4. Define the term *Total Safety Management*.
5. List and briefly describe the fundamental elements of TSM.
6. What is the rationale for TSM?
7. What are the three broad phases that make up the model for implementing TSM?
8. What is the most important step in the TSM implementation model. Why?

ENDNOTES

1. Goetsch, David L., and Davis, Stan. *Introduction to Total Quality: Quality, Productivity, Competitiveness.* (New York: Macmillan College Publishing Company, 1994), p. 6.
2. Ibid., p. 585.
3. White, Eleanor J. "Saturn's Implementation of Deming's Fourteen Points," an undated training document, p. 1.
4. Ibid., p. 3.

CHAPTER TWENTY-NINE

Safety, Health, and Competition in the Global Marketplace

One of the most frequently heard terms in the language of modern business and industry is *competitiveness*. To survive and prosper in today's global marketplace, industrial companies must be competitive. Companies that used to compete only with neighboring firms now find themselves competing against companies from Japan, Germany, Taiwan, Korea, France, Great Britain, China, and many other countries throughout the world.

The global marketplace is intensely competitive. It is like a sports contest that never ends. You may win today, but the race begins again tomorrow. Competing in the global marketplace has been described as the equivalent of running in a race that has no finish line.

The need to achieve peak performance levels day after day puts intense pressure on companies, and pressure runs downhill. This means that all employees from executive-level managers to workers on the shop floor feel it. It is not uncommon for this pressure to create a harried atmosphere that can increase the likelihood of accidents. It can also lead to shortcuts that increase the potential for health hazards (e.g., improper storage, handling, and use of hazardous materials).

This is unfortunate because the most competitive companies typically are also the safest and healthiest. It is critical that safety and health professionals understand the positive relationship between safety/health and competitiveness. It is also extremely important that they be able to articulate that connection effectively.

This chapter provides prospective and practicing safety and health professionals with the information they need to (1) understand the concept of global competitiveness; (2) understand its impact on a company's safety and health program; and (3) understand how to use the global competitiveness issue to gain a commitment to safety and health or to increase an existing commitment.

COMPETITIVENESS DEFINED

The Institute for Corporate Competitiveness defines **competitiveness** as "the ability to consistently succeed and prosper in the marketplace whether it is local, regional, national, or global."[1] The most competitive companies are those that do the following: (1) consistently outperform their competitors in the key areas of quality, productivity, response time, service, cost, and corporate image (Figure 29–1); and (2) continually improve all of these areas. The two key concepts associated with competitiveness are **peak performance** and **continual improvement.**

Figure 29–1
Key areas that result in competitiveness.

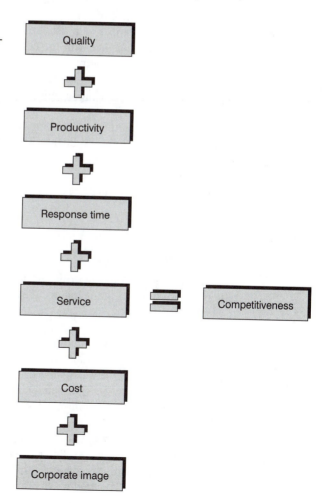

Competing in the **global marketplace** is difficult to do but easy to understand. One need only think of Olympic athletes to gain a perspective on the concept. Olympic athletes train continually to improve their performance to world-class levels. Their goal is to improve continually in practice so that they can achieve peak performance levels in competition. Those who do this consistently win. However, these athletes also know that, if they want to continue competing, there will always be another race. Consequently, the process of improvement must be continual. Winners today who rest on their laurels will be losers tomorrow. Records set today will be broken tomorrow. What is true for the Olympic athlete is also true for the modern industrial company.

The productivity record of today will be broken tomorrow. What is considered world-class quality today will be considered mediocre tomorrow. What is excellent service today will be unsatisfactory service tomorrow. The acceptable costs of today will be considered too high tomorrow. Record response time today will be too slow tomorrow. Finally, an excellent corporate image today can fade quickly and be gone tomorrow.

The marketplace is truly like a never-ending series of athletic contests. In order to compete, industrial companies must continually and consistently combine the best people and the best technologies with the best management strategies. As explained later in this chapter, a safe and healthy workplace is one of the best management strategies in this age of global competitiveness. It is also one of the best strategies for getting and keeping the best people while simultaneously protecting valuable technologies from damage.

Of the various factors influencing a company's competitiveness, the two most important are productivity and quality. The other factors are, to a large extent, functions of these two. Therefore, safety and health managers should be knowledgeable about these two concepts and be able to discuss them on an equal footing with production managers, supervisors, and representatives of higher management.

PRODUCTIVITY AND COMPETITIVENESS

Productivity is the concept of comparing **output** of goods or services to the **input** of resources needed to produce or deliver them. Productivity is typically expressed as the ratio of output to input in the following manner:

$$\text{Output/Input} = \text{Productivity}$$

To fully understand productivity, it is necessary to understand the concept of value added. Converting raw materials into useable products adds value to the materials. **Value added** is the difference between what it costs to produce a product and what it costs to purchase it (Figure 29–2). This difference represents the value that has been added to the product by the production process. Value added is increased when productivity is increased.

In Figure 29–2, Company A and Company B manufacture the same product. Each is able to sell their product for $1,300. However, Company B can manufacture the product for $400 because it is more productive than Company A, which produces the same product at a cost of $700. The productivity difference gives Company B at least two competi-

Figure 29–2
The value-added concept.

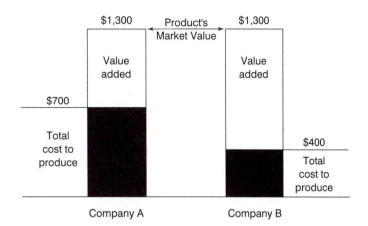

Company A   Company B

tive advantages: (1) It can lower its price for the product below that of Company A and still make a profit; and (2) its greater profit margin gives Company B more capital to reinvest in upgrading its facility, equipment, and personnel, which will, in turn, improve its competitiveness even more.

This productivity difference is important because all competing companies that produce a given product will probably pay approximately the same for the raw materials. Consequently, the winner will most likely be the one that adds the most value to the materials, which means the one that is the most productive (produces the most output with the least input).

Following are some rules-of-thumb that production managers use to monitor productivity in their plants. Safety and health managers should be familiar with these rules.

■ Productivity is declining when (1) output declines, and input is constant; or (2) output is constant, but input increases.

■ Productivity is improving when, (1) output is constant, but input decreases; or (2) output increases, and input is constant.

In considering the concept of productivity, remember that it represents only half of the equation. Nothing has been gained if productivity is improved to the detriment of quality. When productivity is improved, quality must also improve or at least remain constant.

Productivity: A Global View

In the community of industrialized nations, the United States has historically been the most productive. This country's productive capacity peaked during World War II when the United States amazed its enemies and allies alike with its enormous production capabilities.

The perception, however, is that the United States has seen productivity in the industrial sector decline steadily since World War II. Actually, this isn't true. Between

1950 and 1997, productivity in the United States increased by 129 percent.[2] The problem is that productivity in other industrialized countries, particularly Japan, increased even more during this same period (Figure 29–3). As a result, companies in the United States have found it increasingly difficult to compete in the global marketplace, particularly in the production of automobiles, computers, and consumer electronics products.

Pressure to increase productivity often results in actions that are detrimental to the safety and health of workers. In their rush to meet deadlines or quotas, workers may disassemble safeguards, stop taking the time to use appropriate personal protective gear, ignore safety rules, neglect equipment maintenance duties, improperly handle or store toxic substances, and take chances that they wouldn't take under normal conditions.

Such efforts occasionally result in short-term productivity improvements, but they invariably do more harm than good in the long run. The inevitable result is that productivity ultimately suffers, and the company finds that it won a battle only to lose the war.

It's as though a football team is losing late in a game, and the coach is getting desperate. Rather than sticking to what he knows will work best in the long run, the coach gives in to desperation and has the team's star quarterback run several trick plays that leave him vulnerable. His team wins the game, but in the process loses the star quarterback for the rest of the season to a crippling injury. The team's chances of having a championship season are forfeited for the sake of one game.

Production managers often respond to the pressures of competition in much the same way as the coach did and invariably with the same results. Safety and health professionals will probably face similar situations. Strategies for preventing this type of response are discussed later in this chapter.

Figure 29–3
Increases in output per hour of workers (1950–1997).

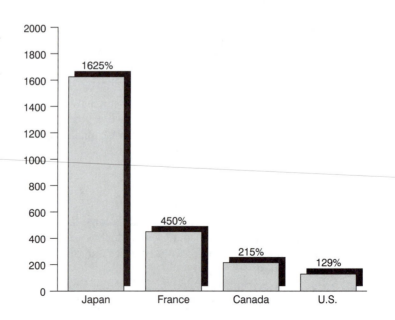

QUALITY AND COMPETITIVENESS

Quality goes hand-in-hand with productivity in the competitiveness equation. Today's production company must have both. Quality without productivity results in costs that are too high to be competitive. Productivity without quality results in a shabby product that quickly tarnishes the corporate image.

Quality is a measure of the extent to which a product or service meets or exceeds customer expectations. According to Peters, it is important to define quality in terms of customer expectations.[3]

Quality: A Global View

The most important consumer product in the global marketplace is the automobile. Therefore, it is often used as the measuring stick for making quality comparisons among the leading industrialized nations. U.S. automobile makers compete with those of Japan, Europe, East Asia, Brazil, and Mexico.

The basis of comparison, used by Congress's Office of Technology Assessment, is assembly defects. Beginning in the 1970s and extending through the 1980s, Japanese automakers held center stage in the worldwide quality drama. Quality in the best U.S. plants equaled that in East Asia, Mexico, and Brazil but did not equal that in Japanese or European plants. The worst plants in the United States allowed more defects than the worst plants of all other competitors except those in Europe.

However, by adopting a variety of quality improvement strategies that, collectively, came to be known as Total Quality Management, automakers in the United States have been able to reemerge as quality leaders. The market share taken away from American automakers by their Japanese competitors began to return in the late 1980s and early 1990s. By working with Dr. W. Edwards Deming—the quality guru who is credited with giving birth to the so-called Japanese miracle—Ford was able to lead Chrysler and General Motors down the path to recovery and reemergence. Competition remains fierce in the global automobile market, and consumers worldwide have benefited as a result. World-class quality is an ever-changing phenomenon; continued improvement has become the norm.

Whereas the pressure to increase productivity can be detrimental to safety and health, the pressure to improve quality generally supports safety and health. In the next section, prospective and practicing safety and health professionals will see how to use productivity and quality to gain a commitment to safety and health or to increase a commitment that already exists.

HOW SAFETY AND HEALTH CAN IMPROVE COMPETITIVENESS

When the pressures of competing become intense, it is not uncommon for safety and health to be given a lower priority. Not only is this wrong from an ethical standpoint, it is also wrong from the perspective of competitiveness and profitability. In the days when a worker's most important qualifications were physical strength and stamina, there were always plenty of applicants in the labor pool. If a worker was injured, several equally qualified applicants were waiting to replace him or her.

However, with the dawning of the age of high technology and the advent of global competition, this situation changed. Mental ability became more important than physical ability; suddenly, the number of qualified applicants got smaller. In fact, while the workplace is becoming increasingly technical and complex, the literacy level of the labor pool is actually declining in the United States. Consider the implications for competitiveness of the following statement:

> The basic skills necessary to be productive in a modern industrial setting are increasing steadily. At the same time the national high school dropout rate continues to increase as does the number of high school graduates who are functionally illiterate in spite of their diploma. This means that while the number of high-skilled jobs in modern industry is increasing, the number of people able to fill them is on the decline. The impact this will have on industry in the United States can be summarized as follows:
>
> 1. Difficulty in filling high-skill jobs.
> 2. Lower levels of productivity and, as a result, a lower level of competitiveness.
> 3. Higher levels of waste.
> 4. Higher potential for damage to sophisticated technological systems.
> 5. Greater numbers of dissatisfied employees in the workplace.[4]

The **illiteracy problem,** coupled with rapid and continual technological change, has serious implications for the global competitiveness of U.S. companies. Modern industrial companies are like modern sports teams. To compete, they must recruit, employ, and develop the best possible personnel. Having done so, they must keep them safe and healthy to derive the benefits of their talents.

Talented people working in a safe and healthy environment will be more competitive than equally talented people who are constantly distracted by concerns for their safety and health. In addition, the most talented employees cannot help a company compete if they are slowed by injuries.

The people side of the competitiveness equation is only part of it. There is also the technology side. With technological change occurring so rapidly, reinvestment in modern equipment is essential. Today's industrial firm must invest a higher percentage of its profits in equipment upgrades and do so more frequently. Money that must be diverted to workers' compensation, medical claims, product liability litigation, and environmental clean-ups is money that could have been invested in technological upgrades.

If they are able to convey these facts effectively, safety and health professionals should be able to secure a strong commitment to safety and health in the workplace. In attempting to do so, these professionals should still use the ethics and compliance arguments because these are proper and compelling. However, health and safety professionals must not stop there. They must convince companies that committing to safety and health is not just the right thing to do ethically or the smart thing to do legally, it is also the profitable thing to do in terms of competitiveness.

Nothing will help safety and health managers make this point more than the competitiveness equation: productivity, quality, response time, cost, service, and corporate image. Each of these ingredients relies directly in some way on having a safe and healthy workplace. Some of the more direct ways are summarized in the following paragraphs.

Productivity and Cost

Productivity is a function of people, technology, and management strategies. Safety and health have a direct effect on the first two. Companies with a reputation for providing a safe and healthy workplace will find it easier to attract and keep the best people. Correspondingly, these talented employees will be able to focus their skills more intently on being productive rather than worrying about accidents and/or health problems. On the **cost** side of the issue, companies with a record of safe and healthy practices will be better able to reinvest in equipment upgrades than those who must divert funds into such nonproductive costs as medical claims, environmental cleanups, and health/safety-related litigation. As shown in Figure 29–2, the more productive a company, the more competitive it can be in setting prices for its products.

Quality

Quality is essential to competitiveness. Fortunately, those practices that enhance quality also tend to enhance safety and health. Quality requires strict adherence to established production practices, attention to detail, and a commitment to doing things the right way. So do safety and health. For this reason, safety and health professionals should also be advocates of quality.

Response Time

Response time is like productivity in that it is a function of people, technology, and management strategies. Therefore, the same arguments that apply to productivity apply to response time. This is particularly important in the age of global competition because response time and quality, taken together, are becoming more important than the old

---

SAFETY MYTH

Safety and Global Competitiveness

Maintaining a safe and healthy workplace cannot be shown to enhance a company's global competitiveness. Right? Not necessarily. Productivity—a key determinant of a company's ability to compete—is measured as a ratio of output to input. The input elements are employees, material, money, and motivation. All four of these factors are affected by the work environment. In fact, more than 90 percent of employees identify *quality of the work environment* as critical in terms of their motivation and performance.[5]

"low bid" approach when it comes to winning contracts. The ability to deliver a quality product on demand is a prerequisite to participating in just-in-time contracts. These are contracts in which suppliers provide their products just in time to be used in the production process. This allows manufacturers to produce their products while carrying little or no inventory, a characteristic of today's most competitive companies.

Service

Service is an important ingredient in competitiveness. With industrial companies, **service** typically means in-field or after-delivery service. The most common example is service provided to people who purchase an automobile. Service is important because it can have a significant impact on customer satisfaction and, in turn, on corporate image. Service is not closely associated with safety and health and is, therefore, not a component to be used when trying to gain a commitment to safety and health.

Image

In a competitive world, industrial companies must be concerned about their **corporate image**. An image of being concerned about employee safety and health will help companies attract and keep the best people. An image of being concerned about product safety will help companies market their products. An image of being concerned about the environment will make a company a welcome neighbor in any community. Correspondingly, a poor image in any of these areas can undermine a company's competitiveness. A product safety problem that becomes a media story can cause a company's sales of a given product to plummet overnight. This happened to a major automobile manufacturer when it was discovered that one of its economy models had a tendency to ignite and explode in rear-end collisions. It also happened to pharmaceutical manufacturers when various headache relief products were tampered with and found to

---

DISCUSSION CASE

What Is Your Opinion?

"Explain this to me again. You say that making all of these modifications to our workstations and implementing these new procedures will make us more competitive. How? These workstation modifications are expensive, and it seems to me that some of these new procedures will just slow down production." This wasn't the first time that Garner Baxter, an independent safety and health consultant, had been asked to explain the strategic advantages of a safe and healthy workplace. Can the quality of the work environment really make an organization more competitive? What is your opinion?

have been laced with poison. Had these manufacturers not shown their concern by immediately recalling all containers that might have been affected and quickly developing tamper-proof substitutes, they could have suffered irreparable damage in the marketplace.

Image problems are not limited to product liability issues. Companies that are not careful about protecting the environment may find themselves the subject of protest demonstrations on the nightly news. Such negative publicity can harm the corporate image and translate very quickly into market losses. As corporate decision makers become more sensitive to these facts, safety and health professionals can use them to gain a commitment to their programs.

SUMMARY

1. Competitiveness is the ability to succeed and prosper in the local, regional, national, and/or global marketplace. The most competitive companies are those that consistently outperform their competitors in the key areas of quality, productivity, response time, service, cost, and image.
2. Productivity is a measure of output in goods and services compared to input of resources needed to produce or deliver them. Part of productivity is the concept of value added, which is measured as the difference between what it costs a company to produce a product and the competitive market price of that product.
3. Quality is a measure of the extent to which a product meets or exceeds customer expectations. It goes hand-in-hand with productivity. Quality without productivity results in costs that are too high to be competitive. Productivity without quality results in an unacceptable product.
4. Safety and health contribute to competitiveness in the following ways: (a) by helping companies attract and keep the best people; (b) by allowing employees to focus on peak performance without being distracted by concerns for their safety and health; (c) by freeing money that can be reinvested in technology updates; and (d) by protecting the corporate image.

KEY TERMS AND CONCEPTS

Competitiveness	Output
Continual improvement	Peak performance
Corporate image	Productivity
Cost	Quality
Global marketplace	Response time
Illiteracy problem	Service
Input	Value added

REVIEW QUESTIONS

1. Explain why and how global competition can have a negative impact on safety and health in the workplace.
2. Define the term *competitiveness*.
3. What are the common characteristics of the most competitive companies?
4. Explain the importance of continual improvement as it relates to global competitiveness.
5. Define the terms *productivity* and *value added*. Explain their relationship.
6. State two rules-of-thumb that explain how to recognize declining productivity.
7. State two rules-of-thumb that explain how to recognize improving productivity.
8. Explain the relationship between productivity and quality as it relates to competitiveness.
9. Define the term *quality*.
10. Write a brief rebuttal to the following statement: "A safety and health program is just a bunch of bureaucratic regulations that get in the way of profits."

ENDNOTES

1. Institute for Corporate Competitiveness. *Manual of Services* (Niceville, FL: Institute for Corporate Competitiveness, 1992), p. 1.
2. NCOE and AACJC. *Productive America: Two-Year Colleges Unite to Improve Productivity in the Nation's Workforce* (Washington, D.C.: 1990), pp. 1–12.
3. Peters, T. *Thriving on Chaos* (New York: Harper & Row, 1987), p. 78.
4. Goetsch, D. L. *Industrial Supervision in the Age of High Technology* (Columbus, OH: Merrill, 1992), p. 413.
5. Institute for Corporate Competitiveness, "Employee Perceptions: Impact of Work Factors on Job Performance," Report 95–6, August 1995.

CHAPTER THIRTY

Violence in the Workplace

"America has been hard at work in the past 10 days and here is what happened: A Federal Express pilot took a claw hammer and attacked three others in the cockpit, forcing one of them to put the fully loaded DC-10 cargo plane through a series of violent rolls and nose dives in a melee that brought the whole crew back bleeding. A purchasing manager in suburban Chicago stabbed his boss to death because, police say, they couldn't agree on how to handle some paperwork. And a technician who quit because he had trouble working for a woman sneaked back inside the fiber optics laboratory, pulled out a 9-mil semiautomatic pistol and started firing at workers, who ducked or fled and curled up in closets and file cabinets. By the time he finished the job, two were dead, two were injured; he walked upstairs to an office and shot himself in the head."[1]

Workplace violence has emerged as a critical safety and health issue. According to the Bureau of Labor statistics, homicide is the second leading cause of death to American workers, "accounting for 16 percent of the 6,588 fatal work injuries in the United States."[2] Although more than 80 percent of workplace homicide victims are men, workplace violence is not just a male problem. In fact, workplace homicide is the leading cause of death on the job for women in the United States.

Almost one million people are injured or killed in workplace-violence incidents every year in the United States, and the number of incidents is on the rise. In fact, according to the U.S. Department of Justice, the workplace is the most dangerous place to be in the United States.[3] Clearly, workplace violence is an issue of concern to safety and health professionals.

OCCUPATIONAL SAFETY AND WORKPLACE VIOLENCE: THE RELATIONSHIP

The prevention of workplace violence is a natural extension of the responsibilities of safety and health professionals. Hazard analysis, records analysis and tracking, trend monitoring, incident analysis, and prevention strategies based on administrative and engineering controls are all fundamental to both concepts. In addition, emergency response and employee training are key elements of both. Consequently, occupational safety and health professionals are well suited to add the prevention of workplace violence to their normal duties.

WORKPLACE VIOLENCE: DEFINITIONS

Safety and health professionals should be familiar with the language that has developed around the issue of workplace violence. This section contains the definitions of several concepts as they relate specifically to workplace violence.

- *Workplace violence.* Violent acts, behavior, or threats that occur in the workplace or are related to it. Such acts are harmful or potentially harmful to people, property, or organizational capabilities.
- *Occupational violent crime (OVC).* Intentional battery, rape, or homicide during the course of employment.[4]
- *Employee.* An individual with an employment-related relationship (present or past) with the victim of a workplace-violence incident.
- *Outsider.* An individual with no relationship of any kind with the victim of a workplace-violence incident or with the victim's employer.
- *Employee-related outsider.* An individual with some type of personal relationship (past or present) with an employee, but who has no work-related relationship with the employee.
- *Customer.* An individual who receives products or services from the victim of a workplace-violence incident or from the victim's employer.

Each of these terms has other definitions. Those presented herein reflect how the terms are used in the language that has evolved around workplace violence.

SAFETY FACT

High-Risk Occupations and People

High-risk occupations in terms of workplace violence are taxicab drivers; retail workers; police/security officers; and finance, insurance, real estate, health-care, and community service employees. Employees 65 years and older are more likely to be victims than are younger employees.

WORKPLACE VIOLENCE: CASES

This section contains numerous cases of workplace violence that occurred in the United States during a one-year period. These cases are provided to give students of occupational safety and health a better understanding of the types of violent incidents that occur frequently in today's workplace. The names of individuals involved have been changed, but the incidents are real.

1. *Jackson, Mississippi.* A 32-year-old fireman shot his wife through the head and then proceeded to a firehouse. Using an assault rifle, the fireman shot six co-workers—all supervisors—killing four of them and seriously wounding another two. He then fled the scene and exchanged fire with a police officer, wounding the officer, before being shot in the head himself and critically injured. The president of the union representing the shooter described him as "a time bomb waiting to go off."

2. *Fort Lauderdale, Florida.* Shouting, "Everyone is going to die," a maintenance employee who had been fired walked into a meeting of his former co-workers and began shooting, killing five people and injuring another. The employees were inside a temporary trailer office when the disgruntled former employee showed up with two handguns. Police say that he chased workers around the office and shot them methodically, pausing only to reload.

3. *Honolulu, Hawaii.* An employee returned to his former workplace after being fired from his job. He held five co-workers hostage—including his former boss—for up to six hours. During the incident, the disgruntled former employee shot and seriously wounded his former supervisor. He was eventually killed by police after he held a shotgun to the head of one co-worker for several hours while negotiating with officers. Prior to being shot by police, the perpetrator threatened to kill the co-worker and started a countdown to pulling the trigger of the gun. The countdown prompted the hostage to grab the barrel of the gun and gave police the opportunity to shoot. The perpetrator died of his wounds.

4. *Waterville, Maine.* A former mental patient accused of beating and stabbing four nuns, killing two of them, had been turned down for a job at their convent the previous week. Police said that they caught the perpetrator in the convent's chapel Saturday evening, standing over one of the nuns and beating her with a religious figurine. Officers

said that they had to pull him off the woman and that he had also beaten and stabbed three others in an adjacent part of the convent. Nuns at the convent of the Servants of the Blessed Sacrament said that the man had applied for a job but had been turned down. Nuns had just finished a prayer service when the perpetrator smashed the glass on a locked door, opened it, and walked inside. The perpetrator was described as an accomplished musician who had played trumpet in a local jazz combo and studied at the University of Maine at Augusta.

5. *Evensdale, Ohio.* A male in his early fifties returned to the offices from which he had recently been fired. Brandishing two pistols, he shot and killed three employees and wounded a fourth. A witness quoted the perpetrator as saying he was "going after someone who had screwed him over." After the murders, he surrendered quietly to authorities.

6. *Harlem, New York.* A man who was apparently involved in protesting the closure of an electronics store entered a clothing store with a gun and arson materials. He immediately set the store on fire and began shooting at employees. Eight individuals were killed (including the perpetrator), and another four were wounded—three seriously—by gunfire.

7. *Columbus, Ohio.* A man was fired from his job at a bank credit card center following charges of sexual harassment against co-workers. A year later, he forced his way into at least two homes of former co-workers, fatally shooting four individuals (including a four-month-old child) and wounding two others. He was captured by police attempting to flee the city in his automobile.

8. *San Jose, California.* A young accountant, on the job for just six weeks, shot and killed his female supervisor, then committed suicide with the same gun. This happened one day after he received his first performance counseling session. The killer was a 28-year-old Asian male. The victim was a 32-year-old Caucasian female.

9. *Palatine, Illinois.* A postal worker reported to work with a handgun and shot two co-workers who he claimed were his friends. The shooter was a 53-year-old Caucasian male who had worked at the same location for almost 20 years and had an exemplary service record.

10. *Los Angeles, California.* A city electrician with a 12-year work history shot and killed four of his supervisors when he learned he was facing possible dismissal for poor performance. After shooting his supervisors, the killer, 42 years old, quietly waited for police to arrive and arrest him. He was heard saying that he had specifically targeted the four murdered individuals because he "felt he was being picked on and singled out" by them.

11. *Industry, California.* A "quiet, unassuming" postal worker who had been on the job for 22 years shot and killed his supervisor. The perpetrator was 58 years old and was easily disarmed by co-workers after the shooting.

12. *Asheville, North Carolina.* A "classic loner" just fired from his job at a machine tool company returned the next day with a rifle and a pistol. The perpetrator, 47 years old, killed three workers and wounded another four before quietly surrendering to police.

13. *Littleton, Colorado.* A man distraught over marital problems opened fire in a crowded grocery store, killing three people (including his wife) before he was subdued by a bystander. The perpetrator was a 35-year-old Caucasian. Among the victims was his wife, age 37, and the store manager, age 39. This type of incident is becoming more common. It is often referred to as *spillover* violence.

14. *Richmond, California.* After a Richmond Housing Authority employee was fired from his job, he went to his automobile, retrieved a handgun, and returned to shoot and kill a supervisor and a co-worker. The perpetrator was 38 years old. Both victims were women, aged 47 and 24 years.

SIZE OF THE PROBLEM

Violence in the workplace no longer amounts to just isolated incidents that are simply aberrations. In fact, workplace violence should be considered a common hazard worthy of the attention of safety and health professionals. In a report on the subject, the U.S. Department of Justice revealed the following information:[5]

- About 1,000,000 individuals are the direct victims of some form of violent crime in the workplace every year. This represents approximately 15 percent of all violent crimes committed annually in America. Approximately 60 percent of these violent crimes were categorized as *simple assaults* by the Department of Justice.

- Of all workplace violent crimes reported, over 80 percent were committed by males; 40 percent were committed by complete strangers to the victims; 35 percent by casual acquaintances, 19 percent by individuals well known to the victims, and 1 percent by relatives of the victims.

- More than half of the incidents (56 percent) were not reported to police, although 26 percent were reported to at least one official in the workplace.

- In 62 percent of violent crimes, the perpetrator was not armed; in 30 percent of the incidents, the perpetrator was armed with a handgun.

- In 84 percent of the incidents, there were no reported injuries; 10 percent required medical intervention.

- More than 60 percent of violent incidents occurred in private companies, 30 percent in government agencies, and 8 percent to self-employed individuals.

- It is estimated that violent crime in the workplace caused 500,000 employees to miss 1,751,000 days of work annually, or an average of 3.5 days per incident. This missed work equates to approximately $55,000,000 in lost wages.

The Society for Human Resource Management (SHRM) periodically surveys its members on the issue of workplace violence. One such survey produced the following results:[6]

Regarding violent incidents in the workplace:

- 33 percent of all managers surveyed experienced at least one violent incident in the workplace.

- 54 percent of these managers reported between two and five acts of violence in the five years prior to the survey.

Regarding the type of violence experienced:

- 75 percent of the reported incidents were fistfights.
- 17 percent of the incidents were shootings.
- 8 percent of the incidents were stabbings.
- 6 percent of the incidents were sexual assaults.

Regarding the victims of the incidents:

- 54 percent of the incidents were employee against employee.
- 13 percent of the incidents were employee against a supervisor.
- 7 percent of the incidents were customer against worker(s).

Regarding the gender of the perpetrator:

- 80 percent of all violent acts were committed by males.

Regarding the injuries sustained by the victims:

- 22 percent of the incidents involved serious harm.
- 42 percent of the incidents required medical intervention.

Regarding the reasons for the violent incidents:

- 38 percent were attributed to personality conflicts.
- 15 percent were attributed to marital or family problems.
- 10 percent were attributed to drug or alcohol abuse.
- 7 percent were nonspecific as to attribution.
- 7 percent were attributed to firings or layoffs.

Regarding crisis management programs:

- 28 percent of the organizations had a crisis management program in place prior to the violent incident.
- 12 percent of the organizations implemented a crisis management program after the violent incident occurred.

Regarding the effect of a violent incident on the workplace:

- 41 percent of the organizations reported increased stress levels in the workplace after a violent incident.
- 20 percent reported higher levels of paranoia.
- 18 percent reported increased distrust among employees.

SAFETY MYTH

Workplace Violence—An Isolated Problem?

Most working Americans are unconcerned about workplace violence. They see it as an isolated problem that happens in other organizations, but not theirs. Right? Not any more. A TIME/CNN poll of the general population in the United States found that 37 percent of respondents viewed workplace violence as a growing problem. Of those responding to the poll, 18 percent had personally witnessed some form of violence at work. The same percentage of respondents (19 percent) said they feared for their own safety while at work.

LEGAL CONSIDERATIONS

Most issues relating to safety and health have legal ramifications, and workplace violence is no exception. The legal aspects of the issue revolve around the competing rights of violent employees and their co-workers (Figure 30-1). These conflicting rights create potential liabilities for employers.

Rights of Violent Employees

It may seem odd to be concerned about the rights of employees who commit violent acts on the job. After all, logic suggests that in such situations the only concern would be the protection of other employees. However, even violent employees have rights. Remember, the first thing that law enforcement officers must do after taking criminals into custody is to read them their rights.

According to Stephen C. Yohay,

> Employee rights are granted by a number of sources, including individual employment contracts, collective bargaining agreements, statutes such as the various state, local, and federal civil rights statutes, and sometimes from policies and procedures utilized by the employer. To the extent possible under the circumstances, the employer should consider following the procedures outlined in applicable contracts or policies before taking any adverse employment action against a threatening or violent employee.[7]

This does not mean that an employer cannot take the immediate action necessary to prevent a violent act or the recurrence of such an act. In fact, failure to act prudently in this regard can subject an employer to charges of negligence. However, before taking

Figure 30-1
Conflicting rights.

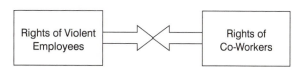

long-term action that will adversely affect the violent individual's employment, employers should follow applicable laws, contracts, policies, and procedures. Failure to do so can serve to exacerbate an already difficult situation.

For example, in *International Union vs. Micro Manufacturing, Inc.,* the employer was required to reinstate an employee who had been fired for assaulting and battering the owner of the company.[8] The reason? The violent employee had been terminated on the spot without a union representative present. This approach violated the terms of the union's collective bargaining agreement.

In addition to complying with all applicable laws, policies, and procedures, it is also important to apply these consistently when dealing with violent employees. Dealing with one violent employee one way while dealing with another in a different way puts the employer at a disadvantage legally. Consequently, it is important for employers to be prepared to deal both promptly and properly with violent employees. According to Yohay,

> The courts and the EEOC as a general matter do not require an employee to retain an employee who is violent or threatening to the other employees. However, . . . the violent and threatening employee does have some competing rights. Employers must check the law in their jurisdictions to ascertain exactly where the line is drawn regarding the rights of the threatening or violent employee.[9]

Employer Liability for Workplace Violence

Having to contend with the rights of both violent employees and their co-workers, employers often feel as if they are caught between a rock and a hard place. Fortunately, the situation is less bleak than it may first appear due primarily to the **exclusivity provision** of workers' compensation laws. This provision makes workers' compensation the employee's exclusive remedy for injuries that are work-related. This means that even in cases of workplace violence, as long as the violence occurs within the scope of the victim's employment, the employer is protected from civil lawsuits and the excessive jury verdicts that have become so common.

The key to enjoying the protection of the exclusivity provision of workers' compensation laws lies in determining that violence-related injuries are within the scope of the victim's employment—a more difficult undertaking than one might expect. For example, if the violent act occurred at work but resulted from a non-work-related dispute, does the exclusivity provision apply? What if the dispute was work-related, but the violent act occurred away from the workplace?

Making Work-Related Determinations

The National Institute for Occupational Safety and Health (NIOSH) developed the following guidelines for categorizing an injury as being work-related:

■   If the violent act occurred on the employer's premises, it is considered an on-the-job event if one of the following criteria apply:

   The victim was engaged in work activity, apprenticeship, or training.

The victim was on break, in hallway, restrooms, cafeteria, or storage areas.

The victim was in the employer's parking lots while working, arriving at, or leaving work.

■ If the violent act occurred off the employer's premises, it is still considered an on-the-job event, if one of the following criteria apply:

The victim was working for pay or compensation at the time, including working at home.

The victim was working as a volunteer, emergency services worker, law enforcement officer, or firefighter.

The victim was working in a profit-oriented family business, including farming.

The victim was traveling on business, including to and from customer–business contacts.

The victim was engaged in work activity in which the vehicle is part of the work environment (e.g., taxi driver, truck driver, and so on).[10]

RISK REDUCTION STRATEGIES

Figure 30-2 is a checklist that can be used by employers to reduce the risk of workplace violence in their facilities. Most of these risk reduction strategies grow out of the philosophy of *Crime Reduction through Environmental Design,* or CRTED.[11] CRTED has the following four major elements, to which the author has added a fifth (administrative controls).

■ Natural surveillance

■ Control of access

■ Establishment of territoriality

■ Activity support

■ Administrative controls

DISCUSSION CASE

What Is Your Opinion?

A man walks into an office building and asks to see his wife. The man is well-known to the other employees, one of whom escorts him to his wife's workstation. Suddenly, the man pulls a gun and shoots his wife and another employee who tries to intervene. Is this an on-the-job event? Is the employer at fault? What is your opinion about this incident involving an **employee-related outsider?**

✓ Identify high-risk areas and make them visible. Secluded areas invite violence.

✓ Install good lighting in parking lots and inside all buildings.

✓ Minimize the handling of cash by employees and the amount of cash available on the premises.

✓ Install silent alarms and surveillance cameras where appropriate.

✓ Control access to all buildings (employee badges, visitor check-in/out procedure, visitor passes, and so on).

✓ Discourage working alone, particularly late at night.

✓ Provide training in conflict resolution as part of a mandatory employee orientation.

✓ Conduct background checks before hiring new employees.

✓ Train employees how to handle themselves/respond when a violent act occurs on the job.

✓ Develop policies that establish ground rules for employee behavior/responses in threatening or violent situations.

✓ Nurture a positive, harmonious work environment.

✓ Encourage employees to report suspicious individuals and activities or potentially threatening situations.

✓ Deal with allegations of harassment or threatened violence promptly before the situation escalates.

✓ Take threats seriously and act appropriately.

✓ Adopt a *zero-tolerance* policy toward threatening or violent behavior.

✓ Establish a *violence hot line* so that employees can report potential problems anonymously.

✓ Establish a *threat-management team* with responsibility for preventing and responding to violence.

✓ Establish an *emergency response team* to deal with the immediate trauma of workplace violence.

Figure 30-2
Checklist for workplace-violence risk reduction.

Natural Surveillance

This strategy involves designing, arranging, and operating the workplace in a way that minimizes secluded areas. Making all areas inside and outside of the facility easily observable allows for **natural surveillance.**

Control of Access

One of the most common occurrences of workplace violence involves an **outsider** entering the workplace and harming employees. The most effective way of stopping this type of incident is to **control access** to the workplace. Channeling the flow of outsiders to an access-control station, requiring visitor's passes, issuing access badges to employees, and isolating pickup and delivery points can minimize the risk of violence perpetrated by outsiders.

Establishment of Territoriality

This strategy involves giving employees control over the workplace. With this approach, employees move freely within their **established territory** but are restricted in other areas. Employees come to know everyone who works in their territory and can, as a result, immediately recognize anyone who shouldn't be there.

Activity Support

Activity support involves organizing work flow and natural traffic patterns in ways that maximize the number of employees conducting natural surveillance. The more employees observing the activity in the workplace, the better.

Administrative Controls

Administrative controls consist of management practices that can reduce the risk of workplace violence. These practices include establishing policies, conducting background checks, and providing training for employees.

CONTRIBUTING SOCIAL/CULTURAL FACTORS

Another way to reduce the risk of workplace violence is to ensure that managers understand the social and cultural factors that can lead to it. These factors fall into two broad categories: individual and environmental factors.

Individual Factors Associated with Violence

The factors explained in this section can be predictors of the potential for violence. Employees and individuals with one or more of the following factors may respond to anger, stress, or anxiety in a violent way.

1. *Record of violence.* Past violent behavior is typically an accurate predictor of future violent behavior. Consequently, thorough background checks should be a normal part of the employment process.

2. *Membership in a hate group.* Hate groups often promote violence against the subjects of their prejudice. Hate-group membership on the part of an employee should raise a red flag in the eyes of management.

3. *Psychotic behavior.* Individuals who incessantly talk to themselves, express fears concerning conspiracies against them, say that they hear voices, and/or become increasingly disheveled over time may be violence-prone.

4. *Romantic obsessions.* Workplace violence is often the result of romantic entanglements or love interests gone awry. Employees who persist in making unwelcome advances may eventually respond to rejection with violence.

5. *Depression.* People who suffer from depression are prone to hurt either themselves or someone else. An employee who becomes increasingly withdrawn or overly stressed may be suffering from depression.

6. *Finger pointers.* Refusal to accept responsibility is a factor often exhibited by perpetrators of workplace violence. An employee's tendency to blame others for his or her own shortcomings should raise the caution flag.

7. *Unusual frustration levels.* The workplace has become a competitive, stressful, and sometimes frustrating place. When frustration reaches the boiling point, the emotional explosion that results can manifest itself in violence.

8. *Obsession with weapons.* Violence in the workplace often involves a weapon (gun, knife, or explosive device). A normal interest in guns used for hunting or target practice need not raise concerns. However, an employee whose interest in weapons is unusually intense and focused is cause for concern.

9. *Drug dependence.* It is common for perpetrators of workplace violence to be drug abusers. Consequently, drug dependence should cause concern not only for all of the usual reasons but also for its association with violence on the job.

10. *Environmental factors associated with violence.* The environment in which employees work can contribute to workplace violence. An environment that produces stress, anger, frustration, feelings of powerlessness, resentment, and feelings of inadequacy can increase the potential for violent behavior. The following factors can result in such an environment:

■ *Dictatorial management.* Dictatorial, overly authoritative management that shuts employees out of the decision-making process can cause them to feel powerless, as if they have little or no control over their jobs. Some people respond to powerlessness by striking out violently—a response that gives them power, if only momentarily.

■ *Role ambiguity.* One of the principal causes of stress and frustration on the job is role ambiguity. Employees need to know for what they are responsible, how they will be held accountable, and how much authority they have. When these questions are not clear, employees become stressed and frustrated, factors often associated with workplace violence.

- *Partial, inconsistent supervision.* Supervisors who play favorites engender resentment in employees who aren't the favorite. Supervisors who treat one employee differently than another or one group of employees differently from another group also cause resentment. Employees who feel that they are being treated unfairly or unequally may show their resentment in violent ways.

- *Unattended hostility.* Supervisors who ignore hostile situations or threatening behavior are unwittingly giving them their tacit approval. An environment that accepts hostile behavior will have hostile behavior.

- *No respect for privacy.* Supervisors and managers who go through the desks, files, tool boxes, and work areas of employees without first getting their permission can make them feel invaded or even violated. Violent behavior is a possible response to these feelings.

- *Insufficient training.* Holding employees accountable for performance on the job without providing the training that they need to perform well can cause them to feel inadequate. People who feel inadequate can turn their frustration inward and become depressed or turn it outward and become violent.

The overriding message in this section is two-fold. First, managers should endeavor to establish and maintain a positive work environment that builds employees up rather than tearing them down. Second, managers should be aware of the individual factors that can contribute to violent behavior and respond promptly if employers show evidence of responding negatively to these factors.

OSHA'S VOLUNTARY GUIDELINES

The U.S. Department of Labor, working through the Occupational Safety and Health Administration (OSHA), has established *advisory* guidelines relating to workplace violence.[12] Two key points to understand about these guidelines are as follows:

- The guidelines are *advisory* in nature and *informational* in content. The guidelines *do not* add to or enhance in any way the requirements of the General Duty clause of the OSHAct.

- The guidelines were developed with night retail establishments in mind. Consequently, they have a service-oriented emphasis. However, much of the advice contained in the guidelines can be adapted for use in manufacturing, processing, and other settings.

Figure 30-3 is a checklist of those elements of the specifications that have broader applications. Any management program relating to safety and health in the workplace should have at least these four elements.

Management Commitment and Employee Involvement

Management commitment and **employee involvement** are fundamental to developing and implementing any safety program, but they are especially important when trying to prevent workplace violence. The effectiveness of a workplace-violence prevention program may be a life-or-death proposition. Figure 30-4 is a checklist that explains what

Figure 30-3
Broadly applicable elements
of OSHA's advisory guide-
lines on workplace violence.

✓ Management commitment and employee involvement

✓ Worksite analysis

✓ Hazard prevention and control

✓ Safety and health training

management commitment means in practical terms. Figure 30-5 describes the practical application of employee involvement. Figure 30-6 is a checklist that can be used to ensure that violence prevention becomes a standard component of organizational plans and operational practices. These three checklists can be used by any type of organization to operationalize the concepts of management commitment and employee involvement.

Workplace Analysis

Workplace analysis is the same process used by safety and health professionals to identify potentially hazardous conditions unrelated to workplace violence. Worksite analysis

✓ Hands-on involvement of executive management in developing and implementing prevention strategies.

✓ Sincere, demonstrated concern for the protection of employees.

✓ Balanced commitment to both employees and customers.

✓ Inclusion of safety, health, and workplace-violence prevention in the job descriptions of all executives, managers, and supervisors.

✓ Inclusion of safety, health, and workplace-violence prevention criteria in the performance evaluations of all executives, managers, and supervisors.

✓ Assignment of responsibility for providing coordination and leadership for safety, health, and workplace-violence prevention to a management-level employee.

✓ Provision of the resources needed to prevent workplace violence effectively.

✓ Provision of or guaranteed access to appropriate medical counseling and trauma-related care for employees affected physically and/or emotionally by workplace violence.

✓ Implementation, as appropriate, of the violence-prevention recommendations of committees, task forces, and safety professionals.

Figure 30-4
Elements of management commitment to workplace-violence prevention.

> ✓ Staying informed concerning all aspects of the organization's safety, health, and workplace-violence program.
>
> ✓ Voluntarily complying—in both letter and spirit—with all applicable workplace-violence prevention strategies adopted by the organization.
>
> ✓ Making recommendations—through proper channels—concerning ways to prevent workplace violence and other hazardous conditions.
>
> ✓ Prompt reporting of all threatening or potentially threatening situations.
>
> ✓ Accurate and immediate reporting of all violent and/or threatening incidents.
>
> ✓ Voluntary participation on committees, task forces, and/or focus groups concerned with preventing workplace violence.
>
> ✓ Voluntary participation in seminars, workshops, and/or other educational programs relating to the prevention of workplace violence.

Figure 30-5
Elements of employee involvement in workplace-violence prevention.

> ✓ Include the prevention of workplace violence in the safety and health component of the organization's strategic plan.
>
> ✓ Adopt, disseminate, and implement a *no-tolerance* policy concerning workplace violence.
>
> ✓ Adopt, disseminate, and implement a policy that protects employees from reprisals when they report violent, threatening, or potentially threatening situations.
>
> ✓ Establish procedures for reporting violent and threatening incidents.
>
> ✓ Establish procedures for making recommendations for preventing workplace violence.
>
> ✓ Establish procedures for monitoring reports of workplace violence so that trends can be identified and incidents predicted and prevented.
>
> ✓ Develop a comprehensive workplace-violence prevention program the contains operational procedures and standard practices.
>
> ✓ Develop a workplace-violence component to the organization's emergency response plan.
>
> ✓ Train all employees in the application of standard procedures relating to workplace violence.
>
> ✓ Conduct periodic emergency-response drills for employees.

Figure 30-6
Checklist for incorporating workplace-violence prevention in strategic plans and operational practices.

should be ongoing and have at least four components (Figure 30-7). An effective way to conduct an ongoing program of workplace analysis is to establish a threat-assessment team with representatives from all departments and led by the organization's chief safety and health professional.

Records Monitoring and Tracking

The purpose of records monitoring and tracking is to identify and chart all incidents of violence and threatening behavior that have occurred within a given time frame. Records to analyze include the following: incident reports, police reports, employee evaluations, and letters of reprimand. Of course, individual employees' records should be analyzed in confidence by the human resources member of the team. The type of information that is pertinent includes the following:

- Where specifically did the incident occur?
- What time of day or night did the incident occur?
- Was the victim an employee? Customer? Outsider?
- Was the incident the result of a work-related grievance? Personal?

Trend Monitoring/Incident Analysis

Trend monitoring and **incident analysis** may prove helpful in determining patterns of violence. If there have been enough incidents to create one or more graphs, the team will want to determine if the graphs suggest a trend or trends. If the organization has experienced only isolated incidents, the team may want to monitor national trends. By

Figure 30-7
Four components of workplace analysis.

analyzing both local and national incidents, the team can generate information that will be helpful in predicting and, thereby, preventing workplace violence. The team should look for trends in severity, frequency, and type of incidents.

Employee Surveys/Focus Groups

Employees are one of the best sources of information concerning workplace hazards. This is also true when it comes to identifying vulnerabilities to workplace violence. Employee input should be solicited periodically through either written **surveys** or **focus groups** or both. Where are we vulnerable? What practices put our employees at risk? These are the types of questions that should be asked of employees. An effective strategy for use with focus groups is to give participants case studies of incidents that occurred in other organizations. Then ask such questions as, "Could this happen here? Why? Or why not? How can we prevent such incidents from occurring here?"

Security Analysis

Is the workplace secure, or could a disgruntled individual simply walk in and harm employees? It is important to ask this question. The team should periodically perform a **security analysis** of the workplace to identify conditions, situations, procedures, and practices that make employees vulnerable. The types of questions to ask include the following:

- Are there physical factors about the facility that make employees vulnerable (e.g., isolated, poorly lighted, infrequently trafficked, or unobservable)?
- Is there a process for handling disgruntled customers? Does it put employees at risk?
- Are the prevention strategies already implemented working?
- Is the training provided to employees having a positive effect? Is more training needed? Who needs the training? What kind of training is needed?
- Are there situations in which employees have substantial amounts of money in their possession, on or off-site?
- Are there situations in which employees are responsible for highly valuable equipment or materials late at night and/or at isolated locations?

Hazard Prevention and Control

Once hazardous conditions have been identified, the strategies and procedures necessary to eliminate them must be put in place. The two broad categories of prevention strategies are engineering controls and administrative controls, just as they are with other safety and health hazards. In addition to these, organizations should adopt post-incident response strategies as a way to prevent future incidents.

Engineering Controls

Engineering controls relating to the prevention of workplace violence serve the same purpose as engineering controls relating to other hazards. They either remove the hazard, or they create a barrier between it and employees. Engineering controls typically involve changes to the workplace. Examples of engineering controls include the following:

■ Installing devices and mechanisms that give employees a complete view of their surroundings (e.g., mirrors, glass or clear plastic partitions, interior windows, and so on)

■ Installing surveillance cameras and television screens that allow for monitoring of the workplace

■ Installing adequate lighting, particularly in parking lots

■ Pruning shrubbery and undergrowth outside and around the facility

■ Installing fencing so that routes of egress and ingress to company property can be channeled and, as a result, better controlled

■ Arranging outdoor sheds, storage facilities, recycling bins, and other outside facilities for maximum visibility

Administrative Controls

Whereas engineering controls involve making changes to the workplace, **administrative controls** involve making changes to how work is done. This amounts to changing work procedures and practices. Administrative controls fall into four categories (Figure 30-8).

■ **Proper work practices** are those that minimize the vulnerability of employees. For example, if a driver has to make deliveries in a high crime area, the company may employ a security guard to go along, change delivery schedules to daylight hours only, or both.

■ **Monitoring and feedback** ensures that proper work practices are being used and that they are having the desired effect. For example, say, a company established a controlled access system in which visitors must check in at a central location and receive a visitor's pass. Is the system being used? Are all employees sticking to specified procedures? Has unauthorized access to the workplace been eliminated?

Figure 30-8
Categories of administrative controls.

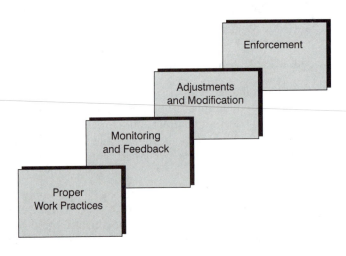

- **Adjustments and modifications** are made to violence prevention practices if it becomes clear from monitoring and feedback that they are not working or that improvements are needed.
- **Enforcement** involves applying meaningful sanctions when employees fail to follow the established and proper work practices. An employee who has been fully informed concerning a given administrative control, has received the training needed to practice it properly, but consciously decides not to follow the procedure should be disciplined appropriately.

Post-Incident Response

Post-incident response relating to workplace violence is the same as post-incident response relating to traumatic accidents. The first step is to provide immediate medical treatment for injured employees. The second step involves providing psychological treatment for traumatized employees. This step is even more important in cases of workplace violence than with accidents. Employees who are present when a violent incident occurs in the workplace, even if they don't witness it, can experience the symptoms of psychological trauma shown in Figure 30-9. Employees experiencing such symptoms or any others growing out of psychological trauma should be treated by professionals such as psychologists, psychiatrists, clinical nurse specialists, or certified social workers. In addition to one-on-one counseling, employees may also be enrolled in support groups. The final aspect of post-incident response is the investigation, analysis, and report. In this step, safety professionals determine how the violent incident occurred and how future incidents may be prevented, just as post-accident investigations are handled.

Training and Education

Training and education are as fundamental to the prevention of workplace violence as they are to the prevention of workplace accidents and health-threatening incidents. A complete safety and health training program should include a comprehensive component covering all aspects of workplace violence (e.g., workplace analysis, hazard preven-

Figure 30-9
Symptoms of psychological trauma in cases of workplace violence.

✓ Fear of returning to work

✓ Problems in relationships with fellow employees and/or family members

✓ Feelings of incompetence

✓ Guilt feelings

✓ Feelings of powerlessness

✓ Fear of criticism by fellow employees, supervisors, and managers tion, proper work practices, and emergency response). Such training should be provided on a mandatory basis for supervisors, managers, and employees.

Record Keeping and Evaluation

Maintaining accurate, comprehensive, up-to-date **records** is just as important when dealing with violent incidents as it is when dealing with accidents and nonviolent incidents. By evaluating records, safety personnel can determine how effective their violence prevention strategies are, where deficiencies exist, and what changes need to be made. Figure 30-10 shows the types of records that should be kept.

- *OSHA log of injury and illness (OSHA 200).* OSHA regulations require inclusion in the Injury and Illness Log of any injury that requires more than first aid, is a lost-time injury, requires modified duty, or causes loss of consciousness. Of course, this applies only to establishments required to keep OSHA logs. Injuries caused by assaults, which are otherwise recordable, also must be included in the log. A fatality or catastrophe that results in the hospitalization of three or more employees must be reported to OSHA within eight hours. This includes those resulting from workplace violence and applies to all establishments.

- *Medical reports.* Medical reports of all work injuries should be maintained. These records should describe the type of assault (e.g., unprovoked sudden attack), who was assaulted, and all other circumstances surrounding the incident. The records

Figure 30-10
Types of records that should be kept.

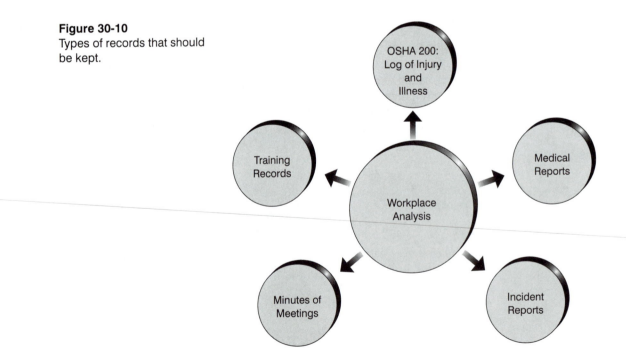

should include a description of the environment or location, potential or actual cost, lost time, and the nature of injuries sustained.

- *Incidents of abuse.* Incidents of abuse, verbal attacks, aggressive behavior—which may be threatening to the employee but do not result in injury, such as pushing, shouting, or acts of aggression—should be evaluated routinely by the affected department.
- *Minutes of safety meetings.* Minutes of safety meetings, records of hazard analyses, and corrective actions recommended and taken should be documented.
- *Records of all training programs.* Records of all training programs, attendees, and qualifications of trainers should be maintained.

As part of its overall program, an employer should regularly evaluate its safety and security measures. Top management should review the program regularly, as well as each incident, to determine the program's effectiveness. Responsible parties (managers, supervisors, and employees) should collectively evaluate policies and procedures on a regular basis. Deficiencies should be identified, and corrective action taken. An evaluation program should involve the following activities:

- Establishing a uniform violence reporting system and regular review of reports
- Reviewing reports and minutes from staff meetings on safety and security issues
- Analyzing trends and rates in illness/injury or fatalities caused by violence relative to initial or *baseline* rates
- Measuring improvements based on lowering the frequency and severity of workplace violence
- Keeping up-to-date records of administrative and work practice changes to prevent workplace violence to evaluate their effectiveness
- Surveying employees before and after making job or workplace changes or installing security measures or new systems to determine their effectiveness
- Keeping abreast of new strategies available to deal with violence as they develop
- Surveying employees who experience hostile situations about the medical treatment they received initially and, again, several weeks afterward, and then several months later
- Complying with OSHA and state requirements for recording and reporting deaths, injuries, and illnesses
- Requesting periodic law enforcement or outside consultant review of the workplace for recommendations on improving employee safety

Management should share violence prevention evaluation reports with all employees. Any changes in the program should be discussed at regular meetings of the safety committee, union representatives, or other employee groups.

CONFLICT RESOLUTION AND WORKPLACE VIOLENCE

When developing a violence prevention program for an organization, the natural tendency is to focus on protecting employees from outsiders. This is important. However, increasingly with workplace violence, the problem is internal. All too often in the modern workplace, conflict between employees is turning violent. Consequently, a violence prevention program is not complete without the following elements: **conflict management** and **anger management.**

Conflict Management Component

Disagreements on the job can generate counterproductive conflict. This is one of the reasons why managers in organizations should do what is necessary to manage conflict properly. However, it is important to distinguish between just conflict and counterproductive conflict. Not all conflict is bad. In fact, properly managed conflict that has the improvement of products, processes, people, and/or the work environment as its source is positive conflict.

Counterproductive conflict—the type associated with workplace violence—occurs when employees behave in ways that work against the interests of the overall organization and its employees. This type of conflict is often characterized by deceitfulness, vindictiveness, personal rancor, and anger. Productive conflict occurs when right-minded, well-meaning people disagree, without being disagreeable, concerning the best way to support the organization's mission. Conflict management has the following components:

- Establishing conflict guidelines
- Helping all employees develop conflict prevention/resolution skills
- Helping all employees develop anger management skills

Establishing Conflict Guidelines

Conflict guidelines establish ground rules for discussing and debating differing points of view, differing ideas, and differing opinions concerning how best to accomplish the organization's vision, mission, and broad objectives. Figure 30-11 is an example of an organization's conflict guidelines. Guidelines such as these should be developed with a broad base of employee involvement from all levels in the organization.

Develop Conflict Prevention/Resolution Skills

If managers are going to expect employees to disagree without being disagreeable, they are going to have to ensure that all employees are skilled in the art and science of conflict resolution. The first guideline in Figure 30-11 is an acknowledgment of human nature. It takes advanced human relation skills and constant effort to disagree without being disagreeable. Few people are born with this ability. Fortunately, it can be learned. The following strategies are based on a three-phase model developed by Tom Rusk and described in his book, *The Power of Ethical Persuasion.*[13]

Explore the Other Person's Viewpoint

Allow the other person to present his or her point of view. The following strategies will help make this phase of the discussion more positive and productive.

Conflict Guidelines
Micro Electronics Manufacturing (MEM)

Micro Electronics Manufacturing encourages discussion and debate among employees at all levels concerning better ways to improve continually the quality of our products, processes, people, and work environment. This type of interaction, if properly handled, will result in better ideas, policies, procedures, practices, and decisions. However, human nature is such that conflict can easily get out of hand, take on personal connotations, and become counterproductive. Consequently, in order to promote productive conflict, MEM has adopted the following guidelines. These guidelines are to be followed by all employees at all levels:

- The criteria to be applied when discussing/debating any point of contention is as follows: Which recommendation is most likely to move our company closer to accomplishing its mission?

- Disagree, but don't be disagreeable. If the debate becomes too hot, stop and give all parties an opportunity to cool down before continuing. Apply your conflict resolution skills and anger management skills. Remember, even when we disagree about how to get there, we are all trying to reach the same destination.

- Justify your point of view by tying it to our mission and require others to do the same.

- In any discussion of differing points of view, ask yourself the following question: "Am I just trying to win the debate for the sake of winning (ego), or is my point of view really the most valid?"

Figure 30-11
Sample conflict guidelines.

1. Establish that your goal at this point is mutual understanding.
2. Elicit the other person's complete point of view.
3. Listen nonjudgmentally and do not interrupt.
4. Ask for clarification if necessary.
5. Paraphrase the other person's point of view and restate it to show that you understand.
6. Ask the other person to correct your understanding if it appears to be incomplete.

Explain Your Viewpoint

After you accurately and fully understand the other person's point of view, present your own. The following strategies will help make this phase of the discussion more positive and productive:

1. Ask for the same type of fair hearing for your point of view that you gave the other party.
2. Describe how the person's point of view affects you. Don't point the finger of blame or be defensive. Explain your reactions objectively, keeping the discussion on a professional level.

3. Explain your point of view accurately and completely.
4. Ask the other party to paraphrase and restate what you have said.
5. Correct the other party's understanding if necessary.
6. Review and compare the two positions (yours and that of the other party). Describe the fundamental differences between the two points of view and ask the other party to do the same.

Agree on a Resolution

Once both viewpoints have been explained and are understood, it is time to move to the resolution phase. This is the phase in which both parties attempt to come to an agreement. Agreeing to disagree—in an agreeable manner—is an acceptable solution. The following strategies will help make this phase of the discussion more positive and productive:

1. Reaffirm the mutual understanding of the situation.
2. Confirm that both parties are ready and willing to consider options for coming to an acceptable solution.
3. If it appears that differences cannot be resolved to the satisfaction of both parties, try one or more of the following strategies:

    Take time out to reflect and try again.

    Agree to third-party arbitration or neutral mediation.

    Agree to a compromise solution.

    Take turns suggesting alternative solutions.

    Yield (this time), once your position has been thoroughly stated and is understood. The eventual result may vindicate your position.

    Agree to disagree while still respecting each other.

Develop Anger Management Skills

It is difficult, if not impossible, to keep conflict positive when anger enters the picture. If individuals in an organization are going to be encouraged to question, discuss, debate, and even disagree, they must know how to manage their anger. Anger is an intense, emotional reaction to conflict in which self-control may be lost. Anger is a major cause of workplace violence. Anger occurs when people feel that one or more of their fundamental needs are being threatened. These needs include the following:

1. Need for approval
2. Need to be valued
3. Need to be appreciated
4. Need to be in control
5. Need for self-esteem

When one or more of these needs is threatened, a normal human response is to become angry. An angry person can respond in one of four ways:

1. *Attacking*. With this response, the source of the threat is attacked. This response often leads to violence, or at least verbal abuse.

2. *Retaliating.* With this response, you fight fire with fire, so to speak. Whatever is given, you give back. For example, if someone calls your suggestion ridiculous (threatens your need to be valued), you may retaliate by calling his or her suggestion dumb. Retaliation can escalate into violence.

3. *Isolating.* This response is the opposite of venting. With the isolation response, you internalize your anger, find a place where you can be alone, and simmer. The childhood version of this response was to go to your room and pout. For example, when someone fails to even acknowledge your suggestion (threatens your need to be appreciated), you may swallow your anger, return to your office, and boil over in private.

4. *Coping.* This is the only positive response to anger. Coping does not mean that you don't become angry. Rather, it means that, even when you do, you control your emotions instead of letting them control you. A person who copes well with anger is a person who, in spite of his or her anger, stays in control. The following strategies will help employees manage their anger by becoming better at coping:

> Avoid the use of anger-inducing words and phrases including the following: *but, you should, you made me, always, never, I can't, you can't,* and so on.

> Admit that others don't make you angry; you allow yourself to become angry. You are responsible for your emotions and your responses to them.

> Don't let pride get in the way of progress. You don't have to be right every time.

> Drop your defenses when dealing with people. Be open and honest.

> Relate to other people as equals. Regardless of position or rank, you are no better than they, and they are no better than you.

> Avoid the human tendency to rationalize your angry responses. You are responsible and accountable for your behavior.

If employees in an organization can learn to manage conflict properly and to deal with anger positively, the potential for workplace violence will be diminished substantially. Conflict/anger management will not prevent violent acts from outsiders. There are other methods for dealing with outsiders. However, properly managing conflict and anger can protect employees from each other.

DOS AND DON'TS FOR SUPERVISORS

Supervisors can play a pivotal role in the prevention of workplace violence. Following are some rules-of-thumb that will enhance the effectiveness of supervisors in this regard:

- *Don't* try to diagnose the personal, emotional, or psychological problems of employees.
- *Don't* discuss an employee's drinking unless it occurs on the job. Restrict comments to performance.
- *Don't* preach to employees. Counsel employees about attendance, tardiness, and job performance, not about how they should live their lives.
- *Don't* cover up for employees or make excuses for inappropriate behavior. Misguided *kindness* may allow problems to escalate and get out of hand.

■ *Don't* create jobs to get problem employees out of the way. Stockpiling an employee simply gives him or her more time to brood and to allow resentment to build.

■ *Don't* ignore the warning signs explained earlier in this chapter. The problems that they represent will not simply go away. Sooner or later, they will have to be handled. Sooner is better.

■ *Do* remember that chemical dependence and emotional problems tend to be progressive. Left untreated, they get worse, not better.

■ *Do* refer problem employees to the Employee Assistance Program or to other mental health service providers.

■ *Do* make it clear to employees that job performance is the key issue. They are expected to do what is necessary to maintain and improve their performance.

■ *Do* make it clear that inappropriate behavior will not be tolerated.[14]

EMERGENCY PREPAREDNESS PLAN[15]

To be prepared for properly handling a violent incident in the workplace, employers should form a crisis management team. The team should have only one mission—immediate response to violent acts on the job—and be chaired by a safety and health professional. Team members should receive special training and be updated regularly. The team's responsibilities should be as follows:

■ Undergo trauma response training

■ Handle media interaction

■ Operate telephone/communication teams

■ Develop and implement, as necessary, an emergency evacuation plan

■ Establish a backup communication system

■ Calm personnel after an incident

■ Debrief witnesses after an incident

■ Ensure that proper security procedures are established, kept up-to-date, and enforced

■ Help employees deal with post-traumatic stress

■ Keep employees informed about workplace violence as an issue, how to respond when it occurs, and how to help prevent it

============= SUMMARY =============

1. Preventing workplace violence is a natural extension of the responsibilities of safety and health professionals. Like the traditional responsibilities of such professionals, dealing with workplace violence involves such activities as hazard analysis, records analysis and tracking, trend monitoring, and incident analysis.

2. Key concepts relating to workplace violence include the following: occupational violent crime (OVC), employee, outsider, employee-related outsider, and customer. These terms have definitions relating specifically to workplace violence.

3. Approximately one million people are victims of workplace violence every year. These incidents result in more than 1.75 million lost days of work annually.

4. Almost 30 percent of all violent acts in the workplace are committed by males. The majority of violent incidents reported each year (75 percent) are fistfights.

5. When dealing with violent incidents on the job, it is important to remember that even the perpetrator has rights. Employee rights are protected by employment contracts, collective bargaining agreements, and various local, state, and federal civil rights statutes. To the extent possible, when dealing with a violent employee, follow the procedures stipulated in contracts, agreements, and statutes.

6. Although it is important that employers consider perpetrators rights when dealing with workplace violence, it is equally important that they act prudently to prevent harm to other employees and/or customers.

7. The exclusivity provision of workers' compensation laws provide employers with some protection from liability in cases of workplace violence, provided that the incident is work-related. When this is the case, workers' compensation is the injured employee's exclusive remedy.

8. A violent act can be considered an on-the-job incident, even if it is committed away from the workplace. Specific guidelines have been established by NIOSH for determining whether a violent act can be classified as an on-the-job incident.

9. The concept of Crime Prevention through Environmental Design (CRTED) has four major elements as follows: natural surveillance, control of access, establishment of territoriality, and activity support. The author has added another: administrative controls.

10. OSHA has produced voluntary advisory guidelines relating to workplace violence. Although the guidelines are aimed specifically at the night retail industry, they provide an excellent framework that can be used in other industries including manufacturing, transportation, and processing. The framework has the following broad elements: management commitment/employee involvement, worksite analysis, hazard prevention control, and safety and health training.

KEY TERMS AND CONCEPTS

Activity support	Customer
Adjustments and modifications	Employee
Administrative controls	Employee-related outsider
Anger management	Employee surveys/focus groups
Conflict guidelines	Enforcement
Conflict management	Engineering controls
Control of access	Establishment of territoriality

Exclusivity provision

Management commitment/employee involvement

Monitoring and feedback

Natural surveillance

Occupational violent crime

Outsider

Post-incident response

Proper work practices

Record keeping and evaluation

Records monitoring and tracking

Security analysis

Trend monitoring/incident analysis

Workplace analysis

Workplace violence

REVIEW QUESTIONS

1. Define the following terms as they relate to violence in the workplace: occupational violent crime, employee, and outsider.
2. Approximately how many people are direct victims of workplace violence annually?
3. Defend or refute the following statement: Employees who commit violent acts forfeit their rights and can be dealt with accordingly.
4. What is the exclusivity provision of workers' compensation laws? Why is this provision significant?
5. Defend or refute the following statement: A violent act that occurs away from the employer's premises cannot be considered work-related.
6. Explain the concept of Crime Reduction through Environmental Design (CRTED).
7. Defend or refute the following statement: A manufacturer must comply with OSHA's guidelines on workplace violence.
8. What elements of OSHA's guidelines on workplace violence can be adapted for future use in most types of business and industrial firms?
9. What are the primary causes of conflict on the job?
10. Explain the five ways in which an angry person may respond in a work setting.

ENDNOTES

1. U.S. Department of Labor. Occupational Safety and Health Administration. *OSHA Workplace Violence Guidelines*—Draft, June 1996, p. 3.
2. Toufexis, Anastasia. "Workers Who Fight Firing with Fire." *Time*, April 25, 1994, Vol. 143, No. 17, p. 34.
3. U.S. Department of Justice. *Violence and Theft in the Workplace* (NCJ-148199) (Annapolis Junction, MD: Bureau of Justice Statistics Clearinghouse, July 1994).
4. Nigro, Lloyd G. "Violence in the American Workplace: Challenges to the Public Employees," *Public Administration Review*, July–August 1996, Vol. 56, No. 4, p. 326.
5. U.S. Department of Justice, *Violence and Theft in the Workplace.*
6. Society for Human Resource Management. "Workplace Violence Survey." Results published in *USA Today Magazine*, April 1994.

7. Yohay, Stephen C., and Peppe, Melissa L. "Workplace Violence: Employer Responsi-
    bilities," *Occupational Hazards*, July 1996, p. 22.
8. International Union vs. Micro Manufacturing, Inc., 895, F. Supp. 170, 171 (E.D.
    Mich. 1995).
9. Yohay and Peppe, "Workplace Violence: Employer Responsibilities," p. 24.
10. National Institute for Occupational Safety and Health and U.S. Department of
    Health and Human Services. *Homicide in U.S. Workplaces: A Strategy for Preven-
    tion and Research* (Washington, D.C.: Centers for Disease Control and Prevention,
    NIOSH) September 1992.
11. Thomas, Janice L. "A Response to Occupational Violent Crime," *Professional
    Safety*, June 1997, pp. 27–31.
12. U.S. Department of Labor, Occupational Safety and Health Administration. *Guide-
    lines for Workplace Violence Prevention Programs for Night Retail
    Establishments*, May 1997.
13. Rusk, Tom. *The Power of Ethical Persuasion* (New York: Penguin Books, 1994), pp.
    xv–xvii.
14. Illinois State Police. *Do's and Don'ts for the Supervisor*, http://www.state.il.us/
    isp/viowkplc/vwpp6c.htm, August 1996.
15. Illinois State Police. *Do's and Don'ts for the Supervisor*, http://www.state.il.us/
    isp/viowkplc/vwpp8.htm, August 1996.

Facility Assessment Checklist

Appendix A is a comprehensive checklist that can be used to help health and safety professionals conduct comprehensive facility assessments. Such assessments should be conducted on a regular basis, and no condition should be presumed safe until it is actually checked. Hazards that are identified should be corrected immediately.

This checklist is designed for use in conducting periodic assessments of the health and safety hazards that may be found in a modern industrial facility. It is a general checklist that can be used in any industrial facility. Supplemental criteria may be needed for highly specialized facilities.

Mechanical Hazards

- Are point-of-operation guards being properly used where appropriate?
- Are point-of-operation devices being properly used where appropriate?
- Are the safest feeding/ejection methods being employed?
- Are lockout/tagout systems being employed as appropriate?
- Are automatic shutdown systems in place, clearly visible, and accessible?
- Are safeguards properly maintained, adjusted, and calibrated as appropriate?

Fall-Related Hazards

- Are foreign objects present on the walking surface or in walking paths?
- Are there design flaws in the walking surface?
- Are there slippery areas on the walking surface?
- Are there raised or lowered sections of the walking surface that might trip a worker?
- Is good housekeeping being practiced?
- Is the walking surface made of or covered with a nonskid material?
- Are employees wearing nonskid footwear as appropriate?
- Are ladders strong enough to support the loads to which they are subjected?
- Are ladders free of cracks, loose rungs, and damaged connections?
- Are ladders free of heat damage and corrosion?
- Are ladders free of moisture that can cause them to conduct electricity?
- Are metal ladders free of burrs and sharp edges?
- Are fiberglass ladders free of blooming damage?
- Are employees wearing head and face protection as appropriate?
- Are employees wearing foot protection as appropriate?

Lifting Hazards

- Are posters that illustrate proper lifting techniques displayed strategically throughout the workplace?
- Are machines and other lifting aids available to assist employees in situations where loads to be lifted are too heavy and/or bulky?
- Are employees who are involved in lifting using personal protective devices?

Heat and Temperature Hazards

- Are workers in hot environments being gradually acclimatized?
- Are workers in hot environments rotating into cooler environments at specified intervals?
- Are workers in hot environments wearing personal protective clothing?
- Are first aid stations and supplies readily available for the treatment of burn victims?
- Are eye wash and emergency shower stations readily available for chemical burn victims?
- Are central or spot heating furnaces, warm air jets, contact warm plates, or radiant heaters being used where employees work in cold environments?
- Are work areas in cold environments shielded from the wind?
- Are the handles of tools used in cold environments covered with insulating material?
- Are metal chairs in cold environments covered with insulating material?
- Are heated tents or shelters provided so that workers in cold environments can take periodic warming breaks?
- Are appropriate types of warm drinks made available to workers in cold environments?
- Are jobs done in cold environments designed to require movement and minimize sitting or standing still?
- Are jobs done in cold environments designed so that as many tasks as possible are performed in a warm environment?

Pressure Hazards

- Are boilers properly installed (level, sufficient room around them for inspecting all sides, and so on)?
- Are control/safety devices on boilers present and in proper working condition?
- Is a schedule of regular inspections for all boilers posted and clearly visible?
- Is a schedule of preventive maintenance for all boilers posted and adhered to?
- Are pulse dampening devices/strategies being employed on high pressure systems?
- Are high pressure systems installed with a minimum of joints?
- Are appropriate pressure gauges in place and working properly?
- Are shields placed around high pressure systems?
- Are remote control and monitoring devices being used with high pressure systems?
- Is access restricted in areas where high pressure systems are present?
- Are leak detection methods being employed with pressurized gas systems?

Electrical Hazards

- Are short circuits present anywhere in the facility (use a professional electrician to run the necessary checks)?
- Are static electricity hazards present anywhere in the facility?
- Are electrical conductors in close enough proximity to cause an arc?
- Are explosive/combustible materials stored or used in proximity to electrical conductors?
- Does the facility have adequate lightning protection?
- Has insulation degradation resulted in bare wires anywhere in the facility?
- Is all electrical equipment in proper working order (use a professional electrician to run the necessary checks)?

Fire Hazards

- Are automatic fire detection systems employed in the facility?
- Are flammable materials and substances stored in a way that isolates them from oxygen and heat?
- Are automatic fire extinguishing systems employed in the facility?
- Are fixed extinguishing systems strategically located throughout the facility?

Toxic Substance Hazards

- Are all hazardous substances clearly identified and marked?
- Are all toxic substances properly stored?
- Are persons handling toxic substances using appropriate personal protective equipment that is properly maintained?
- Is access to areas that contain toxic substances appropriately restricted?
- Is equipment used to produce, process, and/or package toxic substances properly maintained?

Explosive Hazards

- Is smoking prohibited in the vicinity of explosive materials?
- Are methods employed to eliminate static electricity?
- Are spark-resistant tools employed wherever possible?
- Are processes that use explosive substances segregated in stand-alone facilities?
- Are processes and facilities that use explosive materials well ventilated?
- Are automatic shut-down systems employed on all processes that use explosives?
- Are all areas where explosives are present free of ignition sources?
- Is good housekeeping being practiced in all areas where explosives are present?

Radiation Hazards

- Are personal monitoring devices being used where appropriate?
- Are radiation areas clearly marked with caution signs?
- Are high radiation areas clearly marked with caution signs?
- Are airborne radiation areas clearly marked with caution signs?
- Are all containers in which radioactive materials are stored and/or transported clearly marked with caution signs?
- Is an evacuation warning signal system in place and operational?
- Are all containers in which radioactive materials are stored secured against unauthorized removal?

Vibration Hazards

- What are the sound level meter readings taken at different times and locations?
- What are the dosimeter readings for each shift of work?
- Are audiometer tests being conducted at appropriate intervals?
- What are the results of audiometric tests?
- Are records of audiometric tests being properly maintained?
- What is being done to reduce noise at the source?
- What is being done to reduce noise along its path?
- Are workers wearing personal protective devices where appropriate?
- Are low-vibration tools being used wherever possible?
- Are employees who use vibrating tools or equipment doing the following as appropriate?
  Wearing properly fitting thick gloves?
  Taking periodic breaks?
  Using a loose grip on vibrating tools?
  Keeping warm?
  Using vibration-absorbing floor mats and seat covers as appropriate?

Automation Hazards

- Are workers who use VDTs as their primary work tool employing methods to prevent eye strain?
- Are all robot sites well lighted and marked off?
- Are the floors around robot sites clean and properly maintained so that workers won't slip and fall into the robot's work envelope?
- Are all electrical and pneumatic components of all robots equipped with guards and covers?
- Are the work envelopes of all robots equipped with guards and covers?

Automation Hazards, *continued*

- Are automatic shut-off systems in place for all robots and other automated systems?
- Are lockout systems used before workers enter a robot's work envelope?
- Are safety fences erected around all automated systems?
- Are the controllers for all automated systems located outside of work envelopes?

Ergonomic Hazards

- Are tasks being performed that involve unnatural or hazardous movements?
- Are tasks being performed that involve frequent manual lifting?
- Are tasks being performed that involve excessive wasted motion?
- Are tasks being performed that involve unnatural or uncomfortable postures?
- Are tasks being performed that should be automated?

Glossary

abatement period: The amount of time given an employer to correct a hazardous condition that has been cited.

ability to pay: Applies when there are a number of defendants in a case but not all have the ability or means to pay financial damages.

absorption: Passage through the skin and into the bloodstream.

acceleration: Increase in the speed of a falling object before impact.

accident/incident theory: Theory of accident causation in which overload, ergonomic traps, and/or a decision to err lead to human error.

accident prevention: The act of preventing a happening that may cause loss or injury to a person.

accident rate: A fixed ratio between the number of employees in the workforce and the amount that are injured or killed every year.

accident report: Records the findings of an accident investigation, the cause or causes of an accident, and recommendations for corrective action.

accident scene: The area where an accident occurred.

accidents: Unexpected happenings that may cause loss or injuries to people who are not at fault for causing the injuries.

acclimatization: Process by which the body becomes gradually accustomed to heat or cold in a work setting.

accommodation: The ability of the eye to become adjusted after viewing the VDT so as to be able to focus on other objects, particularly objects at a distance.

adjustable guards: A device that provides a barrier against a variety of different hazards associated with different production operations.

administrative controls: Procedures that are adopted to limit employee exposure to hazardous conditions.

aerosols: Liquid or solid particles so small that they can remain suspended in air long enough to be transported over a distance.

aerospace engineering: Program of study incorporating a solid foundation of physical and mathematical fundamentals that provides the basis for the development of the engineering principles essential to the understanding of both atmospheric and extra-atmospheric flight.

agreement settlement: The injured employee and the employer or its insurance company work out an agreement on how much compensation will be paid and for how long.

air dose: Dose measured by an instrument in the air at or near the surface of the body in the area that has received the highest dosage of a toxic substance.

altitude sickness: A form of hypoxia associated with high altitudes.

amperes: The unit of measurement for current.

analysis and evaluation: All potential solutions to a problem are subjected to scientific analysis and careful evaluation.

ancestry: A person's line of descent.

anesthetics: In carefully controlled dosages, anesthetics can inhibit the normal operation of the central nervous system without causing serious or irreversible effects.

appeals process: The process of challenging an OSHA standard as it goes through the adoption process and after formal adoption.

artificial environment: One that is fully created to prevent a definite hazardous condition from affecting people or material.

aseptic necrosis: A delayed effect of decompression sickness.

asphyxiants: Substances that can disrupt breathing so severely as to cause suffocation.

associate degree: A two-year college degree.

assumption of risk: Based on the theory that people who accept a job assume the risks that go with it. It says employees who work voluntarily should accept the consequences of their actions on the job rather than blaming the employer.

audiogram: The results of an audiometric test to determine the noise threshold at which a subject responds to different test frequencies.

audiometric testing: Measures the hearing threshold of employees.

auto-ignition temperature: The lowest temperature at which a vapor-producing substance or a flammable gas will ignite even without the presence of a spark or a flame.

automatic ejection: A system that ejects work pneumatically and mechanically.

automatic feed: A system that feeds stock to the machine from rolls.

baccalaureate degree: A four-year college degree.

barometer: Scientific device for measuring atmospheric pressure.

bends: Common name for decompression sickness.

biological hazards: Come from molds, fungi, bacteria, and insects.

black-and-white approach: Right is right, wrong is wrong, and circumstances are irrelevant.

blackball: To ostracize an employee.

body surface area: The amount of surface area that is covered with burns.

bonding: Used to connect two pieces of equipment by a conductor. Also, involves eliminating the difference in static charge potential between materials.

Boyle's law: States that the product of a given pressure and volume is constant with a constant temperature.

BSDSs: Material safety data sheets. These sheets contain all of the relevant information needed by safety personnel concerning specific hazardous materials.

cancer: A malignant tumor.

carcinogen: Any substance that can cause a malignant tumor or a neoplastic growth.

carpal tunnel syndrome: An injury to the median nerve inside the wrist.

case law: Serves the purpose of establishing precedents that can guide future decisions.

causal relationship: A situation in which an action leads to a certain result.

ceiling: The level of exposure that should not be exceeded at any time for any reason.

central factor: The main issue or factor in a problem or act.

chemical burn injuries: Burn damage to the skin caused by chemicals such as acids and alkalies.

chemical engineer: An engineer concerned with all the physical and chemical changes of matter to produce economically a product or result that is useful to humans.

chemical hazards: Include mists, vapors, gases, dusts, and fumes.

chokes: Coughing and choking, resulting from bubbles in the respiratory system.

circadian rhythm: Biological clock.

circuit tester: An inexpensive piece of test equipment with two wire leads capped by probes and connected to a small bulb.

claim notice: Notice filed to indicate an expectation of workers' compensation benefits owed.

closed environment: One that is completely or almost completely shut off from the natural environment.

closing conference: Involves open discussion between the compliance officer and company/employee representatives.

code: A set of standards, rules, or regulations relating to a specific area.

Code of Hammurabi: Developed around 2000 B.C. during the time of the Babylonians, the ruler Hammurabi developed this code which encompassed all of the laws of the land at that time. The significant aspect is that it contained clauses dealing with injuries, allowable fees for physicians, and monetary damages assessed against those who injured others.

coefficient of friction: A numerical correlation of the resistance of one surface against another surface.

cold stress: Physical and/or mental stress that results from working in cold conditions.

Combination theory: The actual cause of an accident may be explained by combining many models.

combustible substance: Any substance with a flash point of 100°F or higher.

combustion: A chemical reaction between oxygen and a combustible fuel.

combustion point: The temperature at which a given fuel can burst into flame.

common-sense test: Requires person to listen to what their instincts and common sense are telling them.

competitiveness: The ability to succeed and prosper consistently in the marketplace whether it is local, regional, national, or global.

conduction: The transfer of heat between two bodies that are touching or from one location to another within a body.

conductors: Substances that have many free electrons at room temperature and can pass electricity.

confined space: An area with limited means of egress that is large enough for a person to fit into, but is not designed for occupancy.

consistency: The rules are enforced in the same manner every time with no regard to any outside factors.

continuity tester: May be used to determine whether a conductor is properly grounded or has a break in the circuit.

contributory negligence: An injured worker's own negligence contributed to the accident. If the actions of employees contributed to their own injuries, the employer is absolved of any liability.

controlled environment: A natural or induced environment that has been changed in some way to reduce or eliminate potential environmental hazards.

convection: The transfer of heat from one location to another by way of a moving medium.

convergence: The coordinated turning of the eyes inward so as to focus on a nearby point or object.

Cooperative Safety Congress (CSC): The CSC was a result of the Association of Iron and Steel Electrical Engineers' (AISEE) desire to have a national conference on safety. The first meeting took place in Milwaukee in 1912. The meeting planted the seeds for the eventual establishment of the National Safety Council (NSC).

corium: The inner layer of human skin.

cost: The amount of money needed to produce a product, not to be confused with "price," which is the amount of money needed to purchase a product after the cost has been marked up.

cost allocation: Spread the cost of workers' compensation appropriately and proportionately among industries ranging from the most to the least hazardous.

creeps: Caused by bubble formation in the skin, which causes an itchy, crawling, rashy feeling in the skin.

critical burns: Second-degree burns covering more than 30 percent of the body and third-degree burns covering over 10 percent are considered critical.

crushing: Occurs when a part of the body is caught between two hard surfaces that progressively move together, thereby crushing anything between them.

cutis: The inner layer of human skin.

cutting: Occurs when a body part comes in contact with a sharp edge.

Dalton's law of partial pressures: States that in a mixture of theoretically ideal gases, the pressure exerted by the mixture is the sum of the pressures exerted by each component gas of the mixture.

damages: Financial awards assigned to injured parties in a lawsuit.

DBBS: Conducts research in the areas of toxicology, behavioral science, ergonomics, and the health consequences of various physical agents.

death rates: A fixed ratio between the number of employees in the workforce and the number that are killed each year.

decibel: The unit applied when measuring sound. One-tenth of a bel. One decibel is the smallest difference in the level of sound that can be perceived by the human ear.

decompression sickness: Can result from the decompression that accompanies a rapid rise from sea level to at least 18,000 feet or a rapid ascent from around 132 feet to 66 feet underwater.

demonstration method: The instructor shows students how to perform certain skills or tasks.

dermis: The inner layer of human skin.

design flaw: A defect in a product.

design process: A plan of action for reaching a goal.

direct settlement: The employer or its insurance company begins making what it thinks are the prescribed payments.

discovery period: Period in which evidence is collected, depositions are taken, and products are examined.

document and communicate: Engineering drawings, detailed calculations, and written specifications document the design of a product and communicate its various components to interested parties.

Domino theory: Injuries are caused by the action of preceding factors. Removal of the central factor negates the action of the preceding factors and, in so doing, prevents accidents and injuries.

dose: The amount of ionizing radiation absorbed per unit of mass by part of the body or the whole body.

dose threshold: The minimum dose required to produce a measurable effect.

dosimeter: Provides a time-weighted average over a period of time such as one complete work shift.

double insulation: Another means of increasing electrical equipment safety.

drowning: The act of suffocating due to submersion in water.

DTMD: Implements Section 21 of the OSHAct, which sets forth training and education requirements.

dusts: Various types of solid particles that are produced when a given type of organic or inorganic material is scraped, sawed, ground, drilled, heated, crushed, or otherwise deformed.

dysbarism: The formation of gas bubbles due to rapid ambient pressure reduction.

ego strength: An employee's ability to undertake self-directed tasks and to cope with tense situations.

electrical engineering: A science-oriented branch of engineering primarily concerned with all phases and development of the transmission and utilization of electric power and intelligence.

electrical hazards: Potentially dangerous situations related to electricity (e.g., a bare wire).

electrical system grounding: Achieved when one conductor of the circuit is connected to the earth.

electricity: The flow of negatively charged particles called electrons through an electrically conductive material.

electrolytes: Minerals that are needed for the body to maintain the proper metabolism and for cells to produce energy.

electromechanical devices: Contact bars that allow only a specified amount of movement between the worker and the hazard.

electrons: Negatively charged particles.

ELISA: Screening test currently used to ensure the accuracy of an HIV antibody test.

Emergency Action Plan (EAP): A collection of small plans for every anticipated emergency (e.g., fire, hurricane, chemical spill, and so on).

emergency coordinator: A person who is clearly identified in a company Emergency Response Plan as the responsible party for a specific type of emergency situation.

emergency notification: Requires that chemical spills or releases of toxic substances that exceed established allowable limits be reported to appropriate LEPCs and SERCs.

emergency planning: Requires that communities form local emergency planning committees and that states form state emergency response commissions.

emergency response network: A network of emergency response teams that covers a designated geographical area and is typically responsible for a specific type of emergency.

Emergency Response Plan: A written document that identifies the different personnel/groups that respond to various types of emergencies and, in each case, who is in charge.

emergency response team: A special team that responds to general and localized emergencies to facilitate personnel evacuation and safety, shut down building services and utilities as needed, work with responding civil authorities, protect and salvage company property, and evaluate areas for safe reentry.

employee: A person who is on the company's payroll, receives benefits, and has a supervisor.

employee responsibilities: Specific obligations of employees relating to safety and health as set forth in OSHA 2056 (revised).

employee rights: Protections that an employee has against punishment for complaining to an employer union, OSHA, or any governmental agency about hazards on the job.

employer-biased law: A collection of laws that favored employers over employees in establishing a responsibility for workplace safety.

employer liability: In 1877, the Employer's Liability Law was passed and established the potential for employers to be liable for accidents that occurred in the workplace.

employer responsibilities: Specific responsibilities of employers as specified in OSHA 2056 (revised).

employer rights: The rights of an employer as specified in OSHA 2056 (revised).

environment: The aggregate of social and cultural conditions that influence the life of an individual.

environmental engineer: An individual whose job is to protect and preserve human health and the well being of the environment.

environmental factors: Characteristics of the environment in which an employee works that can affect his or her state of mind or physical conditions such as noise or distractions.

environmental heat: Heat that is produced by external sources.

Environmental Management System (EMS): That component of an organization with primary responsibility for leading, planning, organizing, and controlling as these functions relate specifically to an organization's processes, products, and/or services and the impact that they have on the environment.

Epidemiological theory: Holds that the models used for studying and determining epidemiological relationships can also be used to study causal relationships between environmental factors and accidents or diseases.

epidermis: The outer layer of human skin.

ergonomic hazards: Workplace hazards related to the design and condition of the workplace. For example, a workstation that requires constant overhead work is an ergonomic hazard.

ergonomic traps: An unsafe condition unintentionally designed into a workstation.

ergonomics: The science of conforming the workplace and all of its elements to the worker.

ERISA of 1974: Protects the benefits of employees by prohibiting actions taken against them based on their eligibility for benefits.

ethical behavior: That which falls within the limits prescribed by morality.

ethics: The study of morality.

expert system: A computer programmed to solve problems.

expiration: Occurs when air leaves the lungs and the lung volume is less than the relaxed volume, increasing pressure within the lungs.

explosion: A very rapid, contained fire.

explosive range: Defines the concentrations of a vapor or gas in air that can ignite from a source.

exposure ceiling: Refers to the concentration level of a given substance that should not be exceeded at any point during an exposure period.

exposure threshold: A specified limit on the concentration of selected chemicals. Exposure to these chemicals that exceeds the threshold is considered hazardous.

extraterritorial employees: Those who work in one state but live in another.

failure mode and effects analysis: A formal, step-by-step analytical method that is a spin-off of reliability analysis, a method used to analyze complex engineering systems.

falls: Category of accidents in which an employee unintentionally drops under the force of gravity from a surface that is elevated.

false negative: A result that shows no HIV antibodies in people who, in reality, are infected.

false positive: A result that shows the presence of HIV antibodies when, in reality, no such antibodies exist.

fault tree analysis: An analytical methodology that uses a graphic model to display visually the analysis process.

feedback: Employee's opinions of a project or an employer's opinion about an employee's job performance.

fellow servant rule: Employers are not liable for workplace injuries that result from negligence of other employees.

flame-resistant clothing: Special clothing made of materials or coated with materials that are able to resist heat and flames.

flexible hours: A work/time scheduling system in which employees are allowed to adopt a work schedule that more closely matches their personal needs.

fire: A chemical reaction between oxygen and a combustible fuel.

fire hazards: Conditions that favor the ignition and spread of fire.

fire point: The minimum temperature at which the vapors or gas in air can ignite from a source of ignition.

fire-related losses: Loss of life caused by fires and instances related to fires such as burns, asphyxiation, falls, and so on.

first-degree burn: Result in a mild inflammation of the skin known as erythema.

fixed guards: Provide a permanent barrier between workers and the point of operation.

flammable substance: Any substance with a flash point below 100°F and a vapor pressure of less than 40 pounds per square inch at 100°F.

flash point: The lowest temperature for a given fuel at which vapors are produced in sufficient concentrations to flash in the presence of a source of ignition.

flashing back: When a flame consumes a flammable material such as a gas traveling quickly back to its source.

fluid loss: Depletion of necessary body fluids, primarily through perspiration.

foreign object: Any object that is out of place or in a position to trip someone or to cause a slip.

foreseeability: Concept that a person can be held liable for actions that result in damages or injury only when risks could have been reasonably foreseen.

four-step teaching method: Preparation, presentation, application, and evaluation.

four-to-one ratio: Base one foot away from the wall for every four feet between the base and the support point.

free environment: One that does not interfere with the free movement of air.

freeze: The inability to release one's grip voluntarily from a conductor.

frequency: The number of cycles per second.

friable asbestos: Asbestos that is in a state of crumbling deterioration. When asbestos is in this state, it is most dangerous.

front-page test: Encourages persons to make a decision that would not embarrass them if it were printed as a story on the front page of their hometown newspaper.

frostbite: Occurs when there is freezing of the fluids around the cells of the outer body tissues.

frostnip: Less severe than frostbite. It causes the skin to turn white and typically occurs on the face and other exposed parts of the body.

fuses: Consist of a metal strip or wire that will melt if a current above a specific value is conducted through the metal.

gases: Formless fluids that are airborne and can be toxic.

gates: Provide a barrier between the danger zone and workers.

global marketplace: The worldwide economic market in which many companies must compete for business.

good housekeeping: Proper cleaning and maintenance of a work area.

ground fault: When the current flow in the hot wire is greater than the current in the neutral wire.

ground fault circuit interrupter: Can detect the flow of current to the ground and open the circuit, thereby interrupting the flow of current.

grounded conductor: Neutral wire.

grounding conductor: The groundwire.

hand-arm vibration syndrome (HAV): A form of Raynaud's Syndrome that afflicts workers who use vibrating power tools frequently over time.

harmful equipment: A work environment in which exists physical and/or psychological factors that are potentially hazardous.

hazard: A condition with the potential of causing injury to personnel, damage to equipment or structures, loss of material, or lessening of the ability to perform a prescribed function.

hazard analysis: A systematic process for identifying hazards and recommending corrective action.

hazard and operability review: An analysis method that was developed for use with new processes in the chemical industry.

hazardous condition: A condition that exposes a person to risks.

hazardous waste reduction: Reducing the amount of hazardous waste generated and, in turn, the amount introduced into the waste stream through the process of source reduction and recycling.

health physicist: Concerned primarily with radiation in the workplace.

hearing conservation: Systematic procedures designed to reduce the potential for hearing loss in the workspace. Employers are required by OSHA to implement hearing conservation procedures in settings where the noise level exceeds a time-weighted average of 85dBA.

heart disease: An abnormal condition of the heart that impairs its ability to function.

heat burn injuries: Burn damage to the skin caused by high temperatures from some of the following activities: welding, cutting with a torch, and handling tar or asphalt.

heat cramps: A type of heat stress that occurs as a result of salt and potassium depletion.

heat exhaustion: A type of heat stress that occurs as a result of water and/or salt depletion.

heat rash: A type of heat stress that manifests itself as small raised bumps or blisters that cover a portion of the body and give off a prickly sensation that can cause discomfort.

heat stroke: A type of heat stress that occurs as a result of a rapid rise in the body's core temperature.

heat transfer: The spread of heat from a source to surrounding materials/objects by conduction, radiation, or convection.

high-radiation area: Any accessible area in which radiation hazards exist that could deliver a dose in excess of 100 millirem within one hour.

horizontal work area: One that is designed and positioned so that it does not require the worker to bend forward or to twist the body from side to side.

human error: A mistake that is made by a human, not a machine.

human error analysis: Used to predict human error and not as an after-the-fact process.

Human Factors theory: Attributes accidents to a chain of events ultimately caused by human error.

humidification: Adding moisture to the air to reduce electrical static.

hyperoxia: Too much oxygen or oxygen breathed under too high a pressure.

hypothermia: The condition that results when the body's core temperature drops to dangerously low levels.

ignition temperature: The temperature at which a given fuel can burst into flame.

impact accidents: Involve a worker being struck by or against an object.

impulse noise: Consists of transient pulses that can occur repetitively or nonrepetitively.

inanimate power: Power that is lacking life or spirit. During the Industrial Revolution, humans and animals were replaced with inanimate power (e.g., steam power).

inappropriate activities: Activities undertaken with disregard for established safety procedures.

inappropriate response: A response in which a person disregards an established safety procedure.

income replacement: Replacement of current and future income (minus taxes) at a ratio of two-thirds (in most states).

independent contractor: A person who accepts a service contract to perform a specific task or set of tasks and is not directly supervised by the company.

indirect costs: Costs that are not directly identifiable with workplace accidents.

industrial hygiene: An area of specialization in the field of industrial safety and health that is concerned with predicting, recognizing, assessing, controlling, and preventing environmental stressors in the workplace that can cause sickness or serious discomfort. Concerns environmental factors that can lead to sickness, disease, or other forms of impaired health.

industrial hygiene chemist: Often hired by companies that use toxic substances to test the work environment and the people who work in it. Industrial hygiene chemists have the ability to detect hazardous levels of exposure or unsafe conditions so that corrective/preventive measures can be taken at an early stage.

industrial hygienist: A person having a college or university degree(s) in engineering, chemistry, physics, medicine, or related physical or biological sciences who, by virtue of special studies and training, has acquired competence in industrial hygiene.

industrial medicine: A specialized field that is concerned with work-related safety and health issues.

industrial place accidents: Accidents that occur at one's place of work.

industrial safety engineer: An individual who is a safety and health professional with specialized education and training. Responsible for developing and carrying out those aspects of a company's overall safety and health program relating to his or her area of expertise.

industrial safety manager: An individual who is a safety and health generalist with specialized education and training. Responsible for developing and carrying out a company's overall safety and health program including accident prevention, accident investigation, and education/training.

industrial stress: Involves the emotional state resulting from a perceived difference between the level of occupational demand and a person's ability to cope with this demand.

infection: The body's response to contamination by a disease-producing microorganism.

ingestion: Entry through the mouth.

inhalation: Taking gases, vapors, dust, smoke, fumes, aerosols, and/or mists into the body by breathing in.

inspection tour: A facility tour by an OSHA compliance officer to observe, interview, and examine as appropriate.

inspiration: When atmospheric pressure is greater than pressure within the lungs, air flows down this pressure gradient from the outside into the lungs.

instructional approach: A brief action plan for carrying out the instruction.

interlocked guards: Shut down the machine when the guard is not securely in place or is disengaged.

interlocks: Automatically break the circuit when an unsafe situation is detected.

internal factors: Factors that can add a burden on a person and interfere with his or her work, such as personal problems.

ionizers: An ionizer ionizes the air surrounding a charged surface to provide a conductive path for the flow of charges.

ionizing radiation: Radiation that becomes electrically charged or changed into ions.

irritants: Substances that cause irritation to the skin, eyes, and the inner lining of the nose, mouth, throat, and upper respiratory tract.

islands of automation: Individual automated systems lacking electronic communication with other related systems.

job autonomy: Control over one's job.

job descriptions: Written specifications that describe the tasks, duties, reporting requirements, and qualifications for a given job.

job safety analysis: A process through which all of the various steps in a job are identified and listed in order.

job security: A sense—real or imagined—of having the potential for longevity in a job and having a measure of control concerning that longevity.

kinetic energy: The energy resulting from a moving object.

lacrimination: The process of excreting tears.

learning objectives: Specific statements of what the learner should know or be able to do as a result of completing the lesson.

let-go current: The highest current level at which a person in contact with the conductor can release the grasp of the conductor.

liability: A duty to compensate as a result of being held responsible for an act or omission.

lifting hazards: Any factor that if not properly dealt with may lead to an injury from lifting it.

lightning: Static charges from clouds following the path of least resistance to the earth, involving very high voltage and current.

line authority: The safety and health manager has authority over and supervises certain employees.

litigation: The carrying on of legal matters by judicial process.

load: A device that uses currents.

lockout/tagout system: A system for incapacitating a machine until it can be made safe to operate. "Lockout" means physically locking up the machine so that it cannot be used without removing the lock. "Tagout" means applying a tag that warns employees not to operate the machine in question.

locus of control: The perspective of workers concerning who or what controls their behavior.

lost time: The amount of time that an employee was unable to work due to an injury.

lost wages: The amount that an employee could have earned had he or she not been injured.

Machiavellianism: The extent to which an employee will attempt to deceive and confuse others.

malpractice: Negligent or improper practice.

meaninglessness: The feeling that workers get when their jobs become so specialized and so technology-dependent that they cannot see the meaning in their work as it relates to the finished product or service.

means of egress: A route for exiting a building or other structure.

mechanical engineering: The professional field that is concerned with motion and processes whereby other energy forms are converted into motion.

mechanical hazards: Those associated with power-driven machines, whether automated or manually operated.

mechanical injuries: Injuries that have occurred due to misuse of a power-driven machine.

medical expenses: Money paid to cover the costs of emergency medical response and follow-up treatment when an employee is injured.

Merit Program: Seen as a stepping-stone to recognize companies that have made a good start toward Star Program recognition.

metabolic heat: Produced within a body as a result of activity that burns energy.

mindlessness: The result of the process of dumbing down the workplace.

minor burns: All first-degree burns are considered minor as well as second-degree burns covering less than 15 percent of the body.

mists: Tiny liquid droplets suspended in air.

moderate burns: Second-degree burns covering less than 30 percent of the body and third-degree burns covering less than 10 percent are considered moderate.

monetary benefits: Actual money owed to injured employees or their relatives under workers' compensation laws.

NACOSH: Makes recommendations for standards to the secretary of health and human services and to the secretary of labor.

narrow band noise: Noise that is confined to a narrow range of frequencies.

National Council of Industrial Safety: Established in 1913, one year after the first meeting of the CSC. In 1915, the NCIS changed its name to the National Safety Council. It is now the premier safety organization in the U.S.

Natural disasters: Incidents prompted by nature such as earthquakes, hurricanes, floods, and tornadoes.

natural environment: Not man-made. It is the environment that we typically think of as Earth and all of its natural components, including the ground, the water, flora and fauna, and the air.

NEC: Specifies industrial and domestic electrical safety precautions.

negative pressures: Caused by pressures below atmospheric level.

negligence: Means failure to take reasonable care or failure to perform duties in ways that prevent harm to humans or damage to property.

negligent manufacture: The maker of a product can be held liable for its performance from a safety and health perspective.

neoplastic growth: Cancerous tissue or tissue that might become cancerous.

NIOSH: This organization is part of the Centers for Disease Control of the Department of Health and Human Services. It is required to publish annually a comprehensive list of all known toxic substances. It will also provide on-site tests of potentially toxic substances so that companies know what they are handling and what precautions to take.

noise: Unwanted sound.

nonionizing radiation: That radiation on the electromagnetic spectrum that has a frequency of 10^{15} or less and a wavelength in meters of 3×10^{-7} or less.

nonskid footwear: Shoes that have special nonskid soles.

normlessness: The phenomenon in which people working in a highly automated environment can become estranged from society.

notice of contest: A written note stating that an employer does not wish to comply with a citation, an abatement period, and/or a penalty.

notice of proposed rule making: Explains the terms of the new rule, delineates the proposed changes to existing rules, or lists rules that are to be revoked.

nuclear engineering: Concerned with the release, control, and safe utilization of nuclear energy.

objectivity: Rules are enforced equally regardless of who commits an infraction from the newest employee to the chief executive officer.

occupational diseases: Pathological conditions brought about by workplace conditions or factors.

occupational health nurse: One whose job it is to conserve the health of workers in all occupations.

ohms: Measure resistance.

opening conference: An initial meeting in which an OSHA compliance officer informs company officials and employee representatives the reason that an on-site inspection is going to occur and what to expect.

organized labor: A group of employees who joined together to fight for the rights of all employees (i.e., unions).

OSHA: The government's administrative arm for the Occupational Safety and Health Act. It sets and revokes health and safety standards, conducts inspections, investigates problems, issues citations, assesses penalties, petitions the courts to take appropriate action against unsafe employers, provides safety training, provides injury prevention consultation, and maintains a database of safety and health statistics.

OSHAct: Occupational Safety and Health Act. Passed by the United States Congress in 1970 and updated periodically since that time.

OSHA Form 101: Supplementary Record of Occupational Injuries and Illnesses.

OSHA Form 200/300: Log and Summary of Occupational Injuries and Illnesses.

OSHA Form 200S: Annual Survey.

OSHA Poster 2203: Explains employee rights and responsibilities as prescribed in the OSHAct.

OSHRC: An independent board whose members are appointed by the president and given quasi-judicial authority to handle contested OSHA citations.

overexertion: The result of employees working beyond their physical limits.

overload: Amounts to an imbalance between a person's capacity at any given time and the load that person is carrying in a given state.

oxygen limit: The amount of oxygen that must be present in a given substance in order for an explosion to occur.

patent defect: One that occurs in all items in a manufactured batch.

permanent partial disability: The condition that exists when an injured employee is not expected to recover.

permanent total disability: The condition that exists when an injured employee's disability is such that he or she cannot compete in the job market.

permanent variance: An exemption from an OSHA standard awarded to an organization that can show it already exceeds the standards.

personal monitoring devices: Devices worn or carried by an individual to measure radiation doses received.

personal protective equipment (PPE): Any type of clothing or device that puts a barrier between the worker and the hazard (e.g., safety goggles, gloves, boats, hard hats, and so on).

photoelectric devices: Optional devices that shut down the machine any time that the light field is broken.

photoelectric fire sensors: Detect changes in infrared energy that is radiated by smoke, often by the smoke particles obscuring the photoelectric beam.

physical hazards: Include noise, vibration, extremes of temperature, and excessive radiation.

PMA: Available to employers who intend to correct the situation for which a citation was issued, but who need more time.

point-of-operation guards: Machine guards that provide protection right at the point where the user operates the machine.

poisoning: To be injured or killed by a harmful substance.

powerlessness: The feeling that workers have when they are not able to control the work environment.

preceding factors: Factors that led up to an accident.

predispositional characteristic: Human personality characteristics that can have a catalytic effect in causing an accident.

preliminary hazard analysis: Conducted to identify potential hazards and prioritize them according to (1) the likelihood of an accident or injury being caused by the hazard; and (2) the severity of injury, illness, and/or property damage that could result if the hazard caused an accident.

pressure: The force exerted against an opposing fluid or thrust distributed over a surface.

pressure hazard: A hazard caused by a dangerous condition involving pressure.

private insurance: Workers' compensation coverage purchased from a private insurance company.

problem identification: Engineers must draft a description of a problem before anything else can be done.

produce and deliver: Shop or detail drawings are developed, and the design is produced, usually as a prototype. The prototype is analyzed and tested. Design changes are made. The product is then produced and delivered.

product literature: Purpose of this literature is to tell users about hazards that cannot be removed by design or controlled by guards and safety devices.

product safety management program: Purpose is to limit as much as possible a company's exposure to product liability litigation and related problems.

productivity: The concept of comparing output of goods or services to the input of resources needed to produce or deliver them.

professional societies: Typically formed for the purpose of promoting professionalism, adding to the body of knowledge, and forming networks among colleagues in a given field.

proof pressure tests: Tests in which containers are "proofed" by subjecting them to specified pressures for specified periods.

property damage: Facilities, equipment, or other non-personnel items damaged as a result of an accident.

proposed penalty: The initial penalty proposed by the OSHA compliance officer after an inspection tour.

protons: Positively charged particles.

proximate cause: The cause of an injury or damage to property.

PSM auditor: Responsible for evaluating the overall organization and individual departments within it.

PSM coordinator: Responsible for coordinating and facilitating a product safety management program in all departments of a company.

psychophysiological techniques: Require simultaneous measurement of heart rate and brain waves, which are then interpreted as indexes of mental workload and industrial stress.

psychosocial questionnaires: Evaluate workers' emotions about their jobs.

public hearing: If an injured worker feels that he/she has been inadequately compensated or unfairly treated, a public hearing can be requested.

pullback devices: Pull the operator's hands out of the danger zone when the machine starts to cycle.

puncturing: Results when an object penetrates straight into the body and pulls straight out, creating a wound in the shape of the penetrating object.

quality: A measure of the extent to which a product or service meets or exceeds customer expectations.

quality product: One that meets or exceeds customer standards and expectations.

rad: A measure of the dose of ionizing radiation absorbed by body tissue stated in terms of the amount of energy absorbed per unit of mass of tissue.

radiant heat: The result of electromagnetic nonionizing energy that is transmitted through space without the movement of matter within that space.

radiation: Consists of energetic nuclear particles and includes alpha rays, beta rays, gamma rays, x-rays, neutrons, high-speed electrons, and high-speed protons.

radiation area: Any accessible area in which radiation hazards exist that could deliver doses as follows: (1) within one hour, a major portion of the body could receive more than 5 millirem; or (2) within five consecutive days, a major portion of the body could receive more than 100 millirem.

radiation control specialist: Monitors the radiation levels to which workers may be exposed, tests workers for levels of exposure, responds to radiation accidents, develops company-wide plans for handling radiation accidents, and implements decontamination procedures when necessary.

radioactive material: Material that emits corpuscular or electromagnetic emanations as the result of spontaneous nuclear disintegration.

radio-frequency devices: Capacitance devices that brake the machine if the capacitance field is interrupted by a worker's body or another object.

reasonable risk: Exists when consumers (1) understand risk, (2) evaluate the level of risk, (3) know how to deal with the risk, and (4) accept the risk based on reasonable risk/benefit considerations.

receptacle wiring tester: A device with two standard plug probes for insertion into an ordinary 110-volt outlet and a probe for the ground.

reclamation: A process whereby potentially hazardous materials are extracted from the byproducts of a process.

rehabilitation: Designed to provide the needed medical care at no cost to the injured employee until he/she is pronounced fit to return to work.

Rehabilitation Act of 1973: Enacted to give protection to handicapped people, including handicapped workers.

rem: A measure of the dose of ionizing radiation to body tissue stated in terms of its estimated biological effect relative to a dose of one roentgen (r) of x-rays.

repeat violation: A violation of any standard, regulation, rule, or order where, upon reinspection, a substantially similar violation is found.

repetitive motion: Short-cycle motion that is repeated continually.

repetitive strain injury: A broad and generic term that encompasses a variety of injuries resulting from cumulative trauma to the soft tissues of the body.

resistance: A tendency to block flow of electric current.

response time: The amount of time between when an order is placed and the product is delivered.

restraint devices: Hold the operator back from the danger zone.

restricted area: Any area to which access is restricted in an attempt to protect employees from exposure to radiation or radioactive materials.

retrofit: Renovating rather than replacing.

risk analysis: An analytical methodology normally associated with insurance and investments.

role ambiguity: The condition that occurs when an employee is not clear concerning the parameters, reporting requirements, authority, and/or responsibilities of his or her job.

Role-reversal test: Requires a person to trade places with the people affected by the decision that he or she made and to view the decision through their eyes.

safeguarding: Machine safeguarding was designed to minimize the risk of accidents of machine-operator contact.

safety and health professional: An individual whose profession (job) is to be concerned with safety and health measures in the workforce.

safety movement: Began during WWII when all of the various practitioners of occupational safety and health began to see the need for cooperative efforts. This movement is very strong today.

safety policy: A written description of an organization's commitment to maintaining a safe and healthy workplace.

safety trip devices: Include trip wires, trip rods, and body bars that stop the machine when tripped.

SARA: Designed to allow individuals to obtain information about hazardous chemicals in their community so that they are able to protect themselves in case of an emergency.

second-degree burn: Result in blisters forming on the skin.

self-insurance: Workers' compensation coverage in which a company insures itself by building its own fund.

semiautomatic ejection: A system that ejects the work using mechanisms that are activated by the operator.

semiautomatic feed: A system that uses a variety of approaches for feeding stock to the machine.

semiconductors: Substances that are neither conductors nor insulators.

shift work: Employees work at different times of the day instead of during the same hours.

shock: A depression of the nervous system.

short circuit: A circuit in which the load has been removed or bypassed.

short-term exposure limit: The maximum concentration of a given substance to which employees may be safely exposed for up to 15 minutes without suffering irritation, chronic or irreversible tissue change, or narcosis to a degree sufficient to increase the potential for accidental injury, impair the likelihood of self-rescue, or reduce work efficiency.

sick building syndrome: An internal environment that contains unhealthy levels of biological organisms in the air. A common cause is the introduction of unhealthy outdoor air that is brought in and circulated through the cooling system.

simulation: Involves structuring a training activity that simulates a line situation.

situational characteristics: Factors that can change from setting to setting and can have a catalytic effect in causing an accident.

situational factors: Environmental factors that can affect an employee's safety and that can differ from situation to situation.

smoke: The result of the incomplete combustion of carbonaceous materials.

Social environment: The general value system of the society in which an individual lives, works, grows up, and so on.

sound: Any change in pressure that can be detected by the ear.

sound level meter: Produces an immediate reading that represents the noise level at a specific instant in time.

spraining: The result of torn ligaments.

staff authority: The safety and health manager is the staff person responsible for a certain function, but he/she has no line authority over others involved with that function.

standards/testing organizations: Conduct research, run tests, and establish standards that identify the acceptable levels for materials, substances, conditions, and mechanisms to which people may be exposed in the modern workplace.

standpipe and hose systems: Provide the hose and pressurized water for fire fighting.

Star Program: Recognizes companies that have incorporated safety and health into their regular management system so successfully that their injury rates are below the national average for their industry.

state funds: Workers' compensation coverage provided by the state.

static electricity: A surplus or deficiency of electrons on the surface of a material.

step and fall: An accident that occurs when a person's foot encounters an unexpected step down.

straining: The result of overstretched or torn muscles.

stress: A pathological, and therefore generally undesirable, human reaction to psychological, social, occupational, or environmental stimuli.

stress claims: Workers' compensation claims that are based on stress-induced disabilities.

stressors: Stimuli that cause stress.

stump and fall: An accident that occurs when a worker's foot suddenly meets a sticky surface or a defect in the walking surface.

subjective ratings: Ratings that are less than objective and can be affected by emotions, human biases, presumptions, and perceptions.

synthesis: Second step in the design process wherein engineers combine systematic, scientific procedures with creative techniques to develop initial solutions.

Systems theory: Views a situation in which an accident may occur as a system comprised of the following components: person (host), machine (agency), and environment.

technic of operation review: An analysis method that allows supervisors and employees to work together to analyze workplace accidents, failures, and incidents.

technological alienation: The frame of mind that results when employees come to resent technology and the impact that it has on their lives.

technology access: Access to time- and work-saving devices, processes, and/or equipment that are up-to-date.

temporary emergency standards: OSHA standards that can be adopted on a temporary basis without undergoing the normal adoption procedures.

temporary partial disability: The injured worker is incapable of certain work for a period of time but is expected to recover fully.

temporary total disability: The injured worker is incapable of any work for a period of time but is expected to recover fully.

temporary variance: Employers may ask for this when they are unable to comply with a new standard but may be able to if given time.

thermal expansion detectors: Use a heat-sensitive metal link that melts at a predetermined temperature to make contact and ultimately sound an alarm.

third-degree burn: Penetrate through both the epidermis and the dermis. They may be fatal.

Three E's of Safety: Engineering, Education, and Enforcement.

threshold limit values (TLVs): The level of exposure to which all employees may be repeatedly exposed to specified concentrations of airborne substances without fear of adverse effects. Exposure beyond the TLV is considered hazardous.

threshold of hearing: The weakest sound that can be heard by a healthy human ear in a quiet setting.

threshold of pain: The maximum level of sound that can be perceived without experiencing pain.

time-weighted average: The level of exposure to a toxic substance to which a worker can be repeatedly exposed on a daily basis without suffering harmful effects.

tort: An action involving a failure to exercise a reasonable care that may, as a result, lead to civil litigation.

total quality management (TQM): A way of managing a company that revolves around a total and willing commitment of all personnel at all levels to quality.

total safety management (TSM): The principles of total quality management (TQM) applied to safety management.

toxic substance: One that has a negative effect on the health of a person or animal.

trade associations: Promote the trade that they represent.

trenchfoot: A condition that manifests itself as tingling, itching, swelling, and pain.

trip and fall: An accident that occurs when a worker encounters an unseen foreign object in his/her path.

two-hand controls: Require the operator to use both hands concurrently to activate the machine.

UL (Underwriters Laboratory): Determines whether equipment and materials for electrical systems are safe in the various NEC location categories.

ultraviolet detectors: Sound an alarm when the radiation from fire flames are detected.

unreasonable risk: Exists when (1) consumers are not aware that a risk exists; (2) consumers are not able to judge adequately the degree of risk even when they are aware of it; (3) consumers are not able to deal with the risk; and (4) risk could be eliminated at a cost that would not price the product out of the market.

unrestricted area: Any area to which access is not controlled because no radioactivity hazard is present.

unsafe act: An act that is not safe for an employee.

unsafe behavior: The manner in which people conduct themselves that is unsafe to them or another.

useful consciousness: A state of consciousness in which a person is clear-headed and alert enough to make responsible decisions.

vacuum mentality: Workers think that they work in a vacuum and don't realize that their work affects that of other employees and vice versa.

vacuums: Caused by pressures below atmospheric level.

value added: The difference between what it costs to produce a product and the value the marketplace puts on it.

vapor: A mist state into which certain liquids and solids can be converted (e.g., gasoline fumes are vaporized petroleum).

vapors: The state in which certain solids and liquids are transformed into at elevated temperatures and abnormal pressures.

vertical work area: One that is designed and positioned so that workers are not required to lift their hands above their shoulders or bend down in order to perform any task.

vocational rehabilitation: Involves providing the education and training needed to prepare the worker for a new occupation.

volatility: The evaporation capability of a given substance.

voltage: Measures the potential difference between two points in a circuit.

wage-loss theory: Requires a determination of how much the employee could have earned had the injury not occurred.

water hammer: A series of loud noises caused by liquid flow suddenly stopping.

wellness program: Any program designed to help and encourage employees to adopt a healthier lifestyle.

whole-person theory: What the worker can do after recuperating from the injury is determined and subtracted from what he/she could do before the accident.

wide band noise: Noise that is distributed over a wide range of frequencies.

willful/reckless conduct: Involves intentionally neglecting one's responsibility to exercise reasonable care.

wind-chill factor: Wind or air movement causes the body to sense coldness beyond what a thermometer actually registers as the temperature.

work envelope: The total area within which the moving parts of a robot actually move.

work injuries: Injuries that occur while an employee is at work.

work stress: A complex concept involving physiological, psychological, and social factors.

worker negligence: Condition that exists when an employee fails to take necessary and prudent precautions.

workers' compensation: Developed to allow injured employees to be compensated appropriately without having to take their employer to court.

workplace accidents: Accidents that occur at an employee's place of work.

workplace inspection: An on-site inspection conducted by OSHA personnel.

workplace stress: Human reaction to threatening situations at work or related to the workplace.

INDEX